Countryside History

Oliver Rackham and the ancient cypress above Anopolis, Crete, July 2010. (A. Helman-Ważny)

Countryside History

The Life and Legacy of Oliver Rackham

edited by

Ian D. Rotherham and Jennifer A. Moody

PELAGIC PUBLISHING

First published in 2024 by
Pelagic Publishing
20–22 Wenlock Road
London N1 7GU, UK

www.pelagicpublishing.com

Countryside History: The Life and Legacy of Oliver Rackham

https://doi.org/10.53061/PTLZ6584

British Library Cataloguing in Publication Data
A catalogue record for this book is available from the British Library

ISBN 978-1-78427-316-3 Hbk
ISBN 978-1-78427-317-0 ePub
ISBN 978-1-78427-318-7 PDF

Published with the support of:
Jennifer A. Moody
Professor Christopher Kelly, Master of Corpus Christi College, Cambridge
through the Spencer Fund

Contents

Foreword by Peter Grubb vii

Information on Contributors ix

Introduction: An Overview of the Work and Influences of Oliver Rackham 1
Ian D. Rotherham and Jennifer A. Moody

PART I: WOODLAND STUDIES IN ENGLAND 15

1 Concepts of Ancient Woodland 17
George Peterken

2 The Influence of Oliver Rackham in our Understanding of Wooded Landscapes 31
Della Hooke

3 How the Wildwood Worked: Rackham's Contribution to Forest Ecology 49
Adrian C. Newton

4 Stability and Change in Woodland Ground Flora 64
Keith Kirby

5 Echoes of the Wildwood? Investigating the Historical Ecology of some Warwickshire Lime Woodlands 1986–2000 78
David R. Morfitt

PART II: EUROPEAN STUDIES 97

6 On the Shoulders of Oliver Rackham 99
Frans Vera

7 Forest History versus Pseudo-History: The Relevance of Oliver Rackham's Concepts in the Conservation of Białowieża Primeval Forest 115
Tomasz Samojlik, Piotr Daszkiewicz and Aurika Ričkienė

8 Old-Growth Forests in the Eastern Alps: Management and Protection 124
Elisabeth Johann

9 Biocultural Landscapes of Europe: A Journey with Oliver Rackham 138
Gloria Pungetti

PART III: MEDITERRANEAN STUDIES 151

10 Trees Grow Again: Greece and the Mediterranean in Oliver Rackham's Publications 153
J. Donald Hughes

11 Friend or Foe? Oak Agroforestry Systems in the Mediterranean and the Role of Grazing 159
Thanasis Kizos

12 The Irreplaceable Trees of Crete 165
Jennifer A. Moody

13 Walking in Sacred Forests with Oliver Rackham: A Conversation about Relict Landscapes in Epirus, North-West Greece 191
Rigas Tsiakiris, Kalliopi Stara, Valentino Marini Govigli and Jennifer L.G. Wong

14 Historical Ecology and the History of 'Individual Landscapes': Oliver Rackham's Field Visits to Liguria (North-West Italy) 218
Roberta Cevasco, Diego Moreno and Charles Watkins

PART IV: APPROACHES TO COUNTRYSIDE RESEARCH 233

15 Oliver Rackham, Archives and Ancient Woodland Research 235
Melvyn Jones

16 From Household Equipment to Countryside in Eleventh-Century Bavaria 250
Richard Hoffmann

17 It's a Fair Coppice: Methodological Considerations of the History of Woodland Management 258
Péter Szabó

18 Oliver Rackham and Shadow Woods 271
Ian D. Rotherham

19 Oliver Rackham and the Archaeology of Ancient Woods of Norfolk 285
Tom Williamson

PART V: WIDER PERSPECTIVES 297

20 Reflections from the Antipodes 299
Paul Adam

21 Managing Pollards and the Last Forest 320
Vikki Bengtsson

22 The Value and Meaning of Traditional Natural Resource Use Systems in Satoyama Landscapes in Japan 329
Katsue Fukamachi

23 Pollard Beech Trees in Snowy Areas of Japan 340
Tohru Nakashizuka, Hideo Miguchi and Tomohiko Kamitani

24 So Human a Landscape: Oliver Rackham's Influence on a New England Ecologist 350
Henry W. Art

PART VI: LEGACY, ARCHIVE, AND PUBLICATIONS 365

25 Conclusions: The Legacy of Oliver Rackham 367
Jennifer A. Moody and Ian D. Rotherham

An Oliver Rackham Bibliography 382

Acknowledgements from the Editors 396

General Index 397

Index of Persons 420

Foreword

Oliver Rackham was a very special person. Countless readers will know that he combined an ability to exploit historical documents (often in medieval Latin) with a deep knowledge of the ecology of plants and a talent for writing original and inspiring books. He is also well known for the impact of his writings on the management of lowland woods in England: the reversal of coniferization and the culling of deer populations where excessively large. Less known are his command of modern languages, his ability in physics, which was to be his major subject as an undergraduate until he switched to botany, his PhD work on the physiology of plants (establishing the reasons for their not having higher maximum rates of photosynthesis, a problem taken up by others some 30 years later), his ability to make his own apparatus, his appreciation of silver work, his skill as a carpenter, and his knowledge of the history of woodwork and architecture.

I met Oliver when he was a second-year undergraduate. His quick thinking and ability to ask endless questions made it very difficult for him to be given supervisions (Cambridge parlance for tutorials) with less able undergraduates. I was asked to supervise him alone. At that stage he was very keen to learn what the authoritative writers had discovered and the ideas they had established. How different from the mature Oliver who distrusted any supposed fact or received idea! Anything I did to help him was repaid many times over when – on a botanical trip to Munich – his enthusiasm for architectural history inspired me to explore that subject and make it my major outside interest to this day.

Oliver's mistrust of supposed facts and current ideas was strongly reflected in his first books, which inspired and excited so many people: *Trees and Woodland in the British Landscape* (1976), *Ancient Woodland: Its history, vegetation and uses in England* (1980) and *The History of the Countryside* (1986). Most of the authors of this tribute volume acknowledge the great impact that these books had on their thinking and research.

It is important to emphasize that although Oliver first applied his historical approach in a book about a wood (*Hayley Wood* 1975), he was soon applying it to the whole of the countryside. In this tribute volume we meet accounts of woods, forests and the whole landscape in 11 countries apart from Britain and Ireland: Austria, the Czech Republic, Germany, Greece, Italy, the Netherlands, Poland, Sweden, the USA, Japan and Australia. Oliver visited all of these countries bar the Czech Republic and Poland. Some contributors write mainly about woods and forests, while others are concerned throughout with the whole 'biocultural landscape'. Most emphasize that Oliver inspired them not only through his books but also by his showing them in the forest or field what they should record, and what were the features of outstanding interest at a particular site. Not to mention his strong sense of humour – the more preposterous the story the more he liked it.

In the chapters covering Central and Southern Europe we learn about the importance of forests of mountainous areas in protecting towns and villages below them from rock falls, landslides and avalanches, and the problems in their management – a phenomenon rarely faced in Britain. In the chapters on Greece and Japan we learn about the nature and problems of sacred forests, likewise not encountered on these islands.

Some chapters give fascinating accounts of the development and problems of key concepts, notably 'ancient forest' and 'ancient forest indicator plants'. Oliver was keen to be clear about the meaning of the concepts he introduced. In the foreword to his last magnum opus (*Woodlands* 2006) he wrote in one place 'I am a general practitioner of science' but at another 'I have no particular theory to promote'. One contributor to this volume asks, in effect, was Oliver in his mature years a scientist? It is true that he did not introduce new generalizations about any of the questions concerning the workings of forests and grasslands that have intrigued many ecological scientists during his lifetime, such as the features that make plants light-demanding or shade-tolerant, the significance of seed size, the kinds of differences between species that enable several or many to coexist indefinitely in a forest, scrub or grassland, or the reasons why some plant species invest more than others in defence against herbivores (spines and poisonous or deterrent chemicals). Instead, his originality lay in his introduction of generalizations about new ways of interrogating a site, which involved both percipient observations in the forest or field and rigorous use of historical sources to establish how the plant cover at a given site had come to be what it is. That left room for each site to be unique.

I congratulate the editors on bringing together such a wide range of contributors to show in one volume the huge impact that Oliver's research and writings had on the study of countryside history around the world.

Peter J. Grubb
Cambridge, January 2024

Contributors

Professor Paul Adam is a botanist, plant geographer and ecologist, who received his PhD from Cambridge University and now has a senior academic position at the University of New South Wales. A Fellow of the Royal Zoological Society of New South Wales, he was awarded honorary membership in the general division of the Order of Australia in recognition of his contributions to science, biodiversity conservation, and science education. He received the Gold Medal of the Ecological Society of Australia. Paul's particular research interests are in wetlands and rainforest.

Professor Henry W. Art has broad research interests including the investigation of long-term changes in successional relationships among species comprising the various communities in the College-owned Hopkins Forest. In this work he considers the extent to which natural and human-use disturbances have played a role in shaping the present patterns of communities and ecosystems. The study has involved data collection from a permanent plot system initiated in 1935 by the U.S. Forest Service when they operated the facility. The studies involve sources such as deed history, oral history, and other socioeconomic data to complement the ecological databases on the Hopkins Forest.

Vikki Bengtsson, with a background in entomology, botany and ecology, is interested in and working on issues surrounding the management and care of ancient and other veteran trees including old, lapsed pollards. She also works on veteranization and the potential benefits for nature conservation.

Dr Roberta Cevasco, with a PhD in Geography and a first degree in Natural Sciences, is Associate Professor at Università degli Studi di Scienze Gastronomiche, Pollenzo. Her main research themes include historical ecology and geography of landscapes, environmental factors influencing crop growth, local production for the management of environmental-rural heritage and for sustainable local development.

Piotr Daszkiewicz is a research officer at PatriNat, the centre of expertise and data on natural heritage, based in Paris, France. He is especially interested in the European Bison *Bison bonasus*, its portrayal in historical and archival resources, and its relationship to woodland, and has published widely on these subjects. He is also a key developer of TaxRef, the taxonomic repository of fungi, flora and continental and marine fauna for France and its overseas territories – a project sponsored by PatriNat and Muséum National d'Histoire Naturelle.

Dr Katsue Fukamachi was educated at The University of Tokyo and is now Professor (Associate) affiliated to Kyoto University working on environmental science, agricultural science, and landscape science with particular research interests in landscape ecology, forest culture, forest conservation and forest landscape planning.

Dr Valentino Marini Govigli is a Junior Assistant Professor in the Department of Agri-Food Sciences and Technologies at the University of Bologna, Italy. His research interest is at the intersection of environmental economics, human geography and rural sociology, applied to the forest and agriculture sectors. He uses qualitative and quantitative tools in three main fields of research: (i) stakeholder preferences and consumer behaviour analysis in natural resource management and agro-forestry production models, (ii) evaluation of cultural and intangible ecosystem services, (iii) social innovation brokerage and multi-actor engagement.

Professor Peter Grubb is a British ecologist and emeritus professor of botany at Cambridge University. He joined the staff of Magdalene College, later becoming a full professor, and retired in 2001. His early work was influenced by A.S. Watt, and he has worked on diverse botanical and ecological subjects, from physiology to biomes and from chalk grassland to tropical rain forest. His is particularly associated with the concept of the regeneration niche. Peter co-edited the *Journal of Ecology* from 1972 to 1977. President of the British Ecological Society in 1992, he is now an honorary member of the society.

Professor Richard Hoffmann is Professor emeritus and senior scholar in the Department of History, York University and does research in environmental history and pre-modern Europe. After training in History at the University of Wisconsin (1965) and in Medieval Studies at Yale (1970), he practiced the craft of premodern economic and environmental history at York until formal retirement in 2008. His award-winning publications have included *Land, Liberties, and Lordship in a Late Medieval Countryside* (1989), *Economic Development and Aquatic Ecosystems in Medieval Europe* (1996), *Fishers' Craft and Lettered Art* (1997) and *An Environmental History of Medieval Europe* (2014). His latest volume is *An Environmental History of Medieval Fisheries: The Catch* (2021).

Dr Della Hooke is a historian of Anglo Saxon England and Fellow of the Society of Antiquaries of London. She is a member of the Society for Landscape Studies (editor since 1985), Birmingham and Warwickshire Archaeological Society (editor since 1988), History Geography Research Group Institute British Geographers, Society for Name Studies, English Place-Name Society. She has published widely with volumes such as *Trees in Anglo-Saxon England* (2010), *The Anglo-Saxon Landscape: the Kingdom of the Hwicce* (2009), *England's Landscape: The West Midlands* (2006) and many others. She has been an educator at many institutions especially the University of Birmingham where since 1996, she has been Honorary Fellow of the Institute Advisory Research Arts & Social Sciences.

Professor J. Donald Hughes is John Evans Distinguished Professor of History, University of Denver. He is the author of *What is Environmental History?* (2006), *The Mediterranean: An Environmental History* (2005), and *Pan's Travail: Environmental Problems of the Ancient Greeks and Romans* (1994). He is a founding member of the American and European Societies for Environmental History and past editor of *Environmental History.*

Dr Elisabeth Johann has worked as a forester, has taught forest history at the University of Freiburg and is currently a guest lecturer in the Institute for Socioeconomics, University of Natural Resources and Applied Life Sciences, Vienna. She is a leader in the Austrian Forest Association's Working Group on Forest History and Deputy Coordinator in IUFRO's Forest history and traditional knowledge group. She has written papers and book chapters on Austrian forest history, nature-based forestry in Europe, traditional forest knowledge and European spruce forests. Her book on the history of Austrian forests, Österreiches *Wald in Verganheit und Gegenwart,* was published in 1983. In 2013 she co-authored 'Science and Hope: a forest history'.

Professor Melvyn Jones was a highly acclaimed and passionate researcher and writer on matters of woodland history and historic management. He mixed academic outputs at Sheffield Hallam University with popular writing and broadcasts on radio for the BBC, for many years being the 'Radio Sheffield History Man'. His research on the woodland history of his native South Yorkshire led to a million-pound Heritage Lottery Project to implement programmes of traditional woodland management. This was called 'Fuelling the Revolution' as it connected ancient woodland history and survival to the iron and steel industries of the Sheffield region.

Professor Tomohiko Kamitani is Professor Emeritus in the Faculty of Natural Sciences at Niigata University, Japan. He is also on the advisory board of the Center for Toki and Ecological Restoration there and active at the Tadami Beech Center, near Fukushima, Japan. He has published widely in English and Japanese. His research has focused on Forest Ecology and Forest Management, including coppicing, pollarding, seed dispersal and forest restoration.

Professor Keith Kirby joined the Nature Conservancy Council in 1979 as a woodland ecologist. His role was as a specialist in forestry and woodland work (science and policy) across the NCC and subsequently English Nature then Natural England until 2012 when he retired. Since then, he has been a visiting researcher in the Department of Plant Sciences, at the University of Oxford, continuing his research interests particularly in long-term changes in the woodland ground flora. Other recent work has involved woodland grazing and rewilding and exploring the diaries of Charles Elton from 1942 to 1965. In his role with the agencies and subsequently he published widely in both peer-reviewed and more popular journals (such as *British Wildlife* and *Quarterly Journal of Forestry*). In this work he has sought to build bridges through being active in both the British Ecological Society and the Institute of Chartered Foresters.

Dr Thanasis Kizos works at the Department of Geography, University of the Aegean. His background is in geography, food science and environmental science, and his current research is in related fields. Recent projects include Terra Lemnia (https://terra-lemnia.net/en/) and Terracescape (http://www.lifeterracescape.aegean.gr/en/)

Professor Hideo Miguchi has been professor of Production and Environmental Sciences in the Faculty of Agriculture at Niigata University since 2004, and of the Graduate School for Science and Technology since 2010. He is interested in the dynamic between small forest animals and the arboreal environment, especially as it relates to maintaining biodiversity, and has published on this and other ecological subjects in English and Japanese. Hideo has been the principal investigator on several important research projects such as 'Effects of habitat selection of wood mice on the diversity of canopy tree species in mixed forests' (2002–2003) and 'Functional estimations of forest ecosystem dynamics by field manipulate experimentation in a heavy snow fall region' (2013–2015).

Dr Jennifer A. Moody is an Aegean archaeologist at the University of Texas, Austin, specializing in landscape and paleo-climate reconstruction and ceramic fabric analysis. She has worked on the island of Crete for over 40 years, where she has directed four archaeological surveys and consulted for many more. She helped establish the William A. MacDonald Petrography program at the INSTAP East Cretan Study Center in Pachyammos Crete and founded its associated internship for ceramic petrographers. She is an advocate for landscape conservation and the preservation of cultural heritage in Greece and elsewhere. In 1989 she was awarded a MacArthur Fellowship for her research, exploring the interface between environment, climate change and culture in Cretan prehistory. From 1991 to 2001 she was Visiting Professor and Senior lecturer in Anthropology at Baylor University and since 2006 has been a Research Fellow in Classics at the University of Texas at Austin. In 1996 she and Oliver Rackham co-authored *The Making of the Cretan Landscape*, for which they won the Runciman prize. A translation of their book was published in Greek in 2004.

Professor Diego Moreno is former full professor of Geography and currently works at the Laboratorio di Archeologia e Storia Ambientale (Environmental archaeology and History Laboratory) at the University of Genoa. Diego's research covers environmental resources archaeology, historical rural landscapes and historical geography. He employs multi-proxy evidence in the regressive approach of the British site historical ecology.

Dr David R. Morfitt is a historical ecologist who undertook a research doctorate on the ancient woodland of Piles Coppice near Coventry, England. The site was under threat of destruction by the Coventry Eastern By-pass but was saved and then purchased by the Woodland Trust. David's research into this unspoilt wood involved developing an explanation of the patterns of tree and herb distribution in the context of its historical development.

Professor Tohru Nakashizuka is Director General of the Forestry and Forest Products Research Institute in Tsukuba, Japan. For many years he was Professor in the Graduate School of Life Sciences at Tohoku University, Japan. He is widely published in English and Japanese. His publications focus on ecology, biodiversity, botany, forest canopy, and agroforestry. He works on ecological aspects of species diversity, old-growth forest, species richness, temperate forest and deciduous woodlands. His biodiversity research is multidisciplinary, incorporating elements of ecosystem, ecosystem services and environmental resource management. His botanical research is multidisciplinary, involving both agronomy and horticulture to interconnect temperate deciduous forest studies relating to forest canopy issues. He won the Award of the Japanese Forest Society in 2003, the Konosuke Matushita Memorial Award of Flower Exposition in 2004, and the Midori Scientific Award, Cabinet Office, Government of Japan, in 2007.

Professor Adrian C. Newton is a professor at the University of Bournemouth with research examining human impacts on the environment, with a particular focus on biodiversity loss and its consequences. Having worked on his PhD at Cambridge with Oliver Rackham, he has a particular interest in analysing the biodiversity dynamics of fragmented forest landscapes. He also researches societal responses to biodiversity loss, including protected areas, ecological restoration, and development of the green economy. He has coordinated international, collaborative research projects, principally in Latin America, but also in East and West Africa, Central and South-East Asia, and in Britain. Adrian has produced over 150 research publications on conservation science and management, including books on biodiversity loss and conservation, forest ecology, and tropical forest resources. Recent research activities included analysis of the impacts of ecological restoration on biodiversity and provision of ecosystem services, human impacts on forest biodiversity, and the effectiveness of protected areas in reducing biodiversity loss.

Dr George Peterken, OBE, specialized in woodland ecology and conservation for the Nature Conservancy and successor bodies, where he developed the concept of ancient woodland indicators, initiated the ancient woodland inventory, helped negotiate and develop the UK Government's 1985 Broadleaves Policy, and summarized his experiences in the book *Woodland Conservation and Management* (1981). His interest in natural processes led to *Natural Woodland: Ecology and Conservation in North Temperate Regions* (1996). Since 1992 he has worked in woodlands in the Lower Wye Valley and is currently president of the Gwent Wildlife Trust and associate professor of Nottingham University. In 1994, George was awarded an OBE for services to forestry.

Professor Gloria Pungetti is Founder Chair of Biocultural Landscape and Seascape at the University of Sassari, Italy. She is also Founder of the Cambridge Centre for Landscape and People and Chair of the Darwin College Society, University of Cambridge, UK. She coordinates international initiatives and European projects on holistic landscape research, is co-editor-in-chief of IMIC Journal, and sits on boards of journals and international working groups, including Biocultural Landscape by IALE International.

Dr Aurika Ričkienė is a researcher at the Nature Research Centre in Vilnius, Lithuania. From 2018 to 2020, she was project leader in Lithuania of 'Perception of the European Bison and primaeval forest in the 18th–19th centuries: shared cultural and natural heritage of Poland and Lithuania'. She is interested in the history of biological and botanical science using historical and history of science methods. Aurika is also researching innovative solutions for the sustainable use of natural biomass, such as algae and cyanobacteria, in agriculture.

Professor Ian D. Rotherham is an ecologist and landscape historian and the author of over 500 academic research papers, many hundreds of poplar articles, and author or editor of over fifty books. He is Emeritus Professor at Sheffield Hallam University and has chaired meetings and conferences around the world for over twenty years. He has chaired committees for IUFRO, for ESEH, and for BES.

Dr Kalliopi Stara currently works at the Department of Biological Applications and Technology, University of Ioannina. With interests in ethnobotany ecology and cultural anthropology, Kalliopi does research in ethnoecology, cultural anthropology and sacred natural sites. Her current project is INCREdible 'Innovation Networks of Cork, Resin and Edibles in the Mediterranean basin', funded by the European Commission H2020 programme.

Dr Péter Szabó works at the Institute of Botany of the Czech Academy of Sciences in Brno, Czech Republic. He holds MA degrees in history, English, and medieval studies and completed his PhD in medieval studies at the Central European University in Budapest, Hungary. His research interests lie in the long-term interactions between human societies and wooded environments, with a special focus on what historical knowledge can contribute to today's nature conservation. His publications cover issues ranging from prehistoric forest dynamics through medieval woodland management to the use of large databases in history. In 2012–2016, he led an interdisciplinary team in the LONGWOOD project, funded by the European Research Council (ERC). Because his works rely heavily on crossing the 'great divide' between the humanities and the natural sciences, he has also published extensively on the conceptual aspects of connecting history and ecology. Péter serves on the editorial board of *Global Environment* as well as on the editorial advisory committee of Environment and History. From 2017 to 2019, he served as president of the European Society for Environmental History (ESEH).

Dr Rigas Tsiakiris is currently working as a forester in the Department of Forest Management, Forestry Service of Ioannina, Epirus, Greece. He was working until recently as Scientific advisor of the Alternate Minister of Environment and Energy and for the Ministry of Rural Development and Food. He has worked as an independent free launcher researcher at the Forest Research Institute and several Environmental NGOs. He is an external collaborator researcher at the University of Ioannina. His background is in ecology and forestry.

Dr Frans Vera is a Dutch biologist and conservationist. He has played a key part in devising the current ecological strategy for the Netherlands. He is particularly known for his work that hypothesized that Western European primeval forests at the end of the Pleistocene epoch were not only of 'closed-canopy' high-forest but included pastures combined with forests. He worked as director of the Stichting Natuurlijke Processen (Natural Processes Foundation), University of Groningen, as a guest staff member at the University of Groningen's Centre for Ecological and Evolutionary Studies, and as Senior Policy Adviser to the Strategic Policies Division at the Minister's Office in The Hague. He retired in June 2014, but still works as scientist and policy adviser. His book *Grazing Ecology and Forest History* challenged views on the nature of the former natural landscape. Frans played a significant role in the nature conservation project to develop the Oostvaardersplassen nature reserve in southern Flevoland, a site that was reclaimed in 1967.

Professor Charles Watkins is Professor of Rural Geography at the University of Nottingham. He studies land management, history, conservation and the geography of trees and forestry. He works on interdisciplinary, international projects linking cultural studies with natural science. His recent books include *Uvedale Price (1747–1829): Decoding the Picturesque* (2012) with Ben Cowell, *Trees, Woods and Forests: A Social and Cultural History* (2014) and *Europe's Changing Woods and Forests* (2015), edited with Keith Kirby, *Trees in Art* (2018), and *Rediscovering Lost Landscapes* (2021).

Professor Tom Williamson is a landscape historian and landscape archaeologist with wide-ranging interests. His recent research projects have included a GIS-aided study of Agriculture and the Landscape in Midland England, funded by the AHRC; and investigations of the history of tree populations and tree disease in England since 1600, funded by the AHRC, DEFRA, and the Woodland Trust. He has written very widely and has been an inspirational speaker who crosses between history, landscape and ecology to produce a remarkable synthesis of environmental history.

Dr Jennifer L.G. Wong is Managing Director of Wild Resources Limited, a forestry consultancy group based in Bangor, Wales, that provides high-quality applied research and technical services to support sustainable use of wild products from tropical and temperate forests. Jenny has worked in tropical forestry, specializing in inventory, since 1988. Over the past 10 years or so her work has broadened to include inventory of all wild products derived from forests and European forestry. She has a mix of field and academic experience and is uniquely placed to bridge the gap between research and practice. She holds an honorary Lectureship at the School of Environment and Natural Resources, Bangor University.

INTRODUCTION

An Overview of the Work and Influences of Oliver Rackham

Ian D. Rotherham and Jennifer A. Moody

Summary

Professor Oliver Rackham (1939–2015), botanist, historical ecologist and landscape historian, died on 12 February at the age of 75. A brilliantly original researcher and writer, he inspired many people – botanists, landscape historians and local historians (amateur and professional), conservationists, and countryside managers – to look again with new eyes at the landscape. His insight particularly addressed woods, wooded commons and parklands, both as natural ecosystems and as products of long-term human management.

He was a Life Fellow of Corpus Christi College, Cambridge, and elected Master of the College in October 2007 until October 2008. He was also keeper of the College silver. Oliver was awarded the OBE in 1998 for services to nature conservation and in 2000 an Honorary Doctorate from the University of Essex. In 2002, he was elected to the British Academy and in 2006 appointed Honorary Professor of Historical Ecology in the Department of Plant Sciences at the University of Cambridge.

This volume is a small tribute to the life, work and influence of Oliver Rackham, contributed to by scholars from around the world. Among other things, the book helps demonstrate both the breadth and the depth of Oliver's remarkable scholarship. Chapters have been provided by invited authors from all continents and from varying academic disciplines.

We begin the story with a brief background to the man and his work.

Introduction

Oliver Rackham remains one of the most influential and inspiring academics of all time, and for the study of trees and woods, is without peer. However, his roots in Cambridge ecology and woodland study, before he was so well known, are often overlooked. It is thus worthwhile to look back to a seminal visit to Hayley Wood in 1962 with Dr (later Professor) C. Donald Pigott (botanist and Director of the Cambridge Botanical Garden 1984–1996) and Dr Michael Martin and imagine how that event triggered a revolution in how we recognise, think about and understand our remarkable woodland landscapes. From tiny acorns grow mighty oaks.

Ian D. Rotherham and Jennifer A. Moody, 'An Overview of the work and influences of Oliver Rackham' in: *Countryside History: The Life and Legacy of Oliver Rackham*. Pelagic Publishing (2024).
DOI: 10.53061/JDWL9754

Early days at Cambridge

In the words of Professor Donald Pigott:

> Oliver was certainly a 'one-off' as he was a physicist and mathematician as well as a botanist, an unusual combination. This led him to do his PhD at Cambridge under the supervision of Dr Clifford Evans on the energy balance of leaves. This resulted in an excellent thesis, but I do not think the findings were published. He also had remarkable linguistic abilities allowing him to learn rapidly, for example, mediaeval Latin in order to read monastic charters and other documents in college archives. (The Cambridge colleges often owned woodlands on their estates).
>
> In 1962, I was working with Michael Martin on his PhD at Hayley Wood, 10 miles west of Cambridge. The occupants of the 'railway cottage', the cottage at the entrance to the wood, told me that the farmer who owned the site was proposing to sell it to a timber merchant and have it cleared. Armed with this information I visited the owner and managed to persuade him to sell it to the Cambridgeshire Wildlife Trust (CWT) for an asking price of £5,000. Furthermore, I then managed to persuade the CWT, (a more difficult task), to raise the money by an Appeal. For this, I had the energetic support of both Michael Martin and then Oliver Rackham, who visited the wood with me and this, I believe, was probably a life-changing event.
>
> We toured the wood and looked at its flora and I recall being impressed that he was identifying Basidiomycetes, of which I was shamefully ignorant. I then took him to a small area of the wood that is secondary and was partly planted with oaks (*Q. robur*). This is on land cut off from a field by the railway, subsequently cultivated and then abandoned, into which the true woodland flora, bluebells, wood anemone, and oxlips had spread into a fringe only a metre or two wide, after about 50 years, (I am sure it is all in Oliver's book on Hayley Wood), and he became totally fascinated. Was this, I wonder, the event that sparked his change of direction in research? Oliver's first book [in 1975] followed his exploration of this remarkable woodland and interestingly was more hard scientific text than the next. Subsequently, with some supervision by Dr David Combe, Oliver extended his work to East Anglian woods.
>
> Certainly, after the initial visit to Hayley Wood, Oliver and Michael were my energetic support team in collecting information for the Appeal. They were both involved in the historic tour of the wood with over 100 visitors, which ended with a brief vote of thanks and discussion about raising the money. The latter, in the modern jargon, ended with several people saying '£5,000, no problem' and indeed, it was not. In 1964, I moved to Lancaster and Michael to Bristol. Oliver took over the research on Hayley Wood and responsibility for planning the silvicultural policy for the site. In my opinion, Hayley Wood is in part a memorial to him.
>
> C.D. Pigott in Rotherham 2015: 50–51

Oliver's involvement in the conservation of Hayley Wood was a watershed in his career. Indeed, his very first book, written in 1975, was devoted to it: *Hayley Wood: Its History and Ecology*. Impressively, the book's 'Introduction' and 'Foreword' were written by two of the most eminent British ecologists of the time: Sir Harry Godwin (Professor of Botany, University of Cambridge) and Dr S. Max Walters (botanist, Director of the Cambridge Botanical Garden 1973–1984). In the words of Sir Harry Godwin and Dr S. Max Walters:

> It is the good fortune of the Hayley Wood research that it has found, in Dr Rackham, a scholar with the remarkable combination of scientific training

with linguistic and historical aptitude, to exploit and interpret a mass of documentation from the time of Domesday onwards, that by equal good fortune concerns this wood directly or, through parallel research in East Anglia, indirectly. Equally fortunately, it turns out that there has been woodland in the same boundaries throughout the last nine centuries, and Dr Rackham accordingly now presents a most convincing picture of the woodland through all this time as it played its changing but continuing role in the economy of the region.

It has always been the tendency of British ecology to concern itself far less with the establishment of a taxonomic hierarchy of plant communities, than with an examination of the processes and mechanisms operating within the ecosystem, with a demonstration of how these control observed successional changes and how distributional patterns can be attributed to soil and climatic variation through systems of operation ultimately referable to the intimate biology and physiology of the component species. It is an approach lending itself to experimental methods in both field and laboratory and possibly owes its adoption to the high regard for experimental biology, including plant physiology, that has long prevailed in this country. Cambridge has been a strong centre for ecological research of this kind, and Dr Rackham's account of present-day ecology of Hayley Wood is an outstanding exposition of its character. It chiefly subsumes ecological research on the Wood done by Dr A.S. Watt and Professor C.D. Pigott or by their students, Dr P. Wardle, Professor Abeywickrama and Dr M.H. Martin, but many extensions of their studies are in progress, often under Dr Rackham's own hand. The operation of waterlogging in relation to the drainage pattern and through ferrous iron toxicity in the gley soils, like the role of phosphate deficiency, are themes of very general significance. This is like the impact of Dr Rackham's research into historical woodland management, that now appears, surprisingly enough, to illuminate even the long-standing problem of the status of the oxlip and primrose in these boulder-clay woods.

The necessary breadth of synecological study is exemplified by the inclusion of an account of tree-diseases that draws on Cambridge-expertise in plant pathology, and by a section on the bryophyte and lichens reflecting the recently much increased interest in these organisms.

The volume [*Hayley Wood* 1975] finally constitutes reassuring evidence that the rank and file of the Naturalists' Trust indeed supports a broad ecological policy: members have included species lists for various groups of organisms that will be indicators of future change, and a regime of experimental coppicing has already operated for some years. They, like Dr Rackham and the enlightened officers of the Trust, deserve every encouragement; appreciation will certainly follow this timely publicity for their work. It will give all conservationists heart to see the speed and effectiveness of their control, and ecologists within this country and outside will be happy to profit from this account of the manner in which the evolution and extension of their own field of science is illustrated in a single significant woodland site. Not least, ecologists, like local historians and economists, will note the new dimension provided by the combination of ecological expertise and effective use of documentary history. For my own part, I confess to pleasure in having been on hand to follow the changes in attitude and the progression of knowledge to which this book bears witness.

Sir H. Godwin, *Foreword*, in Rackham 1975: xiv–xv

> Although Hayley was sufficiently well known to have been designated a Special Site of Scientific Interest by the Nature Conservancy Council, little was known of its natural history interest, and few published records of plants and animals of the Wood existed before the Trust bought it. One of the very great advantages of nature reserve status is that once a number of interested naturalists have been stimulated to visit the site, the knowledge the flora and fauna soon grows. To the success of this policy in Hayley over twelve years, this book is an eloquent testimony. It is probably true to say that any piece of land intensively studied will prove to have its unexpected interest; but this is abundantly true in the case of Hayley Wood, and there can be little doubt that the interest of the Wood and its plant and animal communities will continue to grow in the years to come. Indeed, the systematic recording of nature reserves is probably one of their most valuable uses; changes of flora and fauna resulting from planned management are of the utmost importance in determining whether a particular policy is successful and is therefore to be continued. Various aspects of management and their consequences are discussed in this book.
>
> S.M. Walters, *Introduction*, in Rackham 1975: 3

Woodlands and the British countryside

Oliver wrote many books (see Rackham's bibliography *this volume*) and articles. Volumes include *The History of the Countryside* (1986) and *The Last Forest* (1989) on Hatfield Forest. Rackham's writing and passion for trees, woods and the countryside affected everyone involved in those fields, and not only in Great Britain, but across Europe and around the world.

In an obituary he wrote for Oliver, Melvyn Jones (Professor of Landscape History, Sheffield Hallam University) notes:

> He [Oliver] first came to notice nationally in Britain with his book *Trees and Woodland in the British Landscape* (Dent 1976). This was re-written to take account of further research for a second edition published in 1990. Praise for the first edition was unanimous, the *New Scientist* proclaiming that he had 'a gift for presenting solid information that kindles the imagination and stimulates the sense of curiosity'. [The reviewer continued, 'Drawing upon years of research, Oliver Rackham is as highly readable as he is informative and topical, concluding this definitive study with a section on the conservation and future of Britain's trees, woodlands, and hedgerows … it is difficult to convey the quality of Dr Rackham's book on the strength of a few quotations. As an aid to understanding the landscape] … I haven't found its equal.' Moreover, *Country Life* said it was 'a masterly account … a classic of recorded fieldwork and meticulous scholarship'. In addition, it was in the second edition of this book that he introduced the concept of *pseudo-history*, which as he [Oliver] pointed out had no connection with the real world and was made up of *factoids* that look like facts, are respected as facts, and have all the properties of facts except they are untrue. Among the factoids he lists are that medieval England was still well-wooded, that coke was used for smelting iron from the eighteenth century because there were no trees left, and that the last remnants of ancient woodland were cut down during the two world wars.
>
> Jones 2015: 48–49

Mel described Rackham's seminal 1980 work *Ancient Woodland: its history, vegetation and uses in England* (Edward Arnold), as his *magnum opus* (also see contributions in this volume by Peterken, Hooke, Newton, Kirby, Morffit, Vera and Williamson). This 400-page volume was re-published in a new edition in 2003 (Castlepoint Press) and launched at the Sheffield

Hallam 2003 conference 'Working and Walking in the Footsteps of Ghosts'. However, Oliver had already produced another ground-breaking book in 1986, the monumental *The History of the Countryside*. It was surely this volume that transcended scientific scholarship and achieved popular acclaim.

Mel Jones further observes:

> Rackham was just as good when researching on a small scale as well as on a national or international scale. This is obvious whether he was compiling and writing a study of an individual wood, as in *Hayley Wood, its history and ecology* (Cambridge and Isle of Ely Naturalists' Trust 1975), a former royal forest such as *Hatfield Forest: the story of Hatfield Forest* (Dent 1989), or a small sub-region like *The Woods of South-East Essex* (Rochford District Council 1986).
>
> It was not just in his major works that he made an outstanding contribution to knowledge and the changing of attitudes; this was also true of the articles that he contributed to journals over the years. For example, in 1979, he wrote a pioneering article in *Landscape History* entitled 'Documentary Evidence for the Historical Ecologist' in which he described at length the various types of documentary evidence that could be used to compile the history of a wood. He made what might now be regarded as the obvious point, but what at that time was quite novel, that: 'Without such information it is impossible adequately to assess the importance of individual woods and woodland types or to draw up rational management schemes'. More recently, in 2004, he wrote a thought-provoking article in the journal *Landscapes* on pre-existing trees and woods in country house parks in which he illustrated the fact that the creation of parks had removed samples of countryside from the normal pressures of agriculture and preserved the trees, other vegetation, and antiquities of the previous landscape.
>
> Jones 2015: 49

Oliver was fascinated with what place-names and fieldnames could reveal about a landscape, and his interpretations were always supported by meticulous work in archives and other documentary sources such as old maps and even oral traditions. Of course, the ability to read medieval Latin was a great help when searching and interpretating old legal papers and documentation. This aspect of Oliver's approach inspired others such as Melvyn Jones (Jones 2009, and *this volume*) – in Mel's case spending months in the Wentworth Muniments in Sheffield. Similarly prolific authors like Hooke (*this volume*) and Hoffmann (*this volume*) also used this rigorous application of diverse sources to the understanding of landscape and countryside history.

Searching for ancient trees and woodland

Oliver was passionate about his trees and woods, still giving lectures, travelling to remote places and even running regular field courses just a few months before he died. In 2013, he was unable to attend the 'Trees Beyond the Wood' conference in Sheffield (Rotherham et al. 2013) because, already into his seventies, he 'had a once in a lifetime chance to visit Ethiopia', where he cast his expert, discerning eye over the tree landscapes of that region.

Although he preferred being in the field, Oliver understood the value of conference-going for disseminating and developing ideas. Over the years he attended a number of conferences at Sheffield – 'Working and Walking in the Footsteps of Ghosts' in 2003 (Rotherham et al. 2012) and 'Animals, Man and Treescapes' in 2011 (Rotherham and Handley 2011) – where he presented papers and contributed important chapters to their associated publications (Rackham 2012b, 2013).

Since the 1970s, Oliver's influence on the study of 'woods' and 'woodland' and his contributions to the emergence of new concepts like 'shadow woods' and 'ghost woods' has been profound (see Rotherham *this volume*). Furthermore, their critical understanding helps drive practical approaches to ancient woods and to the conservation of veteran trees (e.g., Bengtsson *this volume*).

European landscape history and beyond

As evidenced by contributions to this volume, Oliver's interests and influences extended beyond the British Isles.

The Mediterranean

Oliver first encountered the Mediterranean in July 1967 on a Cambridge Botany School field-trip led by S. Max Walters and Peter Grubb (ecologist and botanist, Professor of Investigative Plant Ecology, Cambridge University). It was on this trip that Oliver demonstrated his legendary linguistic gifts, as recently retold by John and Hilary Birks:

> Driving along a twisty steep road on the Croatian island of Rab, the bus driver saw a priest walking along the road. The driver offered a lift to the priest. The only spare seat on the bus was next to Oliver so he sat there. Oliver tried to converse with the priest in English (no success), then French (no success), then German (no success), and finally Italian (still no success). After a short silence, Oliver struck up a conversation with the priest in Latin! Oliver quizzed him about management and use of the local woods. Meanwhile the bus-load of students collapsed with quiet laughter.
>
> J. and H. Birks 2022: 8–9.

A year later (July 1968) he arrived in Crete as the expeditionary botanist on Peter Warren's excavation at Early Minoan Myrtos (Warren 1972). His publications on the project's vegetation and archaeological charcoal continue to be important references today (Rackham 1972a,b).

During that summer, Oliver hiked all over the Myrtos area, including the high Lasithi mountains, where he encountered his first ancient Cretan trees – the impressive Prickly Oaks *Quercus coccifera* of Selakano. Oliver was familiar with ancient woodland and veteran pollards in England, but he did not expect to see them in Crete. This was the beginning of his love affair with the Mediterranean's ancient and venerable trees about which, some years later, he observed:

> Ancient pollards and coppice stools, a metre or more in diameter and several centuries old, are to be found in most Mediterranean countries, supremely in Greece and Crete with their ancient evergreen and deciduous oaks, pines, junipers, planes, cypresses and many others.
>
> Rackham 2001: 56
> (Also see Moody *this volume*)

By the time Oliver returned to Crete in 1981, at Jennifer Moody's invitation, he had worked in other parts of Greece – Boeotia, Sitagroi, Santorini, Chalki – and had garnered a reputation among Aegean archaeologists as a brilliant if eccentric botanist and historical ecologist. Moody and Rackham hit it off and wrote *The Making of the Cretan Landscape* in 1996, which won the Runciman Prize and was translated into Greek in 2004.

In the 1980s and 1990s, Oliver explored parts of Italy in the company of Diego Moreno, Roberta Cevasco, Mauro Agnoletti and others, where he had a profound impact (Cevasco et al. *this volume*). In a letter he wrote to the Master of Corpus Christi College when Oliver died, Mauro Agnoletti (Professor of Landscape and Cultural Heritage, University of Florence) notes:

> Oliver Rackham has been a fundamental source of inspiration for my work. I met him about 25 years ago and had the chance to collaborate with him several times, inviting him to conferences and meetings. I had the privilege not only to learn

> from his academic work, but also to transfer some of his research ideas into laws and political documents for my government, for the European Commission, for UNESCO, CBD and FAO.

But it was Oliver's collaboration with A.T. (Dick) Grove in the 1990s that took him to the remote corners of the Mediterranean and resulted in their groundbreaking book *The Nature of Mediterranean Europe* (2001) (Figure 1). Using examples from western Turkey to the west coast of Portugal, Grove and Rackham hoped that they

> have thoroughly demolished the myth of Ruined Landscape, at least as a general proposition. ... Mediterranean vegetation should be understood on its own terms and using its own categories, rather than misinterpreted as degraded forms of a once universal forest of tall, timber-quality trees. It makes no sense to reproach Greece for not being like northern Europe.
>
> Grove and Rackham 2001: 362

This was then and continues to be today a controversial and debated position (Hughes *this volume*; Kizos *this volume*).

Always keen to explore unfamiliar landscapes, in 2005 Oliver began working in Epirus (Greece) with Kalliopi Stara and Rigas Tsiakiris, both at the University of Ioannina (see Tsiakiris et al. *this volume*). In a letter he wrote to Jennifer Moody when Oliver died, Rigas says:

> I was astonished by his ability to discover the history of every place. To be his companion in the forests of Epirus was a delight, like an Aristotelian philosophical walk: every creature, the shape of every tree, every small hidden plant or fallen log had a history to tell and Oliver had this extraordinary ability to discover or to sense the facts behind every scene. But more importantly, every fact raised a new question, a new field of research, a new hidden mystery.
>
> Oliver was the man who taught us all to escape easy facts and to search on our own for the evolution of cultural landscapes. Therefore, it is not an exaggeration to call him the 'Darwin of landscapes', the one who described the endless evolution of Culture and Nature with the simplicity of a fairy tale.

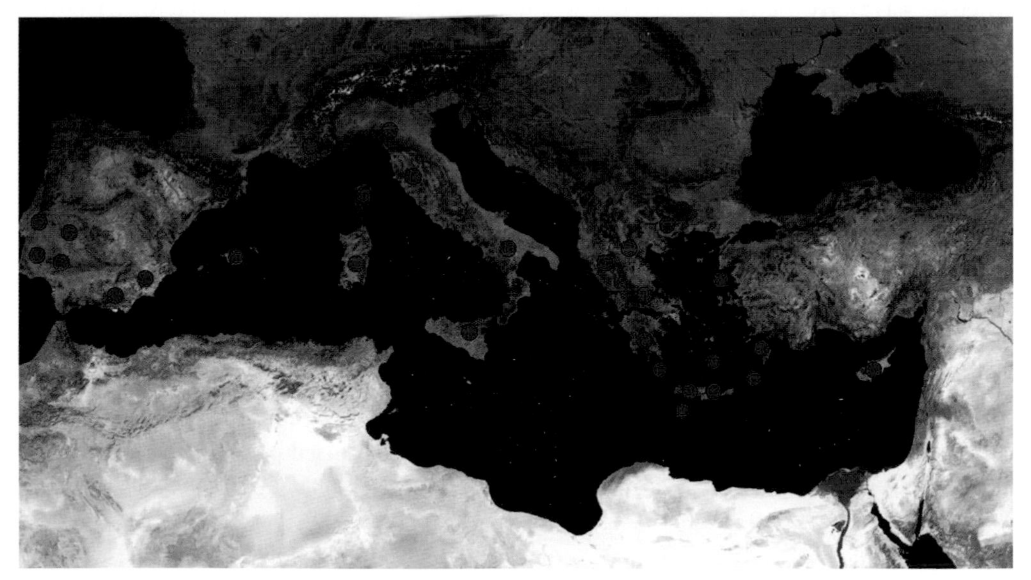

Figure 0.1 Oliver Rackham in the Mediterranean. Compiled from the indices he wrote for his red fieldbooks. (J. Moody 2022)

A few years later, Oliver began collaborating with Gloria Pungetti (Associate Professor, University of Sassari, Sardinia) on several pan-European initiatives concerned with biocultural landscapes (Pungetti *this volume*).

Oliver's impact in the Mediterranean World was not confined to field work. He participated in conferences (Rackham 1978, 1992a, 1992b, 2015), taught classes, and advised Master's and Doctoral students at the University of Crete, the Mediterranean Agronomic Institute Chania (MAICH), the University of the Aegean, Laboratorio di archeologia e storia ambientale at the University of Genoa, the University of Sassari, among others.

In the Mediterranean, as in Britain, Oliver fought to protect heritage landscapes. One example is Cavo Sidero in east Crete. In 2007, Oliver and Jennifer Moody chanced to hear that a monstrous golf resort called Cavo Sidero had been given planning permission on the remote and starkly beautiful Itanos Peninsula, where they had been working. The development was wrong on many levels – the landscape is a museum of antiquities and there are wetlands and local endemic plants – so, they wrote to diplomats and journalists, published articles in local Greek newspapers and set up an online petition that collected over 10,000 signatures. Three years later the permit was revoked (Rackham and Moody 2012).

Oliver's work in the Mediterranean even impacted his British friends and colleagues in unexpected ways. Melvyn Jones writes:

> I remember spending a week in Crete in 1998, supposedly on a birding and wild flowers holiday, but having just devoured Rackham and Moody's book on Crete I spent most days trying to photograph a goat up a prickly-oak 'goat pollard' that was so vividly illustrated in a line drawing on page 113.
>
> Jones 2015: 49

Around the world

With a huge range of books, articles and papers, Oliver's influence was global. Furthermore, over his career he received invitations to speak at meetings and to collaborate with researchers around the world. Increasingly, too, he was able to visit and research far-flung regions and through this to both learn from and influence scholars on their home turf (Figure 2).

Oliver's first trip to the Continent was hiking in the Austrian Alps with his father in 1956, when he was 17. He then proceeded to visit and revisit the French, German and Austrian Alps over the next 20 years. It was not until his third book, *Ancient Woodland* (1980/2003), that Oliver began to call upon his observations on Alpine and other non-British landscapes to illustrate his points. For instance, when discussing lime reproduction and temperature, Oliver notes:

> Lime extends further north than oak into Finland, and I have seen it ascending consistently higher than oak in many Alpine valleys, e.g., in South Tyrol.
>
> Rackham 1980/2003: 243
> (Also see Johann *this volume*)

The country outside of Europe that Oliver most frequented was the United States. Between 1981 and 2014, Oliver travelled to the United States twenty-nine times. His first visits were especially to New England, at the invitation of Henry Art (Professor of Environmental Studies and Biology, Williams College), who says:

> There are many continuing influences of Oliver Rackham on my ways of thinking about and interacting with nature. As a colleague he infused patience in paying attention to details of what one observes, openness to information of all sorts – yet filtered by critical eyes, and an organic sense of humour.
>
> H. Art *this volume*, p. XXX

Oliver's later books are full of references to what he saw and learned in the USA. For instance, his twenty-five trips to Texas informed his ideas on savanna and Vera's proposal that grazing wild beasts propagated prehistoric Europe's mosaic of grassland and wildwood:

> Most savannas in the world today consist of single trees in grassland, which is not (in the main) what Vera had in mind, although one can imagine that some of the groves might never have gotten further than a patch of thorns with one or two oak-trees. A closer modern parallel might be the mott savannas of Texas ... with clonal patches of oaks and elms in grassland, some two or three acres in extent, having woodland shrubs and woody climbers as well as trees.
>
> Rackham 2006: 92

Oliver travelled to Japan five times between 1998 and 2010, filling twenty-seven of his red fieldbooks with notes on the country. His final trip, in 2010, was to work on the Japanese translation of his seminal book, *The History of the Countryside*, which came out in 2012. Japanese timber architecture, ancient trees (Nakashizuka et al. *this volume*), and Satoyama landscapes (Fukamachi *this volume*) fascinated him.

In two visits to Australia (1996–97; 2001) and one to Tasmania (2001), Oliver filled eighteen red fieldbooks. He was fascinated by fire, and Australia's unique fire history and fire-adapted vegetation and wildlife did not disappoint (Adam *this volume*):

> Australia is the Planet of Fire. Except in the small area of rainforest, fire is as necessary to Australian native vegetation as rain to Britain. All the thousand species of *Eucalyptus* that dominate Australian vegetation appear to be fire-adapted. One cannot live long in Sydney without witnessing the awesome spectacle of a eucalyptus crown fire, or the nonchalant way in which the trees carry on growing afterwards.
>
> Rackham 2006: 59

And thus, the Rackham legacy of landscape studies carries on around the world, even in countries he never managed to visit such as Kyrgyzstan (Newton *this volume*), Poland (Samojlik et al. *this volume*) and the Czech Republic (Szabó *this volume*).

Oliver's final exotic excursion was to Ethiopia in the fall of 2012. He filled nearly four red fieldbooks with notes and prepared a PowerPoint presentation with detailed comments on his travels. Sadly, Oliver's insights from this trip, which surely would have been illuminating, never made it into print. His notes, however, are available in his archive at Corpus Christi College (see Moody and Rotherham *this volume*).

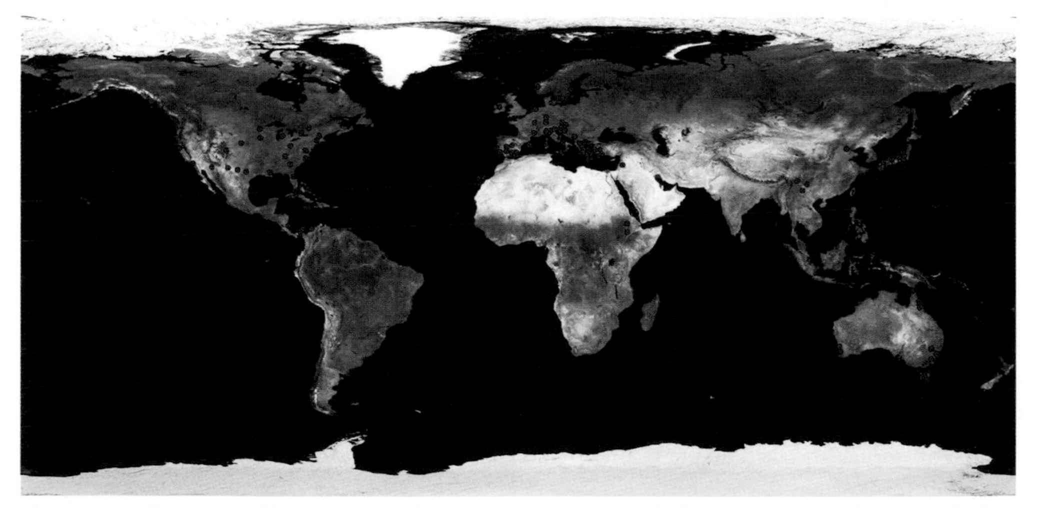

Figure 0.2 Places visited by Oliver Rackham 1956 to 2015. Compiled from the indices he wrote for his red fieldbooks. (J. Moody 2022)

Setting the scene for this volume

In the obituary carried by the *Arboricultural Journal* (Rotherham 2015), Ted Green (MBE, founder of the Ancient Tree Forum) describes Oliver as a 'Field man, observer, ecologist' and a 'scholar of the historical written word', and then writes the following:

> No doubt as a young man wandering the leafy lanes and woods with their ancient coppice stools and pollards, they threw up so many questions for his incredibly enquiring mind; which in turn led him to the vast archives of our historical and cultural history. Thankfully many of these original observations and much of Oliver's research have been passed on to us and provided a springboard for countless researchers past and present. Many of his 'throw-away comments' are well embedded in ecological folk law and are often quoted to illustrate a point and invariably not challenged by the audience. However, one 'Oliver' quote that is still ignored is 'Broad-leaved woodland burns like wet asbestos.' No doubt with time, he will again be recognized for his powers of observation. The debt we owe Oliver cannot be measured but rest assured the references to his works will continue unabated for all time. Oliver will never be forgotten.
>
> Ted Green in Rotherham 2015: 3

Now through this current volume there is an opportunity to write about the influence he had and still has today on the Ancient Tree Forum and more widely on the world of history, culture and conservation. This impact was across Britain, Europe and around the world in North America, Australia and Japan, and of course in the Mediterranean, especially in Crete. By the mid-1980s, inspired by Oliver's early writings, in the conservation and tree worlds, concerns were being raised about the plight and importance of our remaining ancient trees. A small group of highly motivated, idealistic and caring individuals got together and fortunately began to meet in places with concentrations of ancient trees. The discussions that flowed in the debates about the trees under which we stood formed an unbreakable bond of concerned individuals. The dye was set and in 1993 the Ancient Tree Forum was formed. This followed the seminal 1992 meeting of the Landscape Conservation Forum (established 1986) in Sheffield ('Ancient Woodlands their archaeology and ecology a coincidence of interest'), at which Donald Pigott was the keynote presenter along with Chris Baines, and Ted Green spoke passionately about ancient and veteran trees (Parsons et al. 1993). Oliver was unable to attend due to a prior commitment abroad.

However, from those early beginnings at those meetings through his writings Oliver was already 'right there'. In fact, two of Oliver's most quoted assertions concern ancient and veteran trees:

> 10,000 oaks of 100 years old are not a substitute for one 500 year old oak.
>
> Rackham 1986: 152

And its alternative:

> A single oak that is 400 years old is a series of ecosystems in itself, for which a hundred oaks 200 years old are no substitute.
>
> Rackham 2018: 46

In comments sent to Ian Rotherham on Oliver's death, Ted Green observes:

> We still quote him constantly and his name always comes up again and again in debates about the history of the countryside and especially about our trees. We continue to use his historic data especially on the remnants of wood pasture, commons, and our old parks and forests in England. Oliver was perhaps the first prominent countryside historian to have concerns about our ancient trees and the need to accept their importance. He went on to write and express his views and his doubts about the popularly held and entrenched view of British and European

> vegetation cover being largely dense forest before clearance by human settlers. He suggested a savanna-like landscape over large areas. At this point, mention should be made that as early as the 1960s eminent lichenologist Francis Rose was already pondering why many of his lichens only grew on ancient open-growing trees and suggesting that some ancient pastures in Norfolk and Suffolk had never been treed (Rose 1992, 1993). Similarly, entomologist Keith Alexander could not explain why some many of his warmth-loving, sunlight-dependant beetles associated with decaying wood were again only found in large, open-grown trees and not in the dense dark forest or wood (Alexander 1998). A landmark work to endorse these observations arrived with Frans Vera's definitive study on European wood-pastures and the roles of large grazing herbivores (Vera 2000). This was a key event within a debate about European landscape origins and we continue to use Oliver's savanna-like landscape in explanations today.

Ted further suggests that alongside his meticulous intellect and sharp mind, a significant part of Oliver's power and influence was in his ease of description that Ted calls 'Visual English'. This remarkable ability to communicate complex ideas but in uncomplicated, simple terms, means his work is understood by readers from all walks of life. This is quite a quality.

It seems that Oliver came to only a few of the Ancient Tree Forum gatherings and when present was more of a listener and observer, usually only speaking when asked a question. Maybe, like many of us, he was listening to arborists for the very first time talking about their emerging profession which is the care, conservation and study of trees. Oliver's writings had inspired arborists and he was able to discover that his words had not fallen on stony ground. As Ted observed: 'One wonders what his questions and later discoveries would have been if he had been with arborists in his earlier career.'

Some observations on the Ash

Oliver's final book before he died was *The Ash Tree* (Little Toller Books 2014), reflecting his passion for that species. In this context, an example of Oliver writing for a wider audience was a superb article in *The Mail on Sunday*, 4 November 2012. Rackham gave his sensible, pragmatic view that current waves of plant diseases, especially those of trees, are a consequence of globalized international trade and importation of saplings into and around continents and countries. He noted the common-sense observation that at any one location, diseases like Sudden Dieback of ash come into evolutionary balance over centuries of coexistence with tree and disease. Yet as humans move both trees and diseases around the planet, the local variants and balances of evolution are disrupted – with catastrophic results. When asked about this latest disease rampant, in his inimitable fashion, Oliver stated: 'I told you so!' The message, as always, was clear, unambiguous and to the point (Rackham 2012a).

Conclusions

In 2006 Oliver published *Woodlands*, the 100th volume in the New Naturalist Library (Collins) and his last but one book published before he died. In the preface, he stated that he was now a rather old-fashioned botanist and that, 'for good or ill, I haven't a particular theory to promote'. This was not the case, as the volume was as thought-provoking as ever. An example is his cautionary note about how the use of modern machinery should be out of bounds in archaeologically sensitive areas, and furthermore, how ecologists and archaeologists need to collaborate rather than compete.

> The conservation of the archaeology depends on suitably managing the trees. If lime is allowed to grow up to tall trees, this will indeed result in windblow at the next big storm. As has been shown at Overhall Grove, timber can be removed

> without damaging the earthworks by attention to detail, such as not using machines in wet weather and filling ditches temporarily with logs.
>
> The Forestry Commission has recently surveyed its woods in the east Midlands for archaeological features, entering the results on constraint maps that identify areas out of bounds to certain types of operations. This should set an example, ending the tradition among archaeological and biological conservationists of not talking to each other, even when they war against the same foe.
>
> Rackham 2006: 224

Oliver also provided important conceptual insights through his eloquent descriptions of the 'Locust Years' from the 1950s to the 1970s, when British forestry destroyed vast areas of medieval wooded landscapes (see Rackham 2006: ch. 18). Almost certainly, he would be appalled by the 'New Locust Years' of the early twenty-first century, whereby woodland managers, including now conservation bodies, are laying waste to vast tracts of irreplaceable ancient woods under the diktat of government department Defra and its agency, the Forestry Commission. Centuries, if not millennia, of unique heritage are erased from woodlands by heavy, tracked vehicles in just a few days or even hours, and the landscapes which Oliver loved are irretrievably compromised.

As might be expected, *The Ash Tree* (2014) continued to stir things up. Here, among other things, he turned his critical eye on the globalised import–export trade in trees and the likely consequences for the spread of pests and diseases.

Oliver Rackham was a brilliant researcher and extraordinarily influential communicator. He was inspiring, quirky (often in shorts, red socks and sandals), generous and opinionated (but his opinions were always worth listening to). Back in the 1990s, friend and colleague, pioneering urban ecologist, the late Oliver Gilbert, stated that in his opinion, Oliver Rackham was one of the very few academics who has 'changed the way that everybody thinks'. A rare talent indeed.

A lasting memory for Jennifer is her first day in the field with Oliver in June 1981. Exploring a mountain slope in Crete, suddenly she and Oliver fell to the ground in keen excitement and simultaneously whipped out the hand lenses that hung round their necks on a string. As their eyes locked over their respective prey – Jennifer's a tiny pottery sherd and Oliver's a tiny plant – they knew that this was the beginning of great things to come.

A lasting memory for Ian is of Oliver arriving at the 2003 Sheffield conference, 'Working and Walking in the Footsteps of Ghosts', dressed in striped blazer, shorts (it was hot June weather) and straw boater, *de rigeur* for a Cambridge Don perhaps but less common in Sheffield. He strolled up to the reception desk and announced, 'I am Oliver Rackham and I have arrived!'.

Oliver will be greatly missed by all those who care for trees, woods, history and the landscape, and more widely. Not only has he informed and influenced our professions, but he was a great advocate too. His legacy is immense.

References

Alexander, K.N.A. (1998) The links between forest history and biodiversity: The invertebrate fauna of ancient pasture woodland in the British Isles and its conservation. In: K. Kirby and C. Watkins (eds) *The Ecological History of European Forests*, CABI International, Oxon, Wallingford, pp. 73–80.

Birks, J. and Birks, H. (2022) Some memories of Oliver 1963–2015. *Friends of Oliver Rackham Newsletter #17*, pp. 8–11.

Grove, A.T. and Rackham, O. (2003) *The Nature of Mediterranean Europe: An Ecological History*. Yale University Press, New Haven and London.

Jones, M. (2009) *Sheffield's Woodland Heritage*. Fourth Edition, Wildtrack Publishing, Sheffield.

Jones, M. (2015) Oliver Rackham, 1939–2015 Brilliantly original researcher and writer on landscape history. *Arboricultural Journal*, 37(1), pp. 48–49. https://doi.org/10.1080/03071375.2015.1027611

Parsons, J., Beswick, P. and Rotherham, I.D. (eds) (1993) *Ancient Woodlands: Their Archaeology and Ecology; a Coincidence of Interest; the Proceedings of the National Woodlands Conference Held in Sheffield, England 25–26 April 1992*. Landscape Conservation Forum, Sheffield.

Rackham, O. (1972a) Charcoal and plaster impressions. In: P. Warren (ed.) *Myrtos: An Early Bronze Age settlement in Crete*, British School at Athens Supplementary Volume 7. British School at Athens, London, pp. 299–304.

Rackham, O. (1972b) The vegetation of the Myrtos region. In: P. Warren (ed.) *Myrtos: An Early Bronze Age settlement in Crete*, British School at Athens Supplementary Volume 7. British School at Athens, London, pp. 283–298.

Rackham, O. (1975) *Hayley Wood. Its History and Ecology*. Cambridgeshire and Isle of Ely Naturalists' Trust, Cambridge.

Rackham, O. (1976) *Trees and Woodland in the British Landscape*. Archaeology in the Field Series (First edition), J.M. Dent and Sons Ltd, London.

Rackham, Oliver (1978) The flora and vegetation of Thera and Crete before and after the great eruption. In: C. Doumas (ed.) *Thera and the Aegean World I*. Distributed by Aris and Phillips Ltd, London, pp. 755–764.

Rackham, O. (1979) Documentary Evidence for the Historical Ecologist. *Landscape History*, 1(1), pp. 29–33. https://doi.org/10.1080/01433768.1979.10594337

Rackham, O. (1986) *The History of the Countryside*. J.M. Dent and Sons Ltd, London.

Rackham, O. (1989) *The Last Forest: The Story of Hatfield Forest*. J.M. Dent and Sons Ltd, London.

Rackham, O. (1990) *Trees and Woodland in the British Landscape*. Archaeology in the Field Series (Second, revised edition), J.M. Dent and Sons Ltd, London.

Rackham, O. 1992a Conservation in the cultural landscape: the historical context and the story of Crete. In: *So that God's Creation Might Live: The Orthodox Church responds to the ecological crisis*, Proceedings of the Inter-Orthodox Conference on Environmental Protection, The Orthodox Academy of Crete, November 1991). The Ecumenical Patriarchate of Constantinople (with ICOREC, WWF and Syndesmos), Istanbul, pp. 89–94. [Also in Greek, pp. 107–114]

Rackham, O. 1992b Vegetation history of Crete. In: A.T. Grove, J. Moody and O. Rackham (eds) *Stability and Change in the Cretan Landscape*. Cambridge University Geography Department, Cambridge, pp. 29–39.

Rackham, O. (2003) *Ancient Woodland: its history, vegetation and uses in England*. (Second, updated edition), Castlepoint Press, Dalbeattie.

Rackham, O. (2004) Pre-Existing Trees and Woods in Country-House Parks. *Landscapes*, 5(2), pp. 1–17. https://doi.org/10.1179/lan.2004.5.2.1

Rackham, O. (2006) *Woodlands*. (New Naturalist 100). Collins, London.

Rackham, O. (2012a) Comment. What can I tell you about ash disease? Easy – I told you so! *The Mail on Sunday*, 4 November 2012, p. 19. [Last accessed March 2023]

Rackham, O. (2012b) The ghosts at the ends of the earth: tree-land in four hemispheres. In: I.D. Rotherham, M. Jones, and C. Handley (eds) *Working and Walking in the Footsteps of Ghosts – Volume 1: The wooded landscape*. Wildtrack Publishing for the South Yorkshire Biodiversity Research Group and the Landscape Conservation Forum, Sheffield, pp. 118–132.

Rackham, O. (2013) Woodland and wood-pasture. In: I.D. Rotherham (ed.) *Trees, Forested Landscapes and Grazing Animals: A European Perspective on Woodlands and Grazed Treescapes*. Routledge, London, pp. 11–22.

Rackham, O. (2014) *The Ash Tree*. Little Toller Books, Beaminster, Dorset.

Rackham, O. (2015) Greek landscapes: profane and sacred. In: L. Käppel and V. Pothou (eds) *Human Development in Sacred Landscapes*. V&R Unipress, Göttingen, Germany, pp. 35–50. https://doi.org/10.14220/9783737002523.35

Rackham, O. (2018) Archaeology of trees, woodland, and wood-pasture. In: A. Çolak, S. Kirca and I.D. Rotherham (eds) *Ancient Woodlands and Trees: A guide for forest managers and landscape planners*, IUFRO [International Union of Forest Research Organizations] World Series Vol. 37, Vienna, pp. 39–60. https://doi.org/10.53478/TUBA.2020.039

Rackham, O. and Moody, J. (1996) *The Making of the Cretan Landscape*. Manchester University Press, Manchester.

Rackham, O. and Moody, J. (2004) *Η δημιουργία του Κρητικού τοπίου*. Πανεπιστημιακές Εκδόσεις Κρήτης, Heraklion, Greece.

Rackham, O. and Moody, J. (2012) Drivers of change and the landscape history of 'Cavo Sidero'. In: T. Papayannis and P. Howard (eds) *Reclaiming the Greek Landscape*. Med-INA, Athens, pp. 219–232

Rose, F. (1992) Temperate forest management: its effects on bryophytes and lichen floras and habitats. In J.W. Bates and A.M. Farmer (eds) *Bryophytes and Lichens in a Changing Environment*, Clarendon Press, Oxford, pp. 211–233.

Rose, F. (1993) Ancient British woodlands and their epiphytes. *British Wildlife*, 5, pp. 83–93.

Rotherham, I.D. (2015) The passing of a giant in the study of trees and woods – Oliver Rackham OBE (17 October 1939–12 February 2015). *Arboricultural Journal*, 37(1), pp. 1–6. https://doi.org/10.1080/03071375.2015.1027610

Rotherham, I.D. and Handley, C. (eds) (2011) *Animals, Man and Treescapes*. Wildtrack Publishing, Sheffield.

Rotherham, I.D., Handley, C., Agnoletti, M. and Samoljik, T. (eds) (2013) *Trees Beyond the Wood – an exploration of concepts of woods, forests and trees*. Wildtrack Publishing, Sheffield.

Rotherham, I.D., Jones, M. and Handley, C. (eds) (2012) *Working and Walking in the Footsteps of Ghosts. Volume 1: the Wooded Landscape*. Wildtrack Publishing, Sheffield.

Rotherham, I.D. and Pigott, C.D. (2015) A tribute to Oliver Rackham. *Arboricultural Journal*, 37(1), pp. 50–53. https://doi.org/10.1080/03071375.2015.1033168

Warren, P.M. (1972) *Myrtos: An Early Bronze Age settlement in Crete*. British School at Athens Supplementary Volume 7. British School at Athens, London.

Part I
Woodland Studies in England

CHAPTER 1

Concepts of Ancient Woodland

George Peterken

Summary

The background and development of ideas and concepts of 'ancient woodland' are explored and explained with reference to the work and influence of Oliver Rackham. This emergence of the ancient woodland paradigm came about through dialogue and fieldwork between the author and Oliver over many years from the later 1960s and throughout the 1970s and 1980s.

Keywords: ancient woodland, New Forest, wildwood, primary wood, wood-pasture

Introduction

Ancient woodland was part of my upbringing. My mother's family came from the New Forest, so walks over the forest heaths and into the woods were a staple feature of family holidays. By my teenage years, I knew that the main woods were labelled 'Ancient and Ornamental', and when eventually I became a research student, I chose to work in and around these woods. Three years later, I knew most of them in detail and had learned that their designation had been conferred by Victorian legislators when they passed the New Forest Act of 1877. Quaint though it was, the name seemed appropriate: they were indeed attractive, and by the end of my research I had pieced together how they had developed over the last four centuries (Peterken and Tubbs 1965) (Fig. 1.1).

Four years later, I was a newly appointed woodland ecologist at the Nature Conservancy's Monks Wood Experimental Station, Cambridgeshire, charged with completing the woodland section of what was eventually published as the *Nature Conservation Review* (Ratcliffe 1977). At the time, the woodland specialists were split into two groups, the Woodland Management Section at Monks Wood and the Woodland Research Section at Merlewood Research Station. Though the former was tasked with surveying and selecting woods for inclusion in the *Review*, the latter also had opinions on how this should be done, and the differences were radical. In particular, whereas the Monks Wood group relied on traditional judgement, the Merlewood group insisted on objectivity and mathematical approaches to ecology, reinforced by evangelism for computers. Whilst we at Monks Wood aimed to pick the 'best' examples of the various semi-natural woodland types, they thought we should survey a stratified random sample of British woods, classify these by some form of association analysis, then use the analysis to pick the most typical of each of the groups recognized. Indeed, they carried out such an exercise, leading (if I remember rightly) to a selection for East Anglia of just four

George Peterken, 'Concepts of Ancient Woodland' in: *Countryside History: The Life and Legacy of Oliver Rackham.* Pelagic Publishing (2024). © George Peterken. DOI: 10.53061/HADM7492

Figure 1.1 A ride within Swanton Novers Great Wood, one of the ancient coppice-with-standards woods of Norfolk, 2014. The wood includes several woodland types, including mixed woodland with Small-leaved Lime *Tilia cordata*, valley woodland with Bird Cherry *Prunus padus* and Sessile Oak *Quercus petraea* woodland on strongly acid loamy sand. (G. Peterken)

woods, only one of which later attracted a mention in any of Oliver Rackham's writings, and two of which were of fairly recent origin on former agricultural land. In Cumbria, their selection included an oakwood full of caravans.

Facing the debate within the Nature Conservancy between informed judgement and, as we saw it, impersonal objectivity, and convinced our approach was best, I thought more about our own rationale. By then, Oliver Rackham was actively investigating the ecology of the woods on the Boulder Clay plateau west of Cambridge and I had started to use the woods of central Lincolnshire to test the intriguing suggestion made by Sir Hugh Beevor (1925) that the woods mentioned in Domesday Book, 1086, could be identified by the presence of Bluebell *Hyacinthoides non-scripta*. The idea that individual, identifiable woods had survived 900 years of landscape change, and that at least one species could not colonize new woodland was not only fascinating, but also had important implications for nature conservation in woodlands.

Oliver's interest was sparked by Hayley Wood, Cambridgeshire, which he quickly traced back by name to the Ely Coucher Book of 1251 (Rackham 1975). We were both keen to use historical sources to try to understand the origins and development of the woods, and we both soon realized that the woods with richer floras and the highest incidence of rare and local plant species were woods whose origins could be traced back into the Middle Ages, and which might be remnants of pre-Neolithic woodland that had survived because wood and timber had always been useful.

None of this was entirely new. In fact, Donald Pigott (1969), working in the Peak District, had recently found that Large-leaved *Tilia platyphyllos* and Small-leaved Limes *T. cordata*, together with a suite of herbs, were largely confined to what he called 'primary' woodland, and Francis Merton (1970) confirmed that the famous ash-dominated woods had originated recently.

Colin Tubbs (1964) had traced several of the New Forest's 'Ancient and Ornamental' woods back to medieval encoppicements. Somewhat earlier, Alan Carlisle (Steven and Carlisle 1959) had described the main Highland pinewoods and their history back beyond the availability of records.

Initially, following Pigott, I labelled these older woods 'primary', a term that implied they were remnants of pre-Neolithic woodland, and I argued that they were the most important woodlands for nature conservation. Not only were they richer habitats, but they also could not be recreated once lost (Peterken 1974), whereas woods originating in the nineteenth and twentieth centuries were generally species-poor and could be eventually recreated. Only primary woods provided a direct link to the original wildwood.

At that time, I was frequently in touch with Oliver Rackham. He and I were not only involved with managing some woodland nature reserves for the Cambridgeshire Naturalist Trust, but he was also an integral part of the Historical Ecology Discussion Group we ran from Monks Wood. I also chauffeured him out to several woods in East Anglia when my work took me past Cambridge. So it was natural that sometime in 1970, I should ask him to comment on a paper I had written for the debate within the Nature Conservancy, and that was when he came back with the suggestion that we should call the medieval survivals 'Ancient Woods', not 'Primary Woods'. I have lost the letter, but I am sure the main reason was that one could not prove that a wood had existed continuously since the Neolithic, whereas historical sources did enable woods to be proved beyond reasonable doubt to be medieval. Thereafter we always called them 'ancient', but recognized that, if there were any primary woods (*sensu* remnants of pre-Neolithic woodland), they would fall within ancient woods.

At that time, despite my New Forest origins, I thought we had come up with a new idea, but a few years later, Charles Watkins (1988) pointed out that the idea of ancient woodland was current at the turn of the nineteenth century. The Board of Agriculture Reports of 1790–1813 (Jones 1961) consistently distinguished between 'woodland' and 'plantation', which in modern terms would be ancient and recent secondary woodland. At that time, the former were still almost wholly semi-natural, whereas almost all the latter were easily recognized as recent plantations. This concept also appeared in at least some of the Victoria County Histories (e.g. Nisbet and Vellacott 1907). Far from being a new idea in 1970, ancient woodland was an old idea receiving a new impetus.

How old is ancient?

Ancient woods were the woods that had originated before a particular date, but which date? Oliver initially adopted 1700 as the threshold, but I preferred 1600, which was a closer equivalent to 'medieval wood' and pre-dated early Crown plantations in the New Forest under the 1698 Act. In practice it mattered little, and later studies have often adopted the date of the earliest comprehensive source, such as the Roy maps in Scotland (*c.* 1750), the first Ordnance Survey surveyors drawings (1789 onwards) for routine sifting for the Ancient Woodland Inventory. Other countries have used the date of their first comprehensive survey (e.g. Tack, Van den Bremt and Hermy 1993). In Britain, estate maps sometimes carry evidence back to the later sixteenth century. Just occasionally, old trees and giant coppice stools provide supporting evidence on the ground.

Ancient woodland at its most straightforward

In East Anglia and the east Midlands, where modern ideas of ancient woodland were developed, woods tend to be discrete, compact and sharply defined. By examining Ordnance Survey sources back to 1800 and estate maps before 1800, many sites can simply be classified as either ancient or as secondary woodland originating after 1600. Of course, woods were sometimes cleared and restored between cartographic surveys, but this was exceptionally rare. On the whole, a wood that was present on all available maps was likely to be ancient, especially if it also possessed an irregular outline and dominated the adjacent field shapes. Medieval estate records might also refer to woodland in the same place, and one could always check in Domesday Book facsimiles that the relevant manor contained woodland in 1086.

Ancient woods incorporate a variety of earthworks that define their borders and show how they have changed, and tell us much about how they have been used. I had traced the boundary banks of the medieval encoppicements in the New Forest, and quickly adopted Oliver's custom of mapping boundary banks and so on during the course of every woodland survey. We found that the simplest ancient woods were bounded by wide, often sinuous banks with an external ditch and the overgrown remains of hedges composed of several species, most of which were also found in the interior.

Ancient woods and ancient woodland

Most woods were, of course, more complicated, and the complications took a variety of forms that forced us to refine the idea of ancient woodland (Fig. 1.2).

Maps and field surveys quickly revealed that most ancient woods had expanded and/or contracted since they were first delimited. Lengths of major banks would run within a wood, implying that the wood had been subdivided or had expanded onto neighbouring land. Alternatively, part of a perimeter would be marked by a negligible and often straight bank, indicating that part of the ancient wood had been cleared; in such instances the former ancient woodland perimeter bank often continued as a field bank surmounted by a mixed hedge (Pollard 1973). We also found tracts of ridge-and-furrow cultivation remains within the woods, indicating that parts of the woodland interior had been cleared and cultivated before

Figure 1.2 The antithesis of ancient woodland, 1967. Birch woodland on Stanmore Common, Middlesex, which grew naturally from seed after grazing rights were no longer exercised. (G. Peterken)

reverting to woodland (e.g. Monks Wood National Nature Reserve, Cambridgeshire). Some woods contained small enclosures, defined by banks and ditches, and that had once been pastures or meadows (e.g. Bradfield Woods, Suffolk). All but the smallest woods included tracks and rides bordered by ditches, and former tracks that had reverted to woodland, leaving now-functionless ditches. Many of the ancient woods were, therefore, technically amalgams of ancient woodland and post-1600 secondary woodland. Even in eastern and Midland England, individual woods, if not wholly ancient or wholly secondary, could be anything from mainly ancient with small secondary additions to small ancient woods embedded within a much larger tract of post-1600 secondary woodland. In well-wooded districts, large woods proved to be complex patchworks of former fields mixed with patches of woodland that showed no sign of ever having been cleared. The complexities made fieldwork interesting, but they also became a problem when we embarked on an inventory of ancient woods.

Ancient woodland indicators

My interest in ancient woodland started with that comment by Sir Hugh Beevor (1925), that remnants of Domesday Woods in Norfolk could be identified by the presence of Bluebells (Fig. 1.3), and a remark in a Victorian tree book that I have since been unable to trace, that

Figure 1.3 Bluebells *Hyacinthoides non-scripta*, the species that Sir Hugh Beevor thought indicated Domesday woods in Norfolk. (G. Peterken)

Wild Service Trees *Sorbus torminalis* could be found only in old woods. Similar hints had been expressed by the forest ecologists, Eustace Jones and Alex Watt (Pigott 1992). If some species occurred only or mainly in ancient woods, that implied that these species were vulnerable to changes in woodland distribution.

I put this to the test in the lime-dominated woods of central Lincolnshire. Lists of vascular plant species were compiled for 362 woods, 89 of which were demonstrated to be ancient using maps, written records and evidence from on-site archaeological features and topographical relationships. We also looked at numerous hedges and other habitats in the district. Few species proved to be confined to ancient woods, and all were rare. Thus, for example, Herb-paris *Paris quadrifolia* was found in nine ancient woods and nowhere else (Fig. 1.4). The species that came closest to being the perfect indicator, that is, one that occurred in all ancient woods but no secondary woods, was Wood Anemone *Anemone nemorosa*, found in 81 out of 89 ancient woods and just 14 secondary woods, but even this was also found in a meadow and several hedges. Some 62 herb species and a few trees and shrubs were strongly associated with ancient woods: Bluebell only just came into this category, but Small-leaved Lime and Wild Service were very strongly associated.

Woods in central Lincolnshire have long been sharply defined island habitats, now standing in an ocean of arable, so there was a strong simplifying tendency for plant species to be either woodland species or not, but even there it was possible to see that the idea of ancient woodland indicators was far from simple. Wood Anemone was also found in a meadow, road verges and several hedges. Pale Sedge *Carex pallescens* was found only in ancient woodland rides, perhaps by default, because almost all traditional meadows in the neighbourhood had been destroyed. Bracken *Pteridium aquilinum*, remarkably, was a good indicator in the core of the study area, because in ancient woods it responded to the thin skim of cover sand deposited around the last glaciers, whereas in secondary woodland this skim had been ploughed into the underlying clay before the land reverted to woodland.

Figure 1.4 Herb-paris *Paris quadrifolia*, a species that is strongly associated with ancient woodland throughout Europe. (G. Peterken)

Similar experiences in East Anglia (Rackham 1980) and acquaintance with the wild flora of other regions quickly confirmed that many species were strongly associated with ancient woodland, but few were confined to it or indeed to woodland generally. Wood Anemone is a constituent of some meadows and heaths. Oxlip *Primula elatior*, now strongly associated with ancient woodland, was formerly also a meadow species until the meadows were destroyed. In western Britain, Bluebells cover hillsides, whilst they and Primroses *Primula vulgaris* flower conspicuously along sea cliffs – and Bluebells can even be seen on treeless banks in Norfolk. Woodland plants abounded in hedges and along streamsides, especially outside the intensively farmed districts of eastern and Midland England.

Clearly, these were species that could occupy a range of habitats, but which in some districts were now more-or-less confined to woodland because we had destroyed their alternative habitats outside woodland. We nevertheless dubbed them 'ancient woodland indicators' and did in fact use them informally to identify which woods were ancient, but we concluded that it was better to identify ancient woods from other evidence. We could then use knowledge of woodland origins and land-use history to help identify which plants were slow-colonists and thus vulnerable to habitat change. Much the same can be said about the fauna. Indeed, Boycott's (1934) review of the occurrence of slugs and snails effectively anticipated our interest by four decades.

How much ancient woodland is primary?

From the outset, one of the reasons we gave for the ecological importance of ancient woods was that they were direct links with natural woodland that covered the land before the Neolithic. Indeed, I initially assumed that most ancient woods were 'primary' – land continuously covered in trees since shortly after the glaciers retreated – though we appreciated that they must have been modified by millennia of use and exploitation. Was this right, and did the balance between primary and ancient secondary differ between regions?

Archaeological and historical evidence quickly revealed that some ancient woods were secondary, originating on open ground following a medieval or earlier change in land-use ('ancient secondary woodland'). Indeed, there were many indications that most ancient woods on the chalk and oolitic limestone formations of England were secondary. Doles Wood overlies a Romano-British field system (Dewar *c.* 1926). Later excavations in Micheldever Wood, Hampshire, ahead of motorway construction demonstrated extensive pre-woodland activity (Fasham 1983). In Rockingham Forest, Northamptonshire, Roman settlement remains were so dense that archaeologists were reluctant to allow that any wood could have been continuously part of the landscape through thousands of years of settlement and fluctuating land-use, and their arguments were reinforced by spectacular finds, such as the Roman temple in Collyweston Great Wood, Northamptonshire. At least one ecologist agreed: Francis Rose thought that the only truly primary woods on chalk were the ash woods on deep soils, an example being the Alkham valley, Kent. Elsewhere, however, there seemed to be far fewer signs of prior, non-woodland land-use. For example, in central Lincolnshire, few signs of early clearance were revealed on ground occupied until the mid-nineteenth century by extensive ancient woods.

Later evidence has filled in, but not radically changed, this picture. Most of Savernake Forest has proved to be secondary (Crutchley, Small and Bowden 2009). In the densely wooded south-east England, surveys by Nicola Bannister revealed several ancient woods to be secondary, but not others. Likewise in Norfolk, some ancient woods are undoubtedly secondary, but others could be primary (Barnes and Williamson 2015). Light Detection and Ranging (LiDAR) technology has since enabled us to see through woodland and has revealed further evidence of unwoodland activity within ancient woods. The archaeological findings clearly demonstrate that much has happened in what is now ancient woodland, but do not necessarily indicate that all ancient woodland is secondary. Where field systems underlie woodland, most of the land must have been cleared, but trees and woodland plants could have persisted on the margins between fields. Where Roman villas, Iron Age hill forts and such like have been found in woods, one assumes that the immediate sites must have been cleared, but not necessarily the whole woodland. After all, we find houses today hard up against the boundaries of ancient woods.

Woodland pasturage and ancient woods

Ancient woods in England could have either a coppice or a wood-pasture history, but these two traditional forms of management were not entirely separate. Coppices had been incorporated into wood-pastures, and in any case had themselves been used as pastures once the coppice had grown out of reach. Other structures could be found: the ancient pinewoods in the Highlands, with a long history of pasturage, took both high forest and parkland forms. Pollen analyses from within ancient woods have indicated that most, perhaps all, have been through substantial changes, even if some kind of woodland has always been present on or near the site. Sidling's Copse, Oxfordshire, a remnant of the medieval forest of Shotover and Stowood, became woodland after the Romans departed, the earlier lime-dominated woodland having been cleared long before they arrived (Dark 1993). Oak-dominated woods in Eryri (Snowdonia) had a variety of histories, but had invariably been far more open at various times in the past (Edwards 1986). Much the same can be said of oakwoods in Borrowdale, Cumbria, Wistman's Wood, Devon (Bradshaw et al. 2015) and by Loch Awe, Argyll (Sansum 2005), where mixed woodland had become oak-dominated.

If any ancient woods are primary, they will have had to survive through millennia of use in prehistory, when the unclaimed land beyond settlement provided, *inter alia*, timber, small wood, leaf fodder and pasturage. This is de facto wood-pasture, which, unlike historic parklands, common woods and the unenclosed woods of the surviving medieval forests, must have been an informal mixture of lopped trees, coppice, high forest and open pasture. We can see signs of this in the Somerset trackways (Rackham 1977) and in the remnants of Roman woodland management, which probably included coppicing (Dark 2000). Enclosed coppices may still have been post-Roman innovations.

By 1086, coppices (*silva minuta*) were widespread in Lincolnshire and nearby, but most woodland was pasture (*silva pastilis*) or significant mainly for pannage (Darby 1950, and regional volumes in the *Domesday Geography* series; Rackham 1980). The ancient coppices of Norfolk were formed from the eleventh to the thirteenth centuries by enclosing parts of extensive wood-pastures (Barnes and Williamson 2015). Some woods in the Forest of Dean were mixtures of coppice, dottards and great trees in the thirteenth century (Hart 1966). The partition into coppices and wood-pasture around the upper Wye Gorge happened in the sixteenth century, but lingered until the nineteenth century in some common woods (Peterken 2008). In much of the western uplands, oak was widely planted into sparse wood-pastures in the eighteenth and nineteenth centuries to form enclosed coppices and high forest (Tittensor 1970; Smout, MacDonald and Watson 2005). Those parts of the early woodland that remained as wood-pastures steadily lost their trees, especially where they were still subject to common rights, though many still retain a scatter of old trees and raise questions about whether they are woodland, let alone ancient woodland. Ian Rotherham calls them 'shadow woods' (Rotherham 2017), but they take many forms. For instance, this article is written in the Hudnalls, Gloucestershire, a former ancient wood-pasture, where veteran oak *Quercus* spp., European Beech *Fagus sylvatica* and Small-leaved Lime trees have survived in the 200-year-old walls surrounding tiny meadows.

Composition and links to 'wildwood'

Palynology also allows us to understand ancient woods through their composition. Pollen-based studies show that, before the Neolithic, limes were much more abundant in lowland woods than they are now (Birks et al. 1975; Greig 1982), but they were not absolutely dominant. Rather, as Rackham's analysis of pollen-records from Norfolk shows, the pre-Neolithic woodland was by no means a uniform tract of lime-dominated woodland, but rather a mixture of types that could be related to the variety of types in ancient woodland today (Rackham 2003). Since then, lime has certainly decreased at several sites. At Sydlings Copse, Oxfordshire, the woodland that was cleared before the Romans contained lime, but the woodland that returned after they left did not. In Epping Forest (Baker et al. 1978), woodland was never cleared, but lost most of its lime and started its transformation into beech-dominated wood-pastures in Roman times. In the New Forest, limes are exceptionally rare, but their pollen is found in many prehistoric deposits.

Today, both species of lime are strongly associated with ancient woodland throughout their ranges in England and Wales. Combined, they occupy a wide range of soil types, with *Tilia platyphyllos* preferring well-drained calcareous soils and *T. cordata* willing to spread onto a range of strongly acid sites. Both compete strongly with other native tree species at all stages of growth, and individuals of both species continue to resprout vigorously after repeated coppicing, eventually developing into enormous spread stools.

So why do so few ancient woods contain limes? First, cattle find them particularly palatable, so it is no surprise that they declined and largely disappeared from ancient wood-pastures. Second, they demonstrate only a limited ability to colonize nearby secondary woodland, and when they do, the colonizing distances are short. Examples include secondary woodland adjacent to Groton Wood, Suffolk; some small fields embedded within the ancient lime-dominated woods of the Lower Wye Valley; and, until destroyed by archaeologists, on the Silures encampment near Caerwent. It seems that the presence of lime in ancient woods today indicates that the wood is likely to be primary, but the absence of lime does not prove that an ancient wood is secondary.

If at least some ancient woodland is primary, we must also consider how it relates to the pre-Neolithic woodland from which it descended. This, if we follow Vera (2000), was a form of wood-pasture but, if so, why did later woodland pasturage generate such changes (Fig. 1.5)? The climate has changed, which has, for example, restricted lime regeneration at its fringes (Pigott and Huntley 1978, 1980, 1981). Soils have matured with time and probably become more base poor, and this is no longer countered by subsoil disturbance associated with wind-throw. Natural grazing and browsing was probably less intense than grazing by domestic herbivores nursed through winter by leaf and grass hay.

Ancient Woodland Inventory

'Ancient woodland' was never just a concept in landscape history. From the outset, it formed part of the rationale of nature conservation where priorities had to be defined and justified. Ancient woods as a class were demonstrably richer than other kinds of woodland in some wildlife groups, and offered the possibility of direct links to the pre-Neolithic 'wildwood'. They also had historical significance, defined local landscape character and demonstrated the long co-evolution of people and nature. In practice, the great majority of the woods that were already nature reserves and Sites of Special Scientific Interest were, or appeared to be, ancient.

The practical expression of this was the ancient woodland inventory (Goldberg, Peterken and Kirby 2011). This attempt to list all ancient woods started tentatively in Norfolk in 1978 (Goodfellow and Peterken 1981) and was rolled out during the 1980s to the rest of Britain. During this process, we faced many practical and fundamental dilemmas – some outlined earlier – but in 1985 the new Broadleaves Policy accepted the inventory as part of the

Figure 1.5 Staverton Park, Suffolk, a medieval park dominated by Pedunculate Oak and still in its original condition, 2016. The many wood-pastures that have a lower density of trees presented particular problems for the ancient woodland inventory. (G. Peterken)

consultative process. However, we knew that it was an approximation based on the best available evidence, and agreed that it should always be 'provisional', that is, open to revision in the light of new information. We also acknowledged that it was bound to be incomplete – the first lists, completed under pressure, omitted all woods below 2 ha – and accepted that particular examples of other kinds of woodland could have more value for nature conservation than some ancient woods. It formed the basis of a consultation, not a final word.

Reassessing ancient woodland

The idea of ancient woodland has appealed strongly to public imagination. Since 1985, it has been a factor in national forestry policies, a priority for the Woodland Trust and a factor in planning, all facilitated by the inventory of ancient woodlands. Within nature conservation, ancient, semi-natural woodland has been recognized as the most important type of woodland and broadleaved trees are being restored to many of the ancient woods that were replanted with conifers.

All this has happened despite misunderstandings, criticism and revisionist commentary. Initially, ancient woodland was often equated with old trees, rather than long continuity of trees in general, and this still happens occasionally. The threshold of 1600, before which woodland should originate in order to qualify as ancient, was criticized as arbitrary, even though it was based on a step-change in cartographic sources, had been placed well before the onset of widespread tree planting, and in any case some such threshold was required. In practice, it has been varied where sources have been different, notably Scotland (Roberts et al. 1992). Ancient woodland indicators were often taken too literally and accordingly criticized (Day 1992; Barnes and Williamson 2015; Webb and Goodenough 2018), when they were found in manifestly secondary woods, but an alternative, more accurate label, such as 'slow colonizing species', would have caught nobody's imagination. Related to this, ancient woodland has even been criticized as a good marketing device, but this sounds like praise to those concerned for nature conservation.

Other criticisms have required some change in thinking. At the outset, we could be said to have had a simplistic view of ancient woods, based on the ancient coppices of Cambridgeshire, Huntingdonshire and Lincolnshire. We were well aware that some had expanded over adjacent land (e.g. Hayley in Rackham 1975); that others had been reduced or completely cleared, leaving their former boundaries as 'ghosts' in the nearby field patterns (Pollard 1973); that parts of others had been temporarily cleared for agriculture; and that the landscape context and its history, revealed in Hoskins (1955) and several subsequent county volumes, should be taken into account. Even so, as Barnes and Williamson (2015) say, wider awareness of archaeological evidence and landscape history sources would have revealed a more complex history. However, the problems were not just with ecologists: during the 1970s, when we tried to interest archaeologists in boundary banks and other artefacts of woodland management, we discovered that they more or less ignored woodland archaeology and the value it might have for interpreting ecological patterns. Since then, the numerous surveys and writings of Oliver Rackham have long since done much to reveal the complexities.

This issue relates to difference between the legalistic and the biological understanding of ancient woodland. The former draws hard lines on maps and disqualifies any woodland as ancient if any sign is found that its land had once lacked trees. The latter recognizes that plant and animal populations move around, albeit slowly and over short distances in some cases, and accepts that suitable habitats could have been continuously present if the intervening land-use could have supported the species under consideration, or a wooded habitat remained very close. Thus, a secondary wood formed on old meadow might have habitat continuity for the several species that can thrive in both meadows and woods. Many ancient woods contain disused tracks that have been recolonized by trees, which are strips of secondary woodland on a legalistic basis, but merge back into ancient woodland from the ecological perspective. Equally, a secondary wood that incorporates a former ancient woodland boundary (wood-relict hedge) has a degree of woodland habitat continuity. Nevertheless, there is a strong case for recognizing the continuity of semi-natural habitats collectively, rather than continuity of woodland separately, as Barnes and Williamson (2015) suggest, and this has long been central to the interpretation of individual species distributions (Peterken and Game 1984). There has also been a need to reconsider the degree to which ancient woodlands are natural. Ecologists have always distinguished ancient semi-natural woods (ASNW) from plantations on ancient woodland sites (PAWS), and accepted that the two grade one into the other. Indeed, the 'semi-natural' label acknowledges that there is an artificial element to ASNW, notably their structure and the dominance of oak as standards within coppices. Some types of ASNW have long been recognized as more artificial than others, for example, most of the oceanic sessile oakwoods of western Britain, most of the beech stands in the south-east lowlands and many of the coppices of the south-east dominated by hazel, ash, or some other single species, but later study has given us greater understanding of the degree to which traditional woodmanship manipulated composition. Since the 1970s, neglect has allowed many ASNW to develop a more natural structure, but it still seems appropriate to recognize naturalness as a multidimensional continuum grading from 100 per cent natural (the hypothetical virgin forest) to totally artificial (pole stage spruce plantation on a bog). Of course, the character of natural woodland has been much debated since Frans Vera (2000) emphasized the importance of large herbivores, but the general point remains.

Conclusion

Our understanding of ancient woodland has become ever more nuanced. Even in the sharply defined and poorly wooded landscapes of eastern and Midland England, few ancient woods have been stable, unchanging entities. Whilst many have been reduced to small fragments of their medieval extent, others long ago wholly or partially reconstituted themselves on land that was once cleared and used for cultivation or pasturage. Before the eleventh century, most woodland was not partitioned into wood-pastures and coppices, and some remained like this until the nineteenth century. Even after partition in historical times, they overlapped, merged and changed one into the other.

At the same time, our understanding of natural woodland and links to pre-Neolithic woodland has changed. Pre-Neolithic woodland was at least indirectly influenced by people through changes to the populations of large carnivores and herbivores. Herbivores reinforced soil and topographical factors that kept some ground open, and may have promoted constant change in the pattern of open and tree-covered ground. Natural disturbances also ensured that woodland changed in structure and composition. Before ancient woods were defined as discrete entities, Neolithic and later peoples must have changed the woodland to a mosaic of wood-pasture, scrub and coppice.

Demonstrating that any particular patch of ground has been wooded since before the Neolithic may be impossible. However, if one takes the ecological, not the legalistic view of habitat continuity, then direct links remain plausible. In modern times, we can observe woodland species surviving in hedges, shaded banks, tall herb communities, stream sides, meadows, sea cliffs and limestone pavements, none of which will be 'green on the map', and such habitats would have been at least as available as they are now during the medieval period. In prehistoric times, with spring grazing pressures reduced by winter mortality of cattle and the degree of woodland fragmentation much less than it later became, the opportunities for woodland species in 'semi-woodland' and non-woodland habitats would have been greater, and in any case, much of the woodland fauna and flora is associated with edges and open spaces. The contrast between woodland and non-woodland habitats must have been far less than we see now, and slow-colonizing species must have been much more capable of moving around the landscape than the modern landscape of scattered woods in intensively cropped farmland allows.

Against this background, ancient woodland is a useful concept for understanding the ecology of woodlands and communicating with a wider audience. It contains an important truth for nature conservation – that there are many species that are slow to respond to the changes we impose, and which therefore require continuity of habitat in space and time, which may mean minimal spatial or temporal isolation, which could impact on colonisation rates. We accept that this continuity may be partial, in the sense that ancient coppices are more likely to embody continuity of ground flora and ancient wood-pastures are more likely to embody continuity of mature timber habitats. We embrace the long history of human use and exploitation, which is entertaining and allows general audiences to relate to woodland ecology and conservation. Aware of the history, we are alerted to the long timescale of change in woodlands and their ability to recover in time.

Ancient woodland continues to offer a rationale for nature conservation in woodlands. Even if we think only of biodiversity, every woodland and plantation has value of some kind, but ancient woods tend to support more local and vulnerable species than recent secondary woods (especially if they are managed to maintain open spaces) – although secondary woodland may support species associated with woodland open spaces and the precursor habitats, heathland, grassland and mire. In fact, just as the Royal Commission on the Historical Monuments for England repeatedly judged that the medieval church was the most important historical monument in a parish without ignoring the others, so the ancient wood is often the most important habitat. Expressing these priorities through the ancient woodland inventory gives a basis for regulation, provided that it is used flexibly as a basis for discussion.

References

Baker, C.A., Moxey, P.A. and Oxford, P.M. (1978) Woodland continuity and change in Epping Forest. *Field Studies*, 4, pp. 645–669.

Barnes, G. and Williamson, T. (2015) *Rethinking Ancient Woodland. Archaeology and history of woods in Norfolk.* Studies in Regional and Local History 13. University of Hertfordshire Press, Hatfield.

Beevor, H. (1925) Norfolk Woodlands from the evidence of contemporary chronicles. *Quarterly Journal of Forestry*, 19, pp. 87–100.

Birks, H.J.B., Deacon, J. and Peglar, S. (1975) Pollen maps for the British Isles 5000 years ago. *Proceedings of the Royal Society of London*, B189, pp. 87–105. https://doi.org/10.1098/rspb.1975.0044

Boycott, A.E. (1934) The habitats of land *Mollusca* in Britain. *Journal of Ecology*, 22, pp. 1–38. https://doi.org/10.2307/2256094

Bradshaw, R.H.W., Jones, C.S., Edwards, S.J. and Hannon, G.E. (2015) Forest continuity and conservation value in western Europe. *The Holocene*, 25, pp. 194–202. https://doi.org/10.1177/0959683614556378

Darby, H.C. (1950) Domesday woodland. *The Economic History Review*, 3, pp. 21–43. https://doi.org/10.2307/2589941

Dark, S.P. (1993) Woodland origin and 'ancient woodland indicators': a case study from Sidling's Copse, Oxfordshire, UK. *The Holocene*, 3, pp. 45–53. https://doi.org/10.1177/095968369300300105

Dark, S.P. (2000) *The Environment of Britain in the First Millennium AD*. Duckworth, London.

Day, S.P. (1992) Origins of medieval woodland. In: P. Beswick, I.D. Rotherham and J. Parsons (eds) *Ancient Woodlands, their Archaeology and Ecology*. Landscape Conservation Forum, Sheffield, pp. 12–25.

Dewar, H.S.L. (*c.* 1926) The field archaeology of Doles. *Papers and Proceedings of the Hampshire Field Club and Archaeological Society*, 10, pp. 118–126.

Edwards, M.E. (1986) Disturbance histories of four Snowdonia woodlands and their relation to Atlantic bryophyte distributions. *Biological Conservation*, 37, pp. 301–320. https://doi.org/10.1016/0006-3207(86)90075-3

Fasham, P.J. (1983) Fieldwork in and around Micheldever Wood, Hampshire, 1973–1980. *Proceedings of the Hampshire Field Club and Archaeological Society*, 39, pp. 5–45.

Goldberg, E., Peterken, G. and Kirby, K. (2011) Origin and evolution of the Ancient Woodland Inventory. *British Wildlife*, 23, pp. 90–96.

Goodfellow, S. and Peterken, G.F. (1981). A method for survey and assessment of woodlands for nature conservation using maps and species lists: the example of Norfolk woodlands. *Biological Conservation*, 21, pp. 177–195. https://doi.org/10.1016/0006-3207(81)90090-2

Greig, J. (1982) Past and present lime woods of Europe. In: M. Bell and S. Limbrey (eds) *Archaeological Aspects of Woodland Ecology*. British Archaeological Reports, S.146, Oxford. pp. 23–55.

Hart, C.E. (1966) *Royal Forest. A history of Dean's woods as producers of timber*, Clarendon Press, Oxford.

Hoskins, W.G. (1955) *The Making of the English Landscape*. Hodder and Stoughton, London.

Jones, E.W. (1961) British Forestry in 1790–1813. *Quarterly Journal of Forestry*, 55, pp. 36–40, pp. 131–138.

Merton, L.F.H. (1970) The history and status of the woodlands of the Derbyshire limestone. *Journal of Ecology*, 58(3), pp. 723–744. https://doi.org/10.2307/2258532

Nisbet, J. and Vellacott, C.H. (1907) Forestry. In: W. Page (ed.) *Victoria History of the Counties of England, Gloucester*, Volume 2. Constable, Westminster, pp. 263–286.

Peterken, G.F. (1974) Development factors in the management of British woodlands. *Quarterly Journal of Forestry*, 68(2), pp. 141–149.

Peterken, G. (2008) *Wye Valley*. (New Naturalist 105). Collins, London.

Peterken, G.F. and Game, M. (1984) Historical factors affecting the number and distribution of vascular plant species in the woodlands of central Lincolnshire. *Journal of Ecology*, 72, pp. 155–182. https://doi.org/10.2307/2260011

Peterken, G.F. and Tubbs, C.R. (1965) Woodland regeneration in the New Forest, Hampshire, since 1650. *Journal of Applied Ecology*, 2, pp. 159–170. https://doi.org/10.2307/2401702

Pigott, C.D. (1969) The status of *Tilia cordata* and *T. playtyphyllos* on the Derbyshire limestone. *Journal of Ecology*, 57, pp. 491–504. https://doi.org/10.2307/2258394

Pigott, C.D. (1992) The history and ecology of ancient woodlands. In: P. Beswick, I.D. Rotherham and J. Parsons (eds) *Ancient Woodlands, their Archaeology and Ecology*. Landscape Conservation Forum, Sheffield, pp. 1–11.

Pigott, C.D. and Huntley, J.P. (1978) Factors controlling the distribution of *Tilia cordata* at the northern limits of its geographical range. *New Phytologist*, 81, pp. 429–441. https://doi.org/10.1111/j.1469-8137.1978.tb02648.x

Pigott, C.D. and Huntley, J.P. (1980) Factors controlling the distribution of *Tilia cordata* at the northern limits of its geographical range. *New Phytologist*, 84, pp. 145–164. https://doi.org/10.1111/j.1469-8137.1980.tb00757.x

Pigott, C.D. and Huntley, J.P. (1981) Factors controlling the distribution of *Tilia cordata* at the northern limits of its geographical range. *New Phytologist*, 87, pp. 817–839. https://doi.org/10.1111/j.1469-8137.1981.tb01716.x

Pollard, E. (1973) Hedges VII. Wood relic hedges in Huntingdonshire and Peterborough. *Journal of Ecology*, 61, pp. 343–352. https://doi.org/10.2307/2259030

Rackham, O. (1975) *Hayley Wood. Its history and ecology*. Cambridgeshire and Isle of Ely Naturalists' Trust, Cambridge.

Rackham, O. (1977) Neolithic woodland management in the Somerset levels: Garvin's, Walton Heath and Rowland's tracks. *Somerset Levels Papers*, 3, pp. 65–71.

Rackham, O. (1980) *Ancient Woodland*. Arnold, London.

Rackham, O. (2003) *Ancient Woodland*. 2nd edition, Castlepoint Press, Dalbeattie, Kirkcudbrightshire.

Ratcliffe, D.A. (1977) *A Nature Conservation Review*. Cambridge University Press, Cambridge.

Roberts, A.J., Russell, C., Walker, G.J. and Kirby, K.J. 1992. Regional variation in the origin, extent and

composition of Scottish woodland. *Botanical Journal of Scotland*, 46, pp. 167–189. https://doi.org/10.1080/03746600508684786
Rotherham, I.D. (2017) *Shadow Woods. A Search for Lost Landscapes*. Wildtrack Publishing. Sheffield.
Sansum, P. (2005) Argyll oakwoods: use and ecological change, 1000 to 2000 AD – a palynological-historical investigation. *Botanical Journal of Scotland*, 57, pp. 83–97. https://doi.org/10.1080/03746600508685086
Smout, T.C., MacDonald, A.R. and Watson, F. 2005. *A History of the Native Woodlands of Scotland, 1500–1920*. Edinburgh University Press, Edinburgh. https://doi.org/10.3366/edinburgh/9780748612413.001.0001
Steven, H.M. and Carlisle, A. (1959) *The Native Pinewoods of Scotland*. Oliver and Boyd, Edinburgh and London.
Tack, G., Van den Bremt, P. and Hermy, M. (1993) *Bossen van Vlaanderen. Een historische ecologie*. Davidsfonds, Leuven.
Tubbs, C.R. (1964) Early encoppicements in the New Forest. *Forestry*, 37, pp. 95–105. https://doi.org/10.1093/forestry/37.1.95
Vera, F.W.M. (2000) *Grazing Ecology and Forest History*. CABI, Wallingford. https://doi.org/10.1079/9780851994420.0000
Watkins, C. (1988) The idea of ancient woodland in Britain from 1800. In: F. Salbiano (ed.) *Human Influence on Forest Ecosystems Development in Europe*. Pitagora Editrice, Bologna, pp. 237–246.
Webb, J.C. and Goodenough, A.E. (2018) Questioning the reliability of "ancient" woodland indicators: Resilience to interruptions and persistence following deforestation. *Ecological Indicators*, 84, pp. 354–363. https://doi.org/10.1016/j.ecolind.2017.09.010

CHAPTER 2

The Influence of Oliver Rackham in our Understanding of Wooded Landscapes

Della Hooke

Summary

The research and writing of Oliver Rackham transformed understanding and awareness of woods and other wooded landscapes, and this chapter explores his roles and influences. In particular, this account considers the detailed histories of English woods and addresses the often little-known landscape histories of the Anglo-Saxon countryside. Like Rackham's approach to countryside history, the research is based on archival and documentary evidence, for example from the Anglo-Saxon charters, and from meticulous fieldwork and associated place-name analysis.

Keywords: woodland history, Anglo-Saxon charters, archives, fieldwork, wood-pasture

Introduction

Oliver Rackham must take a place in landscape studies alongside such greats as the late W.G. Hoskins, but in the sphere of historical landscape ecology in England his work probably at present reigns supreme. His studies of woodland and wooded landscapes remain a prime source for anyone working in the field. Not only did he investigate the past with an eye more penetrating than that of many a botanist, archaeologist or other landscape historian, but he was also one of the first to recognize the threat of the spread of pathogens to our remaining woodland. For me, the publication of his *Trees and Woodlands in the British Landscape* in 1976 (revised 1990) was a stimulus to much subsequent thought. It has been, however, his *Ancient Woodland, its History, Vegetation and Uses in England,* first published in 1980 but with a new version produced in 2003, which has become one of the most-thumbed books on my bookshelf, frequently consulted in the course of my own work. It would be some years before I was actually to meet Oliver, but his influence on my own research was to be enormous.

Wood-pasture landscapes

It became clear in recent times that the long-held view of early woodland being an unbroken close canopy had to be abandoned. It was certainly not the case in early medieval England when the first documentary evidence appears, for at this time herds of domestic stock,

Della Hooke, 'The Influence of Oliver Rackham in our Understanding of Wooded Landscapes' in: *Countryside History: The Life and Legacy of Oliver Rackham*. Pelagic Publishing (2024). © Della Hooke. DOI: 10.53061/KAFR1092

especially cattle and pigs, were taken seasonally to forage upon woodland resources, resulting in a relatively open type of woodland. It is likely, too, that this had also been the case earlier in prehistoric times, although, as Oliver has shown, the resultant wood-pasture has not necessarily been a static landscape. Indeed, some have queried whether the pollen and snail analysis does not indicate earlier cleared landscapes in certain areas such as the Weald, but pollen analysis is not always a reliable source of evidence and even studies of the type of snails present can be controversial in their interpretation.

Frans Vera's work across Europe was a milestone in understanding the nature of early woodland (e.g. Vera 2000) showing that, even without the influence of man, grazing by herbivorous wild animals would have prevented the formation of close-canopy woodland. After detailed studies, he was forced to conclude that in the lowlands of Europe the original vegetation was more park-like, with the succession of trees determined by the presence of large herbivores and certain birds such as the Jay *Garrulus glandarius* – these birds collect and bury acorns and hazelnuts on the fringes of the scrub and in grassland. Vera argued that grazing wild herbivores or domestic stock was indeed essential for the health of the woodland. It was especially valuable for the regeneration of oak (both Pedunculate/English and Sessile *Quercus robur* and *Q. petraea*) and Common Hazel *Corylus avellana*, trees otherwise crowded out by other more shade-tolerant species such as European Beech *Fagus sylvatica*, Common Hornbeam *Carpinus betulus*, English Elm *Ulmus procera*, and Large-leaved or Small-leaved Lime *Tilia platyphyllos* and *T. cordata*. Grazing helped to produce the open terrain that also provided the right environment for thorny scrub to become established, which in turn protected oak seedlings from browsing animals (Vera 2000: 165–166). He showed how oak was indeed most successful in grazed or park-like landscapes. Vera's conclusion was quite firm: 'The hypothesis that grazing leads to a retrogressive succession from forest to grassland must be rejected' (Vera 2000: 370). This is the landscape that can be characterized as typical of 'a wood-pasture landscape'. Keith Kirby (2003) has followed Vera in arguing that a wood-pasture landscape in England was far more likely to have been a landscape mosaic, with denser woodland surviving only in more inaccessible places, and with large areas of open heathland also to be found on certain soils.

The role of Oliver Rackham in woodland studies

Oliver, based as a Fellow in Corpus Christi College at the University of Cambridge (and Master 2007–8), had already questioned the concept of a pre-Neolithic landscape consisting of an unbroken close-canopy wildwood (Rackham 1998). He did, however, subject Vera's work to critical analysis in his own studies, questioning how far European studies could be related to the British situation where any wildwood had already been greatly altered by man in the prehistoric period. Here, archaeological and earthwork remains indicate that 'most of the big wooded areas of medieval England, and many ordinary wood-lots, did not emerge straight out of wildwood'. He also queried how much royal hunting really affected the landscape and just how dynamic the landscape has really been (Rackham 2003: 501–503).

He was already developing his own work significantly in the 1970s, and it was in 1976 that he published the first edition of his *Trees and Woodland in the British Landscape*. One item that captured the imagination was Oliver's recognition of Britain's landscape regions, and especially his division of lowland England into ancient and planned landscapes. The former was a hedged and walled landscape that drew upon 40 centuries 'between the Bronze Age and the age of the Stuarts' with an irregular field pattern resulting from centuries of 'casual "do-it-yourself" enclosure and piecemeal alterations' (Rackham 1976: 17). The planned landscapes, on the other hand, had been produced mainly by post-1700 enclosure. As a general rule of thumb this was captivating, although possibly somewhat simplistic: 'ancient landscapes' have undergone much change over the centuries, as I'm sure Oliver would have been the first to admit, while 'planned landscapes' have also been influenced by earlier major stages of planning, especially including those as early as the pre-Conquest period, which saw the beginning of the nucleation of settlement and the spread of open-field agriculture. It was in this book too that he 'provided a framework for distinguishing the different ways in which ancient woodland and non-woodland trees had been managed'. Wood-pastures were invaluable for grazing – providing forage by way of acorns and beech-mast and also browse-wood or leaf fodder. Thus it was said that 'He firmly set wood-pastures alongside ancient coppice as worthy of conservation' (Kirby and Perry 2014: 255).

Oliver did not always find face-to-face communication easy, but this trait left him free to delve deeply into his own world of trees and woodland. I vividly remember walking with him across a beach on the island of Limnos in Greece in 2004 (Fig. 2.1) at a conference organized by the Permanent European Conference for the Study of the Rural Landscape. Here we examined together the ancient olive trees; this was an amazing place littered deeply with ancient pottery sherds scrunching underfoot. On the following field trip, however, he was followed eagerly away from the main party and off the route by a small group of fascinated followers. He (and they) remained totally oblivious to the efforts of the trip leader to gather everyone together or the fact that the coach was waiting to take delegates onto the next venue!

Although Oliver produced an impressive volume of publications over the years, much of it inspired by his local Cambridgeshire woodlands such as Hayley Wood (Rackham 1975), it is perhaps his *Ancient Woodland* book that has made the greatest impression. This was first published in 1980 and was one of the first books to explore the complex variety of native woodlands, drawing upon the science of historical ecology; neither did he neglect the trees of parks, commons and present-day forests. He recognized that many books at the time concerning forestry practices were limited to modern plantations, usually of conifers, and his was the first attempt to consider the wide variety of ancient woods and the separate history of each type, together with the role of woods in human affairs.

Figure 2.1 Oliver attends a conference in Limnos, Greece, in 2004. (C. Hooke)

A new edition of this book was published in 2003 by Castlepoint Press some 23 years later, by which time research into woodland history had widened to cover much of Europe, Japan, Tasmania, Texas and West Africa. However, he felt that research still tended to be stifled by the divisions between different disciplines, his conclusion being: 'Those who do no fieldwork, or never study historic buildings, will never appreciate that archives do not tell the whole story. In this field many amateurs perform at least as well as professionals' (Rackham 2003: xvii). Between the two editions, steps had been taken to amass seed banks, classify woodlands (largely by George Peterken, e.g. 1993) and to study the relations between ancient woodland and wildwood. He also tackled the subject of wood-pasture, a source of forage for stock, as distinct from 'ancient woodland': this gave rise to a more open landscape, one in which individual trees might also flourish (Fig. 2.2). It was a source of tree fruits such as acorns or beech-mast and also of leaf fodder as well as grassland pasture: 'Wood-pasture results ... from the fact that cows and sheep cannot climb. Even modern cattle, after hundreds of generations of being bred as grassland animals, still hanker after tree leaves' (Rackham 2003: 496).

Oliver's study of individual tree species has remained one of the most useful available. He showed how the fortunes of many species have fluctuated over the centuries, with Common Holly *Ilex aquifolium* for instance increasing in wood-pasture, where it was a valuable source of leaf fodder. Others, such as the elm, had been affected by disease in earlier centuries but this has probably been worsened by the modern-day transport of timber and bark, which has helped to disperse virulent strains of disease. elm has not been the only species to have suffered in this way: Today the European Ash *Fraxinus excelsior* and Sweet Chestnut *Castanea sativa*, even the oak, may be under serious threat.

Oliver also noted how the conservation scene had changed out of recognition, embodying something that he described as 'applied historical ecology' (2003: xviii), a move that witnessed a growing appreciation of ancient woodlands and the compilation of a register of these, including details of ancient trees. The then Nature Conservancy Council had begun work on producing county-by-county inventories of ancient woodland in 1981, mapping and cataloguing areas of woodland that could be traced back to at least AD 1600, the majority of which they believed to be 'primary', that is 'surviving fragments of primeval forests, the climax vegetation type of this country', but also noting 'Ancient Secondary Woods', mostly comprising small, secondary, semi-natural stands within ancient sites representing regrowth or woods slightly modified by the planting of later tree species (e.g. Worcestershire Inventory of Ancient Woodland: provisional: NCC 1986: 3–4). The Woodland Trust, a private body, had been set up in 1972 with the aim of protecting and campaigning 'on behalf of this country's woods, plant trees, and restore ancient woodland for the benefit of wildlife and people' (www.woodlandtrust.org.uk), launching its Ancient Tree Forum in 2000, a venture 'to secure the long-term future of ancient and veteran trees through best practice in their management' (www.ancienttreeforum.co.uk). Having brought its study of ancient woodland up to date, English Nature (as Natural England) has embarked on a more detailed study of wood-pasture landscapes.

Much of this has been inspired by Oliver's work. His *Ancient Woodland* covers a wide spectrum of woodland history, covering changing management practices and their effects; sources of evidence ranging from pollen and tree-ring analysis, the study of earthworks to surface features and standing buildings; and both written evidence and oral tradition. The effect of different soils and drainage, associations with other types of flora and fungi, and the nature of individual tree species, with what has been probably the most detailed historical study of woodland presently available, have made this an indispensable volume for the landscape historian or anyone else involved in woodland studies.

Few forests have received such detailed attention as the surveys of individual woodlands carried out by Oliver not far distant from Cambridge and of Hatfield Forest in Essex. This latter is a 403.2-hectare Site of Special Scientific Interest now managed in almost pristine condition by the National Trust. As Oliver noted in the title of his book *The Last Forest*,

Figures 2.2 Scenes in the New Forest, Hampshire, June 2008. (D. Hooke)

published in 1989, it is the only remaining intact royal hunting ground established by the Norman kings in the eleventh century. As he noted in 1976:

> Hatfield is of supreme interest in that *all* the elements of a medieval forest survive: deer, cattle, coppice woods, pollards, scrub, timber trees, grassland and fen, plus a seventeenth-century lodge and rabbit warren. As such it is almost certainly unique in England and possibly the world ... The Forest owes very little to the last 250 years ... Hatfield is the only place where one can step back into the Middle Ages to see, with only a small effort of the imagination what a Forest looked like in use.
>
> Rackham 1976: 163–164

But Oliver's work was not here limited to botanical issues. Although he does discuss the present-day woods and trees, including the presence of particular tree species and ancient trees, plus the characteristic flora, he extends his study to the insects, bats, birds and other creatures that make the forest their home. He looks at the practices such as pollarding and coppicing that have shaped the forest over the years, and sets all this within a detailed history that embodies its early archaeology and its subsequent use for hunting, for wood-pasture, even as a place of poachers and hermits. Historical disputes over ownership and usage are outlined, and the structure of the forest buildings portrayed. It is, in other words, a complete 'landscape history'.

Indeed, in recognition of his work in this field, Oliver was made Honorary Professor of Historical Ecology at Cambridge in 2006. At the time, he was widening his field of expertise – at first he was decidedly shaky in his knowledge of Old English terms for some tree species (see below) and the interpretation of place-names (for which he was criticized by the late Margaret Gelling), but this he was remedying (I am honoured that Oliver recognized my own work on the early medieval period in the latest edition of his *Ancient Woodland*). Given all this, few other books offer such a comprehensive guide to the nature of woodland. He had also widened his field geographically and had worked with Jennifer Moody on the Cretan landscape (Rackham and Moody 1996) and elsewhere in the Mediterranean with A.T. Grove (Grove and Rackham 2001).

It was also Oliver who drew attention to the conservation problem caused by the loss of distinction between woodland and wood-pasture, with woods often overgrazed by deer but wood-pastures undergrazed, leading to a suffocating undergrowth of Bracken *Pteridium aquilinum*, or with agricultural-type grassland replacing earlier grassland and heath (Rackham 2013: 20). He claimed that wood-pasture had been neglected by foresters ('despised' is how he put it) and by grassland specialists 'as not being proper grassland' (Rackham 2003: 495–496). He also advised, however, against the attempted recovery of a lost wildwood landscape unaltered by man's activity. Both woodland and wood-pasture sites have been subject to change over time and are today the product of human activity such that each site is unique, formed as ecosystems have adapted over many generations. He declared that conservationists all too frequently 'set up a simplified and idealised model of what an ecosystem ought to be, and to try to make real ecosystems conform to it', often failing to understand and perpetuate the features 'that make each site unique and different from other sites. A site should not be forced to turn into something that it is not' (Rackham 2013: 20–21). He stated that every year was different to every other year, with instability an essential mechanism for maintaining wood-pasture. He was indeed sometimes quite openly critical of conservationists.

An outing with Oliver would always turn into something special. During fieldwork, he would often notice items that others had missed, and he was also able to critically evaluate long-accepted beliefs so that there was always something new and refreshing to entertain and educate his audience. His lectures, too, were always popular (many indeed given at various seminars at Sheffield Hallam University).

The author's own work

I had been working with Anglo-Saxon charters and place-names for many years and interpreting the evidence they provided, especially for the recognition of regional *pays* in the early medieval landscape. This was a subject I had thought about when considering returning to university to carry out postgraduate study after years of teaching and rearing a family. Although charter research had been inspired by a single undergraduate lecture given by the late Harry Thorpe, the latter could not see any potential in this when I approached my old department (where he was now Head of Department), but I received enormous encouragement from Margaret Gelling, who also introduced me to the English Place-Name Society (I later became member of the Council for Name Studies in Great Britain and Ireland). My PhD study was published as a British Archaeological Report, *The Anglo-Saxon Landscapes of the West Midlands: the Charter Evidence*, in 1981, but as time went on university posts remained short-term and unreliable. This had its compensations, because working as a freelance consultant I came to be more and more in touch with ecologists, especially while assessing the Department for Environment, Food and Rural Affairs' Countryside Stewardship Agreements for many years from the late 1990s. I was also inspired by others with similar interests whom I met at European conferences such as those organized by the International Forest Research Organisations Group. Over the years, I became increasingly interested in the survival today of England's traditional landscapes (Hooke 2010a). Some of the most interesting of these were woodland and wood-pasture environments, the latter wooded areas primarily used for seasonal grazing – hence my delight in coming across Oliver's work. Such landscapes usually fell under Oliver's classification of 'Ancient Countryside'.

Here, knowledge of Old English terms came in handy. One suffix that appeared regularly in major and minor place-names was the term lēah. As a result of my research, I was able to show that the original meaning conveyed by this term was 'wood-pasture' rather than the more usual interpretation of 'clearing'. It was more closely related to the Old High German *lōh*, 'copse, grove, woodland, undergrowth, scrub', one associated with 'cutting off' and 'breaking', suggesting the use of available timber or even of intermittent clearing, although the meaning may have changed a little over time (Watts 2004: xlvi; Hooke 2008). Thus *Andredes leag*, a name given to the Kentish Weald, was said in the Anglo-Saxon Chronicle to extend 'a hundred-and-twenty miles long or longer from east to west, and thirty miles broad' (Swanton 1996: 14–15, 84): clearly this could not have referred to a giant clearing; its nature is also described in a charter granting pasture rights to Canterbury minster in the mid-eighth century said to lie *in saltu Andoreda*, 'in the wood-pastures of Andred' (Sawyer 1968: S 25), and in 1018 as *Andredes Wealde*, 'the wood of Andred' (ibid. S 950)

Such landscapes were usually to be found in areas of more marginal soils. An example is offered by Warwickshire documents, where the more acidic clays or infertile drifts of the Arden in the north of the county were of limited agricultural value and carried much woodland and heath. In Anglo-Saxon times this region was known as the Arden, from a British term *arddu* meaning 'high land'. The western part of this region then lay within the kingdom of the Hwicce while the eastern part formed part of Greater Mercia. Even the territories of two Staffordshire folk groups, the Pecsæte and the Tomsæte, reached southwards from their riverine heartlands into the woods and heathlands of the Lickey Hills, which formed the early northern boundary of the Hwiccan kingdom. Indeed, territorial boundaries often ran through these lesser populated marginal regions; another example was that between the Magonsæte and the Wreocensæte in the Welsh borderlands (Hooke 1986). Pre-Conquest charters often show how in England marginal areas of wood-pasture such as these were linked to regions of more intensive agriculture in grants or leases of early medieval estates, links that would often become fossilized or expressed later through manorial or ecclesiastical linkages (but often masked within entries for capital manors in Domesday Book) (Fig. 2.3).

Such regions even carried their own place-name types with Old English *lēah* often indicative of the kind of countryside to be found there. It was usually only from late spring until early

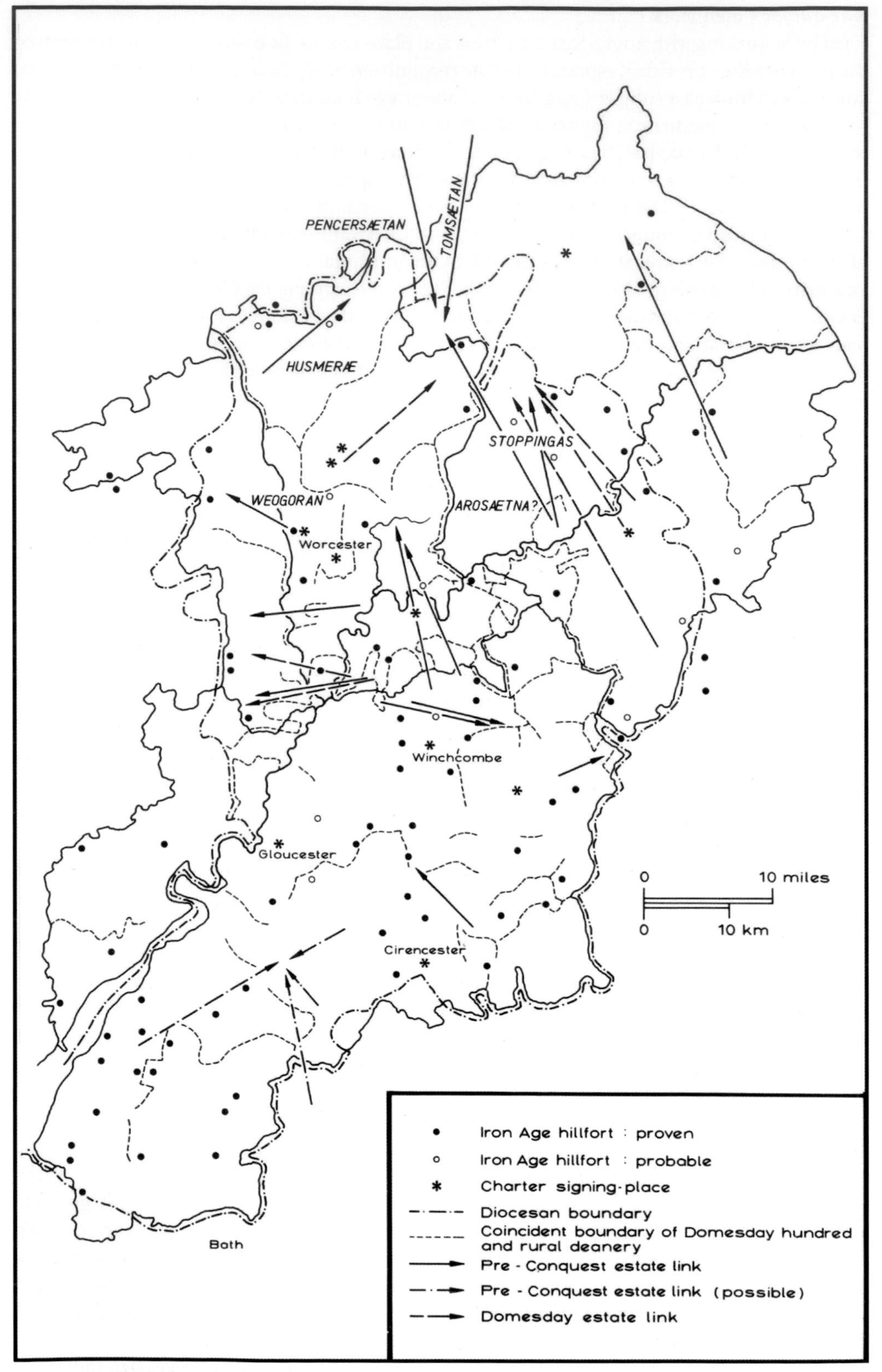

Figure 2.3 Examples of linked estates, these in modern-day Gloucestershire (Hooke 1985: 82, figure 22)

autumn that the pastures were utilized in this way, crops of acorns or beech-mast being especially valuable for fattening swine in late summer and early autumn. The charters of Kent are particularly full of detail concerning this practice, especially within the Weald (then known as *Andredes leag*, as previously mentioned); here the summer settlements and pastures were known as dens. The practice of transhumance – the movement of herds to seasonal pastures – had the added benefit of moving domesticated stock away from crops during the growing season. The animals moved seasonally in this way to forage in the woods were mainly cattle and pigs, but studs for horses were often also found in such areas, and in some woods, such as Needwood in Staffordshire, herds of semi-wild horses were kept from which young animals might be taken to be broken in for riding – the Church especially required many for its preaching duties when priests were visiting their *parochia*. There are many instances of patterns of droveways linking agricultural and wood-pasture regions that have been partly fossilized in later patterns of roads and bridleways (Everitt 1986).

A second term I was able to explore was Old English *haga*, found in Anglo-Saxon charter-bounds specifically in well-wooded regions. Fieldwork across the country showed that this term had described not a normal 'hedge' but a more substantial bank and ditch, the bank probably also topped by a living or a dead thorn hedge, which demarcated areas established chiefly for the hunting of deer – almost the first forests or parks. In Hampshire, some of these extended for over 5 km; one later marked the boundary of a Norman forest and others surrounded whole parishes both there and in Berkshire (Hooke 1989) (Fig. 2.4). A similar term *haie* was used in Domesday Book to describe much smaller enclosures used for the capture of deer, and other similar terms crop up over successive centuries, including the 'ha-ha', used to refer to a hidden boundary step and used by seventeenth- and eighteenth-century landscape designers.

Settlement was expanding into wooded regions before the Norman Conquest, with temporary settlements gradually giving way to more and more permanent settlement as time went by. The wooded region of west Worcestershire lying to the north of the River Teme, known in Anglo-Saxon times as *Weogorenaleage*, was ringed by such settlement by the later Anglo-Saxon period. Here, too, patterns of early droveways can be detected (Hooke 1982: 94–98; 2008: 371, figure 1; 2012) (Fig. 2.5). Such expansion of settlement and associated fields increased in medieval times. In the Warwickshire Arden, surviving documentary evidence shows how manorial lords were deliberately encouraging movement into that area from more heavily populated parts of the county – the Avon Valley and the southern Warwickshire Feldon – in order to raise the revenues from their Arden manors (Roberts 1968). Settlement, however, remained largely dispersed, and many new farmsteads were to be moated – probably more a symbol of status than an actual need for drainage or protection (Hooke 2013a). However, in this region pastoral farming was to prevail, continuing the ancient tradition. Medieval parks, too, commonest in Arden, helped to preserve the historical wood-pasture habitat (Hooke 1993; 2013a).

The pre-Conquest evidence also enables one to study the distribution of particular species of trees. The charter evidence covers only part of the country, being richest for the midlands and southern England. Early place-names (recorded up to 1086) are more ubiquitous, but again are fewer further to the north in regions not covered by Domesday Book. Although neither of these sources offers comprehensive cover, they do much to fill out the general picture (Hooke 2010b). The trees commonly associated with wood-pasture – namely the oak and the ash, but also the Common Yew *Taxus baccata*, holly and lime – help to identify some of the wood-pasture regions, almost always falling into Rackham's 'Ancient Countryside', such as west and central Worcestershire, north-west Wiltshire, west Berkshire and much of Hampshire (Fig. 2.6). Not only was the oak the second-most frequently recorded tree (after the 'thorn') in charters and the third found in early place-names, but it was the most important timber tree for building. In wood-pasture, however, its acorns were a major source of sustenance for the herds of swine taken seasonally into the woods. The leaves of the ash, even commoner than

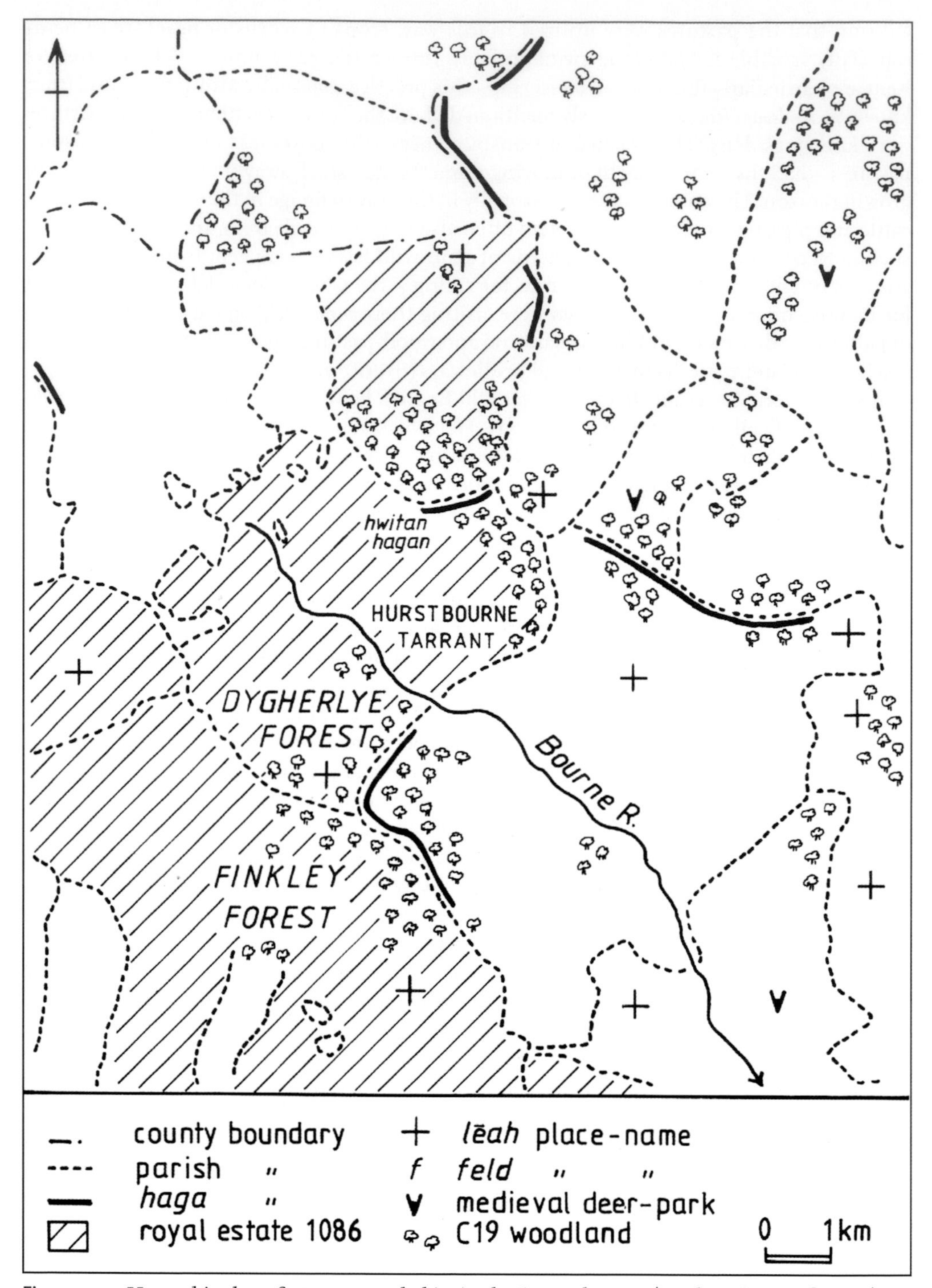

Figure 2.4 Hampshire *haga* features recorded in Anglo-Saxon charters. (Hooke 1989: 127, figure 7)

the oak in early place-names, could also be gathered as animal feed, while holly leaves might also be mashed for animal fodder: it was the more tender new leaves that could be gainfully gathered for this purpose, often stored for the winter season. It is difficult to compile a definitive list of the trees recorded in these sources because a few Old English terms cannot be reliably translated – at first Oliver missed references to beech in his selection of charter

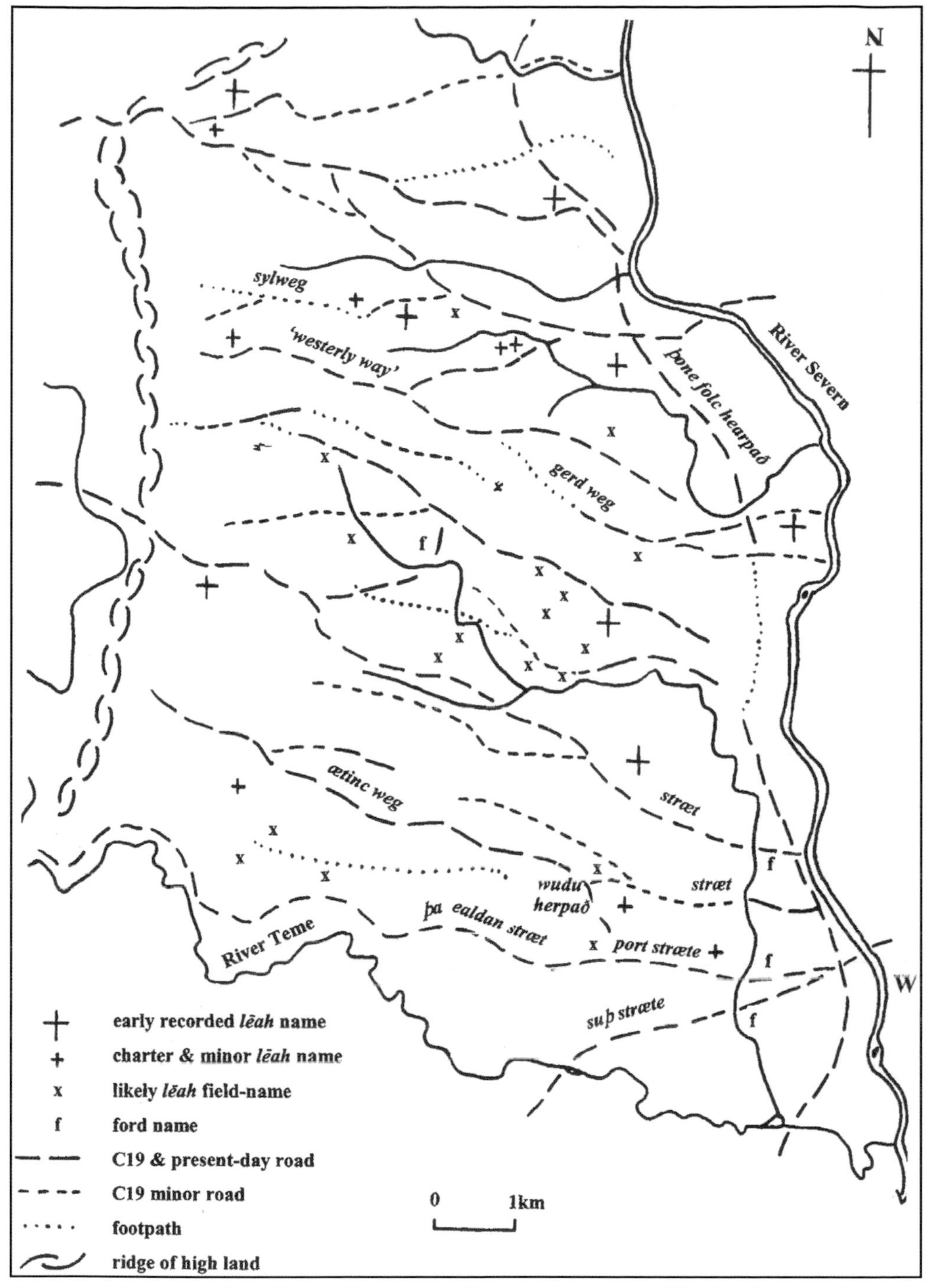

Figure 2.5 Droveways through a wood-pasture region in west Worcestershire. (Hooke 2012: 38, figure 4.2)

references (recorded as Old English *boc* and *bece* in charters and place-names), not recognizing the Old English terms for this tree (Rackham 1976: 54); the yew can also be difficult to recognize with certainty because Old English *ewe* can refer to either a tree or a female sheep. Neither did Oliver (like some others) understand the implications of the Old English terms

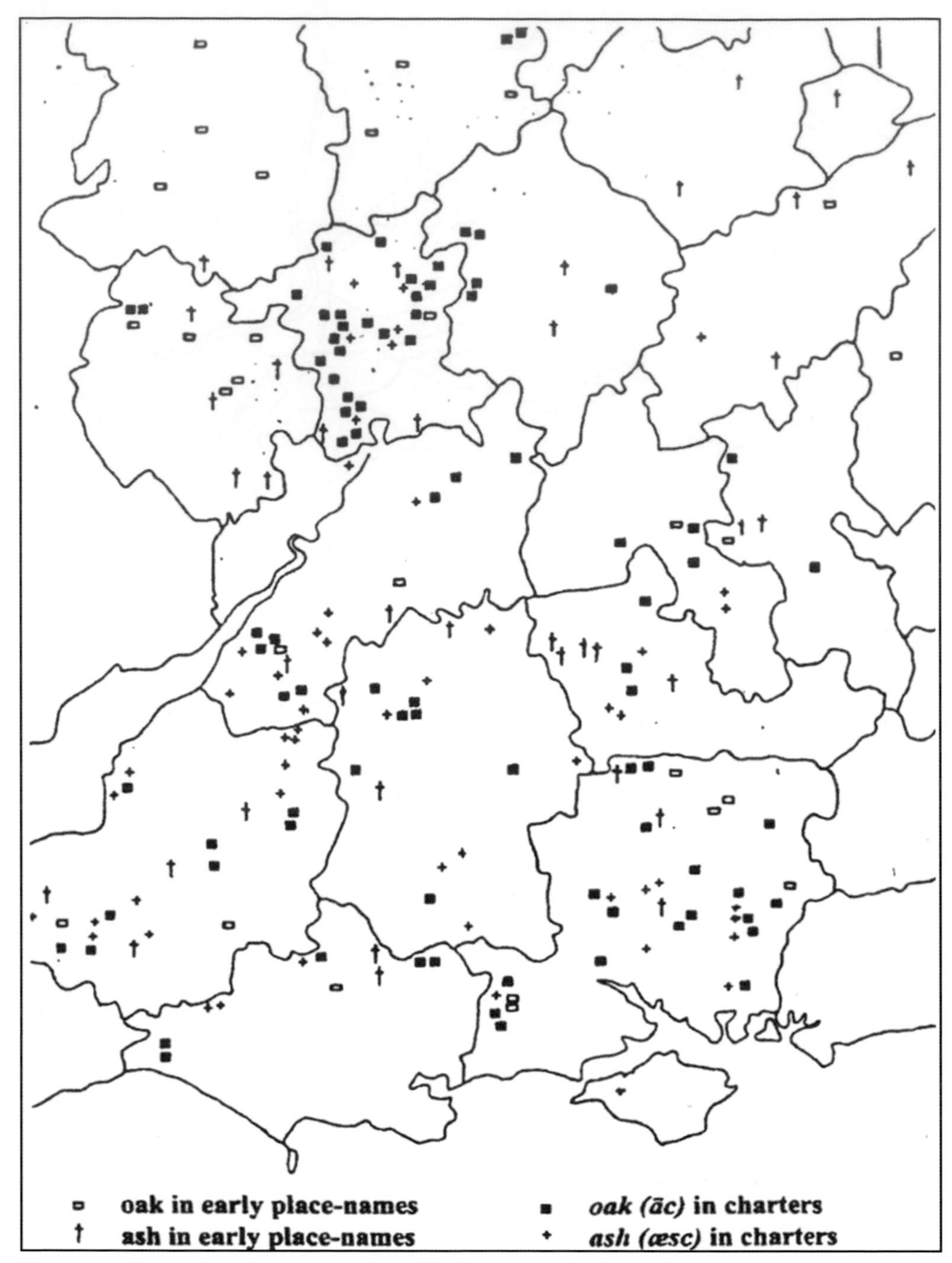

Figure 2.6 Trees of wood-pasture regions. (Hooke 2010a, p. 62, figure 4)

haia and *haga*, which could in some instances relate to very large areas (Hooke 1989; Rackham 2003: 188–189), and he was not at ease with the interpretation of some place-names despite the reliable research carried out by place-name scholars.

Over time, I became increasingly interested in the value of such traditional landscapes and their present-day survival (Hooke 2013b). These have, of course, included wood-pasture landscapes, and more recently, I have traced the history of such wood-pasture landscapes

Figure 2.7 Ancient pollards in Moccas Park, Shropshire, July 2003. (D. Hooke)

throughout history, especially as they have played an ever-decreasing role in farming (Hooke forthcoming). In noting areas of surviving wood-pasture, one is immediately drawn to the parkland of Moccas in the Welsh borderland region of Herefordshire, a landscape partly redesigned by Lancelot 'Capability' Brown in the eighteenth century that may originally have formed part of the fourteenth-century Forest of Dorstone and retains a fair number of ancient pollards. These were described by the Revd Francis Kilvert in 1876 as 'those grey old men of Moccas' (Fig. 2.7) (Harding and Wall c. 2000), and other old limes and yews are still found on the hill-slopes above. The very name Moccas may be a corruption of Welsh *mochros*, 'moor or place of pigs', probably indicating early wood-pasture usage. Ancient wood-pasture trees are limited on Cannock Chase, once part of Cannock Forest in Staffordshire, because both woodland and heathland gave way to conifer planting by the Forestry Commission after the First World War, but old oaks are found, remnants of former wood-pasture usage, in Brocton Coppice (Fig. 2.8). These may have survived because the area was commandeered by the War Office for the construction of a massive training camp during the First World War. Another extensive area of former wood-pasture is that found in Sutton Park near Birmingham, a park enclosed by the fourteenth century out of an extensive chase given to the earls of Warwick by Henry I. But here most of the mature trees were felled, to supply building timber for public buildings such as schools or to raise money from sales of timber, when the park was given by Bishop Veysey to the borough of Sutton Coldfield in the sixteenth century. The only large ones surviving are those that form part of a private estate on the north side of the park. The hollins (Holly groves), however, were particularly valued as a source of fodder for the stock of the townspeople and so escaped destruction, and these are still found, if today overgrown, within their ancient wood banks (Fig. 2.9).

Future directions and Oliver's long-term influence

Oliver has left a huge legacy that will help future practitioners and researchers. He was one of the first to distinguish between different kinds of woodland, and he also discussed the characteristics of wood-pasture landscapes. The latter are now much more highly regarded for their aesthetic, ecological and other environmental benefits. I don't remember him being drawn

Figure 2.8 Ancient oaks in Brocton Coppice on Cannock Chase, Staffordshire, February 2006. (D. Hooke)

Figure 2.9 Holly and oak: wood-pasture trees in Sutton Park near Birmingham, November 2010. (D. Hooke)

into ethical debates, but these will inevitably play a greater part in future planning – and his work will continue to find a place in such debates. As population pressure increases, more and more land is taken up for building and the production of food, and it becomes imperative to preserve or even increase our tree legacy (Hooke forthcoming). However, members of a generally aging public can be averse to change, clinging to the landscapes of their youth: there was an outcry when trees and scrub began to grow again in the Lake District after the foot-and-mouth outbreak in 2001, and others have resisted tree planting in ancient Wychwood in Oxfordshire or in Warwickshire. However, upland sheep farming, which has in many areas scoured the hillsides of trees and shrubs by overgrazing, although it maintains a deeply entrenched way of farming life, can only be upheld by generous government grants. Whether genetically modified foods will help to produce higher yields from current agricultural land without adverse ecological damage remains to be seen, but crop wastage is something that could be more easily remedied. Given the demand by many supermarket chains for fruit and vegetables that are aesthetically pleasing, huge amounts of produce are simply discarded, only a small proportion finding their way into jams, preserves and so on – this at a time when gardens and allotments are getting smaller or being lost to make way for higher building densities. Trees, too, can be disliked by those employed to keep roads, footpaths and railway lines clear of obstacles. This is despite their now recognized role (along with upland peat bogs) in helping to reduce pollution. Fortunately, there is a growing awareness by some members of the public, encouraged by county wildlife trusts and the Woodland Trust, among others, and now being addressed by Natural England, of the benefits of trees, woodlands and traditional wood-pasture landscapes. Well-known regions such as the New Forest are carefully conserved, as are two other smaller areas in the Midlands where wood-pasture landscapes have been preserved or even reinstated – the first on the Cotswold scarp and the second on the Malvern Hills (Figs 2.10 and 2.11).

Figure 2.10 Remnant wood-pastures along the Cotswold scarp on the Worcestershire/ Gloucestershire border, July 2012. (D. Hooke)

Figure 2.11 Woodlands below the Malvern Hills on the Herefordshire/Worcestershire boundary, November 2014. (D. Hooke)

In the latter area, conservators have allowed trees to regrow on the lower slopes, but grazing is maintained to keep not only the upper slopes but also the lower woods sufficiently open for the encouragement of the wild fauna, flora and fungi that form part of this valuable ecological habitat (Duncan et al. 2018; Hooke forthcoming). Many people now volunteer to help with the management of such woodland.

Today, largely owing to Oliver's work, we have a broader understanding of the role of trees and woodland in the landscape. They are part of a wider fragile world of self-sustaining mycological fungi, also creating their own environment and soil conditions through leaf litter and decaying wood, providing rich habitats for wildlife and flora.

Bibliography

Ancient Tree Forum. www.ancienttreeforum.co.uk. [accessed 10 December 2018]

Duncan, I., Garner, P., Comont, R. and Creed, P. (eds) (2018) *The Nature of the Malverns.* Pisces Publications, Newbury.

Everitt, A. (1986) *Continuity and Colonization: The evolution of Kentish settlement.* Leicester University Press, Leicester.

Grove, A.T. and Rackham, O. (2001) *The Nature of Mediterranean Europe: An ecological history.* Yale University Press, New Haven, CT and London.

Harding, P.T. and Wall, T. (eds) (*c.* 2000) *Moccas: An English deer park: The history, wildlife and management of the first parkland nature reserve.* English Nature, Peterborough.

Hooke, D. (1981) *The Anglo-Saxon Landscapes of the West Midlands: The charter evidence.* British Archaeological Reports 95, Oxford. https://doi.org/10.30861/9780860541493

Hooke, D. (1982) The Anglo-Saxon Landscape. In: T.R. Slater and P.J. Jarvis (eds) *Field and Forest: An historical geography of Warwickshire and Worcestershire.* Geo Books, Norwich, pp. 79–103.

Hooke, D. (1985) *The Anglo-Saxon Landscape: The Kingdom of the Hwicce.* Manchester University Press, Manchester.

Hooke, D. (1986) Anglo-Saxon territorial organization: the western margins of Anglo-Saxon Mercia. Dept of Geography Occasional Paper No. 22, University of Birmingham.

Hooke, D. (1989) Pre-Conquest woodland: its distribution and usage. *Agricultural History Review,* 37, pp. 113–129.

Hooke, D. (1993) *Warwickshire's Historical Landscape. 1: The Arden.* School of Geography, University of Birmingham.

Hooke, D. (2008) Early medieval woodland and the place-name term *lēah*. In: O.J. Padel and D.N. Parsons (eds) *A Commodity of Good Names.* Shaun Tyas Publications, Donington, pp. 365–376.

Hooke, D. (2010a) The nature and distribution of early medieval woodland and wood-pasture habitats. In: H. Lewis and S. Semple (eds) *Perspectives in Landscape Archaeology Papers Presented at Oxford 2003–5.* British Archaeological Reports, International series S2103, Archaeopress, Oxford, pp. 55–65.

Hooke, D. (2010b) *Trees in Anglo-Saxon England: Literature, lore and landscape.* Boydell, Woodbridge.

Hooke, D. (2012) 'Wealdbæra and Swina Mæst': wood-pasture in early medieval England. In: S. Turner and B. Silvester (eds) *Life in Medieval Landscapes. People and Places in the Middle Ages.* Windgather, Oxford, pp. 32–49. https://doi.org/10.2307/j.ctv13gvg52.7

Hooke, D. (2013a) The distribution of symbols of status in medieval Warwickshire. *Transactions of the Birmingham and Warwickshire Archaeological Society,* 117, pp. 49–71.

Hooke, D. (2013b) Re-wilding the landscape. Some observations on landscape history. In: I.D Rotherham (ed.) *Trees, Forested Landscapes and Grazing Animals. A European perspective on woodlands and grazed treescapes.* Routledge, Abingdon and New York, pp. 35–50.

Hooke, D. (forthcoming) Historical wood-pasture in England and Wales and its subsequent fate. In: I.D. Rotherham and C. Handley (eds) *Ecology, History and Management of Wood-Meadows and Wood-Pastures.* Wildtrack Publishing, Sheffield.

Kirby, K. (2003) What might a British forest-landscape driven by large herbivores look like? *English Nature Research Report* No. 530, English Nature, Peterborough.

Kirby, K.J. and Perry, S.C. (2014) Institutional arrangements of wood-pasture management. Past and present in the UK. In: T. Hartel and T. Plieninger (eds) *European Wood-Pastures in Transition. A social-ecological approach.* Earthscan/Routledge, London and New York, pp. 254–270.

NCC (Nature Conservancy Council) (1986) *Worcestershire Ancient Inventory of Ancient Woodland.* NCC, Peterborough.

Peterken, G.F. (1993) *Woodland Conservation and Management,* 2nd edition. Chapman and Hall, London.

Rackham, O. (1975) *Hayley Wood: Its history and ecology.* Cambridge and Isle of Ely National Trust, Cambridge.

Rackham, O. (1976) *Trees and Woodlands in the British Landscape.* J.M. Dent and Sons, London.

Rackham, O. (1980) *Ancient Woodland, its History, Vegetation and Uses in England.* Edward Arnold, London.

Rackham, O. (1989) *The Last Forest: The story of Hatfield Forest.* J.M. Dent and Sons, London.

Rackham, O. (1990) *Trees and Woodlands in the British Landscape,* 2nd edition. J.M. Dent and Sons, London.

Rackham, O. (1998) Savanna in Europe. In: K.J. Kirby and C. Watkins (eds) *The Ecological History of European Forests.* CABI, Wallingford, pp. 1–24.

Rackham, O. (2003) *Ancient Woodland, its History, Vegetation and Uses in England,* 2nd edition. Castlepoint Press, Dealbeattie.

Rackham, O. (2013) Woodland and wood-pasture. In: I.D. Rotherham (ed.) *Trees, Forested Landscapes and Grazing Animals: A European Perspective on Woodlands and Grazed Treescapes*. Routledge, Abingdon and New York, pp. 11–22.

Rackham, O. and Moody, J. (1996) *The Making of the Cretan Landscape*. Manchester University Press, Manchester.

Roberts, B.K. (1968) A study of medieval colonization in the Forest of Arden, Warwickshire. *Agricultural History Review*, 16, pp. 101–113.

Sawyer, P.H. (1968) *Anglo-Saxon Charters: An Annotated List and Bibliography*. Royal Historical Society, London.

Swanton, M. (translator and ed.) (1996) *The Anglo-Saxon Chronicle*. J.M. Dent and Sons, London.

Vera, F.W.M. (2000) *Grazing Ecology and Forest History*. CABI, Wallingford. https://doi.org/10.1079/9780851994420.0000

Watts, V. (ed.) (2004) *The Cambridge Dictionary of English Place-Names*. Cambridge University Press, Cambridge.

Woodland Trust. www.ancienttrust.org.uk. [accessed 10 December 2018]

CHAPTER 3

How the Wildwood Worked: Rackham's Contribution to Forest Ecology

Adrian C. Newton

Summary

Oliver Rackham's insights into landscape history have informed woodland and forest managers over several decades. And yet when his great volume *Ancient Woodland* (Rackham 1980) was originally published, it was largely overlooked by the British Forestry Commission. This chapter examines the legacy of Rackham's thinking, observations and written outputs in relation to the communication of countryside history and the shaping of current debates around ideas of authors such as Frans Vera, and to initiatives such as rewilding. This discussion is illustrated by reference to modelling and field studies in the English New Forest and from the Walnut-fruit forests of Kyrgyzstan.

Keywords: forest ecology, woodland composition, structure and dynamics, countryside history, Frans Vera

Introduction

What is Oliver Rackham's legacy to the science of forest ecology? It seems almost unfair to ask this question. For anyone who knew Oliver or enjoys his writings, it is clear that he was a genuine polymath, a kind of Renaissance woodsman, who displayed deep knowledge and understanding of a remarkably wide range of subjects. Partly what excited him were the frontiers between different disciplines, such as between history, ecology and geography. For Oliver these were fertile hybrid zones where intellectual cross-pollination could occur, leading to the development of novel thoughts and ideas. He was an authority in many different areas, but what made his work so distinctive was the innovative way in which he integrated and communicated his transdisciplinary knowledge. From the rich tapestry of his life's work, it seems unreasonable to try to disentangle an individual thread, but that is the task attempted here.

Even within the single field of forest ecology, it is impossible to fully convey Oliver's impact. He taught many valuable lessons to generations of woodland ecologists; above all, the importance of history for understanding woodland composition, structure and dynamics. He left a substantial legacy for the management and conservation of woods, having played a major part in establishing the conservation value of ancient woods, not only in terms of

Adrian C. Newton, 'How the Wildwood Worked: Rackham's Contribution to Forest Ecology' in: *Countryside History: The Life and Legacy of Oliver Rackham*. Pelagic Publishing (2024). © Adrian C. Newton. DOI: 10.53061/CTHR2729

their cultural and historical value, but also in terms of what they can teach us about how woodlands function ecologically. He continually strove to raise awareness of the threats that our woodlands increasingly face, such as browsing by deer, the spread of pests and diseases of tree species, and inappropriate management (Rackham 2008). Yet each of these strands of his endeavour was founded on a deep ecological understanding of how woodlands work as ecological systems, which itself was grounded in the science of forest ecology.

Oliver's own route into this area of science was via physiological ecology, reflecting his initial interest in physics. In Cambridge he encountered the work of researchers such as Clifford Evans, who worked on the measurement of light interception in forest understoreys. This work is illustrated by the classic texts *Light as an Ecological Factor* (vols I and II), both of which were co-edited by Oliver and Clifford (Bainbridge, Evans and Rackham 1967; Evans, Bainbridge and Rackham 1975). As documented elsewhere in this volume, he later shifted his focus to historical ecology, the research area for which he is best known. I would like to be able to say that he never completely forgot his eco-physiological roots and that he retained an understanding of the value of its approaches, including mathematical rigour and experimentation. This is not necessarily the impression one obtains from his writing, however!

What actually do we mean when we speak of the science of forest ecology? And how does this relate to Oliver's own contribution? This is a question well worth asking, as Oliver expressed characteristically forthright opinions on the subject of ecological research as it is currently practised in academia. He suggested that ecology as a science has not fared well:

> particularly at Cambridge which for so long was its capital ... The world contains many more professors of ecology than 30 years ago ... but much of the expertise has been diverted to theory, or the links between ecology and evolution ... Anything long-term, or cheap, or low-technology, or unexpected, or difficult to express in numbers gets neglected ... Ordinary scientists formulate questions and then design experiments and observations to answer them. Their ultimate object is to formulate a theory ... The long-term ecologist cannot do this.
>
> Rackham 2003: 529

It is worth pausing a moment to parse these rather acerbic comments. As a Professor of ecology myself, I would actually celebrate our recent growth in numbers, and also our focus on theory, for what is science if it is not about developing and testing theory (e.g. Peters 1991)? In his somewhat dismissive reference to 'ordinary scientists', Oliver is describing perfectly what most contemporary researchers seek to do: answer questions, ideally with experiments, as these provide the most robust form of scientific evidence. Again, does this not lie at the very core of scientific endeavour (e.g. Ford 2000)? His comments imply that 'long-term ecologists' (such as himself, obviously) represent a superior breed of researcher, operating on some elevated plane, where open-ended, purely qualitative research is not only legitimate, but to be encouraged. Here, Oliver is essentially expressing regret that this more traditional, descriptive approach to ecology, which certainly flourished at Cambridge in the past, is today no longer likely to attract the research funding required to implement it.

For some commentators, the kind of research that Oliver describes is not science at all, but natural history. For example, Peters (1991) states that science, by definition, is explicitly about testing predictions, which are based on theory. Anything else should be classified as natural history. Reading Peters (1991), one might obtain the impression that natural history is somehow inferior to scientific research, and perhaps it is this kind of prejudice that Oliver is complaining about. Yet the science of ecology is grounded in the collection and analysis of open-ended field observations. Studies of the natural world, made simply to describe ecological phenomena, can provide a powerful basis for understanding, as great naturalists such as Charles Darwin and E.O. Wilson have so amply demonstrated. Natural history can provide valuable insights into which ecological questions are worth asking, and where the answers might lie. But it is only through experimentation, and by using allied tools such as modelling and statistical inference, that the mechanisms underlying these natural phenomena can be tested and understood, and robust explanations thereby arrived at (Ford 2000).

If Oliver was more of a natural historian than an ecological scientist, then one might question the relevance of his work to the science of forest ecology, at least in terms of how it is currently practised. My aim here is to demonstrate that his work does, in fact, have value for ecological science. To achieve this, I have not attempted to produce a comprehensive review; rather, I provide some selected illustrations of how Oliver's work has informed recent research in this area. Inevitably, these examples draw heavily on my own experience; this is very much a personal account, and reflects my own particular interests and biases. If the results are somewhat idiosyncratic, then this only seems appropriate, given Oliver's delight in the quirky and unusual. Before exploring these examples, I first provide a brief account of my introduction to the 'Rackhamian' worldview.

Understanding the 'Rackhamian' lens

I first met Oliver during my first week as an undergraduate at Cambridge University, having sought him out as one of the very few members of the academic staff who led field outings for local naturalists. He regularly led fungus forays, and within a year I was co-leading field meetings with him for the local Wildlife Trust, as he asked for some help. Fungus hunting with Oliver was quite an experience: what I remember most clearly was his insistence that any species encountered must be illustrated in the single field guide that he took with him, the venerable Wakefield and Dennis (1950). Of course, with a mega-diverse group such as fungi, any field guide is likely to contain only a small fraction of the species that one might encounter during a typical field visit. Yet Oliver seemed unaware of this. On the basis of this experience, I came to understand that Oliver saw the world in a particular way, as if through a Rackhamian lens. This lens could perhaps sometimes be useful for filtering out unruly observations or 'alternative facts' that did not quite fit the Rackhamian worldview. On the other hand, it undoubtedly also provided him with a tool for discovering many new findings that had escaped the attention of others. He had an outstanding eye for detail when it came to amassing a body of evidence in support of an original idea.

I felt privileged to have been taught by Oliver at Cambridge, even though at the time only a select few attended his classes. I remember my college complaining to me about the expense of all the supervisions I attended with him. Essays were returned with numbered lists of comments, which often extended to several pages themselves. He very generously took a group of us out on field trips to such mythical locations as the Bradfield Woods and the Tolkienesque Staverton Thicks, complete with its giant Common Hollies *Ilex aquifolium* sprouting from shattered veteran Pedunculate Oaks *Quercus robur*. I remember him leaping out of a car while stopped at a traffic light in Lavenham, Suffolk, having seen a small hole in the wall of one of the half-timbered medieval cottages for which the town is renowned. After inspecting some of the wattle and daub he extracted from the hole, Oliver then went on to identify the lichens on the Common Hazel *Corylus avellana* twigs that had been used in its construction. From this, and the size of the twigs, he then deduced the type of woodland management that had been employed where the wattle was sourced. For me, this tour de force of forensic observation and analysis epitomizes how Oliver saw the world, as a rich source of ancient information, much of it long since forgotten or lost, which perhaps only he could decode. He surely found exercising his deductive powers a rich source of enjoyment and fun. Later I carried out an undergraduate research project under his direction, which excitingly involved collecting sediment cores from a woodland pond using a deflating rubber dinghy. Oliver was delighted when I found a single pollen grain of lime *Tilia* at the base of a core, which was evidence enough for him that the woodland site was indeed ancient, and perhaps derived from the original wildwood. That Rackhamian lens again.

As a student, one of the aspects of Oliver's work that I struggled to understand was his frequent reference to meaning. In his best-selling *History of the Countryside*, for example, he states that he is 'specially concerned with the loss of meaning' (Rackham 1986). What did Oliver mean by 'meaning'? Even though he wrote an entire paper about the conservation of meaning (Rackham 1991), he never really defined what he meant by it. After cogitating over this for many years, I began to experience the first glimmerings of insight after attending a workshop organized by a community of makers: people who enjoy creating things with their hands. They taught me a wonderful phrase: 'the making of meaning is the meaning of making'. In other words, the process of creating something with your hands gives an object meaning, as time, care and skill have been devoted to its creation. Among Oliver's many other talents, he was an accomplished carpenter, so he surely appreciated this point; the Rackhamian lens was that of a maker. In relation to landscapes, Oliver understood very profoundly that they are manufactured palimpsests created by centuries of human use. For him, it was perhaps the making of landscapes by people that gave them meaning.

In his paper about landscapes and meaning, when discussing the Small-leaved Lime *Tilia cordata* (or pry), Oliver suggests that 'its meaning lies in being a rare and wonderful tree with a mysterious natural distribution. It is devalued by being made a common tree [by planting]' (Rackham 1991: 911). This indicates that what Oliver meant by meaning is different types of value. Specifically, in the parlance of current research on ecosystem services, what Oliver is referring to here is arguably a type of existence value: enjoyment of the knowledge of wild species (Hirons, Comberti and Dunford 2016), namely the pry in this case. Much of the emphasis in this and his other writings is on cultural and heritage values rather than existence values. For example, when discussing the multiple meanings of Hatfield Forest (which of course merited a book all to itself; Rackham 1989), Oliver referred to the fact that it belonged to 'the wickedest Englishman in history' and that 'five of its owners forfeited it for high treason' (Rackham 1991: 912). These observations provide a perfect example of the kind of lurid historical stories that he particularly delighted in and which enrich his writings. For Oliver, the meaning of ancient woods was arguably more about their cultural and heritage values than any other type of value, such as their existence value.

There is a risk that Oliver's cultural and historical interests, and his knack for telling a good story, could obscure the kernel of scientific value that lies at the core of his work. To highlight this contribution, I focus here on my favourite of his books, *Ancient Woodland* (Rackham 1980). I have always viewed this book as a literary treasure: as a cash-strapped student, it seemed unimaginably expensive, and it was only when the second edition was published many years later (Rackham 2003) that I was finally able to purchase a copy of my own. I was pleased to be able to personally congratulate Oliver when the latter was published; both of us enjoyed the rich irony hidden in the preface. Today, fulsome tributes to the book are provided by the likes of the Forestry Commission and English Nature (now Natural England), but they and their predecessors completely ignored the book when it was first published. In my view this is Oliver's masterpiece, and represents the most important account of British woodland ever written.

In an attempt to illustrate the ongoing value of this book to the science of forest ecology, I focus here on two topics that particularly fired my imagination when I first read it as a student: the nature of wildwood, and the mysterious ecology of rosaceous trees.

The dynamics of wildwood and ancient woods

Oliver did not actually invent the term 'wildwood'. According to the Oxford English Dictionary (http://www.oed.com/), it has antecedents as far back as the Anglo-Saxon Chronicle ('wilde wuda'); as a single non-hyphenated word, the earliest attribution is that of Samuel Taylor Coleridge in 1894. Yet Oliver surely played a major role in popularizing this term: *Ancient Woodland* has a chapter dedicated to it (Rackham 2003). Today the word 'wildwood' is widely used in conservation circles; there are conservation charities named after it, there is a Wildwood safari park and even a restaurant chain has adopted the name. To be honest, Oliver's book chapter is fairly brief, even somewhat cursory. I suspect that for him things got a lot more interesting with the advent of recorded history; significantly, the following chapter on the Domesday Book is much more detailed. There are some tantalizing details here, however, such as the observation that there are sub-fossil forests buried in peat that have no modern analogue, in which there were individual trees of a size now rarely seen in Europe, with oak trunks reaching a length of 27 m to the first branch. Such finds emphasize that the past is, indeed, a foreign country.

So perhaps Oliver did not have a great deal of interest in how the wildwood worked as an ecological system. Yet his work is of clear value to those of us who do. In tracing the various impacts of people on our woodlands over time, Oliver helps us to differentiate between the ecological features that resulted from human activity versus those that are attributable to natural processes. In this way, the study of ancient woods can help us to piece together something of the characteristics of the wildwood from which many of them were derived. For example, the observation that the soils of ancient woodlands can differ significantly from those of other land-use types (see Rackham 2003: ch. 4) is a powerful insight. But why is it important to know how the wildwood worked? Partly, there is a need to understand the ecological processes operating in ancient woodlands in order to conserve and manage them effectively, especially in an era of rapid environmental change. As Oliver noted, academic ecologists also like to identify generalizations and develop theory, which might apply not only to forest ecosystems in the UK, but at a global scale (Newton and Echeverría 2014).

Some conservation practitioners also have a pressing need to understand how the wildwood worked, in particular those involved in rewilding. The term 'rewilding' refers to a form of ecological restoration that seeks to support the recovery of ecosystems primarily through the action of natural ecological processes. While there is little agreement on precisely how rewilding should be defined, there is general consensus that the approach is process- rather than assemblage-oriented; the term has been applied to a wide variety of different conservation practices including species reintroductions (particularly of large vertebrates), taxon substitution, flood pattern restoration and the abandonment of agricultural land (Loth and Newton 2018; Sandom et al. 2019). While rewilding is increasingly being explored in many different parts of the world, in the UK the idea was given particular momentum by publication of George Monbiot's book *Feral* (2014). This has stimulated the creation of a dedicated charity, Rewilding Britain, and the development of a series of new rewilding projects. However, efforts at rewilding have a much longer ancestry than that, with some initiatives now spanning several decades (Newton, Stirling and Crowell 2001).

My own involvement in rewilding began in the early 1990s with development of the Carrifran Wildwood Project in the Scottish Borders, where I lived at the time. This was (and is) a genuinely community-based, grassroots initiative, in which management decisions were taken by consensus following lively discussion among volunteers in a local pub. The idea was to create new woodland in an almost entirely deforested landscape in the Southern Uplands, with a composition and structure that mimicked that of the original wildwood that was once present on the site. The project has been highly successful; it has been referred to as one of the most important conservation projects in the UK. The first trees were established at the beginning of the new millennium, and today the woodland has developed sufficiently to attract woodland birds and fungi, among other components of the ecosystem (Ashmole and Ashmole 2009; Adair 2016). At the outset, identification of which tree species to plant and in what combinations presented a significant scientific challenge, given that the site had been deforested for many centuries (Newton and Ashmole 1998 2000). It is no coincidence that the project incorporated the Rackhamian word 'wildwood' within its name. His *Ancient Woodland* was a key source of guidance, not only as a source of auto-ecological information about individual tree species, but also for insights into how these species combine to form woodland communities. It is partly this ability to span the disciplines of both population and community ecology that makes *Ancient Woodland* such an important text, especially given that the insights it contains are always grounded in Oliver's own copious field observations.

The second edition of *Ancient Woodland*, launched at a Sheffield conference in 2003, also provided a detailed discussion of one of the most influential ideas in woodland ecology of recent years, which itself has played a major role in generating interest in rewilding. This is the cyclical turnover of vegetation theory developed by Frans Vera (Vera 2000). The theory is based on the idea that the original post-glacial vegetation of the lowlands of Europe was a park-like landscape, with areas free of trees as well as patches of woodland. This stands in marked contrast to traditional ideas of the wildwood, which assumed near-continuous forest cover over much of northern and central Europe after about 10,000 years ago. Vera (2000) hypothesizes that successional processes were determined by large herbivorous mammals and birds (such as the Jay *Garrulus glandarius*) that act as seed dispersal agents (Fig. 3.1). Large herbivores, such as Aurochs *Bos primigenius*, Bison and Wild Horses *Equus ferus*, were present at such densities to maintain areas of grassland within a woodland matrix. In areas where grazing intensities were relatively low, thorny shrubs could become established in the grassland, into which species of tree might become established. These were protected from herbivory through a process of facilitation, and subsequently developed into groves of trees. Regeneration of trees within the grove was then prevented because of herbivory, as animals were able to enter the grove as it matured. As a result, the forest grove eventually degenerated into grassland, and the cycle began again (Fig. 3.1).

Vera's theory has stimulated a great deal of debate about the nature of the original wildwood: was it open parkland, as he suggested, or closed forest? What role did large herbivores play in its ecology? Oliver was clearly intrigued by these ideas about the nature of wildwood; he devotes an entire chapter to them (32) in the revised *Ancient Woodland*, where he describes Vera (2000) as a 'remarkable new study' in 'an extraordinary book' that was written with 'panache and imagination' (Rackham 2003: 499). This is high praise indeed, particularly as Vera's book is essentially a reinterpretation of available literature and does not present any original data. The seven pages Oliver devotes to a discussion of this book are, for me, Rackham at his best. One can almost hear his mind grappling with the ideas presented within it, testing it against the unique repository of observation and insight that he had accumulated over the years. There are undoubtedly some features of the theory that appealed to Oliver: 'Vera's theory addresses a number of peculiarities, such as the pre-Neolithic abundance of oak and hazel ... it offers an explanation for such unsolved questions as how people could have made farmland without exhausting their energies on digging up trees' (Rackham 2003: 500).

However, there are many aspects that Oliver takes issue with, such as the omission regarding the suckering behaviour of trees, the lack of consideration of palatability of different tree species and the failure to consider the evidence from medieval timbers. He also finds Vera's proposed dynamics 'not altogether persuasive'; most memorably he asks: 'Has Vera ever seen a dead lime?' (Rackham 2003: 503). In other words, if the forest was dominated by a self-coppicing species such as lime, how was it ever converted to grassland by herbivory? Overall, Rackham concludes with these interesting statements: 'I cannot agree with everything ... [but] I accept with gratitude many of its insights ... The relation between wildwood and the woods or wood pastures of medieval England may be closer than I then thought when writing the original *Ancient Woodland*' (Rackham 2003: 506).

Over a period of some 15 years, my colleagues and I have attempted to test Vera's theory in the ancient pasture woodlands of the New Forest in southern England. This is a royal hunting

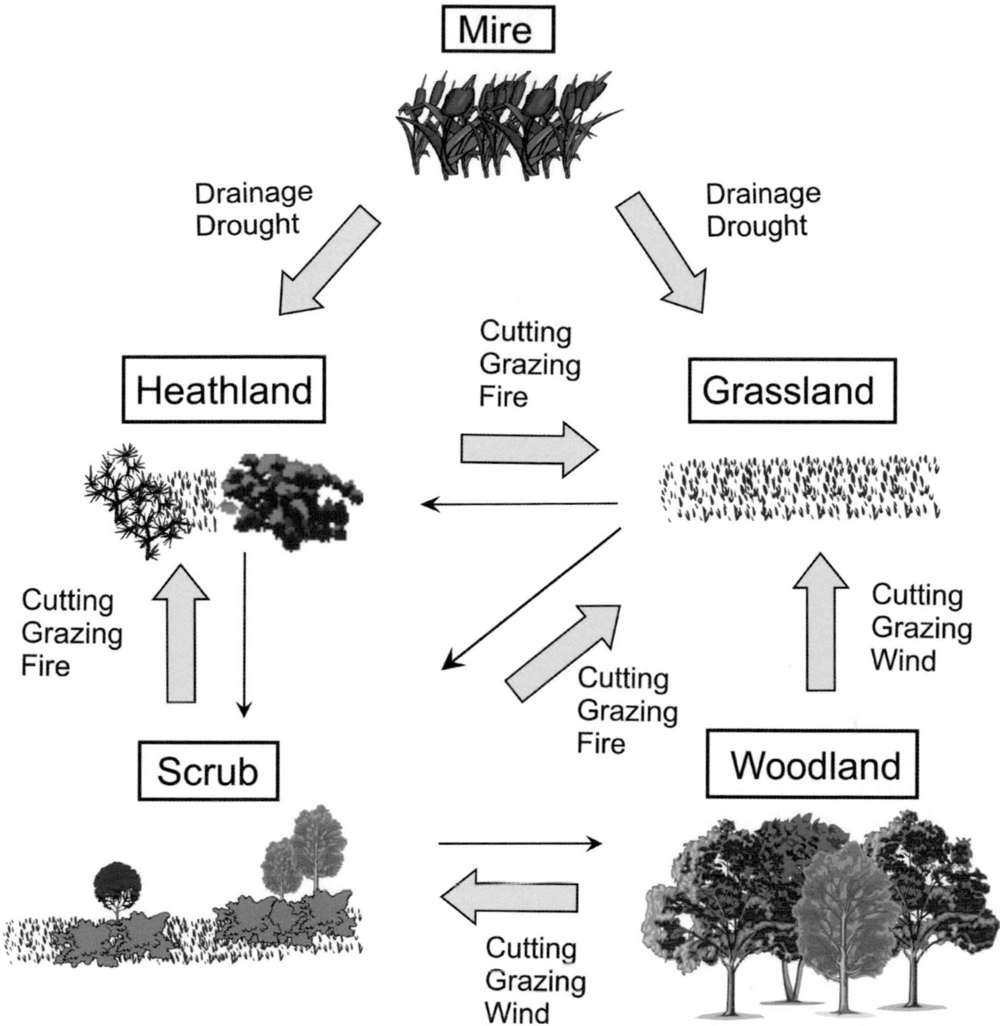

Figure 3.1 Schematic diagram of Vera's cyclical theory of vegetation turnover. The Park phase is a largely open landscape with a thin scatter of trees left from the previous grove; vegetation mainly grassland or heath species. In the Scrub phase, spread of thorny shrubs excludes herbivores; young trees grow up with the shrubs and eventually overtop them. In the Grove phase, which is the tree-dominated phase of the cycle, a closed tree canopy shades out the shrubs, and herbivores return, preventing regeneration. In the Break-up phase, the canopy opens out as trees die; vegetation shifts from woodland to grassland species. (After Newton 2011, based on Vera 2000)

forest that was established more than 900 years ago, making it perhaps the oldest protected area in the world (Newton 2011). Today, owing to the remarkable preservation of a medieval commoning system, it is characterized by a high density of free-ranging large herbivores, including both deer and livestock. This makes it a perfect place to test Vera's ideas, as Oliver noted; in fact, the New Forest was one of the main sources of inspiration for Vera's theory (Vera 2000). To achieve this, our research employed a computer model (LANDIS II) that enabled forest dynamics to be simulated at the scale of the entire landscape and the role of herbivory to be explored. Such models provide a valuable tool for understanding the ecological processes influencing forest dynamics, as they enable experiments to be conducted '*in silico*', involving timescales that would be impossible to implement in the real world. To parameterize the model, we understood a systematic survey of vegetation structure and composition in all 173 woodland patches in the New Forest. Model parameterization also required data on ecological traits of all individual tree species, and for this *Ancient Woodland* again proved to be a valuable source of information (Newton et al. 2013a).

Testing of the model using independent field data found that it was able to accurately predict the abundance and richness of tree species in newly established woodland (Newton et al. 2013a). Having parameterized and tested the model, we were then able to use it to explore the influence of herbivory on long-term woodland dynamics. Results indicated that over the duration of the simulations (300 years), woodland area increased in all of the experimental treatments, with or without herbivory. While woodland area increased more rapidly under conditions of no herbivory, values increased by more than 70 per cent even in the presence of the heavy browsing pressure that currently prevails throughout the New Forest, as indicated by the results of our field survey (Fig. 3.2).

A number of authors have suggested that regeneration of tree species is severely limited or entirely absent in the ancient woodlands of the New Forest as a result of high herbivore pressure (e.g. Peterken and Tubbs 1965; Putman et al. 1989; Mountford et al. 1999; Mountford and Peterken 2003). However, these observations were based on a limited number of sites. In our survey of all New Forest woodlands, we found that tree regeneration was widespread but occurred at low density; for example, oak and European Beech saplings were recorded

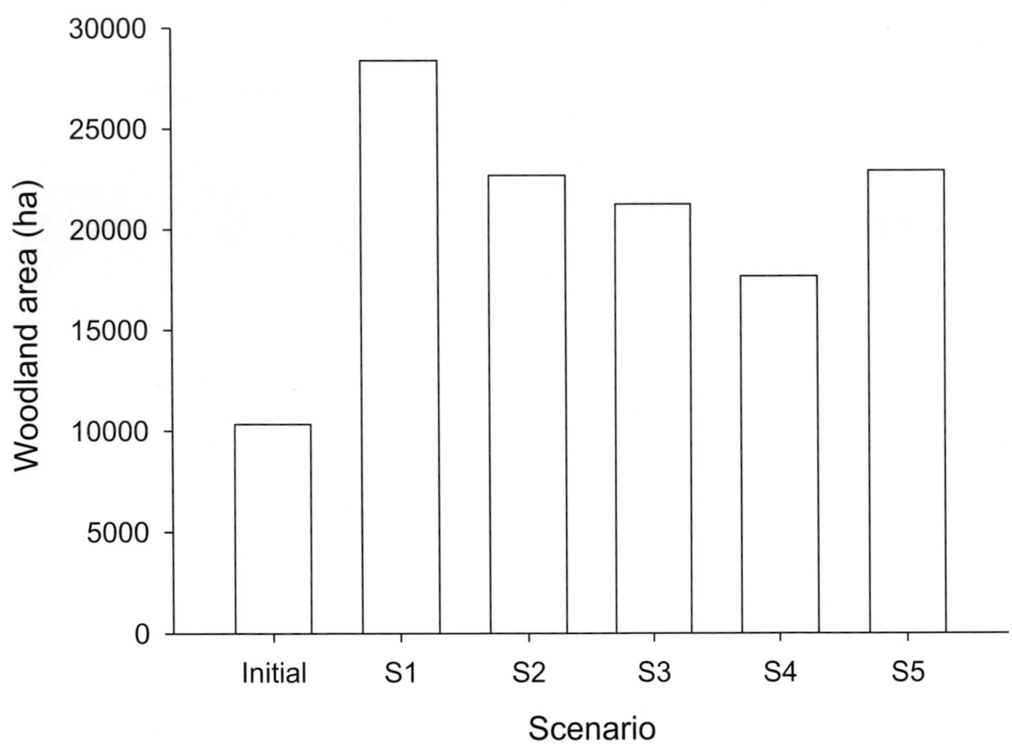

Figure 3.2 Modelling the impact of herbivory on woodland dynamics in the New Forest, using LANDIS II. Scenario 1, no disturbance (neither fire nor browsing); Scenario 2, browsing only; Scenario 3, fire only; Scenario 4, fire plus browsing; Scenario 5, browsing, fire and protection from herbivory by presence of spiny shrubs, where mortality of trees as a result of browsing was set to zero if any of the species Common Hawthorn *Crataegus monogyna*, Blackthorn *Prunus spinosa* or Gorse *Ulex europaeus* of ages 5–100 years old was present within at least 50 per cent of the cells in the stand. Fire was included in some of the scenarios as it is currently used (along with cutting) to manage heathland areas, which would otherwise succeed to woodland. (Newton et al. 2013a)

on 16 per cent and 26 per cent of sites, respectively. The occurrence of regeneration within woodlands may partly be attributed to protection of young trees within protective microsites, such as fallen branches or trunks and spiny shrubs occurring within woodlands, as documented by Morgan (1991) in a single site (Ridley Wood). According to Vera (2000), tree regeneration should occur on the periphery of woodlands in scrub, which results in a concentric expansion of forest. Results from our field surveys (see also Newton et al. 2013b) offered some support for this hypothesis, indicated by the association of young trees with spiny shrubs, but also provided evidence of widespread colonization of both heathland and grassland sites by a range of tree species (especially birch) without such facilitation. This suggests that tree regeneration is restricted neither to the periphery of woodlands, nor to sites where spiny shrubs are present. Facilitation had relatively little impact on the pattern of woodland development; expansion from woodland margins was evident even in the absence of facilitation. As a result of spatial heterogeneity in the suitability of sites for tree establishment, woodland expansion was typically patchy rather than evenly concentric, with or without facilitation (Fig. 3.2).

These results therefore provide limited support for Vera's theory. Most strikingly, none of the model scenarios produced the break-up of woodland groves with maturity and their conversion to either grassland or heathland, as required for cyclic dynamics of vegetation to take place (Vera 2000). Rather, the results indicated the progressive dominance of European Beech and an associated decline of relatively shade-intolerant species such as birch and Scots Pine *Pinus sylvestris*, in a manner consistent with traditional successional theory.

Yet there are ancient woodlands in the New Forest where one can observe stand dieback of beech, which could perhaps equate to the break-up of groves hypothesized by Vera (2000). We examined one such site, Denny Wood, in detail. Here, a permanent transect has been resurveyed repeatedly since the 1950s, affording an outstanding opportunity to study how a beech woodland can progressively die. Based on our resurvey conducted in 2014, we were able to show that since the late 1950s, basal area in the wood has declined by 33 per cent, mostly because of the death of large beech; juvenile tree densities also declined, by approximately 70 per cent. Some of the areas that were closed-canopy beech woodland in 1964 had become relatively open grassland by 2014 (Martin et al. 2015, 2017). This provides a fascinating example of the conditions required to achieve the kind of woodland dieback hypothesized by Vera: high browsing pressure by itself is not necessarily enough. In the case of Denny Wood, we believe that a combination of increased summer droughts and winter waterlogging attributable to anthropogenic climate change have increased beech mortality directly, and have also increased the susceptibility of beech to pathogenic fungi. These factors acting together are progressively killing this woodland, and where browsing pressure is high enough, they are converting it to grassland (Martin et al. 2015, 2017).

Denny Wood is not the only ancient woodland in the New Forest that is dying because of climate change. There are many others. We have examined a series of these sites, to document what the implications of Beech Dieback might be for wildlife, remembering that the New Forest is of exceptional importance for biodiversity conservation (Newton 2010). Our recent field research suggests that the process of stand dieback is not necessarily linear; there appear to be critical thresholds beyond which the rate of biodiversity loss of diverse groups such as ectomycorrhizal fungi and lichens can increase rapidly (Evans et al. 2017; Fig. 3.3). We have also used these field data, together with our LANDIS II model, to examine the different factors that influence the resilience of New Forest woodlands to environmental change (Cantarello et al. 2017). Given the large number of emerging threats that are increasingly affecting our trees and woodlands, as described by Rackham (2008), this type of research, aimed at deepening our understanding of how trees and woodlands can cope with such threats, is surely a high priority for the future. One of our findings is that the longevity of trees may help to buffer forests against environmental change (Martin et al. 2017). The conservation of old, large trees

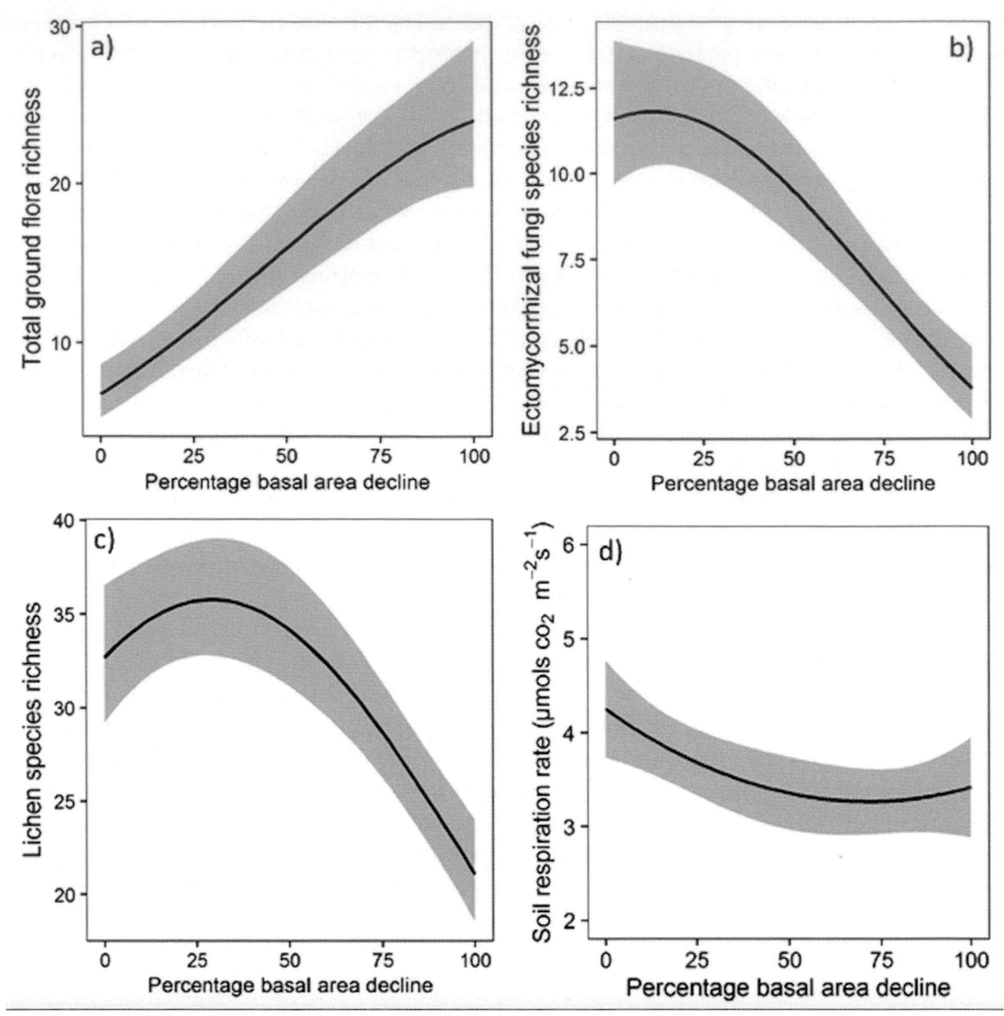

Figure 3.3 Ecological impacts of Beech Dieback in the New Forest. Relationships between degree of Beech Dieback (basal area decline) and species richness of (a) vascular ground flora (n = 60); (b) ectomycorrhizal fungi (n = 60); (c) epiphytic lichen (n = 60); and (d) soil respiration rate (n = 60). The black lines represent prediction using the most parsimonious model coefficients and grey shading the 95 per cent confidence intervals of the coefficients (marginal r2 = 0.60, 0.57, 0.44 and 0.16 for (a–d), respectively). The different coloured points represent the values at each individual site. All species richness values are the number of unique species found in 0.04 ha. (Evans et al. 2017)

is therefore an important priority to strengthen the resilience of woodland stands. This is surely a finding that Oliver would have strongly endorsed.

The mysteries of rosaceous trees

Rosaceous trees provide a further example of how *Ancient Woodland* both inspired and informed some of my research, and also illustrate how the Rackhamian lens can offer ecological insights in parts of the world other than the British Isles. When I first read *Ancient Woodland* as a student, the chapter on rosaceous trees particularly captured my imagination. Here one could read about the mysterious Pear *Pyrus communis* in Hayley Wood, which was a possible rare survivor of the Wild Pear *Pyrus pyraster* recorded in history. Even more wonderfully, its top fell off in 1974 and was sold to make harpsichords. As Oliver put it: 'Historical evidence strongly suggests that there is a native pear, whose fruit is rarely produced and uneatable. If correctly equated with the Hayley pear it is now one of the rarest British plants' (Rackham 1980/ 2003: 356).

Here was an indication that some of our native tree species are deeply enigmatic ecologically: how could this pear colonise colonize woodlands if it hardly ever produced fruit? Rackham (2003) notes that only four individual pear trees are known from Eastern England, all occurring as isolated individual trees in ancient woods. The ecology of this species is perplexing: its ability to survive at very low densities seems reminiscent of many tropical trees, which are similarly poorly understood in terms of their reproductive ecology.

Then there was the remarkable Whitty Pear *Sorbus domestica*, for which ' only one individual has ever been recorded ... Ignored by most British floras ... The earliest written record of any British plant (seventh century) [refers to] one of the rarest British plants' (Rackham 2003: 359). This is surely a species that exemplifies what particularly interested Oliver: a species associated with genuine mystery, history and (dare one say it) meaning. The second edition of *Ancient Woodland* provides a fascinating update (Rackham 2003): when Marc Hampton found the species growing on a sea cliff in Glamorgan in 1983, it represented the first time in three centuries that it had been seen in Britain. Further research established that this population is very likely to be genuinely native (Hampton and Kay 1995). Given that trees are the most conspicuous element of our flora, I find it utterly remarkable that new discoveries of this nature are still possible, and new tree species are still being added to the British list.

Having made a pledge to myself to research some rosaceous trees, I finally got the chance to do so in a project focusing on the *Sorbus* species on the Isle of Arran. The Arran Whitebeams *Sorbus arranensis* form a hybrid complex, and being apomictic, occur in stands of clones. The endemic Arran Whitebeam originates from a cross between Rock Whitebeam *Sorbus rupicola* and Mountain Ash *S. aucuparia*. The similarly endemic Arran Service Tree *Sorbus pseudofennica* arose from a further backcross between the Arran Whitebeam and Mountain Ash. Analysis of the genetic structure of populations of these species using molecular markers confirmed these hybrid origins, but also indicated that there have been at least three origins of Arran Whitebeam on Arran and at least five hybrid origins of Arran Service Tree (Robertson, Newton and Ennos 2004a,b). These results showed that the endemic *Sorbus* taxa on Arran are the products of multiple and ongoing evolutionary events: rather than being frozen in time, these populations are still evolving. Even better, one of the morphotypes we collected turned out to be a new species (Robertson and Sydes 2006), providing further evidence that there are still rosaceous mysteries to be solved. Despite Oliver's protestations to the contrary, this example shows us that links between ecology and evolution are, in fact, fascinating – particularly when one considers that the genetic structure of these species is largely created by the behaviour of migratory thrushes (Vickery, Rivington and Newton 1998).

Some years later I even got to work on a Wild Pear species. This came about after conducting a conservation assessment of tree species in Central Asia (Eastwood, Laskov and Newton 2009). This region is remarkably rich in fruit and nut tree species, which occur in a mixed forest often dominated by Walnut *Juglans regia*. In total, more than 300 wild fruit and nut species occur in the region, including wild species of apple (four species), almond (8–10 species), cherry (8–10 species), plum (4–5 species), and walnut (one species), as well as many domesticated varieties. There are at least three pear species (*Pyrus tadzhikistana*, *P. cajon* and *P. korshinskyi*), all of which are endangered (Eastwood, Lazkov and Newton 2009); like the Hayley Pear, they tend to occur at very low densities, and their reproductive ecology is largely unknown. The rich diversity of fruit and nut species in the region led the Russian geneticist and plant breeder N.I. Vavilov to propose it as one of the world's eight centres of crop origin and domestication. As Oliver so memorably put it: 'This paradise on earth is haunted by the shade of Nikolai Ivanovich Vavilov, greatest of all Soviet scientists, who suspected the apple and walnut story, but never told the world, because he was contaminated by the bourgeois doctrines of Darwin and Mendel, and Stalin murdered him' (Rackham 2007: 47).

Oliver used the word 'paradise' advisedly; the region was described as 'Eden' in a popular account of the forests written by Roger Deakin (Deakin 2007), reflecting its garden-like abundance of species. The reality is somewhat more prosaic; these forests are severely degraded and are threatened by land-use change, fragmentation and overharvesting. I subsequently developed a research project to investigate the conservation ecology of these forests in greater depth, in collaboration with researchers from the region. Our project focused on Walnut-fruit forests in Kyrgyzstan, where remaining forests are most extensive. We conducted a field survey throughout the range of Walnut-fruit forests in this country (Fig. 3.4), supported by a socio-economic survey. Results showed that species differed markedly in abundance. The Wild Apple *Malus sieversii* was relatively common and widespread. This species is now believed to be the wild progenitor of the domestic apple (Juniper and Mabberley 2006), a view supported by molecular evidence (Harris, Robinson and Juniper 2002). Apple was domesticated and transported along the Silk Road in antiquity, eventually reaching Europe, where it was apparently used by Alexander the Great for target practice (Deakin 2007), an anecdote that could hardly be any more Rackhamesque.

In our survey, Pear *Pyrus korshinskyi* was a much rarer species than apple, as were other mysterious rosaceous trees such as *Crataegus pontica* and *Sorbus persica*. Our results showed that unsustainable land-use practices are impacting negatively on populations of threatened fruit tree species, as indicated by a decline in the availability of fruits for harvesting and an increase in time required for collecting fuelwood as reported by local people (Orozumbekov, Cantarello and Newton 2015). Yet, surprisingly, we found that browsing by livestock did not have the negative impact on these species that we anticipated. Use of LANDIS II to model the dynamics of these forests confirmed the negative impact of fuelwood cutting on maintenance of species such as Walnut, Apple and Wild Apricot *Armeniaca vulgaris*, which in our simulations were largely eliminated from the landscape after 50–150 years. In the absence of disturbance or in the presence of grazing only, decline of these species was projected to occur at a much lower rate, owing to competitive interactions between tree species (Cantarello et al. 2014). This research revealed something odd about these forests: although Walnut is dominant in the canopy, regeneration of this species is very sparse, and tends to occur on forest margins. Sound familiar?

Our survey and modelling results supported our field observations, that this is a forest ecosystem that is not at equilibrium. Currently, livestock appear to be playing a positive role in supporting regeneration of tree species, for example by acting as seed dispersers. Individual fruit trees can often be seen in protective, thorny thickets; yet regeneration under the woodland canopy is scant. Could this be an example of a Vera-type system? Within the region, based on research conducted by eminent Russians such as Vavilov, the forests are seen as an example of near-pristine wildwood: a relic of a forest that was much more widespread in the Tertiary era. Yet as a relatively shade-intolerant species, Walnut cannot maintain itself without significant canopy disturbance. Our ecological interpretation was supported by a recent palynological study, which showed that the natural forests in this region were once dominated not by Walnut but by junipers *Juniperus* together with birches *Betula*, ashes *Fraxinus*, roses Rosaceae (hurray!) and maples *Acer*. Pollen evidence for Walnut indicated that stands of this species are mostly around 1,000 years old, and are probably anthropogenic in origin (Beer et al. 2008). Remarkably, this is consistent with a local legend that is still orally transmitted in Kyrgyzstan, which states that Walnut and other fruit species were established under the leadership of Arslan-Bop, the legendary founder of the Uzbek village that still bears his name. He is reputed to have been a respected scholar and leader who died in AD 1120, roughly coinciding with the increase in Walnut detected in the pollen diagrams (Beer et al. 2008). This provides an outstanding example of how history and ecological science can usefully intersect;

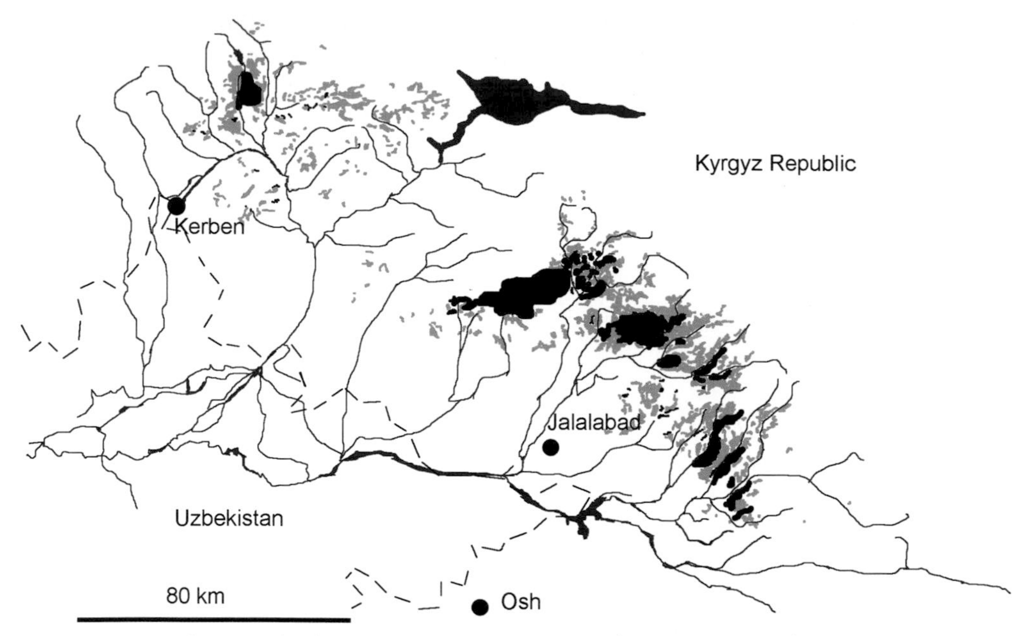

Figure 3.4 Distribution of Walnut-fruit forest in Kyrgyzstan (dark shaded areas), and other broadleaved forest (pale shaded areas). Adapted from a forest map of the Kyrgyz Republic produced by Intercooperation in 2009, based on interpretation of Aster imagery. (Orozumbekov, Cantarello and Newton 2015)

here is a story full of cultural and ecological meaning that Oliver would surely have appreciated, especially given that the wildwood imagined by Vavilov turned out to be an orchard. Eden was, in fact, a garden.

Conclusions

Oliver Rackham is best known for his writing and teaching, through which he reached a wide and diverse audience. For someone who spent his professional life in that most ivory of towers, a Cambridge college, he displayed an outstanding ability to communicate his academic ideas to the general public. A trip into the field with Oliver was something never forgotten. Yet Oliver has also left an exceptional legacy for forest ecologists, which continues to be of value for contemporary ecological research, even if this is to be undertaken by what he referred to as 'ordinary scientists'. By highlighting ecological mysteries and historical curiosities, in his magnificent idiosyncratic style, Oliver created a research agenda that is far from being fully completed. As I have demonstrated in the examples presented here, which each involved teams of researchers, a single Rackhamian anecdote or remark can launch the career of many a young scientist. His work also has direct relevance to contemporary ecological concerns, including biodiversity conservation, sustainability and resilience to environmental change.

My example from the Walnut-fruit forests of Central Asia shows how Oliver's ideas can be relevant in places far from the East Anglian woods in which they were gestated. Our field surveys and modelling analyses suggested that here was a putative wildwood that 'did not work': only through incorporating a historical perspective did it begin to make ecological sense. Surely the need to observe woodlands through a historical lens is Oliver's greatest lesson. Did the wildwood work as Vera would have us believe? I don't think so. Vera's ideas might help us understand how silvo-pastoral orchards work, however. Like the New Forest, and many other ancient pasture woods, the Walnut-fruit forests provide a poor analogue for wildwood – or for rewilding.

In closing, I offer six aphorisms that attempt to capture some of the key lessons I have learned from Oliver's work over the years. One of Oliver's greatest attributes was that he would readily share his knowledge and ideas with anyone who would listen. Now that he has left us, we need to keep listening.

Six Rackhamian lessons for forest ecologists are as follows:

- History underpins ecology
- Making makes meaning
- Species really matter
- Ecology is particular
- Continuity is valuable
- Stasis is interesting

Acknowledgements

Many thanks to Lynn Davy (another one of the select few) for helpful discussions about Oliver's legacy, and for comments on the manuscript.

References

Adair, S. (2016) Carrifran: ecological restoration in the Southern Uplands new native woodland and vegetation succession in the Moffat Hills. *Scottish Forestry*, 70(1), pp. 30–40.

Ashmole, M. and Ashmole, P. (2009) *The Carrifran Wildwood Story*. Borders Forest Trust, Jedburgh.

Bainbridge, R., Evans, G.C. and Rackham, O. (eds) (1967) *Light as an Ecological Factor*. British Ecological Society Symposium No. 6. Wiley, New York.

Beer, R., Kaiser, F., Schmidt, K., Ammann, B., Carraro, G., Grisa, E. and Tinner, W. (2008) Vegetation history of the walnut forests in Kyrgyzstan (Central Asia): natural or anthropogenic origin? *Quaternary Science Review*, 27, pp. 621–632. https://doi.org/10.1016/j.quascirev.2007.11.012

Cantarello, E., Lovegrove, A., Orozumbekov, A., Birch, J., Brouwers, N. and Newton, A.C. (2014) Human impacts on forest biodiversity in protected walnut-fruit forests in Kyrgyzstan. *Journal of Sustainable Forestry*, 33, pp. 454–481. https://doi.org/10.1080/10549811.2014.901918

Cantarello, E., Newton, A.C., Martin, P.A., Evans, P.M., Gosal, A. and Lucash, M.S. (2017) Quantifying resilience of multiple ecosystem services and biodiversity in a temperate forest landscape. *Ecology and Evolution*, 7(22), pp. 9661–9675. https://doi.org/10.1002/ece3.3491

Deakin, R. (2007) *Wildwood: A Journey through Trees*. Penguin Books, London.

Eastwood, A., Lazkov, G. and Newton, A.C. (2009) *The Red List of Trees of Central Asia*. Fauna and Flora International (with IUCN and BGCI), Cambridge.

Evans, G.C., Bainbridge, R. and Rackham, O. (eds) (1975) *Light as an Ecological Factor II*. British Ecological Society Symposium No. 16. Blackwell, New York.

Evans, P.M., Newton, A.C., Cantarello, E., Martin, P., Sanderson, N., Jones, D.L., Barsoum, N., Cottrell, J.E., A'Hara, S.W. and Fuller, L. (2017) Thresholds of biodiversity and ecosystem function in a forest ecosystem undergoing dieback. *Scientific Reports* 7, Article no. 6775. doi: 10.1038/s41598-017-06082-6. https://doi.org/10.1038/s41598-017-06082-6

Ford, E.D. (2000) *Scientific Method for Ecological Research*. Cambridge University Press, Cambridge. https://doi.org/10.1017/CBO9780511612558

Hampton, M. and Kay, Q.O.N. (1995) *Sorbus domestica* L., new to Wales and the British Isles. *Watsonia*, 20(4), pp. 379–384.

Harris, S.A., Robinson, J.P. and Juniper, B.E. (2002) Genetic clues to the origin of the apple. *Trends in Genetics*, 18(8), pp. 426–430. https://doi.org/10.1016/S0168-9525(02)02689-6

Hirons, M., Comberti, C. and Dunford, R. (2016) Valuing cultural ecosystem services. *Annual Review of Environment and Resources*, 41, pp. 545–574. https://doi.org/10.1146/annurev-environ-110615-085831

Juniper, B.E. and Mabberley, D.J. (2006) *The Story of the Apple*. Timber Press, Oregon.

Loth, A.F. and Newton, A.C. (2018) Rewilding as a restoration strategy for lowland agricultural landscapes: stakeholder-assisted multi-criteria analysis in Dorset, UK. *Journal for Nature Conservation*, 46, pp. 110–120. https://doi.org/10.1016/j.jnc.2018.10.003

Martin, P., Newton, A.C., Evans, P. and Cantarello, E. (2015) Stand collapse in a temperate forest and its impact on forest structure and biodiversity. *Forest Ecology and Management*, 358, pp. 130–138. https://doi.org/10.1016/j.foreco.2015.08.033

Martin, P., Newton, A.C., Cantarello, E. and Evans, P.M. (2017) Analysis of ecological thresholds in a temperate forest undergoing dieback. *PLoS ONE*, 12(12), e0189578. https://doi.org/10.1371/journal.pone.0189578.

Monbiot, G. (2014) *Feral: Rewilding the Land, Sea and Human Life*. Penguin, London. https://doi.org/10.7208/chicago/9780226205694.001.0001

Morgan, R.K. (1991) The role of protective understorey in the regeneration system of a heavily browsed woodland. *Vegetatio*, 92, pp. 119–132. https://doi.org/10.1007/BF00036033

Mountford, E.P., Peterken, G.F., Edwards, P. and Manners, J.G. (1999) Long-term change in growth, mortality and regeneration of trees in Denny Wood, an old-growth wood-pasture in the New Forest. *Perspectives in Plant Ecology, Evolution and Systematics*, 2, pp. 223–272. https://doi.org/10.1078/1433-8319-00072

Mountford, E.P. and Peterken, G.F. (2003) Long term change and implications for the management of woodpastures: experience over 40 years from Denny Wood, New Forest. *Forestry*, 76, pp. 19–43. https://doi.org/10.1093/forestry/76.1.19

Newton, A.C. (ed.) (2010) *Biodiversity in the New Forest*. Newbury: Pisces Publications.

Newton, A.C. (2011) Social-ecological resilience and biodiversity conservation in a 900-year-old protected area. *Ecology and Society*, 16(4), art. 13. http://dx.doi.org/10.5751/ES-04308-160413

Newton, A.C. and Ashmole, P. (1998) How may native woodland be restored to southern Scotland? *Scottish Forestry*, 52, pp. 168–171.

Newton, A.C. and Ashmole, P. (eds) (2000) *Carrifran Wildwood Project: Management Plan*. Wildwood Group of Borders Forest Trust, Jedburgh.

Newton, A.C., Cantarello, E., Tejedor, N. and Myers, G. (2013a) Dynamics and conservation management of a wooded landscape under high herbivore pressure. *International Journal of Biodiversity*, Article ID 273948. https://doi.org/10.1155/2013/273948

Newton, A.C., Cantarello, E., Lovegrove, A., Appiah, D. and Perrella, L. (2013b) The influence of grazing animals on tree regeneration and woodland dynamics in the New Forest, England. In: I.D. Rotherham (ed.) *Trees, Forested Landscapes and Grazing Animals – A European Perspective on Woodlands and Grazed Treescapes*. Routledge, Oxford, pp. 163–179. https://doi.org/10.4324/9780203102909-21

Newton, A.C. and Echeverría, C. (2014) Analysis of anthropogenic impacts on forest biodiversity as a contribution to empirical theory. BES Symposium volume. In: D.A. Coomes, D.F.R.P. Burslem and W.D. Simonson (eds) *Forests and Global Change*. Cambridge University Press, Cambridge, pp. 417–446. https://doi.org/10.1017/CBO9781107323506.019

Newton, A.C., Stirling, M. and Crowell, M. (2001) Current approaches to native woodland restoration in Scotland. *Botanical Journal of Scotland*, 53(2), pp. 169–196. https://doi.org/10.1080/03746600108685021

Orozumbekov, A., Cantarello, E. and Newton, A.C. (2015) Status, distribution and use of threatened tree species in the walnut-fruit forests of Kyrgyzstan. *Forests, Trees and Livelihoods*, 24(1), pp. 1–17. https://doi.org/10.1080/14728028.2014.928604

Peterken, G.F. and Tubbs, C.R. (1965) Woodland regeneration in the New Forest, Hampshire, since 1650. *Journal of Applied Ecology*, 2, pp. 159–170. https://doi.org/10.2307/2401702

Peters, R.H. (1991). *A Critique for Ecology*. Cambridge University Press, Cambridge.

Putman, R.J., Edwards, P.J., Mann J.C.E, How, R.C. and Hill, S.D. (1989) Vegetational and faunal changes in an area of heavily grazed woodland following relief of grazing. *Biological Conservation*, 47, pp. 13–32. https://doi.org/10.1016/0006-3207(89)90017-7

Rackham, O. (1980) *Ancient Woodland. Its history, vegetation and uses in England*. Edward Arnold, London.

Rackham, O. (1986) *The History of the Countryside*. J.M. Dent and Sons, London.

Rackham, O. (1989) *The Last Forest: The story of Hatfield Forest*. J.M. Dent and Sons, London.

Rackham, O. (1991) Landscape and the conservation of meaning. *RSA Journal*, 139(5414), pp. 903–915.

Rackham, O. (2003) *Ancient Woodland: Its History, Vegetation and Uses in England*, 2nd edition. Castlepoint Press, Dalbeattie.

Rackham, O. (2007) Child of the New Forest. A review of *Wildwood: A Journey Through Trees* by Roger Deakin. *The Spectator Magazine*. 11 August.

Rackham, O. (2008) Ancient woodlands: modern threat. *New Phytologist*, 180, pp. 571–586. https://doi.org/10.1111/j.1469-8137.2008.02579.x

Robertson, A., Newton, A.C. and Ennos, R.A. (2004a) Breeding systems and continuing evolution in the endemic *Sorbus* taxa on Arran. *Heredity*, 93(5), pp. 487–495. https://doi.org/10.1038/sj.hdy.6800528

Robertson, A., Newton, A.C. and Ennos, R.A. (2004b) Multiple hybrid origins, genetic diversity and population genetic structure of two endemic *Sorbus* taxa on the Isle of Arran, Scotland. *Molecular Ecology*, 13(1), pp. 123–134. https://doi.org/10.1046/j.1365-294X.2003.02025.x

Robertson, A. and Sydes, C. (2006) *Sorbus pseudomeinichii*, a new endemic *Sorbus* (Rosaceae) microspecies from Arran, Scotland. *Watsonia*, 26, pp. 9–14.

Sandom, C.J., Dempsey, B., Bullock, D., Ely, A., Jepson, P., Jimenez-Wisler, S., Newton, A., Pettorelli, N. and Senior, R.A. (2019) Rewilding in the English uplands: policy and practice. *Journal of Applied Ecology*, 56, pp. 266–273. https://doi.org/10.1111/1365-2664.13276

Vera, F.W.M. (2000) *Grazing Ecology and Forest History*. CABI, Wallingford. https://doi.org/10.1079/9780851994420.0000

Vickery, J., Rivington, M. and Newton, A.C. (1998) Reproductive ecology of the endemic *Sorbus* species on the Isle of Arran. In: A. Jackson and M. Flanagan (eds) *The Conservation Status of Sorbus in the UK*. Royal Botanic Garden (Kew), Richmond, pp. 25–31.

Wakefield, E.M. and Dennis, R.W.G. (1950) *Common British Fungi*. P.R. Gawthorn, London.

CHAPTER 4

Stability and Change in Woodland Ground Flora

Keith Kirby

Summary

Oliver Rackham did not introduce any major new theory with respect to woodland ground flora, but through careful observation he tested and sometimes refuted assumptions about why particular plants grow where they do. He encouraged us to look at woods and woodland plants in a different way. His insights have been invaluable in subsequent work, for example in looking at the characteristics of ancient woodland indicators, the impact of deer on the flora and the recolonization of replanted ancient woods as conifers have been removed. Through building on his work, we have a better appreciation of both stability and change in flora over time, of the sort that Rackham elegantly described for sites such as Hayley Wood and Gamlingay Wood in Cambridgeshire.

Keywords: woodland ground flora, ancient woodland, botanical indicators

Introduction

My DPhil supervisor, Phillip Lloyd, had been a friend of Oliver Rackham's at Cambridge and recommended I look at his work. When I got a job in the Lake District, my introduction to woodland conservation was through *Trees and Woodland in the British Landscape* and H.H. Symonds's 1936 polemic on *Afforestation in the Lake District* (Rackham 1990; Symonds 1936). I was taken on by the Nature Conservancy Council as apprentice to George Peterken and was more in his sphere of influence on woodland conservation in the 1970s and 1980s. However, I was involved in setting up Rackham's contract to survey woods in South Wales, and was pleased to help in a small way with the Woodland Trust's posthumous publication of his South Wales studies.

In 1996, Charles Watkins and I organized a meeting on woodland history in Nottingham to mark 20 years since *Trees and Woodland in the British Landscape* had first been published. We left Rackham to choose his own subject, and were surprised when this proved to be about savanna in Europe (Rackham 1998). Had he already picked up on the work that Frans Vera was doing in the Netherlands, some four years before it became widely known in Britain (Vera 2000)? There is nothing in the references or acknowledgements to suggest this; rather,

Keith Kirby, 'Stability and Change in Woodland Ground Flora' in: *Countryside History: The Life and Legacy of Oliver Rackham*. Pelagic Publishing (2024). © Keith Kirby. DOI: 10.53061/HARR7482

his ideas seem to have developed independently – from his work in the Mediterranean and observations in the USA (Grove and Rackham 2001).

His magnum opus *Ancient Woodland* in 1980 had been an eagerly awaited Christmas present when it was first published, so I was pleased that English Nature helped fund the second, extended edition (Rackham 2003). Meanwhile, I would meet Rackham from time to time, sometimes in the field, sometimes at conferences, sometimes in unlikely places, such as on the bridge at Paddington station. I could also trace his activities and influence from the accounts of others who had been on his courses or had him visit their woods.

For several years, we had talked about him coming to look round Wytham Woods, just outside Oxford, which he did in March 2014. As always, he spent much of the time quietly making notes as we walked around looking at historical and ecological features. These included an area of ancient woodland that had burnt not once but twice, in 1955 and again in 1965, notwithstanding his scepticism as to the combustibility of broadleaved woodland. Of course, it was a special case – an open stand with much grass and bracken. The fire scars on the large oaks are still visible, and by poking around under the overgrowing bark, he came up with some charcoal.

Charles Watkins and I put together a new set of papers to update the book that had come from the 1996 Nottingham conference (Kirby and Watkins 2015). As we were doing the final proof checks in early 2015, news came through of Rackham's death. It seemed appropriate that we dedicated the book to his memory, not least because references to his work pervaded all the contributions.

Rackham and studies of the ground flora

Much of Rackham's writing is about trees and shrubs, their variation across the country and their treatment at the hands of conservationists and foresters. Nevertheless, in most woods about 80 per cent of the vascular plant species are in the ground flora, and many of the critical issues affecting ground flora and its conservation are discussed in his works.

He took George Peterken's initial work in Lincolnshire (Peterken 1974) on ancient woodland indicators and looked at how these species behaved in the woods of Eastern England, highlighting both similarities and differences in their general ability to colonize new sites. Wood Millet *Milium effusum* and Wood Sorrel *Oxalis acetosella* (ancient woodland indicators in Lincolnshire) were also strongly associated with ancient woods in Essex and East Anglia, but Goldilocks Buttercup *Ranunculus auricomus*, Wild Garlic *Allium ursinum*, Sanicle *Sanicula europaea* and Early Dog-violet *Viola reichenbachiana* were not such good indicators (Rackham 2003). He then illustrated the differential spread of three species (Bluebell *Hyacinthoides non-scripta*, Dog's Mercury *Mercurialis perennis*, Oxlip *Primula elatior*) into the triangle of new woodland that had developed since the 1920s alongside Hayley Wood (Rackham 2003) (Fig. 4.1).

Oliver was a champion especially of the Oxlip. His account of the impact of deer grazing on its flowers in Hayley Wood in 1975 is one of the first to highlight the threat that deer could pose for British woodland conservation, beyond their impacts on tree regeneration (Rackham 1990). Later, he suggested that while ancient woods would generally survive, their heart might be eaten out by deer. The Oxlip also acted as an exemplar for his concerns about the impact of coniferization of ancient woodland in eastern England where it often died out under the shade of the conifers. He was very critical of the actions of the Forestry Commission in planting up large areas of ancient woodland, with Hempstead and Chalkney woods in Essex being particular examples (Fig. 4.2). However, he was generous in acknowledging their subsequent efforts to restore the native tree and shrub community to aid in the ground flora recovery.

Rackham strongly supported the maintenance and restoration of coppice management in ancient woodland as an important part of the meaning attached to sites (Fig. 4.3). Through his observations at Hayley Wood (Rackham 1975) he grouped ground flora species according to their

Figure 4.1 Dog's Mercury *Mercurialis perennis* and Bluebell *Hyacinthoides non-scripta* growing in the recent woodland triangle in Hayley Wood, 2017. (K.J. Kirby)

Figure 4.2 Restored area in the north of Chalkney Wood showing the stumps from the planted conifers, 2017. (K.J. Kirby)

strategies for surviving through the period of dense shade in mature coppice, building on the earlier work of Adamson (1912) and Salisbury (1924). Rackham distinguished spring-flowering persistent perennials such as Oxlip, summer-flowering persisters such as Meadowsweet *Filipendula ulmaria*; seed-bank species such as Wood Spurge *Euphorbia amygdaloides*; mobile plants that move around the wood from one coppice area to another such as Plumed Thistles *Cirsium* spp.; casual species that come in sporadically from outside and were not sustainable within the wood, such as Borage *Borago officinalis*; and species that do not respond to coppicing and may even decline after cutting, such as Herb-paris *Paris quadrifolia*.

Palynological records for ground flora species that seldom flower in shade lay at the heart of Rackham's 1996 suggestions that prehistoric woodland in Europe may at least locally have been more like savanna. He also made a strong plea for the importance of woodland grassland (Rackham 2003). However, he seems not to have been convinced that the wildwood was as open as Frans Vera later proposed (Vera 2000). Nor was he keen on the current interest in rewilding or restoring wildwood if this meant abandoning historic management and doing nothing. 'Whatever wildwood in England was once like, woodland ecosystems have become adjusted over centuries to periodic felling and lack of grazing animals ... Each wood-lot now has its own individual guild of seed-bank plants ... most of which do not adapt to lack of felling combined with excessive grazing' (Rackham 2008: 582).

Ironically, this lack of recent management in many woods in south-east England meant that Rackham was an early champion of leaving areas to recover naturally after the 1987 storm (Kirby and Buckley 1994) (Fig. 4.4). He stressed that the gaps created would be likely to provide opportunities for the woodland ground flora, as proved to be the case (Smart et al. 2014).

Rackham recognized the potential threats from pollution (although he considered the effects of 'acid rain' in Britain had been overstated), from climate change and increasingly from introduced pests and diseases (Rackham 2008). In his view, however, there was also a threat to the historical significance of ancient woodland from conservation generalizations and fashions. He considered that the species–area relationship shown for woodland ground flora was more easily explained by large woods tending to have more habitats than by the extinction/colonization processes of island biogeography theory. He did not see a lot of value in creating new woodland links between sites in terms of reducing ground flora extinction risks (Rackham 2003). Indeed, he tended to be scathing about the value of new woodland in general, particularly where this was created by planting.

Figure 4.3 Recently cut coppice at Roudsea Wood National Nature, 2018. (K.J. Kirby)

Figure 4.4 Stour Wood, Essex, showing windblown trees and Wood Anemone *Anemone nemorosa*, 2017. (K.J. Kirby)

He was clear that tree and shrubs formed apparently distinct societies reflecting local conditions but was more equivocal about attempts to define communities based on the ground flora. He felt the National Vegetation Classification, in which ground flora is an important diagnostic element, provided too coarse a definition of woodland types (Rodwell 1991). Partly this may be because when it came to the ground flora he preferred to work at a much finer scale. This is illustrated by his critical use of the detailed surveys from past botanists at Hayley Wood and at Gamlingay to explore long-term changes in the ground flora (Adamson 1912; Abeywickrama 1949; Rackham 2003).

Building on Rackham's legacy

Ancient woodland indicators

There has been a massive proliferation of ancient woodland indicator lists both in Britain and on the continent (Hermy et al. 1999; Glaves et al. 2009). Trait analysis suggests that these plants are likely to be more shade-tolerant and stress-tolerant than generalist woodland species (Kirby, Pyatt and Rodwell 2012; Kimberley et al. 2013) (Fig. 4.5). The early assumptions that they are mainly limited by colonizing ability are coming into question, with, for example, various indicators being found in recent woods in Norfolk (Barnes and Willliamson 2015). This perhaps reflects a more nuanced approach to considering woodland origins that is now possible because of increased availability of maps online and a much wider spread of fieldwork. A simple binary division of ancient/recent woodland is neither always necessary, nor helpful (Goldberg et al. 2007); rather, we can view woodland development, including colonization by ground flora, as more of a gradation in some situations.

The impact of increased deer populations

The increased abundance and wider distribution of deer since the Second World War are now well-documented (Ward 2005). Their impact on woodland flora, noted by Rackham in Hayley

Figure 4.5 *Primula elatior* – an ancient woodland indicator, 2017. (K.J. Kirby)

Wood in the 1970s, has been described from a wide range of sites (Kirby 2001). A reduction in Bramble *Rubus fruticosus* and increase in graminoid species are common outcomes, although the influence on ground flora richness is more variable, with both increases and decreases reported (Fig. 4.6). Much of the evidence is based on exclosure studies such as those of Morecroft et al. (2001) or Putman et al. (1989). Less work has been done on determining different levels of impact when the deer numbers are reduced but not completely eliminated (Cooke 2006). There is also relatively little evidence on how far the changes induced are reversible: do grasses decline again when grazing pressures are reduced; can herbaceous species recover where their diversity has been reduced? There is some indication from research at Wytham that this may be the case, but more data are required from longer-term studies.

Changes in forestry policy and management

In 1985, British forestry policy with respect to broadleaved woodland was transformed (Forestry Commission 1985). Since then, broadleaved woodland has been expected to be maintained as such, not replanted with conifers. The special characteristics of ancient semi-natural woodland are to be maintained. Following the policy change, there was a trend to restore plantations on ancient woodland sites to native broadleaves (or native Scots Pine *Pinus sylvestris* in Scotland) (Fig. 4.7). Extensive progress has been made, particularly by the Woodland Trust and other non-governmental organizations, but also by the Forestry Commission (Pryor, Curtis and Peterken 2002; Goldberg 2003; Thompson et al. 2003). The work to remove the unwanted tree species and to diversify the structure of the stands can be done quite quickly, but inevitably, it takes longer for the ground-flora vegetation to recover. There are now, though, studies showing that species richness and composition in restored stands may be similar to those of remnant broadleaved areas in the same forests (Kirby and May 1989; Brown, Curtis and Adams 2015; Kirby and Thomas 2017; Kirby, Goldberg and Orchard 2017).

Traditional coppice management of ancient semi-natural woodland has proved more difficult to promote on a large scale, although the improved market for firewood in recent

Figure 4.6 Bramble inside, grass outside the deer fence at Roudsea Wood, Cumbria, 2014. (K.J. Kirby)

Figure 4.7 Restored oak stand, following conifer removal at Shabbington Woods, Buckinghamshire, 2015. (K.J. Kirby)

years has helped. Coppice-associated species have generally continued to decline (Buckley and Mills 2015). Increased shading, as woods last felled in the Second World War have grown up as high forest, has contributed to this reduction in species-richness in the ground flora of many woods (Kirby et al. 2005). There has been less research done on the flora of high-forest broadleaved stands and, in particular, on alternatives to clear-fell/replant systems (Kerr 1999; Brang et al. 2014). While there is considerable interest currently in continuous cover forestry techniques, it is not clear what the implications for the ground flora will be in the long term.

A changing atmosphere

Climate change and increased nitrogen depositions have been shown to have affected woodland ground flora both in Britain and across Europe generally. The changes in ground flora composition follow from the cumulative effects of changes in individual species performance and survival (Pearce-Higgins et al. 2017). There is, though, concern that these can lead to greater homogenization of the woodland flora (Keith et al. 2009) (Fig. 4.8). To date, the impacts on woodland communities have tended to be less apparent than for some open habitats, because of the buffering effects of forest canopies, examples being described in De Frenne et al. (2013) and Verheyen et al. (2012). However, sudden changes may occur in future, where the canopy becomes more open through forest management or the effects of a disease such as Ash Dieback, caused by *Hymenoscyphus fraxineus.*

The structure of the wildwood and future rewilding

Frans Vera's ideas on the open nature of the prehistoric landscape (Vera 2000) continue to be debated. Evidence that the landscape was generally half-open and that this openness was

Figure 4.8 Recent spread of Wild Garlic *Allium ursinum* may be linked to increased nitrogen deposition in Wytham Woods, Oxfordshire, 2015. (K.J. Kirby)

Figure 4.9 Effects of free-ranging pigs and cattle in a rewilded area on a woodland flora, on the Knepp Estate, Sussex, 2018. (K.J. Kirby)

created by the impact of large herbivores remains elusive (Mitchell 2005; Sandom et al. 2014; Allen 2017). However, the concept has been taken up strongly in many conservation circles and used to support rewilding initiatives across Europe (Navarro and Pereira 2012; Lorimer et al. 2015; Tree 2018).

Reintroduction of large herbivores and large-scale extensive grazing, as in wood-pasture systems, can create and maintain rich and varied landscapes (Plieninger et al. 2015; Rotherham 2013). What are less clear are the implications for the specialist woodland ground flora assemblages that we have come to associate with small fragmented ancient woods (Fig. 4.9). The component species might survive in rewilded landscapes, but with reductions in the massed displays of Bluebells *Hyacinthoides non-scripta*, Wood Anemones *Anemone nemorosa* or Common Primroses *Primula vulgaris* as a consequence of grazing by cattle and the disturbance from wild boar or free-range pigs (Sims, John and Stewart 2014). Conservation based on continued management of some sites will continue to be needed if we wish to maintain such assemblages.

New woods, new meanings?

Woodland cover has continued to increase across Britain, although at a decreasing rate in the last few years (Forestry Commission 2012), both through planting and natural regeneration (Fig. 4.10). Such sites can take decades, if not centuries to accumulate a varied ground flora, even when right next to an ancient woodland (Peterken and Game 1984; Rackham 2003), but the process can be speeded up (Rodwell and Patterson 1994; Francis and Morton 2001; Highways Agency 2005; Craig, Buckley and Howell 2015; Worrell et al. 2016; Buckley et al. 2017). These assemblages do not now have the same meaning as those in an ancient woodland, but new interests and meanings will get attached to them in time (Rotherham 2017). Meanwhile, a more varied, if artificial, ground flora will at the least permit a wider range of invertebrates and fungi to colonize the sites.

Figure 4.10 Yellow Archangel *Lamiastrum galeobdolon* introduced via translocated woodland soil to a new woodland site, New Biggin Wood, Kent, 2012. (K.J. Kirby)

New methods and revisiting old ones

One of the attractions, for many people, of Rackham's approach to studying woods (and the countryside more generally) is its accessibility. It does not need sophisticated or expensive equipment or deep experience before you can try it yourself. The availability, via the internet, of historical maps and documents, aerial photographs and satellite images, species records and past site descriptions has made past landscapes even easier to explore. Field methods have not changed as much as in some other areas of ecology and plant sciences: a pen and clipboard of paper can be sufficient, although increasingly, there is direct field record entry to data loggers, plots are located (and relocated) by GPS, digital photographs replace sketch maps and written descriptions, and capture images of unknown species (Kirby and Hall 2019).

This observational and descriptive ecology, which formed the core of Rackham's work on woodland ground flora, is out of fashion with academic journals. They prefer hypothesis-driven quantitative studies, generally focused on quite narrow questions. Yet if we are to understand how our woods and landscapes will change over the next 50 years, possibly quite radically, we must start with knowing how they have come to be as they are now. This means going back to and interpreting past records in the light of current knowledge and priorities.

I was fortunate to inherit results from a set of 164 permanent plots established (1973–6) in Wytham Woods. We have re-recorded them several times since. The data show little change in mean species richness over a 40-year period, but individual plots show substantial changes in the numbers of species they contain. There have been major changes in ground flora structure: from dominance by Bramble to an abundance of the grass False Wood-brome *Brachypodium sylvaticum*. Individual species show contrasting patterns – from an almost

Figure 4.11 Will Ash Dieback sites come to look like this dead elm grove in Overhall Grove, Cambridgeshire, 2018? (K.J. Kirby)

complete loss of Rosebay Willowherb *Chamaenerion angustifolium* to a substantial spread of Pendulous Sedge *Carex pendula* (Kirby 2010; Kirby and Thomas 2000). In a separate project, the diaries of Charles Elton (1942–65) reveal how Wytham Woods and many of the sites in the surrounding county changed in the decades preceding the first recording of the plots (Kirby 2016; 2017): they explain why some stands are now plantations, but others not; where wartime fellings occurred; and suggest that the woods were generally more open and wetter in the past. These studies now form the baseline for assessing the impact of the disease Ash Dieback in the woods (Fig. 4.11). As European Ash *Fraxinus excelsior* contributes about 25 per cent to the canopy overall, was the fastest growing tree in recent decades and showed the best regeneration, the effects of the disease could be substantial.

Conclusions

Oliver Rackham advanced no great new theory with respect to woodland ground flora, but through careful observation helped test and refine a wide range of ideas about it. When we look at an Oxlip, we now see it in a different light. The digitization of Rackham's notebooks, described in a later chapter, will allow future surveyors literally to follow in his footsteps as he built on the likes of Abeywickrama (1949) and Adamson (1912).

Acknowledgements

The ideas in this chapter draw on insights from the work of Oliver Rackham himself, my mentor George Peterken, and many other foresters and woodland ecologists whom I have had the honour of working with over the years. Parts of this chapter are based on a book on woodland flowers published by British Wildlife Publishing (2020). Any errors and misinterpretations are, however, my own.

References

Abeywickrama, B.S. (1949) A study of the variation in the field layer vegetation of two Cambridgeshire woods. Unpublished PhD thesis, University of Cambridge.

Adamson, R.S. (1912) An ecological study of a Cambridgeshire woodland. *Journal of the Linnean Society of London, Botany*, 40(276), pp. 339–387. https://doi.org/10.1111/j.1095-8339.1912.tb00613.x

Allen, M.J. (2017) The southern English chalklands: molluscan evidence for the nature of the post-glacial woodland cover. In: M.J. Allen (ed.) *Molluscs in Archaeology: Methods, approaches and applications*. Oxbow Books, Oxford, pp. 144–164. https://doi.org/10.2307/j.ctvh1dk5s.15

Barnes, G. and Willliamson, T. (2015) *Rethinking Ancient Woodland: The Archaeology and History of Woods in Norfolk*. University of Hertfordshire Press, Hatfield.

Brang, P., Spathelf, P., Larsen, J.B., Bauhus, J., Bončina, A., Chauvin, C., Drössler, L., García-Güemes, C., Heiri, C., Kerr, G., Lexer, M. J., Mason, B., Mohren, F., Mühlethaler, U., Nocentini, S. and Svoboda, M. (2014) Suitability of close-to-nature silviculture for adapting temperate European forests to climate change. *Forestry*, 87(4), pp. 492–503. https://doi.org/10.1093/forestry/cpu018

Brown, N.D., Curtis, T. and Adams, E.C. (2015) Effects of clear-felling versus gradual removal of conifer trees on the survival of understorey plants during the restoration of ancient woodlands. *Forest Ecology and Management*, 348(1), pp. 15–22. https://doi.org/10.1016/j.foreco.2015.03.030

Buckley, G.P. and Mills, J. (2015) The flora and fauna of coppice woods: winners and losers of active management or neglect. In: K.J. Kirby and C. Watkins (eds) *Europe's Changing Woods and Forests: from Wildwood to Managed Landscapes*. CABI, Wallingford, pp. 129–139. https://doi.org/10.1079/9781780643373.0129

Buckley, P., Helliwell, D.R., Milne, S. and Howell, R. (2017) Twenty-five years on – vegetation succession on a translocated ancient woodland soil at Biggins Wood, Kent, UK. *Forestry*, 90(4), pp. 561–572. https://doi.org/10.1093/forestry/cpx015

Cooke, A.S. (2006) Monitoring Muntjac deer *Muntiacus reevesi* and their impacts in Monks Wood National Nature Reserve. *English Nature Research Report*. English Nature, Peterborough.

Craig, M., Buckley, P. and Howell, R. (2015) Responses of an ancient woodland field layer to soil translocation: methods and timing. *Applied Vegetation Science*, 18(4), pp. 579–590. https://doi.org/10.1111/avsc.12170

De Frenne, P., Rodríguez-Sánchez, F., Coomes, D.A., Baeten, L., Verstraeten, G., Vellend, M., Bernhardt-Römermann, M., Brown, C.D., Brunet, J., Cornelis, J., Decocq, G.M., Dierschke, H., Eriksson, O., Gilliam, F.S., Hédl, R., Heinken, T., Hermy, M., Hommel, P., Jenkins, M.A., Kelly, D.L., Kirby, K.J., Mitchell, F.J.G., Naaf, T., Newman, M., Peterken, G., Petřík, P., Schultz, J., Sonnier, G., Van Calster, H., Waller, D.M., Walther, G.-R., White, P.S., Woods, K.D., Wulf, M., Graae, B.J. and Verheyen, K. (2013) Microclimate moderates plant responses to macroclimate warming. *Proceedings of the National Academy of Sciences*, 110(46), pp. 18561–18565. https://doi.org/10.1073/pnas.1311190110

Forestry Commission (1985) *The Policy for Broadleaved Woodland*. Forestry Commission, Edinburgh.

Forestry Commission (2012) *NFI Preliminary Estimates of Quantities of Broadleaved Species in British Woodlands, with Special Focus on Ash*. Forestry Commission, Edinburgh.

Francis, J.L. and Morton, A. (2001) Enhancement of amenity woodland field layers in Milton Keynes. *British Wildlife*, 12(4), pp. 244–251.

Glaves, P., Handley, C., Birbeck, J., Rotherham, I.D. and Wright, B. (2009) *A Survey of the Coverage, Use and Application of Ancient Woodland Indicator Lists in the UK*. Woodland Trust, Grantham.

Goldberg, E.A. (2003) Plantations on ancient woodland sites. *Quarterly Journal of Forestry*, 97(2), pp. 133–138.

Goldberg, E.A., Kirby, K.J., Hall, J.E. and Latham, J. (2007) The ancient woodland concept as a practical conservation tool in Great Britain. *Journal of Nature Conservation*, 15(2), pp. 109–119. https://doi.org/10.1016/j.jnc.2007.04.001

Grove, A.T. and Rackham, O. (2001) *The Nature of Mediterranean Europe: An Ecological History*. Yale University Press, New Haven, CT and London.

Hermy, M., Honnay, O., Firbank, L., Grashof-Bokdam, C. and Lawesson, J.E. (1999) An ecological comparison between ancient and other forest plant species of Europe, and the implications for forest conservation. *Biological Conservation*, 91(1), pp. 9–22. https://doi.org/10.1016/S0006-3207(99)00045-2

Highways Agency (2005) The establishment of an herbaceous plant layer in roadside woodland. In: *Design Manual for Roads and Bridges*. Highways Agency, London.

Keith, S.A., Newton, A.C., Morecroft, M.D., Bealey, C.E. and Bullock, J.M. (2009) Taxonomic homogenization of woodland plant communities over 70 years. *Proceedings of the Royal Society B: Biological Sciences*, 276(1672), pp. 3539–3544. https://doi.org/10.1098/rspb.2009.0938

Kerr, G. (1999) The use of silvicultural systems to enhance the biological diversity of plantation forests in Britain. *Forestry*, 72(3), pp. 191–205. https://doi.org/10.1093/forestry/72.3.191

Kimberley, A., Blackburn, G.A., Whyatt, J.D., Kirby, K. and Smart, S.M. (2013) Identifying the trait syndromes of conservation indicator species: how distinct are British ancient woodland indicator plants from other woodland species? *Applied Vegetation Science*, 16(3), pp. 667–675. https://doi.org/10.1111/avsc.12047

Kirby, K.J. (2001) The impact of deer on the ground flora of British broadleaved woodland. *Forestry*, 74(3), pp. 219–229. https://doi.org/10.1093/forestry/74.3.219

Kirby, K.J. (2010) The flowers of the forest. In: P.S. Savill, C. Perrins, K.J. Kirby and N. Fisher (eds) *Wytham Woods, Oxford's Ecological Laboratory*. Oxford University Press, Oxford, pp. 75–89. https://doi.org/10.1093/acprof:osobl/9780199605187.003.0006

Kirby, K.J. (2016) The transition of Wytham Woods from a working estate to unique research site (1943–1965). *Landscape History*, 37(2), pp. 79–92. https://doi.org/10.1080/01433768.2016.1249725

Kirby, K.J. (2017) Walking back in time: conservation in Berkshire, Buckinghamshire and Oxfordshire, 1942–65, from the diaries of Charles Elton. *Fritillary*, 7, pp. 49-78.

Kirby, K.J. and Buckley, G.P. (1994) Ecological responses to the 1987 Great Storm in the woods of south-east England. *English Nature Science* 23. English Nature, Peterborough.

Kirby, K.J., Goldberg, E.A. and Orchard, N. (2017) Long-term changes in the flora of oak forests and of oak: spruce mixtures following removal of conifers. *Forestry*, 90(1), pp. 136–147. https://doi.org/10.1093/forestry/cpw049

Kirby, K.J. and Hall, J.E. (2019) *Woodland Survey Handbook: Collecting data for conservation in British woodland.* Pelagic Publishing, Exeter. https://doi.org/10.53061/BPFE8441

Kirby, K.J. and May, J. (1989) The effects of enclosure, conifer planting and the subsequent removal of conifers in Dalavich oakwood (Argyll). *Scottish Forestry*, 43(4), pp. 280–288.

Kirby, K.J., Pyatt, D.G. and Rodwell, J.S. (2012) Characterization of the woodland flora and woodland communities in Britain using Ellenberg Values and Functional Analysis. In: I.D. Rotherham, M. Jones and C. Handley (eds) *Working and Walking in the Footsteps of Ghosts: Volume 1 the wooded landscape.* Wildtrack Publishing, Sheffield, pp. 66–86.

Kirby, K.J., Smart, S.M., Black, H.J., Bunce, R.G.H., Corney, P.M. and Smithers, R.J. (2005) Long-term ecological changes in British woodland (1971–2001). *English Nature Research Report 653*. English Nature, Peterborough.

Kirby, K.J. and Thomas, R.C. (2000) Changes in the ground flora in Wytham Woods, southern England from 1974 to 1991 – implications for nature conservation. *Journal of Vegetation Science*, 11(6), pp. 871–880. https://doi.org/10.2307/3236557

Kirby, K.J. and Thomas, R.C. (2017) Restoration of broadleaved woodland under the 1985 Broadleaves Policy stimulates ground flora recovery at Shabbington Woods, southern England. *New Journal of Botany*, 7(2), pp. 125–135. https://doi.org/10.1080/20423489.2017.1408177

Kirby, K.J. and Watkins, C. (2015) *Europe's Changing Woods and Forests: From wildwood to managed landscapes.* CABI, Wallingford. https://doi.org/10.1079/9781780643373.0000

Lorimer, J., Sandom, C., Jepson, P., Doughty, C., Barua, M. and Kirby, K.J. (2015) Rewilding: science, practice, and politics. *Annual Review of Environment and Resources*, 40, pp. 39–62. https://doi.org/10.1146/annurev-environ-102014-021406

Mitchell, F.J.G. (2005) How open were European primeval forests? Hypothesis testing using palaeoecological data. *Journal of Ecology*, 93(1), pp. 168–177. https://doi.org/10.1111/j.1365-2745.2004.00964.x

Morecroft, M.D., Taylor, M.E., Ellwood, S.A. and Quinn, S.A. (2001) Impacts of deer herbivory on ground vegetation at Wytham Woods, central England. *Forestry*, 74(3), pp. 251–257. https://doi.org/10.1093/forestry/74.3.251

Navarro, L. and Pereira, H. (2012) Rewilding abandoned landscapes in Europe. *Ecosystems*, 15(6), pp. 900–912. https://doi.org/10.1007/s10021-012-9558-7

Pearce-Higgins, J.W., Beale, C.M., Oliver, T.H., August, T.A., Carroll, M., Massimino, D., Ockendon, N., Savage, J., Wheatley, C.J., Ausden, M.A., Bradbury, R.B., Duffield, S.J., Macgregor, N.A., Mcclean, C.J., Morecroft, M.D., Thomas, C.D., Watts, O., Beckmann, B.C., Fox, R., Roy, H.E., Sutton, P.G., Walker, K.J. and Crick, H.Q.P. (2017) A national-scale assessment of climate change impacts on species: Assessing the balance of risks and opportunities for multiple taxa. *Biological Conservation*, 213(part A), pp. 124–134. https://doi.org/10.1016/j.biocon.2017.06.035

Peterken, G.F. (1974) A method for assessing woodland flora for conservation using indicator species. *Biological Conservation*, 6(4), pp. 239–245. https://doi.org/10.1016/0006-3207(74)90001-9

Peterken, G.F. and Game, M. (1984) Historical factors affecting the number and distribution of vascular plant species in the woodlands of central Lincolnshire. *Journal of Ecology*, 72(1), pp. 155–182. https://doi.org/10.2307/2260011

Plieninger, T., Hartel, T., Martín-López, B., Beaufoy, G., Bergmeier, E., Kirby, K., Montero, M.J., Moreno, G., Oteros-Rozas, E. and Van Uytvanck, J. (2015). Wood-pastures of Europe: geographic coverage, social–ecological values, conservation management, and policy implications. *Biological Conservation*, 190, pp. 70–79. https://doi.org/10.1016/j.biocon.2015.05.014

Pryor, S.N., Curtis, T.A. and Peterken, G.F. (2002) *Restoring Plantations on Ancient Woodland Sites.* The Woodland Trust, Grantham.

Putman, R.J., Edwards, P.J., Mann, J.C.E., How, R.C. and Hill, S.D. (1989) Vegetational and faunal changes in an area of heavily grazed woodland following relief of grazing. *Biological Conservation*, 47(1), pp. 13–32. https://doi.org/10.1016/0006-3207(89)90017-7

Rackham, O. (1975) *Hayley Wood.* Cambridgeshire and Isle of Ely Naturalists' Trust, Cambridge.

Rackham, O. (1990) *Trees and Woodland in the British Landscape*, revised edition. J.M. Dent and Sons, London.

Rackham, O. (1998) Savanna in Europe. In: K.J. Kirby and C. Watkins (eds) *The Ecological History of European Forests.* CABI, Wallingford, pp. 1–24.

Rackham, O. (2003) *Ancient Woodland: Its history, vegetation and uses in England*, revised edition. Castlepoint Press, Dalbeattie, Scotland.

Rackham, O. (2008) Ancient woodlands: modern threats. *New Phytologist*, 180(3), pp. 571–586. https://doi.org/10.1111/j.1469-8137.2008.02579.x

Rodwell, J.S. (1991) *British Plant Communities: 1 Woodlands and scrub.* Cambridge University Press, Cambridge. https://doi.org/10.1017/9780521235587

Rodwell, J.S. and Patterson, G. (1994) Creating new native woodland. *Forestry Commission Bulletin.* Forestry Commission, London.

Rotherham, I.D. (2013) *Trees, Forested Landscapes and Grazing Animals.* Routledge, Abingdon. https://doi.org/10.4324/9780203102909

Rotherham, I.D. (2017) *Recombinant Ecology – A Hybrid Future?* Springer, Cham. https://doi.org/10.1007/978-3-319-49797-6

Salisbury, E.J. (1924) The effects of coppicing as illustrated by the woods of Hertfordshire. *Transactions of the Hertfordshire Natural History Society*, 18, pp. 1–21.

Sandom, C.J., Ejrnæs, R., Hansen, M.D.D. and Svenning, J.-C. (2014) High herbivore density associated with vegetation diversity in interglacial ecosystems. *Proceedings of the National Academy of Sciences*, 111(11), pp. 4162–4167. https://doi.org/10.1073/pnas.1311014111

Sims, N.K., John, E.A. and Stewart, A.J.A. (2014) Short-term response and recovery of bluebells (Hyacinthoides non-scripta) after rooting by wild boar (*Sus scrofa*). *Plant Ecology*, 215(12), pp. 1409–1416. https://doi.org/10.1007/s11258-014-0397-9

Smart, S.M., Ellison, A.M., Bunce, R.G.H., Marrs, R.H., Kirby, K.J., Kimberley, A., Scott, A.W. and Foster, D.R. (2014) Quantifying the impact of an extreme climate event on species diversity in fragmented temperate forests: the effect of the October 1987 storm on British broadleaved woodlands. *Journal of Ecology*, 102(5), pp. 1273–1287. https://doi.org/10.1111/1365-2745.12291

Symonds, H.H. (1936) *Afforestation in the Lake District*. J.M. Dent and Sons, London.

Thompson, R., Humphrey, J.W., Harmer, R. and Ferris, R. (2003) Restoration of native woodland on ancient woodland sites. *Forest Practice Guide*. Forestry Commission, Edinburgh.

Tree, I. (2018) *Wilding: The Return of Nature to a British Farm*. Pan Macmillan, London.

Vera, F.W.M. (2000) *Grazing Ecology and Forest History*. CABI, Wallingford. https://doi.org/10.1079/9780851994420.0000

Verheyen, K., Baeten, L., De Frenne, P., Bernhard-Romermann, M., Brunet, J., Cornelis, J., Decoq, G., Dierschke, H., Eriksson, O., Hedl, R., Heinken, T., Hermy, M., Hommel, P., Kirby, K.J., Naaf, T., Peterken, G.F., Petrik, P., Pfadenhauer, J., Van Calster, H., Walther, G.-R., Wulf, M. and Verstraeten, G. (2012) Driving factors behind the eutrophication signal in understorey plant communities of deciduous temperate forests. *Journal of Ecology*, 100(2), pp. 352–365. https://doi.org/10.1111/j.1365-2745.2011.01928.x

Ward, A.I. (2005) Expanding ranges of wild and feral deer in Great Britain. *Mammal Review*, 35(2), pp. 165–173. https://doi.org/10.1111/j.1365-2907.2005.00060.x

Worrell, R., Long, D., Laverack, G., Edwards, C. and Holl, K. (2016) *The Introduction of Woodland Plants into Broadleaved Woods for Conservation Purposes: Best practice guidance*. Plantlife Scotland, Scottish Natural Heritage, Forestry Commission Scotland and Scotia Seeds, Edinburgh.

CHAPTER 5

Echoes of the Wildwood? Investigating the Historical Ecology of some Warwickshire Lime Woodlands 1986–2000

David R. Morfitt

Summary

After discovering, and being inspired by, Oliver's work, I became involved in researching the ancient woods of Warwickshire. Here I concentrate on the ancient limewoods, a little known component of the woods of Warwickshire. As in much of England, limewoods in Warwickshire are relict and a reliable indicator of the ancient status of woods. My PhD on the woods of Binley east of Coventry, with unusually informative medieval documentation for the county, also suggests that grazing may have been especially destructive of lime in ancient woods; ordinarily, short of grubbing, lime is very difficult to destroy. In the study area none of the many thousands of lime seedlings produced in the 1990s has apparently survived to grow into saplings (examination of Piles Coppice, the main limewood in the study, confirmed this in 2017). I saw evidence of extensive Small-leaved Lime *Tilia cordata* reproduction to sapling stage in the Marks Hall Woods in Essex in 2017; will global warming improve Small-leaved Lime reproduction nationally? I discussed global warming with Oliver in March 2011, and we agreed that it would probably benefit most of our native tree and shrub species as they are currently growing at or near the northern limit of their ranges.

Keywords: ancient woodlands, Warwickshire, historical ecology, Piles Coppice, Small-leaved Lime, ageing coppice

Introduction

I began researching the historical ecology of some Warwickshire woods in January 1986, inspired by Oliver Rackham's *Trees and Woodland* (1976) (T&W) and *Ancient Woodland* (1980). I had first read T&W in 1981 and, as for so many people, it was a revelation. As an historian I was most excited by the antiquity of ancient woods, so eloquently evoked by Richard Mabey in his book with Tony Evans, *The Flowering of Britain*: 'These old woods, weather-beaten, hard-worked, spun about with legend and history, each one stocked with its own exclusive cargo of flowers, are life-rafts out of the past' (Mabey 1980: 51).

David R. Morfitt, 'Echoes of the Wildwood? Investigating the Historical Ecology of some Warwickshire Lime Woodlands 1986–2000' in: *Countryside History: The Life and Legacy of Oliver Rackham*. Pelagic Publishing (2024).
 DOI: 10.53061/YDWO7681

I had little opportunity to do any original fieldwork until I moved to Warwickshire from Cambridge; Oliver had, of course, rather cornered the market in woodland research around there, and I had no transport, so working farther afield was difficult. I had met him a number of times, as I was a member of the Hayley Wood Work Party, and on woodland open days as a volunteer at the Cambridgeshire and Isle of Ely Naturalists Trust,[1] which owned Hayley Wood and other ancient woods in Cambridgeshire; but Oliver seemed then more driven and less approachable than he became later.

Early work in Warwickshire

Warwickshire in 1986 was *terra incognita* for the historical ecologist. Warwickshire Wildlife Trust (WWT) had recently acquired Ryton Wood, a large 86 ha wood that begged to be investigated. Such a large site was daunting for a beginner but, as fieldwork showed, it was an interesting wood with intriguing features. Oliver's methodology for researching ancient woods with 'vegetation as a third dimension in a historical and archaeological synthesis' (Rackham 1976; 1980; 1990) frequently proved hard to follow as documentation for Warwickshire woods is often sparse or non-existent. For an historian, this was one of the most frustrating aspects of the work (Fig. 5.1).

Ryton Wood, Warwickshire (UK Grid Reference SP 381 725)

Ryton Wood lies south-east of Coventry and mostly consists of Hazel–Pedunculate Oak woodland (Peterken (1993) Stand Type 6Dc) and birch–Pedunculate Oak woodland (6Cc), but with patches of Small-leaved Lime *Tilia cordata* woodland (5A), on acid clay, sand and silt over sand and gravels.[2] The accessible medieval documentation for Ryton parish woodland was sparse and uninformative (Morfitt 1988; Wager 1998).[3] The records of the Fetherston-Dilke family, who had owned the wood since the sixteenth century and which include Warwickshire deeds, manorial records, legal, estate and other papers from the thirteenth to the twentieth century, were at the family home at Maxstoke Castle in the 1980s. As a beginner researching old records, I lacked credibility with the staff of Warwick County Record Office in Warwick (WCRO) so, as they were not willing to send for those records, I was unfortunately unable to research the documentation of the wood in any satisfactory way. The Fetherston-Dilke records are now at last in the WCRO,[4] and a future project would be to investigate them for material on Ryton Wood.

The wood itself has much to tell about its history (Fig. 5.2), even though interpretation of the evidence is difficult. Areas A and B are surrounded by large woodbanks, with those round A more sinuous than those round B. The breadth of the woodbanks are generally 7–10 m, taking together the bank and the external ditch. Smaller banks and ditches surround much of the

1 Now part of the Wildlife Trust for Bedfordshire, Cambridgeshire and Northamptonshire.

2 Although the National Vegetation Classification (NVC, see Rodwell 1991) has now been available for many years and is used by many nature conservation organizations, the stand-type system, as defined by Peterken (1981, 1993) and Rackham (1980, 1990, 2003, 2006) still has advantages for the study and recording of ancient woodland for historical ecology. The stand-type system emphasizes tree communities, the only aspect of woods for which we have detailed historical records (Rackham 1990); it recognizes many more ancient tree communities than NVC (Rackham 1990), including useful categories such as limewoods ignored by NVC and which have historical meaning (Peterken 1991, 1993; Kirby 1988); and is in general more useful and economical to record (Kirby 1988). Oliver was always sceptical of the value of the NVC for researching ancient woodlands (Rackham 2003; 2006).

3 This includes other documentary references and discussion of the earthworks and plants; a copy is in the Warwickshire Wildlife Trust's archives.

4 WCRO Document reference CR 2981.

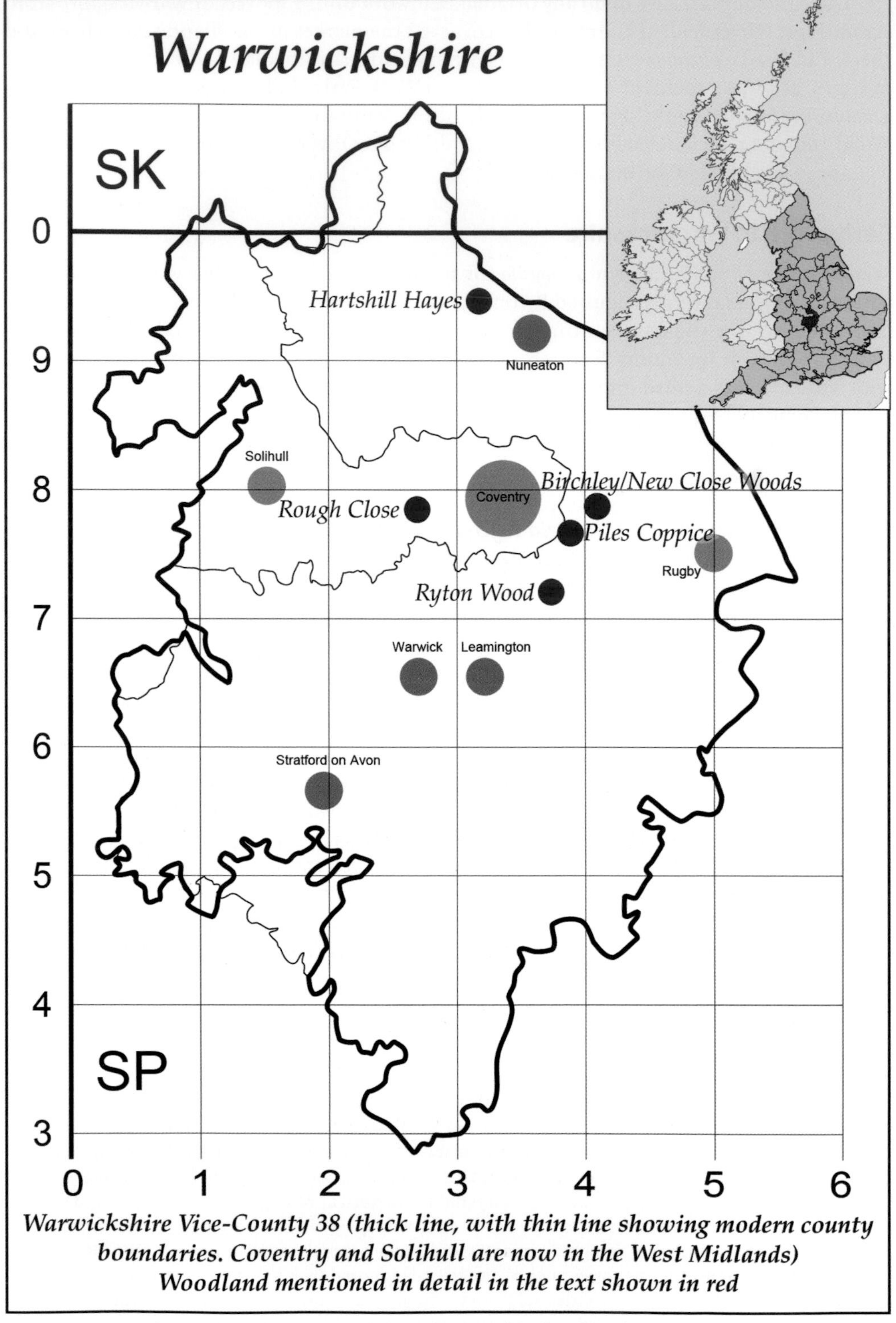

Figure 5.1 Locations of woods in Warwickshire discussed in detail in the text. SP and SK are UK Grid References, each double letter representing a unique 100-km square. The grid is of 10-km squares. (D. Morfitt 2019)

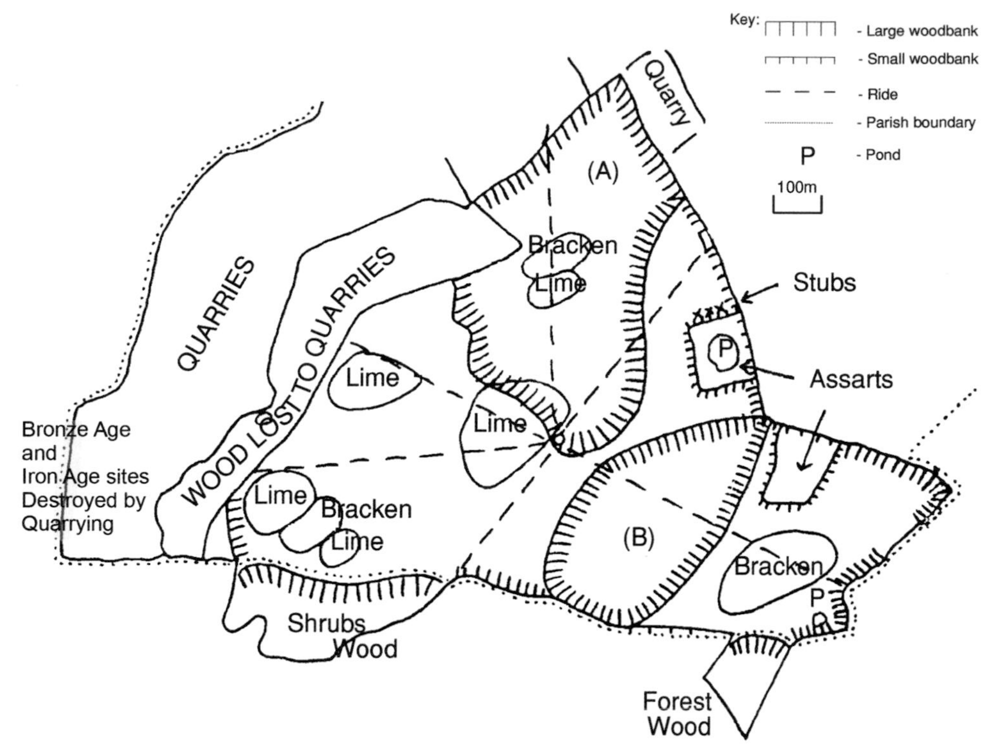

Figure 5.2 Sketch map of lime woodland and archaeological features at Ryton Wood. (D. Morfitt 1988)

rest of the wood. Adjoining Shrubs and Forest Woods, in different parishes from Ryton, have large woodbanks on their side of the parish boundary. A particularly small and acute bank marks the wood edge and parish boundary between the southern edge of area B and Forest Wood. The two assarts (or formerly cleared areas, now wooded again) are surrounded by small, relatively straight banks; the northern assart has stubs or short pollarded trees on its northern boundary, perhaps from a former hedge planted to demarcate the assart.[5] As is frequently the case, accurate dating of these banks is currently impossible.[6] Work elsewhere does suggest that the largest are likely to be of medieval or possibly earlier origin (Rackham 1990). This observation and their interconnection also imply a sequence of construction, with either A (of about 60–70 acres (24–28 ha)) followed by B (of about 40 acres (16 ha)), or A and B together coming first, and the smaller, less sinuous banks in the rest of the wood coming later. The assarts represent efforts to create fields out of parts of the wood at some unknown date. The relatively small size and straightness of the banks may indicate that they are late, perhaps post-medieval (Rackham 1990). The very small acute bank between the edge of area B and Forest Wood is almost certainly eighteenth or early nineteenth century; a map of 1763 shows a large wood in Stretton parish extending south from this edge of Ryton Wood, which

5 See the discussion of stubs as boundary markers in Rackham (1990: 8, 123, 149).

6 Morfitt (2000) – the author's PhD thesis – presented evidence for dating woodbanks in Binley and Brinklow. Extrapolation from this evidence would suggest that the large woodbanks at Ryton may be at least as old as the fourteenth century. Morfitt (2000) is available on loan from the British Library or Coventry University, on CD at the Warwick Record Office, Warwick Museum and Coventry Local Studies Library, and the Bodleian Library (Oxford) or from the author on application.

is the parish boundary.[7] By around 1830, much of that wood had been grubbed out (according to the first edition 1-inch Ordnance Survey Sheet 53 Daventry published October 1834 but surveyed earlier),[8] leaving only a much-truncated Forest Wood connected to Ryton Wood at its eastern end, as it is today. The small acute bank therefore dates from sometime between 1763 and at the latest about 1830.

The evidence of the vegetation complicates this reasoning. As Small-leaved Lime is a relict species, which does not colonize well from seed but persists unless destroyed, we would expect its large old coppice stools of diameter 3 m plus to be restricted to the areas of the wood demarcated by the largest, and therefore probably oldest, banks and ditches – but they also appear elsewhere (see Fig. 5.2). Much evidence has probably also been lost by the quarrying away for sand and gravel of the western edge of the wood (Fig. 5.2). It is inevitable that complex woods such as this, with the evidence of the the woodbanks showing that changes took place unrecorded many centuries ago, will have histories that will be difficult to unravel.

Despite the inconclusive work at Ryton, it was exciting to find patches of ancient Small-leaved Lime in the wood. Of native trees, none is more fascinating and important. In 1981, George Peterken said: 'It is the mixed broadleaved woodlands containing lime ... that probably comprise the least modified remnants of the original forests' (Peterken 1981/1993: 315). He restates this in *Wye Valley* (2008), where he notes lime as an indicator of possible primary woodland. Oliver Rackham said:

> The presence of *Tilia cordata* is strong evidence that a wood is ancient, and is almost conclusive if the lime forms giant stools ... Pry indicates ancient woodland throughout its English range ...
>
> The pry tree, *Tilia cordata*, is a living link with Mesolithic times ... Some of the woods [in South Suffolk] have increased in size, and lime faithfully picks out the parts of the wood inside the original woodbank.
>
> Rackham 1986: 108, 106

Small-leaved Lime was the commonest tree throughout much of the prehistoric forest of lowland England and parts of Wales (Greig 1982; Rackham 1990). In Warwickshire, pollen analysis of sites at Shustoke (Kelly and Osbourne 1964) and Moreton Morrell (Shotton 1967) shows that Small-leaved Lime was common, if not dominant, in the prehistoric forests of parts of north and south Warwickshire. At Shustoke there was also some Large-leaved Lime *Tilia platyphyllos*. Although usually a relict species, defined by Oliver as 'maintaining itself in its old localities but unable to colonize new ground' (Rackham 1980: 243), the behaviour of lime in some sites seems to indicate that it has been capable of at least local expansion in the past millennium; at Swithland Wood, Leicestershire, some Small-leaved Lime overlies the ridge and furrow of probable medieval ploughing. But even this limited colonisation was only possible because lime survived from early woodland clearance in hedges on the site (Woodward 1992). The occurrence of lime in woods often seems to be associated with early encoppicement in areas, such as Suffolk or Lincolnshire, where woodland early acquired scarcity value (Rackham 1980).

Lime species quite often been missed or misidentified in woodland surveys in Warwickshire up to 1985. The Computer Mapped Flora of Warwickshire of 1971 misidentifies most of the lime in the county as Common Lime *Tilia* x *europaea*; much of the small amount of Small-leaved Lime recorded is ascribed to scrub and hedgerows, and Large-leaved Lime is assumed generally to have been planted (Cadbury, Hawkes and Readett 1971). In my researches I discovered that a large coppice stool of Small-leaved Lime at the edge of Hanging Wood,

7 WCRO Lord John Scott Estate 1763 Z8 22/1-3.

8 J.B. Harley, discussion of the origins of Sheet 53 on the David and Charles reprint, renumbered 52, 1970.

Claverdon (GR SP 187643) had been misidentified as a large Common Lime in a Warwickshire Wildlife Trust survey of 1985 and that the approximately 3-ha stand of Small-leaved Lime, straddling a main ride in the middle of Birchley Wood, Brinklow, was missed entirely in the WWT survey of the wood in the same year.

Further work in Warwickshire on limewoods:

Rough Close, Berkswell (GR SP 268787)

Over the next few years, I investigated a number of Warwickshire woods. Rough Close, Berkswell, just to the west of Coventry, was the site of a proposed new coal 'Superpit', and I examined the wood and surrounding landscape, especially Hawkhurst Moor, for the Colliery Opposition Group, eventually writing a long submission for the public enquiry in 1989. A summary of my work appeared in 1987 (Morfitt 1987).[9] The wood has been badly damaged by unsympathetic management since it was acquired by the Scouting Association in 1946, including the creation of large permanent clearings. The *Coventry Evening Telegraph* of 18 August 1950 records the 'digging up of more than 400 roots and bolls of trees ranging from 4–6 feet (1.2–1.8 m) in diameter' to create the camping field, presumably the large west clearing. In the late 1980s, scouts were receiving their forestry badges for planting conifers in the wood.[10] However, the wood still has coppiced Small-leaved Lime, with some coppice stools 4–7 ft (1.2–2.1 m) in diameter, and an impressive woodbank around more than half the periphery. It is largely an acid Pedunculate Oak–Hazel wood on acid loamy clay (Peterken Stand Type 6Dc).

The seventeenth- and eighteenth-century records contain much contradictory and difficult material on the woods of the parish,[11] and suggest, for instance, that Rough Close had a number of different names. It was possibly Hawkers Moore Wood in the early eighteenth century and on the 1802 Berkswell Enclosure Plan is shown as Lewis Wood.[12] The lack of consistency in the names makes tracking the wood in the records difficult, in the absence of clear topographical information. This was a lesson in how unhelpful even relatively full and modern records can be. As with Ryton Wood, the main evidence for Rough Close as an ancient wood is its woodbanks, the structure of the wood as an old coppice wood and the presence of ancient woodland species including lime (Fig. 5.3). In the event the Superpit did not go ahead, for political reasons, and the wood survives to this day.

One interesting landscape feature nearby that was identified by the research was a woodland relict hedge – the linear remnant or *ghost* of a grubbed ancient wood (Rackham 1990). It consists of a hedge including a 40-yd (36-m) stretch of Small-leaved Lime stools, with woodland plants such as Pignut *Conopodium majus*. This is at GR SP 271792, alongside the former Massey Ferguson factory, and is a miniature version of the ghost woodland hedge identified by Oliver Rackham at Shelley in Suffolk, the 600-yd (548 m) remnant of Wither's Wood (see Rackham 1986). The 1841 Tithe Map shows two Wood Fields at this point adjoining Wasp Ford Close.[13] The Berkswell Court Rolls for 1665 refer to a wood in Beechend abutted by Waspern Fields on the west and Banner Lane on the east and for 1711 to a wood of 16 acres (6.5 ha) (which almost exactly fits the two Wood Fields) adjoining Banner Lane and *Waspern*

9 This is a brief summary of some of the evidence collected by the author to produce a report for the Colliery Opposition Group in the 1989 Public Inquiry about the proposed Hawkhurst Moor Superpit. Much of the data remains unpublished.

10 Discussion with scout leader at Rough Close in 1987, and the evidence of young conifers on site.

11 WCRO Berkswell Glebe Terriers DR72A: Terriers of 1612, 1685, 1701, 1722 and 1733. The Berkswell Court Rolls WCRO MR21/1-20 also contain relevant material on the woods of Berkswell.

12 WCRO QS/75/12.

13 WCRO CR569-29.

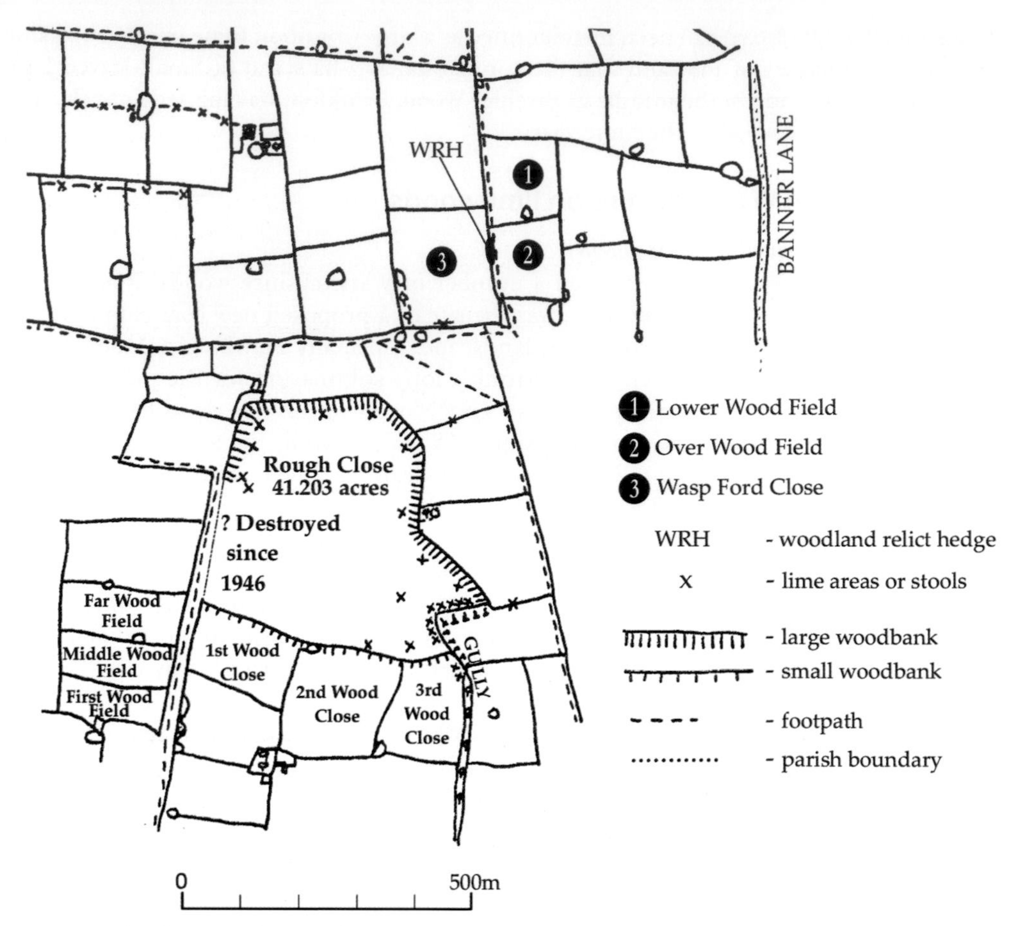

Figure 5.3 Sketch map of lime and woodbanks at Rough Close and woodland relict hedge nearby. Base plan is from the 1902–3 six-inch OS Map. Many hedges have been destroyed since then. Woodbanks and lime are shown as in 1987. Field names are from the 1841 Tithe Map. (D. Morfitt 2019)

Field in Beechend.[14] It was grubbed out some time later, and possibly not long after 1711, as the 1711 reference to the wood appears to be the last in the records. Other hedges in the area that do not appear to be woodland relict have an occasional Small-leaved Lime, probably representing the very occasional successful seedling that has survived from the millions produced over the centuries by Rough Close and other former limewoods nearby (Fig. 5.3).

Hartshill Hayes, Hartshill, North Warwickshire (GR SP 320945)

Space precludes dealing in detail with this large and complex wood, but mention should be made of Hartshill Hayes, approximately 46 ha, which has probably the largest area of ancient limewood in Warwickshire, some 16 ha. According to the current owners (Warwickshire County Council), the wood was acquired by the Forestry Commission (FC) in 1951–2, when much of it was an acid Sessile Oak-limewood 5B and some pure Sessile Oakwood (probably type 6Cb) on mostly acid sandy loam soils. In the late 1950s, the FC began clear felling and coniferizing the wood. Areas of pure Sessile Oakwood have not recovered, although occasional trees do survive. But limewood is hard to kill, as Oliver was latterly delighted to report about many replanted limewoods (see, for instance, Rackham 2003; 2006), and, as I have observed here, the limewood has frequently recovered by outcompeting the planted trees. The site

14 Berkswell Court Rolls for 1665 and 1711 WCRO MR21/1-20.

is topographically complex with deep internal streams and valleys, with alder lining the streams and limewood on the slopes and ridges. As a country park since 1983, the wood has suffered some abuse for amenity, including damage to woodbanks and the intrusion of tarmac pathways, but I have observed that much of the wood away from the entrance and car park is relatively intact. This wood is another example of a site where the early documentation is very sparse (Wager 1998) and the main evidence of the wood's antiquity is its woodbanks, its structure, with many lime stools of at least 3 m diameter, and ancient woodland species such as lime. At Hartshill Hayes, my unpublished survey data shows that, unusually, there are both native species of lime, Small and Large-leaved, with many apparent hybrids. Hartshill Hayes also has large clones of Wild Service Tree *Sorbus torminalis*, especially around the edges. This is another ancient woodland indicator species throughout its English range (Rackham 1980). My unpublished survey work in Warwickshire suggests that Wild Service Tree is indeed an indicator of ancient woods (or woodland relict hedges) in the county.

Allen's local history (Allen 1982) says that the lime in Hartshill Hayes was planted in the late eighteenth century for use by the hat-makers of Atherstone, although the lime stools must then have been large and long-established; this is a classic factoid. According to Oliver Rackham, 'A factoid looks like a fact, is respected as a fact, and has all the properties of a fact except that it is not true' (Rackham 1990: 23).

Woodland work 1990–2000; my PhD on the historical ecology of the woods of Binley, Warwickshire

By 1990, I had been in regular contact with Oliver for some years. I had contacted him in 1986 for permission to use his Woodland Record Card. He sent me one with permission to copy it for my field use. I had visited him in Cambridge, taking examples of woodbank and other plans I had made. He came to Warwickshire at my invitation in March 1990 to visit several woods, and to give a talk and lead a discussion, which raised funds for WWT. One of the woods he visited was Piles Coppice, a wood acquired by the Woodland Trust in 1987 which I had known since its acquisition. Woodland Trust staff and other conservationists in Warwickshire at that time believed that this was a wood of little ecological or historical interest. I was told by a member of WWT staff that WWT consequently had no interest in it when it came up for sale (Morfitt 2000). A member of Woodland Trust staff told me in 1989 that the wood was probably not ancient at all but a plantation in origin; if it was ancient, he thought it might be a 'medieval plantation' (quoted in Morfitt 2000). But what of the wide woodbank around much of its periphery, the impressive spring flora and the imposing large coppice stools of Small-leaved Lime, all features which suggest it is probably ancient? Oliver agreed that field evidence suggested that this *was* an ancient wood, especially interesting for its Small-leaved Lime, not widely known as a probable ancient component of Warwickshire woods (Oliver Rackham, pers. comm.). I suggested to the late Dr Humphrey Smith at Coventry Polytechnic (now Coventry University) that this was an ideal site for detailed historical-ecological investigation. With his support, I applied in the summer of 1990 to do a part-time MPhil/PhD on the woods of Binley with Piles Coppice as the central site (Fig. 5.4).

I eventually realized I had underestimated the need for an active rather than a nominal supervisor, but fortunately Oliver offered to step in as unofficial supervisor of the thesis. Without his help, I might well not have completed the PhD.

The former parish of Binley is on the eastern edge of Coventry, and now straddles the West Midlands/Warwickshire boundary. As well as Piles Coppice, the woods include Binley Little Wood, Binley Common Wood, Big Rough and Little Rough, as well as New Close Wood and The Grove, part of the complex including Birchley Wood, now considered part of Brinklow parish.[15]

15 Ordnance Survey maps nineteenth–twentieth century.

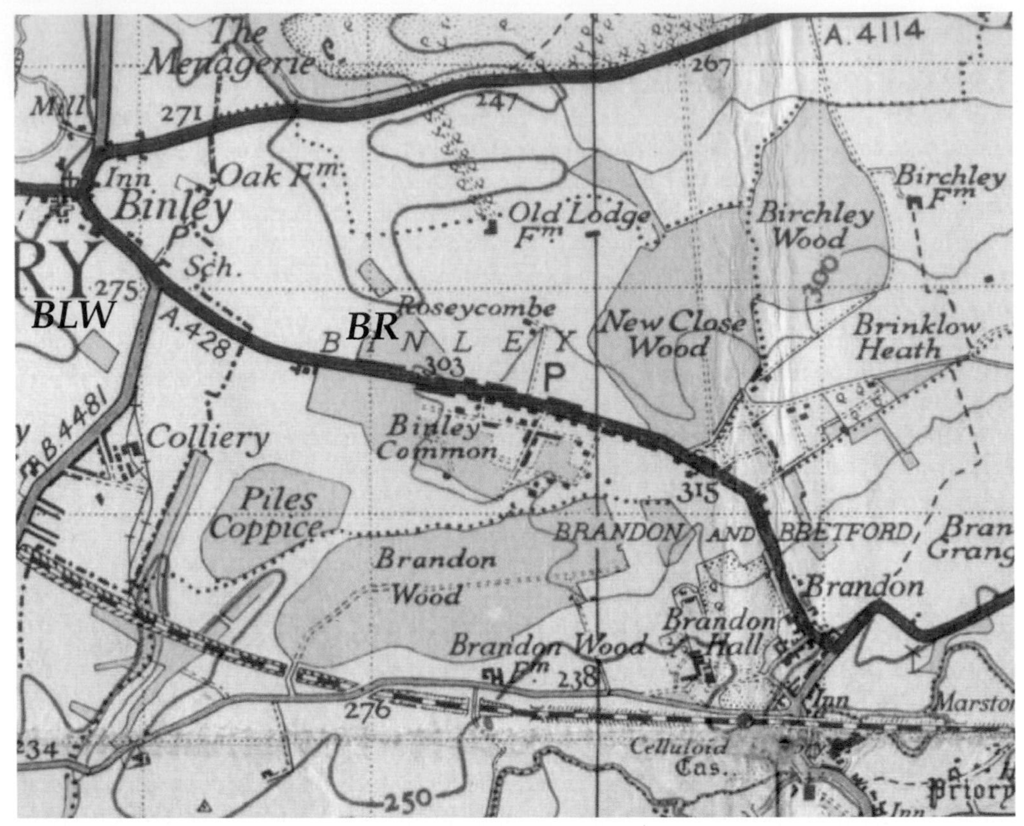

Figure 5.4 Part of the 1946 one-inch Ordnance Survey map Sheet 132 showing the Binley area east of Coventry and woods studied in my PhD. BLW is Binley Little Wood. BR is Big Rough. The map captures the ongoing destruction of Binley Common Wood in the twentieth century; now the only substantial portion of the wood that survives is the western end.

The Bodleian Library in Oxford holds the records for the Craven Estate, which owned the manor of Binley, so the records are publicly accessible.[16] The estate and manor coincidentally included most of the parish from the sixteenth century until the 1920s.

The underlying solid geology of the study area is Mercia Mudstone (Keuper Marl) (Beard 1984). Over much of the area there is an overlay of drift deposits. The woods lie mostly on the Wolston Clay which forms a layer of variable depth, typically 5–10 m deep; there are further sand and gravel deposits beneath the clay (Old, Bridges and Rees 1990). Dunsmore Gravel extends under the southern edge of Birchley and New Close Woods and under much of the eastern half of the former Binley Common Wood. Baginton Sand and Gravel lies under the southern edge of Piles Coppice and the westernmost tip of Binley Common Wood. Thrussington Till, pebbly boulder clay, extends under the northern edge of Birchley Wood, the western half of The Grove, the north-western edge of Big Rough, the western edge of Binley Common Wood and down the western edge and through the southern half of Piles Coppice.

Figures 5.5 and 5.6 show the tree communities and earthworks of Piles Coppice, Birchley and New Close Woods, and The Grove. The soils over most of Piles Coppice appear to be sandy loam, under all tree community types. They are generally strongly acid, light soils, ranging from pH 3.4 under Hazel–Sessile Oakwood 6Db and birch–Sessile Oakwood 6Cb to 4.6 under limewood 5B. At Binley Common Wood the soils are similarly mostly sandy loam,

16 Craven Papers, Bodleian Library, Oxford.

with pHs mostly 4.1–4.6 under the dominant Hazel–Pedunculate Oakwood 6Dc (Morfitt 2000). Although it was not possible to take soil samples from Birchley and New Close Woods because of restricted access, the tree community types and vegetation recorded suggest the pHs will be similar.

I shall discuss three aspects of my research into the woods of Binley: the medieval documentation, which unusually proved both relatively rich and able to produce a consistent story, with invaluable topographical nuggets; the relationship between grazing and Small-leaved Lime; and if lime here is a relict species, defined by Oliver as 'maintaining itself in its old localities but unable to colonize new ground' (Rackham 1980: 243). (For all unreferenced statements that follow, see Morfitt 2000.)

Medieval documentation for Binley (Birchley/New Close Woods were not then in Binley)

As so often, Domesday Book (1086) has the first documentary evidence for woodland in Binley (Morris and Plaister 1976). Two small woods are recorded: 'wood(land) half a league long, and 1 furlong broad',[17] held by the abbey of Coventry and 'wood(land) 4 furlongs long and 2 furlongs broad', held by Hadulf from Turchil of Warwick. Using Rackham's calculation (Rackham 1980/2003), length by breadth by a form factor of 0.7, we reach a rough figure in modern acres of 42 (17 ha) for the Coventry Abbey wood and 56 acres (23 ha) for the wood owned by Hadulf. Adding up all the other land-uses in Domesday does not leave much space for unrecorded woodland in Binley, compared with the known acreage of the later manor and parish.[18] The Combe Cartulary of about 1255 records the gradual acquisition of much of the manor and parish of Binley by Combe Abbey, which was founded in 1150.[19] The woodland references in the Cartulary seem logically, from internal evidence, to fall into two sets (the Victoria County History also detects this pattern).[20] This, of course, mirrors the evidence from Domesday Book. The Hadulf wood is named as Munechet, with a name ending that is almost certainly the Old Welsh *-cêd* (confirmed by Oliver Padel, pers. comm.), meaning a wood. Any wood so named must be pre-Saxon. The Coventry Abbey wood descended through Joeslin, son of Ralph of Billneya; Robert, son of Ralph, brother of Joiilinus; Robert, son of Robert of Billneia; and finally Gaufr[idus], son of Robert of Billneia. The gift of Robert, son of Robert, includes a perambulation, defining a piece of land by its boundaries (Rackham 1986), which places their wood somewhere within the north-east part of the parish on the site of the historic Binley Common Wood, a portion of which has survived to the present.

The Hundred Rolls of 1279 (John 1992)[21] also record two woods for Binley (my translation):

> The aforesaid abbot [of Combe] has a certain foreign wood containing 42 acres of land of which 2 acres are included in the park of Brandon by payment of 2 shillings and a buck per annum.
>
> The aforesaid abbot has in the same place a certain foreign wood containing 40 acres of land.

By this time, Combe Abbey had apparently acquired most of the former Coventry Abbey land holdings in Binley and so it is likely that these two woods still represent the total of woodland in Binley. The correspondence with the Domesday Book entries is striking and offers corroboration of my interpretation of the cryptic Combe Cartulary entries. The term 'foreign' wood

17 According to Rackham, the Domesday league was probably 12 furlongs, and the furlong 220 yards (see Rackham 1980 and 2003: 113, 115).

18 Craven Papers, Bodleian Library, Oxford.

19 Dugdale Manuscripts, Bodleian Library, Oxford.

20 *Victoria County History: Warwickshire*, Vol. 6.

21 WCRO MI 252 and 278: WCRO Microfilm 252 and 278 of the original Rolls and a 16th-century copy give xl = 40; John gives xi = 11. (T. John 1992, p. 128.)

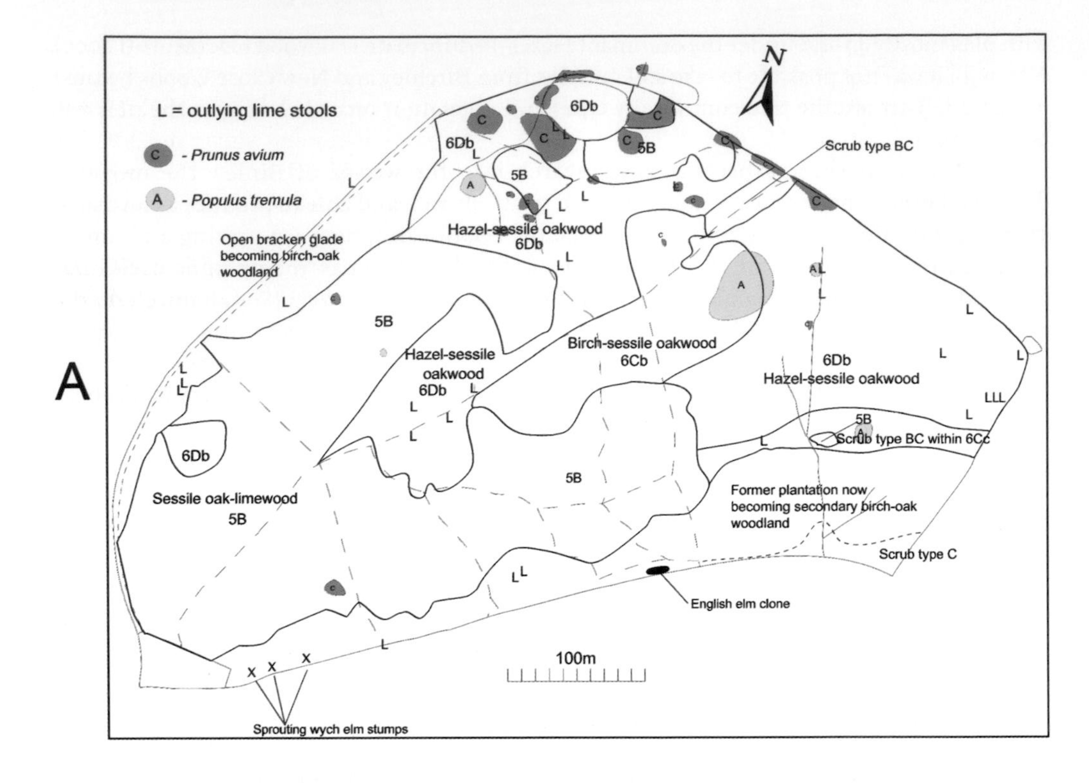

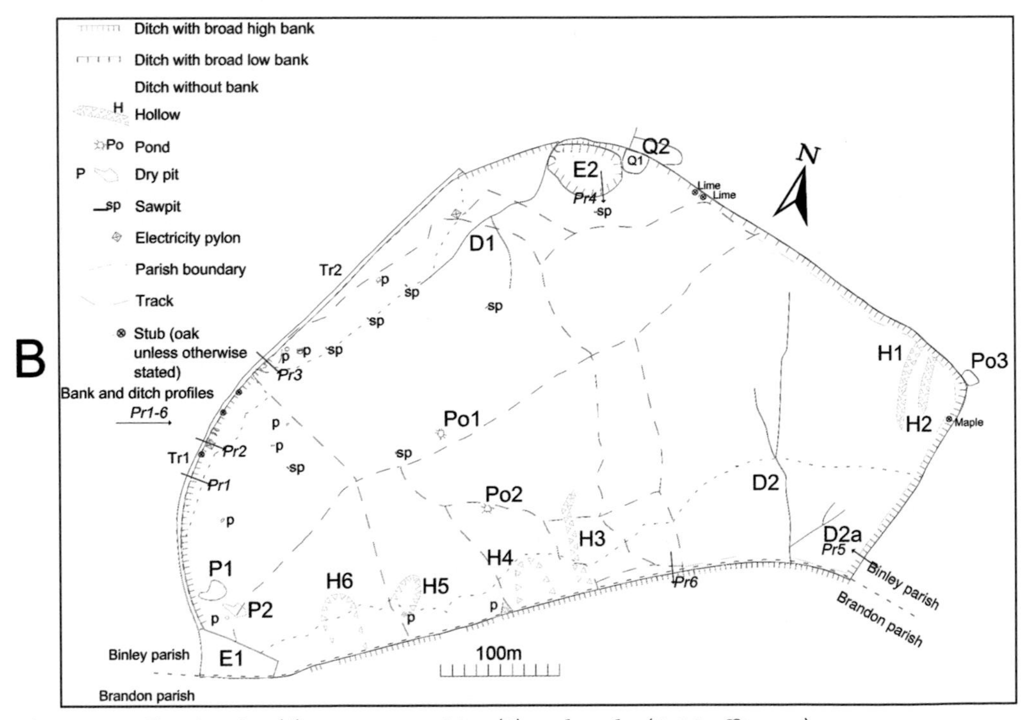

Figure 5.5 Piles Coppice: (A) tree communities; (B) earthworks. (D. Morffit 2000)

boscus forinsecus means a wood outside a park, but in this case it is not clear to which park they refer. Brandon Park is alongside both but is in a separate parish and ownership, which would be unusual (Oliver Rackham, pers. comm.). Medieval woodland areas are often underestimates, possibly because of some systematic error in the method of calculation (Rackham 1968). Before the nineteenth century, a larger perch was often used for woodland acres,[22] so it is likely that these two acreages are smaller than the actual sizes of the two woods. A 42-acre wood calculated by the eighteenth-century perch would be just under 50 modern acres.

The reference in the first entry to the park of Brandon is useful. A map of Brandon and Bretford of about 1630 shows us 'Spyers parke',[23] with two adjacent laundes alongside the Binley boundary at the place where Piles Coppice now is;[24] the map book of the Craven Estates of 1746,[25] as well as that of 1778,[26] also call this part of Brandon 'Brandon Old Park'. The first Hundred Rolls reference specifically locates the wood of 42 acres next to the deer park of Brandon, a park to which this is the first known reference (Cantor 1983). This is an unusually precise topographical reference, and strong evidence that the Hundred Rolls is probably here referring to what later became Piles Coppice. Thus, one of the two woods noted in the Hundred Rolls is almost certainly the present Piles Coppice and is likely to have been approximately the same size as the present and the Domesday Book wood. The Cartulary evidence suggests that Piles Coppice may be the wood called Munechet.

The documentary evidence from 1086 to 1279 is remarkably consistent in recording two woods in Binley. The topographical information in the documents also points to woods on the sites of the present Piles Coppice and Binley Common Wood.

Piles Coppice, defined by woodbanks that from their size are probably ancient (Morfitt 2000) (Fig. 5.5), is very close to its apparent Domesday Book and probable Hundred Rolls size now, at 52 acres (>21 ha); 4 by 2 furlongs also matches the wood in roughly north–south and east–west dimensions: 'We strongly suspect that they are thinking of shape as well as size, and may be giving us the extreme diameters of the wood or some diameters that they guess to be near the mean' (Maitland 1897: 498).

The half a league cited for Binley Common Wood probably equals 0.75 miles (1.2 km); the former Binley Common Wood, and the wooded portion of the common as shown in 1746 before it became a coppice,[27] was close to 0.75 miles (1.2 km) long. The woodbanks within the surviving portion are close to 1 furlong apart. Unfortunately, the destruction of the remainder of the wood has destroyed any evidence of internal earthworks that might indicate its medieval size.

In Binley (and Brinklow parish, in which Birchley Wood now lies), Small-leaved Lime was recorded in recent surveys only in woods for which documents and earthworks indicate they existed in the early medieval period; these were Binley Common Wood, Birchley and New Close Woods, and Piles Coppice. Small-leaved Lime was not recorded from any wood with less than a medieval pedigree. Lime is thus confirmed as a probable relict species here.

Relationship between lime and grazing

Lime is highly susceptible to grazing (Rackham 1980, 1982; Pigott 1991), and it has been suggested that the relative absence of lime in ancient woodland south of the Thames is directly

22 The perch was a traditional unit of length. The woodland perch could vary from 15½ to 30 ft (Rackham 1986). The 18-ft perch was commonly used for woodland throughout England and is recorded before 1600 (Rackham 1980). In Binley, Warwickshire, it was still used in the eighteenth century (Craven Papers 320, Bodleian Library, Oxford (Survey of 1746)).

23 British Museum Add. MS 48181; copy in WCRO Z203U.

24 A launde was a defined grass plain within a park.

25 Craven Papers 320, Bodleian Library, Oxford.

26 WCRO CR8/184.

27 Craven Papers 320, Bodleian Library, Oxford.

related to the high proportion of woodland used for long periods as wood-pasture in that area (Rackham 1980). Grazing may therefore be one of the most significant influences on the composition of woods that formerly had lime. In the light of this, do the amounts of lime in the woods of Binley (including New Close and Birchley) reflect the known history of each site?

Three stools were recorded in the 4-ha remnant of Binley Common Wood. As this is only a small remnant of the original wood, the previous distribution of lime elsewhere in the wood, if it was there, is not known. It is possible that there was no lime outside this surviving remnant of the wood, at least in recent centuries. If lime had been extensive elsewhere on the site, some at least might be expected to have survived in gardens, but none was recorded in my survey of trees which probably survive from the former wood within the housing estate. Binley Common Wood was a grazed common, but the sixteenth- and seventeenth-century references suggest it was still wooded. Then in the middle of the eighteenth century, it became an enclosed coppice wood, protected from grazing.

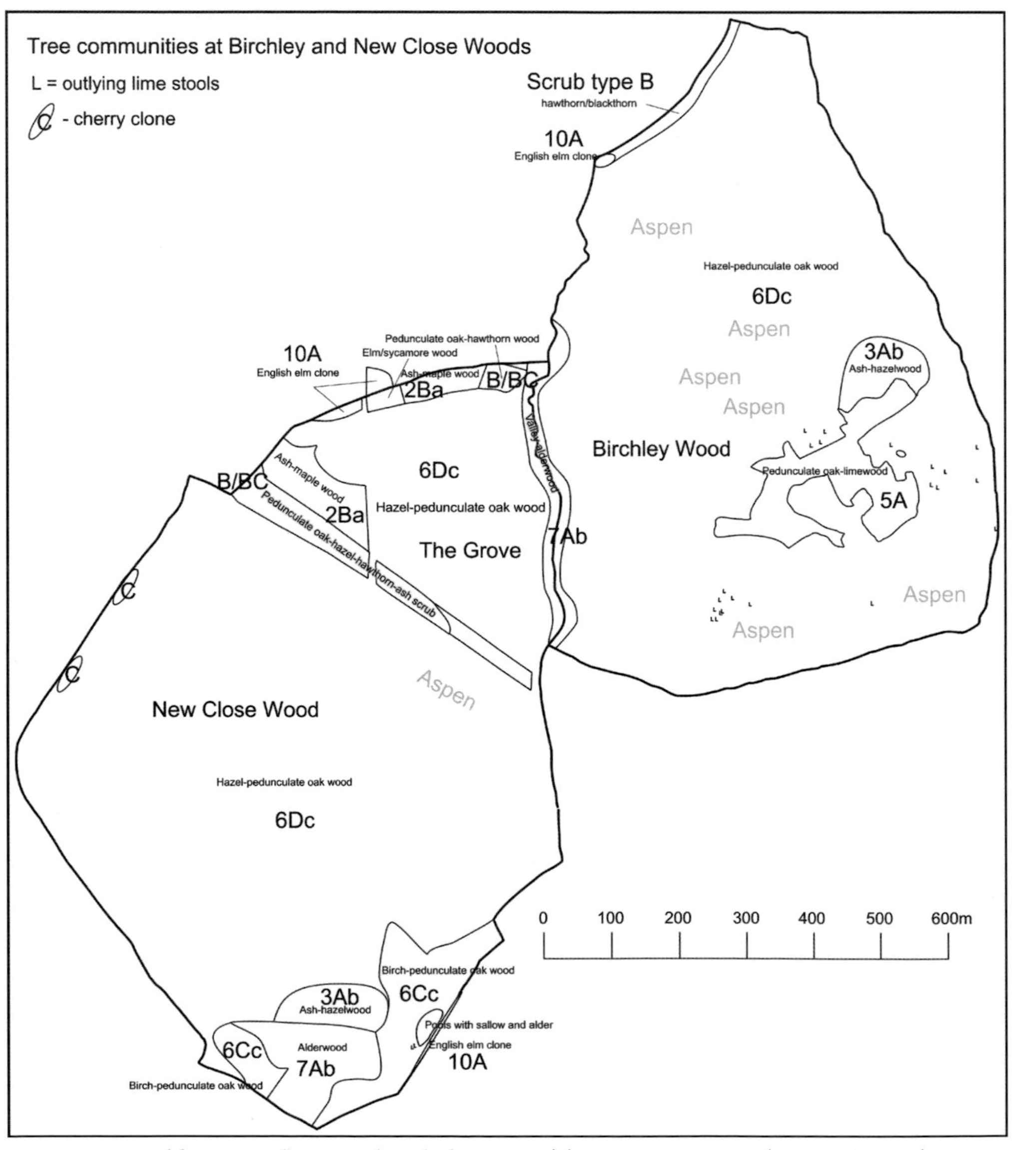

Figure 5.6 Birchley, New Close Wood, and The Grove: (A) tree communities. (D. Morffit 2000)

Only two lime stools were recorded in New Close Wood but around 3 ha in Birchley Wood. The Birchley and New Close Wood complex was grazed by commoners' livestock from at least the twelfth century.[28] Large-scale and possibly effective exclusion of livestock appears to have begun from 1355, when New Close was enclosed against grazing, presumably for coppicing; in 1500–1, Birchley Wood was made into a coppice and access to commoners' animals was restricted. By 1652, the period of exclusion of animals after coppicing had been lengthened from five to seven years. The Grove (formerly Swinestye, implying perhaps the presence of a pigsty) has no lime; it also has evidence of intensive disturbance, with many pits, possibly from a brickworks.

28 WCRO: Gregory Hood papers, catalogue and editor's translation, has the documentation for grazing in Birchley, New Close Wood and The Grove complex.

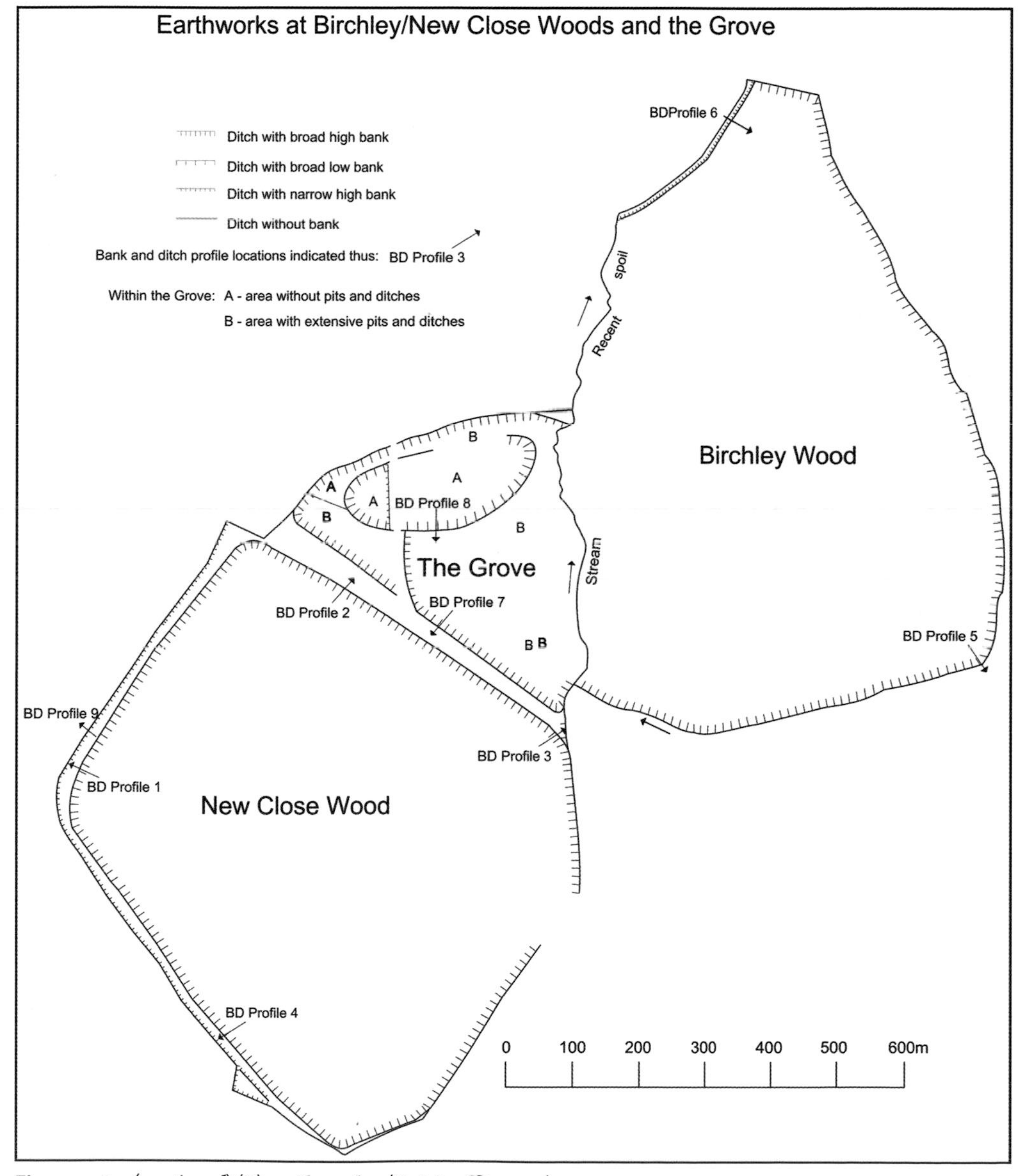

Figure 5.6 (*continued*) (B) earthworks. (D. Morffit 2000)

Why Birchley should have almost all the lime yet have been enclosed 150 years later is not clear. Perhaps New Close never had much lime; perhaps it had all been grazed out by the time the wood was enclosed against grazing in 1355. At present it is impossible to know. Pollen analysis on some of the ponds in the woods or perhaps humus pollen analysis (Bradshaw 1981a, 1981b; Watkins 1990) might offer information on the fluctuations of tree species in the woods through the medieval period.

The shape of the lime area in Birchley Wood is highly suggestive. Unlike most lime stands, which, when mapped, have gently rounded shapes (see, for example, Chalkney Wood (Rackham 1980) and Piles Coppice, Fig. 5.5) the lime stand at Birchley is raggedly indented and has notable re-entrants (angular cuts into the body of the woodland) into it. Could this represent the shrinking remnant of limewood being progressively eaten away by stock but then left as a fossilized shape after effective enclosure had removed or limited the impact of grazing animals and protected the lime from further destruction?

Piles Coppice has the most extensive ancient limewood in the area; nearly 8 ha of the surviving approximately 17 ha of old woodland in the site is lime woodland. Its history and earthworks suggest that the wood has been the same size for 900 years or more; if, as is possible, this wood was the Munechet of the Combe Cartulary the name alone takes its history back a further 600 years or more. The argument from lack of evidence is a potentially hazardous one, but the apparent absence of medieval and later documentation may be an argument in favour of the wood's stability and continuity and probable absence of grazing. There is no suggestion in any of the surviving documentation that there were ever common rights over the wood. Disputes and conflicting demands on a wood from users generate documentation, as with Hatfield Forest (Rackham 1989) or the common woods of West Yorkshire (Faull and Morehouse 1981) or New Close and Birchley Woods (Morfitt 2000).

There is thus a highly suggestive relationship between amounts of lime and the known histories of these woods; much grazing goes with little lime.

Soil variation does not provide an alternative explanation for the differing amounts of lime in these woods. Lime is tolerant of a wide range of soil types (Rackham 1980; Pigott 1991). Much of the soil in these woods is sandy loam; areas of different soils within the woods are marked by variation in woodland types, yet much of the soils of the surviving portion of Binley Common Wood and New Close and Birchley Woods areas that lack lime would appear to be suitable for it.

Lime age and regeneration:

Stool size and age can vary widely from wood to wood, depending on soils and species (Rackham 1990). A very approximate estimate of stool age is that 30 cm of diameter equals 50 years of age (Rackham 1990). On this basis, the biggest stools at Piles Coppice, up to 5 m in diameter, could be at least 800 years old (Fig. 5.7). They might be very much older (Rackham 1975). Ages over 1,000 years are suggested for some large stools of Small-leaved Lime at the northern limit for lime in Britain and Scandinavia (Pigott 1991). The very large circle of Small-leaved Lime at Silk Wood, Gloucestershire, nearly 15 m in diameter with 60 stems, shown by DNA analysis to be one clonal ring, may be one of the oldest known living trees in Britain, estimated at a minimum age of 2,000 years (Mabey 1996; Oliver Rackham pers. comm.).

Lime reproduction:

1) ***Vegetative reproduction:*** Of the 364 limes examined in detail in Piles Coppice, 20 or 5.5 per cent showed active vegetative reproduction, either by dropping major stems at a distance from the stool or by small basal sprouts rooting close to the stool. This feature of lime is well documented (Rackham 1990; Pigott 1991); it may explain how lime at Swithland Wood was able to recolonize the site after it was apparently reduced to remnants in hedges (Woodward 1992).

Figure 5.7 Large Small-leaved Lime stools at Piles Coppice, September 2017. (D. Morfitt)

The rooting of small basal sprouts may also have allowed lime to expand vegetatively even under a regime of short rotation coppicing.

2) ***Lime seedlings:*** The production of fertile seed in lime is very dependent on summer temperatures, especially in July (Pigott 1991). Between 1991 and 1999, many thousands of lime seedlings grew in Piles Coppice. Casual record was made of lime seedlings throughout the wood over a number of years. Seedlings were also recorded beneath and around lime stools during the detailed lime stool recording of winter/spring 1992; these were re-recorded in the summer of 1992 and the autumn of 1999. A detailed record was also made for several years of the lime seedlings growing in one area of the wood where large numbers of lime seedlings were first noticed in the spring of 1991.

The overall conclusion was that, as appears to be usual in English limewoods (Pigott 1991), lime seedlings may be produced in abundance but rarely survive beyond their first year.[29] This may be due to a number of factors, among which shade is likely to be significant (Rackham 1980). Predation of lime seedlings by Bank Vole *Clethrionomys glareolus* has also been shown to be significant elsewhere (Pigott 1985). Bank Voles are present at Piles Coppice (Matthes 1993).

The seeds do not disperse far, despite the wing on the inflorescence (Rackham 1980; Pigott 1991). Seedlings were rarely recorded at Piles Coppice more than 20 m from the edge of the limewood.

3) ***Saplings:*** During the course of the survey, several probable saplings were found, but all disappeared within a few years.

The relative shade tolerance of Small-leaved Lime produces seedlings and consequently saplings, but apparently increased light is needed to allow the sapling to survive beyond 8–12 years (Pigott 1991). At Piles Coppice, very few seedlings developed into saplings at all

29 Woodward (1992) also notes this for Swithland Wood.

and none survived beyond a few years; this reflected the national trend in the 1990s (Pigott 1991). This confirms the status of lime as a relict in Piles Coppice. Although the lime was not recorded in detail in Birchley Wood, as access was very restricted, no saplings were seen in that wood during the necessarily limited survey work done, so lime in that wood is also probably relict.

Afternote:
I visited the Marks Hall Woods in Essex in April 2017, much of which is old Small-leaved Limewood, now recovering under sympathetic management after attempts to 'coniferize' them in the 1960s (Rackham 1980, 2003). I found much evidence of lime regeneration, with millions of lime seedlings and many apparent saplings of probably up to four or five years' growth (Fig. 5.8).

I explored Piles Coppice again in September 2017 to see if there had been any parallel expansion of lime there, but did not find any evidence of successful regeneration from seed. Not a single sapling or surviving seedling could be found. Is this cause for concern? No, of course, there is no reason for panic as long-lived native trees such as lime, especially in the form of coppice, need to reproduce themselves only once in their lifetimes, and the lifetime of a lime coppice stool can be counted in many centuries, if not almost indefinitely (Rackham 1980). It still behaves very much as a relict species in Piles Coppice. It will be interesting to see how lime is faring both nationally and in its other Warwickshire sites.

One very noticeable major change in Piles Coppice since the year 2000 has been a great decline in Bramble *Rubus fruticosus* agg. I had noted in my PhD thesis that Bramble and Bracken covered 80 per cent of the site in the 1990s, but now Bramble probably covers less than 20 per cent. Possible explanations are the increase in deer (I observed increasing numbers of Muntjac *Muntiacus reevesi* throughout the 1990s) and also the dense canopy of the tree communities in Piles Coppice, lime in particular, reaching a 'tipping point' where too little light reaches the ground (see, for example, Morecroft et al. 2001).

Figure 5.8 Small-leaved Lime seedlings in the Marks Hall Woods, April 2017. (D. Morfitt)

As a final (and very topical) note on lime regeneration, I had a discussion with Oliver in March 2011 about the likely effects of global warming on native British trees and shrubs. He pointed out that almost all our trees are at the northern edge of their ranges and confirmed my opinion that they will almost certainly benefit, rather than decline, as a consequence of global warming, as many scaremongers have suggested. (I was at a conference in Leamington in 2006 on native trees and woodland, and the foresters there were adamant that we would need many more exotic species of tree to be planted in the UK to make up for the decline in native trees that they predicted as a consequence of global warming.) Lime seedlings will still have all their associated hazards to survival, such as failure to disperse far from the parent tree, shade, competition from tall herbs and inability to penetrate mor humus (Rackham 1980), so lime is unlikely to become such a universal tree as birch now is, and will almost certainly remain an especially interesting relict species confined to ancient woods such as Piles Coppice.

References

Allen, J. (1982) *Heardred's Hill: A History of Hartshill and Oldbury.* Joan Allen, Nuneaton.

Beard, G.R. (1984) *Soil Survey Record No. 81: Soils in Warwickshire V.* Lawes Agricultural Trust, Harpenden.

Bradshaw, H.W.B. (1981a) Modern pollen-representation factors for woods in south-east England. *Journal of Ecology*, 69(1), pp. 45–70. https://doi.org/10.2307/2259815

Bradshaw, H.W.B. (1981b) Quantitative reconstruction of local woodland vegetation using pollen analysis from a small basin in Norfolk, England. *Journal of Ecology*, 69(3), pp. 941–956. https://doi.org/10.2307/2259646

Cadbury, D.A., Hawkes, J.G. and Readett, R.C. (1971) *A Computer-Mapped Flora: A study of the county of Warwickshire.* Published for the Birmingham Natural History Society by Academic Press, London.

Cantor, L. (1983) *The Medieval Parks of England: A gazetteer.* Department of Education at Loughborough University of Technology, Loughborough.

Faull, M.L. and Moorhouse, S.A. (1981) Woodland. In: *West Yorkshire: An archaeological survey to A.D. 1500*, Vol. 3. West Yorkshire Metropolitan County Council, Wakefield.

Greig, J. (1982) Past and present lime woods of Europe. In: M. Bell and S. Limbrey (eds) *Archaeological Aspects of Woodland Ecology.* Symposia of the Association for Environmental Archaeology No. 2, BAR International Series 146, pp. 23–56.

John, T. (1992) *The Warwickshire Hundred Rolls of 1279–80 Stoneleigh and Kineton Hundreds.* Records of Social and Economic History, New Series XIX. Oxford University Press for the British Academy, Oxford.

Kelly, M. and Osborne, P. J. (1964) Two Faunas and Floras from the alluvium at Shustoke, Warwickshire. *Proceedings of the Linnaean Society of London*, 176(1), pp. 37–65. https://doi.org/10.1111/j.1095-8312.1965.tb00935.x

Kirby, K.J. (1988) *A Woodland Survey Handbook.* Nature Conservancy Council, Peterborough.

Mabey, R. and Evans, T. (1980) *The Flowering of Britain.* Hutchinson, London.

Mabey, R. (1996) *Flora Britannica.* Chatto and Windus, London.

Maitland, F.W. (1897) *Domesday Book and Beyond: Three essays in the early history of England.* Reprint, Fontana Library, 1969. https://doi.org/10.2307/780663

Matthes, G. (1993) A study of small mammals at Piles Coppice. A report submitted to the Department of Biological Sciences towards a Degree of Bachelor of Science with Honours, May 1993. Copy held at Coventry University.

Morecroft, M.D., Taylor, M.E., Ellwood, S.A. and Quinn, S.A. (2001) Impacts of deer herbivory on ground vegetation at Wytham Woods, central England. *Forestry*, 74, pp. 251–257. https://doi.org/10.1093/forestry/74.3.251

Morfitt, D. (1987) *Ecology of the Moor.* Berkswell Miscellany, Vol. 3, no page numbers.

Morfitt, D. (1988) A short summary of evidence for the 'ancient' status of Ryton Wood – documentary, archaeological and ecological. Report written for the owners of the wood, the Warwickshire Wildlife Trust.

Morfitt, D. (2000) The historical ecology of the woods of Binley, Warwickshire. Unpublished PhD thesis. Coventry University.

Morris, J. and Plaister, J. (eds) (1976) *Domesday Book 23 Warwickshire* [edited from a draft translation prepared by Judy Plaister]. Phillimore, Chichester.

Old, R.A., Bridge, D. McC. and Rees, J.G. (1990) *Geology of the Coventry Area: Technical Report WA/89/29.* Onshore Geology Series, British Geological Survey, Nottingham.

Peterken, G.F. (1981) *Woodland Conservation and Management.* Chapman and Hall, London and New York. https://doi.org/10.1007/978-1-4899-2857-3

Peterken, G.F. (1991) Review of *British Plant Communities*, Vol. 1, *Woodlands* by J.S. Rodwell. *Journal of Ecology*, 79 (3), pp. 873–875. https://doi.org/10.2307/2260679

Peterken, G.F. (1993) *Woodland Conservation and Management*, 2nd edition. Chapman and Hall, London and New York.

Peterken, G.F. (2008) *Wye Valley* (New Naturalist 105), Collins, London.

Pigott, C.D. (1985) Selective damage to tree seedlings by bank voles (*Clethrionomys glareolus*). *Oecologia*, 67, pp. 367–371. https://doi.org/10.1007/BF00384942

Pigott, C.D. (1991) Biological Flora of the British Isles: *Tilia Cordata* Miller. *Journal of Ecology*, 79, pp. 1147–1207. https://doi.org/10.2307/2261105

Rackham, O. (1968) Medieval woodland areas. *Nature in Cambridgeshire*, 11, pp. 22–25.

Rackham, O. (1975) *Hayley Wood: Its history and ecology*. Cambridgeshire and Isle of Ely Naturalist's Trust, Cambridge.

Rackham, O. (1976) *Trees and Woodland in the British Landscape*. J.M. Dent and Sons, London.

Rackham, O. (1980) *Ancient Woodland: Its history, vegetation and uses in England*. Edward Arnold, London.

Rackham, O. (1982) The Avon Gorge and Leigh Woods. In: M. Bell and S. Limbrey (eds) *Archaeological Aspects of Woodland Ecology*, Symposia of the Association for Environmental Archaeology No. 2, BAR International Series 146, pp. 23–56.

Rackham, O. (1986) *The History of the Countryside*. J.M. Dent and Sons, London.

Rackham, O. (1989) *The Last Forest: The story of Hatfield Forest*. J.M. Dent and Sons, London.

Rackham, O. (1990) *Trees and Woodland in the British Landscape*, Revised edition. Dent, London.

Rackham, O. (2003) *Ancient Woodland: Its history, vegetation and uses in England*, 2nd edition. Castlepoint Press, Dalbeattie.

Rackham, O. (2006) *Woodlands* (New Naturalist 100), Collins, London.

Rodwell, J.S. (ed.) (1991) *British Plant Communities*. Vol. 1. *Woodlands and scrub*. Cambridge University Press, Cambridge.

Shotton, F.W. (1967) Investigation of an old peat moor at Moreton Morrell, Warwickshire. *Proceedings of the Coventry and District Natural History and Scientific Society*, 4, pp. 13–16.

Wager, S. (1998) *Woods, Wolds and Groves: The woodland of medieval Warwickshire*. British Archaeological Reports, British Series 269, Archaeopress, Oxford. https://doi.org/10.30861/9780860549338

Watkins, C. (1990) *Woodland Management and Conservation*. David and Charles, Newton Abbot.

Woodward, S.F. (1992) *Swithland Wood: A study of its history and vegetation*. Leicestershire Museums publications 116, Leicester.

Part II

European Studies

CHAPTER 6

On the Shoulders of Oliver Rackham

Frans Vera

Abstract

There is a saying in Dutch that as a scientist you are always standing on the shoulders of others. That means that you always build on what others have discovered and published before you. In that sense, I am on Oliver Rackham's shoulders, particularly with regard to his book *Ancient Woodland*. As an ecologist, that volume was a revelation to me. It helped enormously with the writing of my PhD thesis and then my book *Grazing Ecology and Forest History*.

Keywords: ancient woodland, succession theory, climax vegetation, Oostvaardersplassen, moulting greylag geese, grazing large herbivores, landscape engineers, wood-pasture

Clements, Tansley and the succession to the climax-vegetation forest

In this contribution to the memorial volume in honour of Oliver Rackham, I explain how in my own work I climbed upon the shoulders of others and ended up particularly astride those of Oliver Rackham. It was a big detour, and indeed a route that did not begin with trees, woods and forests. Indeed, the journey started with their opposite, namely open grassland. It was at a time when trees were seen only in the context of forest and grassland was perceived as a human-made habitat, which it was believed must have arisen with the advent of agriculture around 6,000 years ago. The first farmers were said to have cut down the forest that was present as the undisturbed vegetation, and the cattle they brought with them degenerated the landscape by trampling and eating seedlings from the trees. This impact caused the primeval forest to degenerate into grassland or heathland. Scientifically, this process was called retrogressive succession.

That natural vegetation was a forest was first and foremost based on the work of the American scientist Frederic Edwards Clements, who formulated the so-called succession theory (Clements 1916). In this model, vegetation developed by a linear succession of plant growth from bare soil until it was in balance with the climate, so-called climax vegetation. In places where trees could grow on the basis of climate and water management, the climax vegetation was supposed to be forest (Fig. 6.1). For Europe, the theory was adopted and modified by the British scientist Arthur Tansley: 'The "climax" represents the highest stage of integration and the nearest approach to perfect dynamic equilibrium that can be attained in a system developed under the given conditions and with the available components'

Frans Vera, 'On the Shoulders of Oliver Rackham' in: *Countryside History: The Life and Legacy of Oliver Rackham*. Pelagic Publishing (2024). © Frans Vera. DOI: 10.53061/PHPS7491

Figure 6.1 The forest of Białowieża, Poland, as *the* example of the closed-canopy forest as the natural, climax-vegetation in temperate Europe, October 2016. (F. Vera)

(Tansley 1935: 300). As early as 1911, he was of the opinion that Britain for example, was originally covered with forest: 'by eating and trampling, it degenerated into grassland or heathland' (Tansley 1911: 65). In the same publication, he writes: 'grassland or heathland have no doubt originated mainly from the clearing of the woodland, and the pasturing of sheep and cattle ... In some cases where grassland is not pastured, the shrubs and trees of the formation recolonize the open land, and woodland is regenerated (7).

In 1935, Tansley described the change from forest to grassland as a result of the continuous effect of grazing animals as a retrogressive succession: 'Here I should include the continuous effect of grazing animals which may gradually reduce forest to grassland' (Tansley 1935: 288). Although he acknowledged that animals could have a significant impact on vegetation they certainly did not change the structure of climax vegetation. According to him, vegetation controlled the development of the vegetation, because animals were completely dependent on it:

> The primary importance of vegetation is what we should expect when considering complete dependence, direct or indirect, of animals upon plants. This fact cannot be altered or gainsaid, however loud the trumpets of the 'biotic community' are blown. This is not to say that animals may not have important effects on the vegetation and thus on the whole organism complex. They may just change the primary structure of the climax vegetation, but usually they certainly do not.
>
> Tansley 1935: 301

A few pages later, he writes about human-introduced grazing animals, which are 'supposed to be merely destructive in their effects, and play no part in any successional or developmental process' (303). He concludes: 'if pasturing were withdrawn their areas would be invaded and occupied, as they were originally occupied, by shrubs and trees (487).

Where does grassland come from?

It was only when I became a nature conservationist that I began to wonder where grassland came from. Like all ecologists, I was educated with succession theory, partly based on the interpretation of pollen diagrams, which stated that Europe was naturally covered with a closed forest. This implied, bearing in mind Tansley (1935: 288, 300, 301, 303, 487), that as steppe tundra developed into a forest the large wild herbivores native to Europe such as primeval cattle, the Aurochs *Bos primigenius*, Wisent or European Bison *Bison bonasus*, Moose *Alces alces*, Red Deer *Cervus elaphus* and Tarpan *Equus ferus ferus*, the European wild horse after the end of the last Ice Age, were all forest animals. As I have indicated, grassland was a human-made habitat, created by the change from climax forest into grassland through the grazing with livestock that was introduced by the advent of agriculture. But where did my question about the origin of grassland come from? It was from my involvement with a marshy area in a drained part of Lake Ijsselmeer, a polder, in Holland. This was the newest polder in the Netherlands, named South Flevoland. There my climbing expedition onto the shoulders of others began; a journey that ended finally upon those of Oliver Rackham thanks to his book *Ancient Woodland* (1980).

My question came about through observations in a vast area of marshland named the Oostvaardersplassen in the polder South-Flevoland, located in the centre of the Netherlands. This was a new nature reserve by 1982 of around 6,000 ha, consisting of 3,600 ha marsh and 2,400 ha of extensive wet and dry grasslands with shallow pools, and further channels and ditches dating back to the time of the reclamation of the polder. The area was created in 1968 when the polder was constructed. Permanent water remained in the lowest part and a huge marshy area developed there, consisting of large-scale shallow open water surrounded by extensive fields of Common Reed *Phragmites australis* and Great Reedmace (or Broadleaf Cattail) *Typha latifolia*. Initially, it was part of a vast reedbed with reeds up to 3 m high that covered the entire polder of 42,000 ha. After the polder was drained, reed seed was sown from small airplanes and helicopters over the clay soil that was still saturated with water. Experience with previously drained polders had shown that the shade cast by the reeds prevented the growth of all kinds of herbs that could hinder further cultivation. This reed facilitated the further extraction of water from the soil by evaporation. The cultivation of the polder was done by the Rijksdienst voor de IJsselmeerpolders (RIJP), the service also responsible for its design. They sprayed the reed with danapon to kill it, then burnt the remaining vegetation before constructing ditches. Subsequently, so-called exploitation agriculture was carried out by RIJP for five years, after which the definitive land-use of the site followed. This was not nature, but mainly agriculture, involving further urban planning and forestry. The lowest part of the polder, where the Oostvaardersplassen is now located, would be the last to become cultivated. Even before the polder was created, the ultimate intention was to establish heavy industry on this site. Yet, because this was not needed immediately, the area was left alone, and quietly developed informally into a fantastic nature area that was enormously rich in birds. Nevertheless, at this time, the outside world and the nature conservation community in general knew nothing about it. However, RIJP biologists recognized the importance of the site, and they worked to preserve at least part of the marsh. But because the drainage and cultivation continued in surrounding areas, the marsh began to become drier. Internally within RIJP, it was decided to create a small dike around the marsh. This structure was completed in 1975, and in 1976 the level of the water in the swamp was raised with pumps. This created a marsh of 3,600 ha, which was designated as a temporary nature reserve within the RIJP (Vera 1988). In 1978, I became directly involved with the Oostvaardersplassen.

Greylag Geese, but moulting rather than breeding

In 1978, the reclamation and layout of the polder for agriculture, urban planning and forestry had already progressed to such an extent that only about 6,000 ha of uncultivated land remained. This consisted of two parts. The first was a drained, dry area overgrown by dry reedbeds, Stinging Nettles *Urtica dioica*, and Common Elder *Sambucus nigra* bushes. These latter formed a rectangular pattern of elongated closed thickets that had formed on mineral soil released during the digging of channels and ditches for the purpose of dewatering the site. The soil that had been thrown up was deposited alongside, forming linear strips of bare mineral soil that created a germination environment for the elder – with the seeds being brought in by birds feeding on the berries elsewhere. This created a kind of hedge-like landscape in the uncultivated part of the Oostvaardersplassen. The second part was the 3,600-ha marsh with extensive reedbeds and large-scale zones of shallow open water.

During the cultivation of the polder, in the early 1980s, non-breeding Greylag Geese *Anser anser* arrived on the marsh to moult their wing feathers. They came from all over Central, Northern and Western Europe every May and June. When these birds moult their wing feathers they lose their primaries all at once, and are then unable to fly for about thirty days. At this time they are very vulnerable to predators, including humans, which is why they look for areas that offer them safety and food for the moult-period. They found this in the marsh of the Oostvaardersplassen, where they withdrew into the vast reedbeds. For the most part, these individuals were immature birds; it takes four years for them to mature, start to form a pair-bond and begin to breed. Any adult birds in the flock were geese that for some reason had not bred. From a few thousand individuals, their numbers grew over the years to a maximum of 60,000 in 1991, Oostvaardersplassen thus becoming the largest moulting area for greylag geese in Europe.

For food, the geese ate Great Reedmace and reeds (Fig. 6.2), digging out the roots of the former and eating the young sprouts and leaves of the latter. In order to get to the leaves of

Figure 6.2 Moulting, temporary flightless Greylag Geese, grazing vast reedbeds in the Oostvaardersplassen, steering the succession towards a mosaic of shallow water and marsh vegetation, in this way preventing it from developing into a fen forest, June 1986. (V. Wigbels)

Figure 6.3 Aerial photograph of the marshy part of the Oostvaardersplassen, May 1986. The channels in the foreground are created by moulting Greylag Geese. The mosaic of shallow water and marsh vegetation is the habitat of many marsh-inhabiting bird species, created by these geese. (F. Vera)

the reeds they broke the stem, and consequently the reeds grew more and more in dense tussocks. By constantly grazing the young reed sprouts that emerged from the rhizomes in May and June, the geese exhausted the plants' rhizomes, causing the reed to disappear. With their grazing, they created a mosaic landscape of shallow open water and marsh vegetation, which became the habitat for huge numbers and a wide variety of species of marsh-dwelling birds (Fig. 6.3). The geese also prevented the marsh area from developing ultimately into a fen forest.

Do grazing animals steer or follow succession?

Marshes such as the Oostvaardersplassen have arisen twice before in two previously drained polders in the IJsselmeer. Both of these polders initially developed into paradise for birds, until exploitation of the area put an end to it. The Oost-Flevoland polder, which dried out in 1957, was a famous site among birdwatchers. Along one of the dikes of this polder, the Knardijk, lay some 1,000s' of ha of bird-rich marsh over a length of more than 20 km and easily viewed from the dike. Yet when the end came, there was no call to preserve it – with very few exceptions among nature conservationists. It seems that the main reason for the lack of outcry was to be found in the theories of Clements and Tansley. Nature conservationists were of the opinion that such areas could not be maintained, because the soil was extremely fertile calcareous clay. This, it was assumed, meant that a vast linear succession would occur, and in a short time climax-forest vegetation would be the result. In order to prevent this, and to preserve the wetland birds' habitat, human intervention in the form of management was necessary, with the reed being mown as in other marsh areas in the Netherlands. However, mowing was and still is a very expensive way of managing nature. Maintaining such a large marsh-area would therefore be unaffordable and therefore uncontrollable; areas such as the Oostvaardersplassen were therefore very interesting for say ten years, but after that it was over. The exception to this view was expressed in *Wild Geese in the Netherlands* (Lebret et al. 1976), although I was unfamiliar with this book when I became involved with the Oostvaardersplassen. It states:

> The main question however is this: under what conditions are areas with low and open vegetation maintained? If we want to understand the place of the wild geese in the ecosystem, we have to think about this. The availability of large areas with low and open vegetation is not self-evident. We do know such types of vegetation as pioneer vegetation, but in general it is precisely these that tend to develop into higher and denser forms of vegetation, a development process that we call succession. However, all sorts of forces work in nature to stop this development. We summarize these forces with environmental dynamics. [...] In the 'wet' ecosystems inhabited by the geese, the following forms of environmental dynamics are important: all forms of water change, namely ebb and flow, dynamic turbulence, changing river discharge, precipitation and evaporation, furthermore frost, thaw and ice, silting up and erosion and finally the salts of seawater and - sometimes deep inland - of groundwater. In addition to these forces from inanimate nature, the so-called abiotic factors, biotic factors also play a role. Living creatures, for example, influence vegetation through feeding and fertilization. Geese play a clear role in the areas they inhabit. Here too they are on a par with grazing ungulates.
>
> Lebret et al. 1976: 24

About the Greylag Geese, the book comments on their southern breeding, as opposed to those species of geese that breed in the tundra:

> Here the harsh cold is missing as a source of environmental dynamics and instead of this the biotic environmental dynamics, the geese themselves, play a major role. With a slight exaggeration one could say that it is not the climate, but the Greylag geese themselves that maintain their breeding habitat. Incidentally, they are of course also helped by grazing ungulates.
>
> Lebret et al. 1976: 28

Later in the book, it is noted that:

> However, the relationship with the biotope is not a one-way street for Greylag geese: not only does the plant growth ensure the 'conservation' of the Greylag geese, but these birds also contribute to the maintenance of the marsh vegetation, by slowing down or bending the succession towards a cyclic ('cycle') succession. This implies that marsh areas where Greylag geese are missing due to human disturbance have no natural vegetation. When managing nature reserves in wet areas, one must therefore ensure that there is a large population of Greylag geese. The density thereof is verse two. But it is precisely in the vegetation period – roughly March-September – that the Greylag geese must help keep the vegetation in check. [...] Not only the vegetation benefits from the goose eater, but also other bird species and probably the entire ecosystem. So the Spoonbills in the Neusiedlersee area benefit greatly.
>
> Lebret et al. 1976: 52–3

Further on, with regard to the new polders, such as South Flevoland, the authors note:

> Large-scale land reclamation – Northeast Polder, Eastern Flevoland and Southern Flevoland – where cultivation can take five to eight years, have temporarily very suitable Greylag Geese biotope, but then it is over: they are 'discarded biotopes'. Only the establishment of a permanent reserve in Southern Flevoland can benefit. In addition, the right species and the right amount of environmental dynamics must then be 'added' by adequate management techniques, with biotic and abiotic

> environmental dynamics being considered. The first supply the geese themselves, but they cannot do it alone and need hoofed animals grazing next to them. The second requires artificial expansion of the water level change, at least outside the breeding season.
>
> Lebret et al. 1976: 55

The Oostvaardersplassen area: unique ecological experiment

In retrospect, these words have turned out to be almost prophetic, but in 1976 they apparently fell on deaf ears. However, when thousands of moulting, non-breeding, grazing Greylag Geese suddenly arrived on the marsh, they prevented it from turning into fen forest. I read about this in a paper by Ernst Poorter (1979), a biologist employed by the RIJP. He stated that in addition to the constant loss of nature that was being experienced more widely, a gain could also be made, and he cited the Oostvaardersplassen as a new nature reserve. Two points intrigued me in this article. The first was the breeding of a pair of Great Egrets *Ardea alba* in the area in 1978. The second was the role of the non-breeding, moulting Greylag Geese as they broke open vast reedbeds to create a habitat suitable for large numbers of birds. He therefore labelled the Greylag Geese as the natural managers of the marsh. Remarkably, he did not indicate whether or not the area should be preserved. I was still an unemployed, recently-graduated biologist at this stage, and I contacted him. I told him that in my opinion the breeding of the Great Egrets was important and significant: they had vanished from the Netherlands in the mid-nineteenth century with the destruction of expansive marshlands through drainage and cultivation. I also mentioned that his description of the role of the Greylag Geese had caused a paradigm shift for me. Instead of grazing animals following the succession, as in the Clements and Tansley paradigms, in the case of the Oostvaardersplassen they appeared to steer it. When I asked him why he did not call in his paper for the Oostvaardersplassen to be protected as a nature reserve, he replied that this was prohibited by his superiors at RIJP. This prompted me to reply that I wanted to devote a paper to this issue, setting out the importance of this area for nature conservation, and therefore the need to safeguard it. For me, the area was primarily an example that showed that nature was resilient; that what was lost could return if we humans wanted it. The Oostvaardersplassen nature reserve, as I wrote later in an article entitled 'Het Oostvaardersplassen area, a unique ecological experiment' (Vera 1979), had the potential to rekindle an ecosystem that once existed in large parts of the Netherlands as a delta of the major rivers Rhine, Waal and Maas. This had the potential to grow into a nature reserve in which breeding birds lost from the Netherlands over the centuries such as Great White Egret *Ardea alba*, Sea Eagle *Haliaeetus albicilla* and Osprey *Pandion haliaetus* could return as breeding birds. It could develop as an unprecedented complete ecosystem for the Netherlands and Western Europe and show that what was lost was not gone forever. This site showed that if you created conditions analogous to complete natural ecosystems, a modern analogy could develop of what was originally present, including lost ecosystems, and this could even be at 4 m below sea level. Essential for this was the presence of tens of thousands of Greylag Geese. For this to happen, it was necessary for them to continue visiting the marsh in the Oostvaardersplassen where they could moult and graze the reedbeds, and therefore grassland was needed. Why was this?

The habitat of moulting Greylag geese: a combination of marsh and grassland

In April 1979, I entered the Staatsbosbeheer (the government organization for forestry and the management of nature reserves) and produced a report (Vera 1980) based on the results of my research. I concluded that in order to bind the Greylag Geese to the marsh, grasslands were needed on the adjacent land.

At that time, the geese gathered before and after the moult on grasslands and arable land close to the marsh, land that was then used and managed by RIJP. Grasslands are of great importance for Greylag Geese, especially before and after the moulting season. It is true that the reeds offer them food while they are moulting, but this is not optimal food. During their time in the swamp, they therefore consume their body-fat reserves. The grasslands are important for building up these reserves both before and after they moult (Zijlstra et al. 1991; van Eerden, Loonen and Zijlstra 1998). That is why grasslands had to be provided at the Oostvaardersplassen within the nature reserve. Without these grasslands, there would be no moulting, grazing Greylag geese in the marsh. Besides providing suitable habitat for the Greylag Geese, grasslands would provide feeding opportunities (in the form of pools with shallow open water) for Great White Egrets and Spoonbills *Platalea leucorodia* breeding in the marsh.

Farmers, cows and grassland

Now the question arose, how do we get grasslands? The general view was very simple. If you want grasslands, you need cows, and if you need cows, you need farmers. So the land directly adjacent to the swamp should be granted to farmers with or without restrictions. I did not think this was a good idea, because those farmers would have to deal with tens of thousands of geese all the year-round. In the months of May, June and July, there would be with tens of thousands of moulting Greylag Geese, then during the migration in the autumn and again in the spring tens of thousands of Greylag Geese, and in the winter tens of thousands of wintering Barnacle Geese and White-fronted Geese. The RIJP tolerated the tens of thousands of geese year-round on their land, but this tolerance would change if the land was to become owned by farmers. This would lead to a major conflict – as happened in the second half of the twentieth century in many places in the Netherlands).

This led to a claim that 6,000 ha should be provided as a nature reserve. Instead of the intended 3,600 ha marshland an area of 2,400 ha of grassland with pools with shallow open water should be added to it. Discussions at government level followed, and it was finally decided to designate 5,600 ha as a nature reserve, rerouting a planned railway line so it passed outside the reserve. But the question still remained: how do you create and maintain grasslands?

However, there was my paradigm shift again, namely that instead of grazing animals following succession, as Tansley claimed, grazing animals could direct succession, as the moulting Greylag Geese in the marsh in the Oostvaardersplassen demonstrated. It was another biologist, Harm van de Veen, who in response to my paper about the Greylag Geese being the marsh's natural managers, pointed out that what they were doing in the marsh was done in many places in the world by large wild herbivores. He pointed me to *The Serengeti: Dynamics of an Ecosystem* (Sinclair and Norton-Griffiths 1979). This book tells how, after a decimation of the large wild ungulates in this area by the rinderpest, their numbers increased enormously after the disappearance of this European, and therefore exotic, cattle disease, resulting in an enormous change in the landscape (Fig. 6.4). Harm was a member of a group called Critical Forest Management, which devoted itself to more natural forests and forest management, and emphasized the role that was played by large herbivores. Although they noted this important role, they still assumed that forest was the natural vegetation. The role of large grazing animals was limited to areas where there were gaps in the canopy, where trees could regenerate in spiny shrubs, as could be seen in wood-pastures. However, even with these large grazers, eventually the gaps would become closed, forming a closed canopy. Forest remained forest. Harm was an exception to this view. He wrote that all over the world, through their grazing, large wild herbivores created the landscape and consequently the habitats of other plant and animal species. He wrote a pamphlet about how the Veluwe, an area in the

Figure 6.4 The Serengeti in Tanzania, February 2012. An example of an area where large, wild living ungulates in large numbers are the engineers of the landscape. (F. Vera)

Central Netherlands consisting of forests and heaths totalling 90,000 ha and home to Red Deer, Wild Boar *Sus scrofa*, Fallow Deer *Dama dama* and Roe Deer *Capreolus capreolus*, could become more natural with, among other things, the reintroduction of European Bison and Grey Wolves *Canis lupus*. There was a lot of opposition to his ideas, especially from hunters. Eventually, through his initiative, Scottish Highland Cattle were introduced to the area as wild animals. Sadly, Harm died on 21 January 1991 at the age of 45 – never to see the pair of wolves has been living in the Veluwe since 2018; in 2019 they had five pups and in 2020 four. Albeit in a limited area, there are also European Bison in the Veluwe.

The people from Critical Forest Management got their inspiration for the role of large herbivores from wood-pastures, such as the Borkener Paradise in Germany and the New Forest in southern England. That is how I came in contact with wood-pasture. But as I explained earlier, their starting point was that the natural vegetation was forest and even with large grazers it remained forest. Grassland, such as the moulting Greylag Geese needed, was not part of it.

If tame cows can create grassland, can't wild cows do the same?

Harm's comparison of the geese with large herbivores led me to the conclusion that if domestic cows can make grassland for geese, then wild cows must also be able to do so (Vera 2009). Furthermore, this would have the advantage that we would not need farmers – with all the risks of a constant conflict between farmers and geese in the nature reserve. Wild cows meant wild cattle. I then wrote a note to propose that for the Oostvaardersplassen there should be the introduction not only of cattle as wild animals, but also other specialized grass-eaters such as the wild horse (Fig. 6.5). My eye fell on the Polish konik pony. There had already been

Figure 6.5 Hecke cattle, red deer and koniks, as wild ungulates, create vast open grasslands on the very productive calcareous clay-soil of the Oostvaardersplassen, a natural habitat of open landscape bird species, July 2008. (F. Vera)

three koniks bought in Poland by the Tarpan Foundation and brought to the Netherlands to be released in a nature reserve in the north of the country.

When I came up with the proposal that wild cattle instead of domestic cattle should develop grasslands for the moulting Greylag Geese, scientists invoked the Tansley mantra. In chorus, they stated that this was impossible, because everything would become forest. We know that when primeval cattle, the Aurochs, were still roaming in Europe, everything was forest. However, I had problems with the ecology of that forest. From all sorts of sources, both ecological and forestry experts, it emerged that oaks *Quercus* did not rejuvenate spontaneously in such forest. They were out-competed by shade-bearing trees such as the European Beech *Fagus sylvatica*, Common Hornbeam *Carpinus betulus* and lime *Tilia*. But nevertheless the oak survived for thousands of years in the presence of shade-tolerant tree species, which competed with it in modern times by not allowing seedlings to grow up in their presence and even killing oak trees by growing over them. Pollen diagrams also indicated that oaks apparently survived for thousands of years in the presence of shade-tolerant tree species. How was that possible? I therefore wondered whether you could deduce the natural vegetation by considering the ecology of the different types of trees. My experience with the moulting Greylag Geese in the Oostvaardersplassen led this thought process. As a nature conservationist, I wanted to know how nature can survive intact with the species currently in it.

On the shoulders of Oliver Rackham

This line of reasoning brought me into conflict with the scientific concept of forest as the natural vegetation. If this had been the case, then all types of trees and shrubs that were part of it would survive there – and the Pedunculate Oak *Quercus robur* and Sessile Oak *Q. petraea* could not survive, as I read in many publications, being out-competed by species such as beech, hornbeam and lime (or linden). But pollen diagrams show that oaks did occur in the presence of those tree species over periods of many thousands of years, so what is the explanation? It was precisely on the basis of these diagrams that scientists concluded that the oak could regenerate successfully in a forest. But pollen diagrams involve the question of so-called proxy data – data that are interpreted on the basis of certain presuppositions. Therefore I began to look into these presuppositions: that the naturally present vegetation was forest, in which oak appeared alongside hornbeam, beech and linden; and therefore that both oak species rejuvenate spontaneously in closed forest and so formed part of the prehistoric forest.

But in reality, based on the ecology of the oak, the reverse was the case. It was assumed that oaks rejuvenate spontaneously in forest, because pollen diagrams show that they were present. However, this is indirect evidence. Evidence for oak regeneration in a closed-canopy forest was supported by forestry experts, who wrote in their manuals that both oak species naturally rejuvenate perfectly well in the forest (see, inter alia, Dengler and Rohrig 1980; Dengler, Rohrig and Gussone 1990). As an ecologist, I interpret 'naturally' as 'in nature'; but in forestry literature, as I discovered after a long search, having to go back to a handbook dated 1922 (Bühler 1922; see Vera 2000: 190ff.), it does not mean that. It is actually about rejuvenating oak trees with the help of all kinds of artificial means, as long as it does not involve sowing or planting them, which is called artificial regeneration. For example, natural regeneration can involve making a seedbed with the help of a plough and then combating, by hand or chemically, all other types of plants that were a threat to oak seedlings. These might include coarse grasses and herbs, and tree species such as hornbeam, beech, lime and elm *Ulmus*. Without all these measures, every regeneration of oak trees failed because the seedlings were killed. The real facts, therefore, contrast with the conclusions that were drawn on the basis of proxy data. In addition, those conclusions were supposed to be supported by historical texts, therefore speaking of the harmful effects of large grazing mammals such as cattle on forests. As a result of these animals, the forest that was naturally present would disappear over the course of 100 years. But then, amid all this confusion, I came across Oliver Rackham's book *Ancient Woodland* (1980).

Ancient Woodland; a revelation

Ancient Woodland was a revelation to me, and I devoured it. Indeed, this volume was the reason why I delved more deeply into books and articles about what was then called the history of the forest. I did not own Oliver's book, but have to admit I photocopied a library copy – which is now full of highlighting, exclamation marks and ticks, especially in the chapters about wood-pastures, Chapter 12 (pp. 173–202) and the oak, Chapter 17 (pp. 283–304). This book revealed to me all kinds of ecological processes that clarified the relationship between large grazers and the vegetation, and not least the answer to my question about grassland. It illustrated how important historical descriptions can be when reconstructing ecological processes that help to answer the questions of what nature looked like naturally before human intervention (see Higgs et al. 2014) and what processes determined the presence of various wild plant and animal species. Eventually, Oliver's book gave me the answer to my question about grassland with wild cattle, and guided me to the final conclusion in my PhD thesis and subsequent book *Grazing Ecology and Forest History* (Vera 2000). This was namely that large herbivores do not follow the succession, as Tansley claimed, but they direct it. They are the architects of the natural landscape and above all creators of grasslands, a natural part of the natural landscape together with large native wild herbivores (Fig. 6.6).

What I present here is mainly based on how I read Oliver's book. I will use quotes from this seminal volume, but cannot, of course, claim to be exhaustive. However, the summary here serves to indicate what Oliver meant to me and my theory formation – and that is a lot.

Figure 6.6 The wood-pasture ecosystem Borkener Paradise in Germany, grazed by livestock, April 2015. In the foreground is an oak protected by hawthorn acting as natural barbed wire. The oak is open grown, that is with a large crown low at the trunk. On the right there is a spiny scrub of blackthorn, also a light-demanding species. It advances in a convex form into the grassland by rootstocks. In the fringe of the scrub, trees become established, protected by this natural barbed wire, growing into trees with a 'forest'-grown small crown. The grove eventually disintegrates in the centre, forming grazed grassland again and closing the circular succession, steered by large herbivores, of grassland, scrub, grove and grassland again. (F. Vera)

One of my very first exclamation points appears under the heading 'Wood-pasture': 'Wood-pastures differ widely in appearance according to the amount of grazing and the method chosen to control it: they can vary from almost open grassland to a close approximation to woodland' (Rackham 1980: 5). Then, in the chapter 'Wood-pasture systems and products', the opening sentence under the heading 'Trees and grazing' struck me: 'Wood-pasture necessarily involves grassland as well as trees' (Rackham 1980: 173). This was a first step towards the answer to my question about grassland. On the same page, Oliver describes how those trees rejuvenated in a wood-pasture:

> Hawthorn or holly scrub was a natural and essential feature of uncompartmented wood-pasture: they protected new oaks growing up in their midst, as well known, for instance, in the New Forest. Under a statute of 1768 it was possible to get up to three months hard labour with a whipping in each month for damaging hollies or thorns.
>
> Rackham 1980: 173

I could have cheered when later in the same chapter under the heading 'Uncompartmented forests' I read: 'Grazing did not prevent the regeneration of trees, which sprang up infrequently but in great numbers' (Rackham 1980: 187). So, grazing did not prevent the regeneration of trees in so-called uncompartmented forests, by which Oliver meant the fully grazed hunting grounds that were under royal jurisdiction, the *ius forestis*.

Oliver's description of the regeneration of trees in uncompartmented grazed wood-pastures by means of bushes illustrated the importance in the past that was attached to thorn bushes as so-called nurse species, which protected seedlings and young trees in grazed areas as natural barbed wire. This also shed light on the meaning of the words 'wood', 'woodland' and 'forest', in texts about their history in English, German, French or Dutch. Their description as such related to thorns rather than meaning forests in the modern sense. It became clear to me that these were wood-pastures. What I found surprising in Oliver's book was that the use of the word 'thorn' in two Anglo-Saxon charters was more common than the second most frequent, 'oak' (Rackham 1980: 17). Oliver says about the Domesday Book: 'The composition of woods is occasionally mentioned. The commonest are spineys and cars (spineta, alneta), i.e. woods of thorn or alder' (Rackham 1980: 19). He also cites a description of the twelfth-century Chilterns as non-cultivated that cannot be read as referring to a closed forest: 'On the north are fields and pastures and a pleasant stretch of meadows ... Nearby lie a huge Forest (*ingens foresta*), wooded glades, and lairs of wild beasts: harts, does, boars and bulls [or, in other MSS, bears]' (Rackham 1980: 123). Oliver also mentions under 'Hawthorn' that in historical sources 'Thorn is more noticed than any other tree' (Rackham 1980: 352). Referring to blackthorn, Oliver mentions that this small tree is numerous in forests and hedges throughout Britain, except the Scottish Highlands: 'It is strongly suckering and forms wickedly spiny thickets through which nobody larger than a rabbit may pass' (Rackham 1980: 351). However, 'The written record is poor because blackthorn was not always distinguished from hawthorn' (Rackham 1980: 352).

The general appearance of this landscape became clear to me when Oliver described the shape of the commons and thus the origin of the coppice. Their external borders were concave: 'I have drawn attention elsewhere to the specific straggling shapes of commons, with concave outlines funnelling out into roads and often enclaves or private land in the interior. This is partly because commons were residues of unenclosed land, left over after a succession of encroachments' (Rackham 1980: 174).

That put me on the trail of the bushes that arose in the thorn. It also brought me to how these bushes' settled convex shape expanded through the encroachments of blackthorn in the grassland by way of underground rhizomes. I understood that the groves became enclosures that then developed into coppices, thus gaining a convex shape and giving the remaining common its concave shape (see Vera 2000: 159). So the picture arose of groves expanding to become grassland. The comment from Oliver on the blackthorn also helped: 'blackthorn was a favourite wood for faggots for heating ovens' (Rackham 1980: 352). I interpreted this as the beginning of coppicing being cutting blackthorn, with young trees being spared. If oaks were present, these were thinned, serving to create oaks with broad crowns that were able to produce many acorns for pannaging pigs (Vera 2000: 143ff.).

Oliver notes a risk with too many trees:

> Despite its suckering habit in the wildwood it persists, being easily overtopped and killed by other trees; timber fellings of the 1920s are sometimes now marked, as in Hatfield Forest, by tracts of dying blackthorn. It is a permanent inhabitant of woodland margins and ride sides. In the wildwood it presumably found opportunities to colonize glades and gaps.
>
> Rackham 1980: 352

What he writes about the encroachment of blackthorn in grassland put me on the trail of how regulated coppice developed in wood-pastures.

With regard to shade in coppices, Oliver's words about hazel were a revelation to me. This turned out to be a very important species in the coppice, explaining the enormously high percentages of pollen from hazel in pollen diagrams. The general explanation was that hazel was an understorey in the prehistoric forest. But, based on his observations, Oliver expresses doubts about this:

> Early pollen analysts thought it was an understory shrub and established a convention for not including hazel in the total of tree pollen. This was a mistake: as anyone can see in a neglected coppice-wood, hazel is not very tolerant of shade; if overtopped by neighbouring trees it may linger for many years but flowers poorly if at all ... The abundance of hazel pollen in most prehistoric deposits implies that the tree was not shaded; we do not know whether it grew taller than it does now or the wildwood as a whole was less tall than is usually supposed.
>
> Rackham 1980: 203

In the chapter 'The prehistory of the woodland', Oliver asks: 'Why should hazel, which is a modern arctic species and does not readily form secondary woodland, have been so prominent at such an early date?' (Rackham 1980: 98). He continues: 'Casual observation suggests that hazel is very productive, probably comparable to oak, if it is not shaded' (Rackham 1980: 101). As already indicated, Oliver argues that because of the high percentages of pollen from hazel, it could never have been a scrub in the prehistoric forest: the bushes must have been in the open. This means, in my opinion, that prehistoric vegetation could never have been a closed forest. Oliver's comments that 'Hazel, ash, and maple are among the commonest underwood species in sixteenth-century surveys or woods' (Rackham 1980: 205) and that 'Woodland management continued to expand by the conversion of wood-pasture to underwood long after 1350' (Rackham 1980: 136) are important. Hazel as part of the mantle and fringe vegetation in a wood-pasture stood together with blackthorn at the cradle of the regulated coppice. This, I felt, originated from the grove being surrounded by a mantle and fringe vegetation (see Vera 2000: paragraph 4.8, 132ff.).

Oliver Rackham on the shoulders of others

Of course, Oliver did not stand alone, as evidenced by his discussion of natural vegetation, which he characterized as wildwood. He based this on pollen diagrams and interpretation by palynologists (Rackham 1980: 100, 102). But the big difference between him and other authors is that the theory of the forest and the destructive effect of cattle, as Tansley put it, did not influence his interpretations of historic texts. He stayed true to their original meanings, even if these were contrary to prevailing theories. If his research differed, he simply noted this – something that makes his book so incredibly valuable. For example, with regard to oaks in relation to other tree species, he mentions in various places how they were out-competed by other species such as beech, hornbeam and ash: 'Invasion by oak with a holly understory, followed by beech – which in places is overtopping and killing the oak – appears to be characteristic of wood-pasture, particularly declining wood-pasture, on almost any soil within the native range of beech' (Rackham 1980: 175). Then he goes on to observe: 'In a neglected wood that ash underwood often overtops and kills the standard oak trees' (1980: 208). About the supposed return to a natural state with the decline of grazing, he remarks: 'It has often been claimed that the Forest is returning to a "natural aspect" of which it has been deprived by centuries of misuse, but this is now known to be an illusion: it is tending to a closed canopy of beech or hornbeam unlike anything in its history' (1980: 202). At the end of Chapter 17, 'Oakwoods and the general history of oak', Oliver criticizes a lack of management:

> D.T. Streeter, concluding the Oak Symposium in 1973, proposed that stands of oak be set aside without management, other than preventing excessive grazing, in order to recover the natural age structure of the climax forest and the later stages now missing from the life cycle. I hesitate to recommend such a policy generally, even where conservation is the chief objective: it would be a serious break with the history of most woods, in some places it would probably lead to the disappearance of oak.
>
> Rackham 1980: 304

As I have already indicated, it is grazing that ensures oak can continue to exist in the presence of other tree species. Without grazing, it is out-competed.

Oliver's book also provided me with astonishing data on the effect of grazing and different animal densities on the regeneration of trees, in particular oak. He writes that 'Oak regenerates well in the less-wooded parts or most wood-pastures' (Rackham 1980: 294). About the failure of oaks to regenerate, he continues: 'The extent of the problem is thus reduced to woods and the wooded parts of wood pastures' (Rackham 1980: 295), and that 'Since about 1850 something has happened to prevent this turnover [the replacement of old oak by young], but without affecting the ability of oak to reproduce outside of woodland' (Rackham 1980: 296). I added many exclamation marks here: 'Wood-pastures encourage oak by eliminating other species more sensitive to grazing and by providing glades and scrub areas which *Q. robur* could colonize' (Rackham 1980: 300).

This gave me the connection between grazing and the regeneration of trees in wood-pastures, ultimately explaining why oak as a light-demanding tree species can continue in the presence of shade-tolerant tree species. The absolute prerequisite for this was grazing by large herbivores, with cattle in forest meadows being a modern equivalent of wild grazing herbivores. Especially important to me in *Ancient Woodland* is what it says about the numbers and densities of animals where regeneration took place. It asserts that on 6,000 acres (2,400 ha) in Epping Forest:

> a) The king's deer were estimated at 3,000 fallow and a few dozen of red;
> b) The grazing rights of commoners and landowners included at least 2,000 sheep and probably over 1,000 cattle;
> c) There were large numbers of goats and other animals illicitly grazed.
>
> Rackham 1980: 185

A few pages later under the heading 'Uncompartmented forests', Oliver observes that in Epping Forest 'Grazing did not prevent the regeneration of trees, which sprang up infrequently but in great number' (Rackham 1980: 187). In my book, I cite other publications that substantiate this (see Vera 2000: paragraph 4.10, 144ff.).

Nevertheless, Oliver also remained critical about the oak:

> No civilized man can be unmoved by the legends, poetry, and song that celebrate the oak – by Abraham's Oak, the oaks of Dodona, the druids, the Gospel Oaks, Herne's Oak, the Major Oak, the Oaks of the reformation, the Oak of Guernica – but these reflect people's attitudes to oaks and not the place of oak in vegetation. Many factors have conspired to over-represent in the record. It is a non-woodland as well as a woodland tree and hence familiar to artists, poets, and the public as well as to the woodmen; the tree, its leaves, acorns timber, and pollen are easily identified; it is variable, and individual trees can be recognized and named.
>
> Rackham 1980: 17

> The over-representation of oak continues back into prehistory, both in peat deposits – for it produces more pollen per tree than most other species – and hence its rot resistant timber is more likely to be preserved in archaeological contexts.
>
> Rackham 1980: 8

Oliver warns against drawing hasty conclusions: 'We must avoid assuming that abundant evidence for oak means that the tree itself was proportionately abundant; greater weight must often be given to less abundant evidence to other trees' (Rackham 1980: 8). In other places too, he warns against the view that the history of British trees is just the history of the oak. I respect his caution. Nevertheless, he led me to regarding the oak as a metaphor for the natural landscape.

I could quote so much more from Oliver' book that inspired me, but that would take too much space. It is with great respect, deep admiration and great gratitude that I think of Oliver Rackham, who explained to me the system of the wood-pasture and the great importance of historical texts in discovering how the natural landscape looked and functioned. For me, his book *Ancient Woodland* was the window onto the natural landscape, and particularly the wood-pasture as a modern analogy for the natural landscape. This provided the answer to my question about where grassland came from. It was from large, wild herbivores – the architects that steered the development of that landscape.

References

Bühler, A. (1922) *Der Waldbau nach wissenschaftlicher Forschunbg und praktischer Erfahrung*. II Band. Eugen Ulmer, Stuttgart.

Clements, F.E. (1916) *Plant Succession: An analysis of the development of vegetation*. Publication No. 242. Carnegie Institution, Washington DC. https://doi.org/10.5962/bhl.title.56234

Dengler, A. and Rohrig, E. (1980) *Silviculture on an Ecological Basis. Vol. I. Forest as a vegetation type and its importance to man*. Verlag Paul Parey, Hamburg.

Dengler, A., Rohrig, E. and Gussone, H.A. (1990) *Silviculture on an Ecological Basis. Vol. II. Choice of species, stand establishment and stand tending*. Verlag Paul Parey, Hamburg.

van Eerden, M.R., Loonen, J.J.E. and Zijlstra, M. (1998) Moulting Greylag Geese *Anser anser* defoliating a reed marsh *Phragmites australis*: seasonal constraints versus long-term commensalism between plants and herbivores. In: M.R. van Eerden (ed.) *Patchwork: Patch use, habitat exploitation and carrying capacity for birds in Dutch freshwater wetlands*. Published PhD thesis, Rijkswaterstaat, Lelystad, The Netherlands, 239–264.

Higgs, E., Falk, D.A., Guerrini, A., Hall, M., Harris, J., Hobbs, R.J., Jackson, S.T., Rhemtulla, J.M. and Throop, W. (2014) The changing role of history in restoration ecology. *Frontiers in Ecology and the Environment*, 12(9), pp. 499–506. https://doi.org/10.1890/110267

Lebret, T., Mulder, T., Philippona, J. and Timmerman, A. (1976) *Wilde ganzen in Nederland*. Thieme, Zutphen.

Norton-Griffiths, M. and Sinclair, A.R. (1979) *Serengeti: Dynamics of an ecosystem*. University of Chicago Press, Chicago.

Poorter, E.P.R. (1979) De Oostvaardersplassen: een nieuw natuurgebied in Nederland. *Vogels*, January/February, 60, pp. 36–39.

Rackham, O. (1980) *Ancient Woodland: Its history, vegetation and uses in England*. Edward Arnold, London.

Tansley, A.G. (ed.) (1911) *Types of British Vegetation*. Cambridge University Press, Cambridge. https://doi.org/10.5962/bhl.title.55266

Tansley, A.G. (1935) The use and abuse of vegetational concepts and terms. *Ecology*, 16, pp. 284–307. https://doi.org/10.2307/1930070

Vera, F. (1979) Het Oostvaardersplassengebied; uniek oecologisch experiment. *Natuur en Milieu*, 79(3), pp. 3–12.

Vera, F.W.M. (1980) De Oostvaardersplassen, de mogelijkheden tot behoud en verdere ontwikkeling van de levensgemeenschap. *Staatsbosbeheer-rapport 1980–1*, Staatsbosbeheer, Inspectie Natuurbehoud, Utrecht.

Vera, F.W.M. (1988) *De Oostvaardersplassen. Van spontane natuuruitbarsting tot gerichte natuurontwikkeling*. IVN/Grasduinen-Oberon, Haarlem.

Vera, F.W.M. (2000) *Grazing Ecology and Forest History*. CABI Publishing, Oxford. https://doi.org/10.1079/9780851994420.0000

Vera, F.W.M. (2009) Large-scale nature development – the Oostvaardersplassen. *British Wildlife*, June, pp. 28–36.

Zijlstra, M., Loonen, M.J.E., Van Eerden, M.R. and Dubbeldam, W. (1991) The Oostvaardersplassen as a key moulting site for Greylag Geese *Anser* in Western Europe. *Wildfowl*, 42, pp. 45–52.

CHAPTER 7

Forest History versus Pseudo-History: The Relevance of Oliver Rackham's Concepts in the Conservation of Białowieża Primeval Forest

Tomasz Samojlik, Piotr Daszkiewicz and Aurika Ričkienė

Summary

This chapter takes the case-study of Białowieża Primeval Forest and applies Oliver Rackham's idea of 'factoids', observing their potential influences on site conservation management. The study applies the findings of archaeological and palynological studies along with cultural sources of information and interrogation of archival and literary sources to challenge some current policies in the management of the forest. The work concludes by noting the potential role of this important site as a test-bed for natural forest processes to inform future forest management across Europe.

Keywords: Białowieża Primeval Forest, factoids, forest history, hunting garden

Introduction: setting the scene at Białowieża Primeval Forest

Białowieża Primeval Forest (BPF) enjoys well-deserved fame as one of the best-preserved temperate forests in the European lowlands. The site survives with a variety of animal, plant and fungi species, diverse forest environments with an abundance of dead and decaying trees, and natural processes that drive the ecology of the forest. Straddling the border between Poland and Belarus, the 1,450-km^2-BPF is a closed-canopy mosaic of different forest habitats with a small number of open areas made up of forest gaps, river valleys and marshes. Most of the 600 km^2 on the Polish side is managed commercially under the state forestry. The protected areas consist of Białowieża National Park (105 km^2 with 47.5 km^2 of old-growth forest strictly protected since 1921) and 21 smaller nature reserves (in total a little over 120 km^2) located within the managed area. Since 1976, the Polish part of BPF has been a UNESCO Biosphere Reserve, and since 2004 it has been covered by a Natura 2000 network of nature protection areas. The Belarussian part is under protection as a Belarusian State National Park, Belovezhskaya Pushcha. The entire forest is a trans-boundary UNESCO World Heritage Site, Białowieża Forest (established in 1992) (Jędrzejewska and Jędrzejewski 1998; Kavalenia et al. 2009; Pabian and Jaroszewicz 2009).

Tomasz Samojlik, Piotr Daszkiewicz and Aurika Ričkiene, 'Forest History versus Pseudo-History: The Relevance of Oliver Rackham's Concepts in the Conservation of Białowieża Primeval Forest' in: *Countryside History: The Life and Legacy of Oliver Rackham*. Pelagic Publishing (2024).
DOI: 10.53061/JELX1549

Apart from natural features, the forest holds an abundance of traces of centuries-old human presence. These traces became the subject of interest of naturalists, travellers and artists as early as the nineteenth century, leading to the creation of many local legends about the forest's past. In modern versions of those legends we see, paradoxically, connections between these works, the ideas of the late Oliver Rackham and historical research linked with BPF. Rackham's works are at first glance set in a completely different world to BPF's environmental history. The English countryside has been well documented, being rich in historical monuments and abundant in ancient trees. In comparison, BPF has seemed to be a '*pushcha*' (as in its original Polish name): empty land, rarely appearing in historical sources and located on the margin of actual history. What is present, though, is a set of well-known 'truths', statements that have been repeated from one book to another and noted in a series of tourist guides. These are 'factoids'. In Rackham's definition, a factoid 'looks like a fact, is respected as a fact, and has all the properties of a fact except that it is not true' (Rackham 2001: 23).

In the case of BPF, those factoids constituted a coherent and widely popular version of the forest's history that seemed to suffice as an explanation of everything that has happened there (both a century and a millennium ago), forming the foundation for all future management decisions. The presence of this 'pseudo-history' has hampered the development of new research into BPF's environmental history, thanks to a simple but daunting question: why study something that is already well known?

Oliver Rackham's approach has therefore been a strong motivation to confront these obstacles. With an international group of co-workers, we have set out to put BPF's factoids to the test, learning that most of them are all but giants with clay feet, and that the myths can be countered with relative ease by close examination of written sources, archaeological excavations and palaeo-ecological studies. The myths have been discounted but not erased. As debate over the conservation and exploitation of BPF continues, old factoids and new pseudo-facts concerning the environmental history of the forest are coming into play. This is even before we address the plethora of pseudo-facts that are connected with the recent (since 2016) bark beetle outbreak or the presence of deadwood in the forest. (This chapter focuses only on historical factoids and so will not consider these contemporary issues further.) However, this is a timely moment to remind all who are interested in BPF about Rackham's approach to forest pseudo-history.

Factoids and Białowieża Primeval Forest

There is a substantial list of factoids connected to BPF's history. Maybe the most understandable are those connected with ancient times – with the lack of reliable archaeological sources leading some authors to believe that the forest was home to Celts or Yotvingians (later proved wrong by archaeological excavations; see Samojlik et al. 2013). The approximately 1,000 artificial mounds (barrows) in the forest were generally regarded as originating in the same medieval epoch. This conclusion was reached despite visible differences in their construction, grouping and distribution, and led to the widespread assumption that BPF was put under heavy pressure by a wave of medieval settlers. This has again been debunked by archaeological and historical studies (Samojlik et al. 2013). Next, there was the idea that the forest's name came from the white stone wall of a donjon or a keep (Białowieża when spoken in Polish sounds very similar to Biała Wieża, White Tower), located by many authors in the forest backwoods and traditionally called Zamczysko (Old Castle Place). Zamczysko turned out to be a medieval Slavonic cemetery, with no traces of an old castle to be found (Krasodębski et al. 2005). Another factoid is an ungrounded belief that BPF was the favourite hunting place of Polish kings and Lithuanian grand dukes, leading to their frequent visits, constant hunting and consequent depletion of game resources. Careful study of historical sources, especially itineraries and court accounts, show a different picture – a hunting ground located in a remote part of the Polish-Lithuanian Commonwealth (similarly distant from both capitals, Kraków and Vilnius), and rarely visited by monarchs for hunting purposes (Samojlik 2006).

In addition to these errors and misinterpretations, nineteenth-century authors reported the presence of species that have never been confirmed as living in BPF, such as Aurochs *Bos primigenius*, Wolverine *Gulo gulo*, Sable *Martes zibellina*, European Wildcat *Felis silvestris*, Old World Flying Squirrel *Pteromys* spp. and Steppe Polecat *Mustela eversmanii*. These errors most probably had biogeographic origins, with species known in other areas of the Polish-Lithuanian Commonwealth being ascribed to BPF (Samojlik et al. 2020). Some of these inaccuracies were clearly based on accidental occurrences, imports or even the fur trade; nevertheless, they were persistently repeated in later periods. The list goes on: some factoids touch upon the great forest fires of 1819 and 1834, described in several books but never confirmed in dendrological studies of BPF's fire history (Niklasson et al. 2010). There is also the idea that European Bison *Bison bonasus* were dependent on the forest environment of BPF, being unable to survive outside it. This was popularized particularly in the nineteenth century as the reason why bison persisted in BPF and disappeared from their wider historical range, and became one of the keystones of modern European bison management and conservation in BPF. Recent research, however, has presented a different picture of a species that is artificially kept confined to suboptimal forest habitat when its original range covered vast open areas (Kerley, Kowalczyk and Cromsigt 2012; see also Samojlik et al. 2019a).

An 'artificial forest': fact or factoid?

Here we focus on the most widespread and deeply rooted factoid that touches upon the history of BPF and strongly influences much of the ongoing debate on the future of the forest's management and protection. This is the conviction that BPF is artificial and has been shaped by management, especially since the emergence of 'scientific' or 'rational' forestry, with the aim of maximizing the forest's production output. Such an outlook is represented both in scientific works connected with forestry (e.g., Hilszczański and Jaworski 2018; Brzeziecki, Andrzejczyk and Żybura 2018) and by popular materials such as tourist guides, leaflets, informational materials printed by State Forests (e.g., CILP 2017a 2017b). This argument was used in an official statement that was issued by the Polish Ministry of Environment in 2017, defending the decision to log bark-beetle-infected spruces in BPF:

> the state and percentage of cover of habitats [...] as well as of the species existing on those sites [...] resulted from the use of the Forest in the past (obtaining wood from stands planted in the past). This has been thoroughly documented in many documents [...] which has been fully ignored by the European Commission, according to whom the Forest constitutes a primeval forest untouched by man.
>
> Ministry of Environment 2017: 1–25

This view of BPF as shaped by long-lasting human presence and utilization, and over the last two centuries by clear-cuts and plantations, is held in high regard by those local inhabitants who are connected with the timber industry and forestry. It is best represented by a touching quote on a tombstone in the cemetery at Hajnówka, a small city on the forest's border. Under the name of a woman who died in 2010 aged 103 is the phrase 'Ten las ja sadziłam', which translates as 'I have planted this forest' (Fig. 7.1). One could believe that this is absolutely true (given that, since 1945, the forest has undergone massive cutting and planting of trees), but at the same time it creates a completely false picture of the forest's history. It sums up an argument made by forestry specialists that the forest was designed by human hands (and that it owes its rich biological diversity to this fact). Therefore, it should be always managed with respect so that biodiversity is not lost, so that forest-building tree species thrive and so the entire set of 'pushcha' features flourish. Since this discussion goes beyond environmental

Figure 7.1 'Ten las ja sadziłam' ('I have planted this forest'): an engraving on a tombstone in cemetery in Hajnówka on the border of Białowieża Primeval Forest. The photograph has been cropped to avoid showing personal data, 24 February 2020. (T. Samojlik)

history, we will not follow it further here. Rather, we will present evidence that rebuts the 'artificial forest' factoid.

Most European forests have experienced at least one phase of complete deforestation in the ancient, medieval or modern period (Williams 2003). In this context, BPF is unique as the only European lowland forest that has been continuously covered by forest vegetation since the last glaciation (about 11,000 to 12,000 years ago; Latałowa et al. 2016), while at the same time being constantly connected with human presence. This historical relationship of people with the forest is of particular interest to interdisciplinary research in environmental history that aims to establish the role of people in changing BPF's natural environment. Recent palynological research has confirmed not only the continuous afforestation of BPF since the end of the last glaciation, and a very low level of anthropogenic indicators compared with other parts of lowland Europe (Latałowa et al. 2016), but also five phases of human presence. Those phases, visible also in archaeological and historical sources, are:

1. An ancient phase, with traces of settlements, cereal cultivation and animal husbandry (lasting until the fifth century AD; Krasnodębski et al. 2008; Olczak et al. 2018).
2. The early medieval to early modern period, from the eighth to the sixteenth century. From the palynological point of view, this phase is characterized by modest settlement and economic activity. Slavonic settlers came to the forest in two waves, the first in the eighth and ninth centuries (evidenced by cremation burials; Krasnodębski et al. 2011), and the second in the eleventh to the thirteenth centuries (evidenced by inhumation graves; Krasnodębski et al. 2005). In the late medieval period, BPF was a hereditary property of the Lithuanian grand dukes, and after the Polish-Lithuanian union in 1385 it became a royal forest, for the next four centuries serving mainly as a hunting

ground for Polish and Lithuanian monarchs. The royal status was connected with a system of conservation, including the prohibition of logging and illegal hunting, and the establishment of a ring of guard villages around the forest; this involved several hundred men in daily control and protection (Hedemann 1939). At the same time, local communities were given the right to enter the forest and use it in a non-destructive way (e.g., by collecting hay, producing honey, fishing in forest rivers and pasturing cattle on forest clearings; Samojlik and Jędrzejewska 2004).

3. The modern phase (seventeenth and eighteenth centuries) was characterized by an increase in human activity indicators in palynological data; it saw the introduction of more invasive, and also more profitable, uses: burning potash, wood tar, birch tar and charcoal. These activities coincided with a rise in the number of small-scale, low-intensity forest fires (Niklasson et al. 2010). In the second half of the eighteenth century, the first attempts at commercial logging were introduced. Nevertheless, the status of royal hunting ground was sustained, along with strengthened protection, which allowed the forest to survive until the end of the eighteenth century. The major part of the woodland was now in a close to natural state with no visible traces of human destruction (Samojlik, Rotherham and Jędrzejewska 2013a).
4. The majority of the fourth phase (nineteenth to the mid-twentieth century) covers the period of Russian rule over BPF. It is commonly known that this time was connected with increased anthropogenic pressure on the forest, with modern forest management being based on the so called German or rational school of forestry. However, environmental history studies lead to a different conclusion. There were several factors, including the presence of European Bison, a lack of good transportation routes for the extraction of timber, overabundance of deadwood and the persistence of traditional types of non-timber forest uses, which hindered and eventually stopped attempts at introducing timber-oriented forest management. Instead, the forest once again became a royal hunting ground, this time belonging to the Russian tsars (Samojlik et al. 2019a, 2019b, 2020). The first time BPF became subject to mass-scale theft of timber was during the First World War, when the forest was occupied and exploited by the German army (1915–17). However, the Germans mainly used selective cutting to remove the most valuable timber. Clear-cuts were introduced after the war, when the newly reinstated Polish government contracted BPF to a British company, the Century European Timber Corporation. However, the government soon terminated this contract. Around the same time, in 1921, the first reserve was created, later being transformed into the Białowieża National Park. This designation protected the best-preserved and most valuable parts of the forest (Więcko 1984: 204).
5. The fifth phase encompasses the most modern part of BPF's history, with its dual approach – partially managed for timber, partially protected – up to the most recent debate on the future of the forest.

All the information here – despite documenting a long-lasting human presence in BPF, a plethora of uses of forest resources and drastic political and administrative changes that shaped the approach to the forest – still evidences the unique preservation of BPF's environment. The forest was never destroyed, and was able to fully regenerate even after prolonged periods of anthropogenic pressure thanks to long gaps between different waves of settlers or phases of human presence (Fig. 7.2). In the modern period (sixteenth to eighteenth centuries), human impact on the natural environment was much lower than in most other areas of Poland and the rest of Europe. This allowed BPF to avoid the fate of neighbouring forests such as Bielska, Kamieniecka, Tokarzewska, Pużycka and Narewka, known only from historical documents. In the nineteenth and early twentieth centuries, the model of forest use did not prevent the natural processes of succession and regeneration, and did not disturb the continuity of forest habitat. However, traditional utilization of the forest resources led to the

Figure 7.2 The best-preserved fragments of old growths in Białowieża Primeval Forest make it especially difficult to pass over the 'artificial forest' factoid lightly, 6 November 2018. (T. Kamiński, with his permission)

creation of traditional or cultural landscapes in BPF. These combined natural habitats with anthropogenically modified elements, such as the landscape of a hunting garden (Samojlik, Rotherham and Jędrzejewska 2013b). The destruction inflicted in the twentieth century is undisputed; but it was not severe enough to disturb the natural processes that govern the entire forest ecosystem, such as cycles of tree seeds production, insect population outbreaks, rodent population peaks and the effects of a large number of decaying trees (Wesołowski et al. 2016). BPF is considered to still have fragments of habitat that are as close to primeval European lowland forest as one can find in our times, and it is used as a benchmark by conservation science, ecology, forestry and evolutionary sciences (Jaroszewicz et al. 2019). At the same time, each year during which BPF is treated as a regular managed forest, each season of timber felling, especially mass extraction after insect outbreaks, threatens the connectivity of the forest – and thus the persistence of natural processes (Mikusiński et al. 2018).

The danger of factoids

As may be seen here, factoids can do a lot of damage. They serve as an easy explanation, especially as they are often backed up by the authority of prominent figures, but they present a false version of reality, leading to irresponsible decisions concerning forest management. But also, paradoxically, they sometimes serve as a flashpoint for research that checks the reliability of 'popular truths' that are repeated over and over again. From this perspective, each factoid is an opportunity to utilize scientific approaches that attempt to show the real situation, as confirmed by research and analysis.

There are also occasions when propositions supposedly based on local tradition, lore or legends have turned out to contain a degree of truth. Such is the case of a very popular tourist

Figure 7.3 One of the oldest oaks in Stara Białowieża (Old Białowieża), 16 September 2008. A group of such oaks has grown on the ruins of a sixteenth-century hunting manor of Polish kings and Lithuanian grand dukes. (T. Samojlik)

trail in BPF, 'The Royal Oaks Route', which was created by a forester, Jacek Wysmułek, in the 1970s. Located in a picturesque part of managed forest, 22 large oaks, with an estimated age between 150 and 400 years, constitute one of the most popular tourist attractions of the forest. The factoid in play here, mentioned earlier, states that BPF was frequented by almost every Polish king and Lithuanian grand duke, whose frequent hunts in Białowieża's woodland led to the depletion of game. Mirroring this, each oak on the trail received a name of a king or a queen, with an explanation of their visit to the forest. The place itself, already called Stara Białowieża (Old Białowieża) in the eighteenth century, was described by the trail's creator: 'It is a place from which the current Białowieża most probably originated. Probably ruins of a hunting manor of [king] Władysław Jagiełło were present there. Passing through one should pay attention to several oaks – natural monuments, which perhaps remember these times' (Wysmułek 1977: 127–34). Remarkably, when archaeological excavations were undertaken after one of the oaks collapsed, the remains of a sixteenth-century hunting manor belonging to Polish kings was revealed; it was most probably used during the hunt of Zygmunt August in 1546 (Fig. 7.3) (Samojlik 2006). In this case, at least part of the fabled hunting took place on this site.

The 'artificial forest' factoid serves as an excuse for many management decisions that are undertaken in BPF, but is also an inspiration for new research. The authors of this chapter are involved in retracing the origins of the connection between the concept of 'primeval forest' and BPF. By studying Europe-wide perceptions of BPF in natural science studies, and in literary descriptions or works of art, we aim to elucidate the role of this place in shaping the idea of primeval forest and, indirectly, influencing the development of modern forestry, phytosociology, biogeography and nature conservation.

Conclusions

Oliver Rackham was famous for being able to find evidence of ancient treescapes in even a small patch of woodland or hedgerow, or even in a single tree. Ironically, in the case of BPF, some observers seem to be unable to find the ancient forest in the pristine, natural woodland. The factoids support their case for intervention and justify drastic and hasty actions when natural disasters occur, bark beetle outbreaks being just one example. Again, Oliver Rackham's words should be heard by decision-makers, as they fit BPF's situation surprisingly well:

> In the 1970s trees and plants seemed to be in such a precarious state that only immediate action could save them: tree-planting – any trees – was thought to be needed in a hurry [...] People now should stop and think and get the details right. This may involve waiting a year or two, or planting fewer but better-chosen trees; or doing nothing and letting natural succession do the job. The time for playing God is over.
>
> Rackham 2006: 538

Hopefully, BPF will soon be treated with the respect it deserves and set apart from further human interference. It can then provide a model from which we can learn how natural processes shaped forest ecosystems across Europe. The knowledge gained from such a case-study might then be the key to elaborate new models of forest conservation that are better suited for upcoming global challenges.

Acknowledgements

This chapter was prepared as a part of the project 'Perception of European bison and primeval forest in the eighteenth and nineteenth centuries: shared cultural and natural heritage of Poland and Lithuania' (UMO-2017/27/L/HS3/031870) financed by the National Science Centre, Poland, and by (S-LL-18-6) Research Council of Lithuania.

References

Brzeziecki, B., Andrzejczyk, T. and Żybura, H. (2018) Natural regeneration of trees in the Białowieża Forest. *Sylwan*, 162(11), pp. 883–896. (in Polish with English summary)

CILP (2017a) *25 pytań o Puszczę Białowieską* (*25 questions about Białowieża Primeval Forest*). CILP, Warsaw. (in Polish)

CILP (2017b) *Niech żyje Puszcza. 100 lat pod opieką leśników* (*Long live the Forest. 100 years under the care of foresters*). CILP, Warsaw. (In Polish)

Hedemann, O. (1939) *Dzieje Puszczy Białowieskiej w Polsce przedrozbiorowej (w okresie do 1798 roku)* (*History of Białowieża Primeval Forest in pre-partition Poland (in the period until 1798)*). Instytut Badawczy Lasów Państwowych, Rozprawy I Sprawozdania Seria A, Nr 41, Warsaw. (in Polish with French summary)

Hilszczański, J. and Jaworski, T. (2018) Biodiversity conservation in the Białowieża Forest in the context of natural and anthropogenic disturbances dynamics. *Sylwan*, 162(11), pp. 927–932. (in Polish with English summary)

Jaroszewicz, B., Cholewińska, O., Gutowski, J. M., Samojlik, T., Zimny, M. and Latałowa, M. (2019) Białowieża Forest – a relic of the high naturalness of European forests. *Forests*, 10(10), p. 849. https://doi.org/10.3390/f10100849

Jędrzejewska, B. and Jędrzejewski, W. (1998) *Predation in Vertebrate Communities. The Białowieża Primeval Forest as a Case Study.* (Ecological Studies 135) Springer-Verlag, Berlin, Heidelberg, New York.

Kavalenia, A.A., Danilovich, V.V., Dounar, A.B. et al. (eds) (2009) *Belavezhskaia Pushcha: vytoki zapavednastsi, gistoria I suchasnasts* (*Białowieża Primeval Forest: the origins of wilderness protection, history and modernity*). Belaruskaia Navuka, Minsk. (in Belorussian and Russian)

Kerley, G.I.H., Kowalczyk, R. and Cromsigt, J.P.G.M. (2012) Conservation implications of the refugee species concept and the European bison: king of the forest or refugee in a marginal habitat? *Ecography*, 35(6), pp. 519–529. https://doi.org/10.1111/j.1600-0587.2011.07146.x

Krasnodębski, D., Samojlik, T., Olczak, H. and Jędrzejewska, B. (2005) Early mediaeval cemetery in the Zamczysko Range, Białowieża Primeval Forest. *Sprawozdania Archeologiczne*, 57, pp. 555–583.

Krasnodębski, D., Dulinicz, M., Samojlik, T., Olczak, H. and Jędrzejewska, B. (2008) A cremation cemetery of the Wielbark culture in Kletna range (Białowieża National Park, Podlasie Province). *Wiadomości Archeologiczne*, 60, pp. 361–376. (in Polish with English summary)

Krasnodębski, D., Olczak, H. and Samojlik, T. (2011) Early medieval cemeteries in Białowieża Forest. In: S. Cygan, M. Glinianowicz and P. Kotowicz (eds) *In silvis, campis ... et urbe. średniowieczny obrządek pogrzebowy na pograniczu polsko-ruskim*. Instytut Archeologii Uniwersytetu Rzeszowskiego, Rzeszów-Sanok, pp. 144–174. (in Polish with English summary)

Latałowa, M., Zimny, M., Pędziszewska, A. and Kuprjanowicz, M (2016) Postglacjalna historia Puszczy Białowieskiej – roślinność, klimat I działalność człowieka (Postglacial history of Białowieża Forest – vegetation, climate and human activity). *Parki Narodowe I Rezerwaty Przyrody*, 35(1): 3–49. (in Polish with English summary)

Mikusiński, G., Bubnicki, J.W., Churski, M. et al. (2018) Is the impact of logging in the last primeval lowland forest in Europe underestimated? The conservation issues of Białowieża Forest. *Biological Conservation*, 227, pp. 266–227. https://doi.org/10.1016/j.biocon.2018.09.001

Ministry of Environment (2017) *Statement of the Ministry of Environment*, 13 October 2017 (https://archiwum.mos.gov.pl/en/news/details/news/statement-of-the-ministry-of-environment-1) [accessed 20 February 2020].

Niklasson, M., Zin, E., Zielonka, T., et al. (2010) A 350-year tree-ring fire record from Białowieża Primeval Forest, Poland: implications for Central European lowland fire history. *Journal of Ecology*, 98(6), pp. 1319–1329. https://doi.org/10.1111/j.1365-2745.2010.01710.x

Olczak, H., Krasnodębski, D., Samojlik, T. and Jędrzejewska, B. (2018) An iron producing settlement of the Stroked Pottery Culture at the Berezowo clearing in the Białowieża Forest. *Wiadomości Archeologiczne*, 69, pp. 149–176. (in Polish with English summary)

Pabian, O. and Jaroszewicz, B. (2009) Assessing socio-economic benefits of Natura 2000 – a case study on the ecosystem service provided by Białowieża Forest. Output of the project Financing Natura 2000: Cost estimate and benefits of Natura 2000 (Contract No.: 070307/2007/484403/MAR/B2), pp. 1–69.

Rackham, O. (2001) *Trees and Woodland in the British Landscape*. Phoenix Press, London.

Rackham, O. (2006) *Woodlands*. (New Naturalist 100) Collins, London.

Samojlik, T. (2006) Łowy I inne pobyty królów polskich I wielkich książąt litewskich w Puszczy Białowieskiej w XV–XVI wieku (Hunts and stays of Polish kings and grand dukes of Lithuania in Białowieża Primeval Forest in the 15–16th century). *Kwartalnik Historii Kultury Materialnej*, 54(3–4), pp. 293–305. (in Polish with English summary)

Samojlik, T. and Jędrzejewska, B. (2004) Użytkowanie Puszczy Białowieskiej w czasach Jagiellonów I jego ślady we współczesnym środowisku leśnym (Utilisation of Białowieża Forest in the times of Jagiellonian dynasty and its traces in the contemporary forest environment). *Sylwan*, 11, pp. 37–50. (in Polish with English summary)

Samojlik, T., Jędrzejewska, B., Michniewicz, M., Krasnodębski, D., Dulinicz, M., Olczak, H., Karczewski, A. and Rotherham, I.D. (2013) Tree species used for low-intensity production of charcoal and wood-tar in the 18th-century Białowieża Primeval Forest, Poland. *Phytocoenologia*, 43(1–2), pp. 1–12. https://doi.org/10.1127/0340-269X/2013/0043-0511

Samojlik, T., Rotherham, I.D. and Jędrzejewska, B. (2013a) Quantifying historic human impacts on forest environments: a case study in Białowieża Forest, Poland. *Environmental History*, 18(3), pp. 576–602. https://doi.org/10.1093/envhis/emt039

Samojlik, T., Rotherham, I.D. and Jędrzejewska, B. (2013b) The cultural landscape of royal hunting gardens from the fifteenth to the eighteenth century in Białowieża Primeval Forest. In: I.D. Rotherham (ed.) *Cultural Severance and the Environment*. Springer, Dordrecht, pp. 191–204. https://doi.org/10.1007/978-94-007-6159-9_13

Samojlik, T., Fedotova, A., Borowik, T. and Kowalczyk, R. (2019a) Historical data on European bison management in Białowieża Primeval Forest can contribute to a better contemporary conservation of the species. *Mammal Research*, 64(4), pp. 543–557. https://doi.org/10.1007/s13364-019-00437-2

Samojlik, T., Fedotova, A., Niechoda, T. and Rotherham, I.D. (2019b) Culturally modified trees or wasted timber: different approaches to marked trees in Poland's Białowieża Forest. *PLoS ONE*, 14(1), e0211025. https://doi.org/10.1371/journal.pone.0211025

Samojlik, T., Fedotova, A., Daszkiewicz, P. and Rotherham, I.D. (2020) *Białowieża Primeval Forest: Nature and culture in the 19th century*. (Environmental History, Vol. 11) Springer, pp. 1–223. https://doi.org/10.1007/978-3-030-33479-6_2

Wesołowski, T., Kujawa, A., Bobiec, A., et al. (2016) Dispute over the future of the Białowieża Forest: myths and facts. A voice in the debate. *www.forestbiology.org* Article 2: 1-19. [accessed 20 February 2020]

Więcko, E. (1984) *Puszcza Białowieska (Białowieża Primeval Forest)*. Państwowe Wydawnictwo Naukowe, Warsaw. (in Polish)

Williams, M. (2006) *Deforesting the Earth: From prehistory to global crisis, an abridgment*. The University of Chicago Press, Chicago, pp. xv–xviii. https://doi.org/10.7208/chicago/9780226899053.001.0001

Wysmułek, J. (1977) Szlaki turystyczne Puszczy Białowieskiej (Tourist trails in Białowieża Primeval Forest). In: J.J. Karpiński (ed.) *Puszcza Białowieska*, 3rd edition. Wiedza Powszechna, Warsaw, pp. 127–134. (in Polish)

CHAPTER 8

Old-Growth Forests in the Eastern Alps: Management and Protection

Elisabeth Johann

Summary

The Alpine region and its ecological function are increasingly endangered by the continuously mounting pressure of human beings. By signing the Framework-Agreement of the Alpine Convention in November 1991, the Alpine states and the EU accepted the fact that the Alpine arc is an indispensable habitat and a last resort for many endangered species of flora and fauna. Provisions were to be made concerning various different ecological aspects, and for their execution the so-called protocols were stipulated. The protocol 'Mountain Forest' (1996) was of great importance. It includes a commitment by the signatory parties to designate 'old-growth forests' of sufficient size and number and to create the necessary apparatus to finance their promotion and compensation arrangements. This chapter investigates how these old woods have survived in some Alpine regions in spite of the pressure of exploitation. It analyses the factors that have contributed to some forest stands staying largely close to their original state in terms of the composition and age pattern of their tree species. In the context of climate change, an open discussion is required to consider the role these old-growth forests play in the upkeep of mountain habitats, the necessary provisions of active forest management and the prevention of logging.

Keywords: forest management, Alpine, nature protection, farm forests, protection forests, naturalness

Introduction

Since the early 1980s, there has been an increased interest in the management and fate of our remaining forests. However, there is a lack of understanding or agreement on what is meant by the various terms that describe the condition of a forest. The term 'old-growth forest' or 'ancient forest' has broadly been used to indicate stands in a developmental phase characterized by high structural heterogeneity. I first became acquainted with this term on the occasion of the conference 'European Woods and Forests' organized by Charles Watkins, Department of Geography, University of Nottingham, in 1997. There I met Oliver Rackham for the first time and, coming from a country with a long tradition in forest management and a forest surface of nearly 50 per cent of the whole territory, I became

Elisabeth Johann, 'Old-Growth Forests in the Eastern Alps: Management and Protection' in: *Countryside History: The Life and Legacy of Oliver Rackham*. Pelagic Publishing (2024). © Elisabeth Johann. DOI: 10.53061/ZOMX6997

somewhat confused. What did he mean by this term, and what was his goal? At this time even the term 'ancient woodland' in German, *Alte Wälder*, was unknown in Austria and only fragmentarily known in Germany among forest historians. Rackham was one of the first to explore woods from a variety of different perspectives. Even though his concept of ancient woodland sites was developed for coppice forests or coppice with standards, his arguments with regard to the natural regeneration of trees could also be applied to high forest stands. Thus, his ideas meshed with those of close-to-nature forest management, which was developed at the same time by the silviculturist Dusan Mlinsek and his European colleagues in the Pro Silva-movement, starting in 1989 in Slovenia. Gradually, foresters also became aware of the value of old and dead trunks, and of the necessity to safeguard the remaining natural forests.

Since then, various attempts have been made to define the term 'old-growth' more accurately. Some definitions that refer to temperate forests have been cited by the Food and Agriculture Organization, United Nations (FAO) (2002). Old-growth forests are those that have originated through natural succession and have not experienced significant human impact over a long period of time. Lund (2002), at the request of Dr Anatoly Shvidenko (International Institute for Applied Systems Analysis), initiated a list of definitions for the International Scientific Conference 'The World's Natural Forests and Their Role in Global Processes', 15–20 August 1999, held in Khabarovsk, Russia.

The Alps are certainly not the largest or highest mountains in the world, but they are characterized by an immense diversity. Their history can be viewed in many different ways: geologically, economically, ethnologically and politically, for example. In each Alpine country, development has not been homogeneous: although the population tripled between 1500 and 1900, from 2.9 to 7.9 million, there has not been continuous growth. From the eighteenth century onwards, there has been a slower population rise in the Alpine region than on the plains. As in the rest of Europe, there is an urbanized region with good economic development and a rural region with considerable structural problems (Norer 2002).

Only relatively recently, with the signing of the Framework-Agreement of the Alpine Convention in November 1991 by the Alpine states, has the Alpine Arc, for the first time in its history, become a region – with attempts being made to create a mutual political and administrative structure. The nine signatories to the agreement (Germany, France, Italy, Liechtenstein, Monaco, Austria, Switzerland, Slovenia and the European Union) accept the fact that the area is an indispensable habitat and a last resort for many endangered species of flora and fauna. The participating parties are therefore aiming to comprehensively protect and conserve the Alps (Mörschel 2004), making provisions concerning population and culture, land-use planning, air pollution control, soil protection, hydrological balance, nature conservation and landscape maintenance, farming, mountain forest, tourism and recreation, traffic, energy and waste management, and for their execution so-called protocols were stipulated. In connection with forests, the protocols 'Alpine Farming' (1994), 'Nature Conservation and Landscape Maintenance' (1994) and 'Mountain Forest' (1996) are of most importance. The last focuses on the protective and economic functions of the mountain forest, and also includes a commitment by the signatory states to designate old-growth forests of sufficient size and number, creating the necessary apparatus to finance their promotion and compensation arrangements (Fig. 8.1).

How is it possible that these old woods have survived in some Alpine regions in spite of the pressure of exploitation? What factors have contributed to some forest stands remaining close to their natural state in terms of the composition and age pattern of tree species? The first is their continued use in rural subsistence agriculture; the second is that they have only been used very carefully if at all, so their protective function has not been affected. These two categories, small-scale farm forest and protected forest, are the subjects of this study.

Study area and sources

The Alpine landscape

The Eastern Alps extend across six European countries: Germany, Italy, Liechtenstein, Austria, Switzerland and Slovenia. They are a labyrinth of valleys and mountain ranges, which rise up to 4,046 m above sea level (Piz Bernina, Switzerland) and show a great number of rock types and microclimates: this is the main reason why there is such an amazing diversity of life. The number of species is remarkable. The World Wildlife Fund (WWF) and the International Union for Conservation of Nature, in their study 'Centers of Plant Variety', named the Alps as one of 234 regions that show the highest plant diversity worldwide. The Alps are also the home of 14 million people across eight countries, with a great number of different cultures and languages. According to the Alpine Convention, they extend over an area of 191,000 sq. km.

Forests with mixed species, dominated by European Beech *Fagus sylvatica*, Silver Fir *Albies alba* and Norway Spruce *Picea abies*, were once widely distributed across the whole Alpine region, and generally occur within an elevation range of 800–1,500 m (Firm, Nagel and Diaci 2005). On average, the highest timberline lies at 2,000–2,200 m, the highest boundary for commercial forest being between 1,700 and 1,800 m. At 2,500 m, it is the highest in the central-Alpine valleys of the Wallis and the Engadin, where the insolation and the relationship between warmed soil and surrounding air are optimal. On the exposed peaks of the pre-Alps, already at 1,800 m, the trees get too little warmth in summer. Because of the close relationship between temperature and tree-growth, an ascent of the timberline correlated with the warming of the climate can be assumed.

Figure 8.1 Old-growth forest with protective and ecological functions, Southern Limestone Alps early 2000s. (M. Johann)

Sources

The idea of an untouched nature, the testimony of a vanished world, has shaped our perception of the Alps throughout history (Guérin 2008). The relationships between human beings and nature, especially in the Alps' extreme environmental conditions, have stimulated much research interest (Koller 1957a; 1970; 1975b; Johann 2004). Jon Mathieu describes the deep bond between human beings and nature in this unique cultural sphere (Mathieu 2015). Great parts of the Alps were used very early, including for intensive grazing. For hundreds of years, the upper timberline in the Alpine region has been influenced by human activity. Remnants of wood emerging from retreating glaciers indicate that in the past the timberline was higher than today (Hagedorn, Rigling and Bebi 2006). Historical agriculture and livestock farming in many parts of the Alps have led to the characteristic cultural landscape that is so important for the preservation of biodiversity. Werner Bätzing (2005) describes how the Alps came into being as a living and economic zone, and as a cultural landscape. In the nineteenth and twentieth centuries, they were changed completely by tourism, industry, growth of towns, traffic and the collapse of Alpine farming. Bätzing provides an overview to show the current situation and problems, finally exploring how modern forms of economy and lifestyle can be connected with the traditional Alpine environmental experience.

Nature conservation has a long tradition in the Alps. Many ecologically important areas are protected: about 20 to 25 per cent of the Alps is protected by law. The differing approaches to conservation are shown by Straubinger (2009). The reports in Frank et al. (2007) give clear insights into the protection of forest biodiversity for individual countries.

Every strategy to maintain biological diversity that has a reasonable chance of success considers the economic, social and political development of the region. With the area's cultural diversity, it is particularly difficult to generalize socio-economic and political trends. Many aspects differ from place to place: while some, such as agriculture and tourism, do not affect the region as a whole, others, such as traffic and climate change, do.

The forests

Origin and development

Silviculture is the second-largest category of land-use in the Alpine region. The forests are for the most part managed, with the maintenance of their protective role in terms of snow and boulder avalanches always having been the priority. Concerning the question of age of forests and human influence on them, especially in connection with nature conservation, the analyses of pollen by Kral (1994) offer good evidence in connection with their most recent history. According to Kral's findings, before the beginning of human influence fir held a position of primacy in montane Alpine mixed forests (700–1,500 m above sea level). In the northern Alps, spruce generally outranked beech, while in the south-east the percentage of beech in the mixture was equal to or higher than that of spruce. The kind of bedrock was also of importance, with limestone promoting the growth of deciduous trees whereas silicate bedrocks supported conifers. In the great majority of commercial forests, the composition of tree species differs greatly from those growing under natural conditions. Owing to long-lasting anthropogenic influences, the natural forest with its great diversity has gradually reduced.

Until about AD 800, the development of forests was mainly controlled by climatic factors and the natural forest was influenced only locally by people. The upper timberline was significantly higher than today, and there was a relatively homogeneous spruce forest with only a little Swiss Pine *Pinus cembra* and larch and even less fir. There is evidence of large pastures and grain cultivation, and also slash-and-burn activities. Following these impacts, the effects of anthropogenic influences came more and more to the fore. In the ninth and tenth centuries, there was a gradual thinning of the spruce forest, as seen by the occurrence of vegetation indicators such as alder and juniper. These point to pastoral farming and grain cultivation, but there were still large forest resources left. In the twelfth and thirteenth centuries, the forest area was heavily reduced owing to the impact of settlers immigrating when the climate was optimum. From 1300 until 1600, intensive farming increased. In terms of tree species, spruce decreased in favour of Swiss Pine, and beech grew up to 1,290 m. Other tree species were birch, alder, willow and larch, and on the areas without forest cover, Green Alder *Alnus viridis* and juniper occurred. The timberline was lowered, in some cases dramatically. From 1600 to 1900, there was a general decrease of the forest area in favour of grazing pastures, and in the lowlands in favour of enclosed fields. People were dependent on the forest as a source of firewood, construction timber, fertilizer and animal food, and as a place of multiple non-timber uses. As a result of permanent pasturing and litter use over hundreds of years, the original forests were in many areas transformed into open, park-like landscapes. Many forest ecosystems have not recovered from that intensive agricultural exploitation. During the nineteenth century, a period of forest regeneration occurred, with spruce becoming dominant. Pastoral farming decreased, and because of changing methods in silviculture, fir disappeared. In higher regions, the percentage of larch increased.

In most of the Alps, extensive exploitation of the forests came to an end in the 1970s and 1980s. Factors included the globalization of the timber market and the scarcity of cheap labour resources. In Alpine countries in the 2020s, woodland covers more than 40 per cent of the landscape (Fig. 8.2). Conifers are the prevailing trees, with the main species including Silver Fir *Abies alba*, Norway Spruce *Picea abies*, European Larch *Larix decidua*, Scots Pine *Pinus sylvestris*, Alpine Pine *P. cembra*, Dwarf Mountain Pine *P. mugo* and Black Pine *P. nigra*. In addition, the region hosts some 40 species of deciduous trees, among them European Beech *Fagus sylvatica*, Common Hazel *Corylus avellana*, European Ash *Fraxinus excelsior*, Sycamore Maple *Acer pseudoplatanus*, and Grey Alder *Alnus incana* and Green Alder (EEA 2002). The distribution of the tree species in the respective Alpine countries can be seen in Figure 8.3.

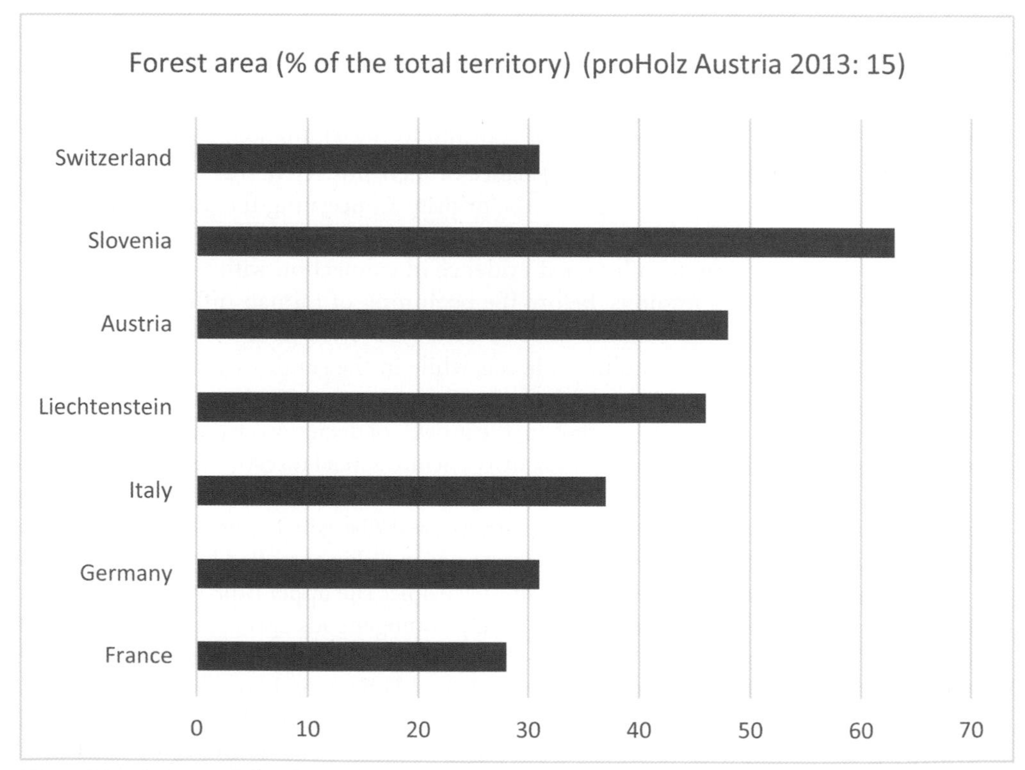

Figure 8.2 Forest area. (proHolz Austria (2013) *Der Wald*. Zuschnitt 51. September 2013, p. 15)

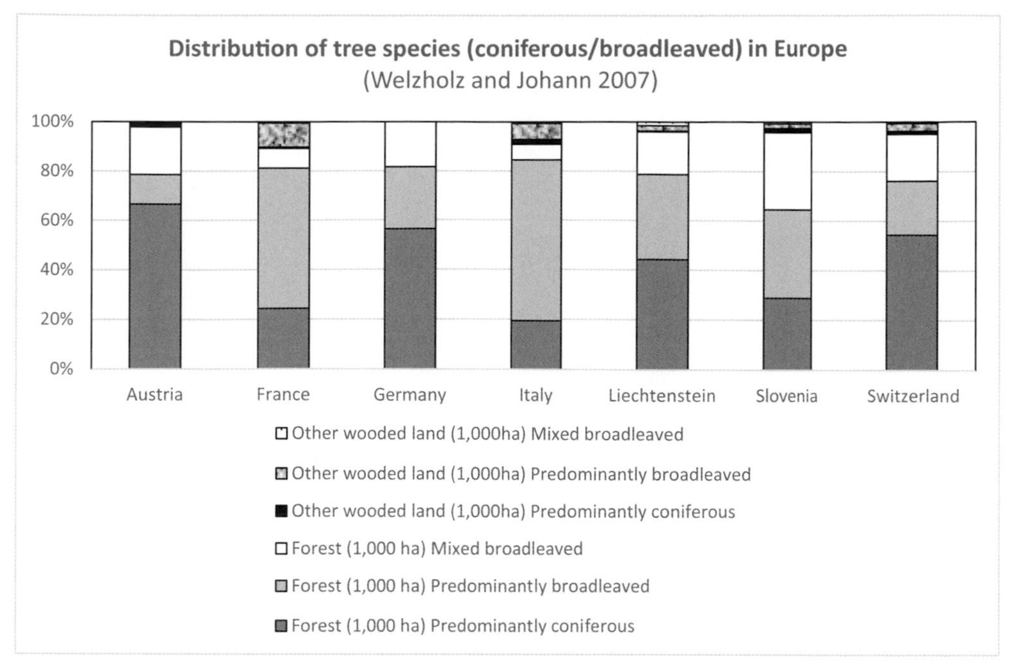

Figure 8.3 Distribution and breakdown of tree species in Europe. (Welzholz and Johann 2007)

Today, the remaining forests showing relatively natural conditions can be found mostly on mountain slopes, where they still cover large areas (Mörschel 2004). However, virgin forests without anthropogenic impacts have survived only in areas that are either absolutely inaccessible or not well suited for agricultural use because of difficult terrain and soil conditions. They are limited to a few hundred hectares (Diaci and Frank 2001). One of these is the Urwald Rothwald in Lower Austria, covering about 500 ha between 900 and 1,600 m above sea level. It is the largest remaining area of virgin forest in Central Europe.

As can be seen in Figure 8.4, a significant part of the mountain forests (on average 60 per cent) were the property of the rural population, either as part of individual farmsteads or as villages' communal property. Management methods focused on the need to supply the multiple demands of a farm, and this is the ultimate reason why these mountain forests remained largely close to nature, include a variety of structures, consist of different species and contain deadwood as well as ancient trees. These features still survive in managed forests as well as in protection forests.

Farm forest utilization and management

In the second half of the nineteenth century, the Alpine regions of Switzerland, Vorarlberg, Tyrol, Salzburg, Carinthia, Styria, Upper Austria and Carniola still had a high portion of their forests, especially in sites that were undeveloped in terms of traffic. Because of the topographical conditions (steepness of the terrain, Karst) and the lack of links to the railway network in these parts, obtaining any income from selling wood was unlikely. Moreover, because extensive privately owned forests were rare, the newest results of forest science (monocultures of coniferous trees, artificial regeneration, etc.) were not adopted. There was also an absence of well-educated forestry staff.

Wood-pastures were used everywhere, being regarded as indispensable for Alpine farmers. Closely connected to livestock farming was the usage of ground and branch litter, which was used in winter as livestock bedding, and was also indispensable as manure. Clearings (slash-and-burn fields), regulated by special laws, were used wherever arable land was scarce. These meant

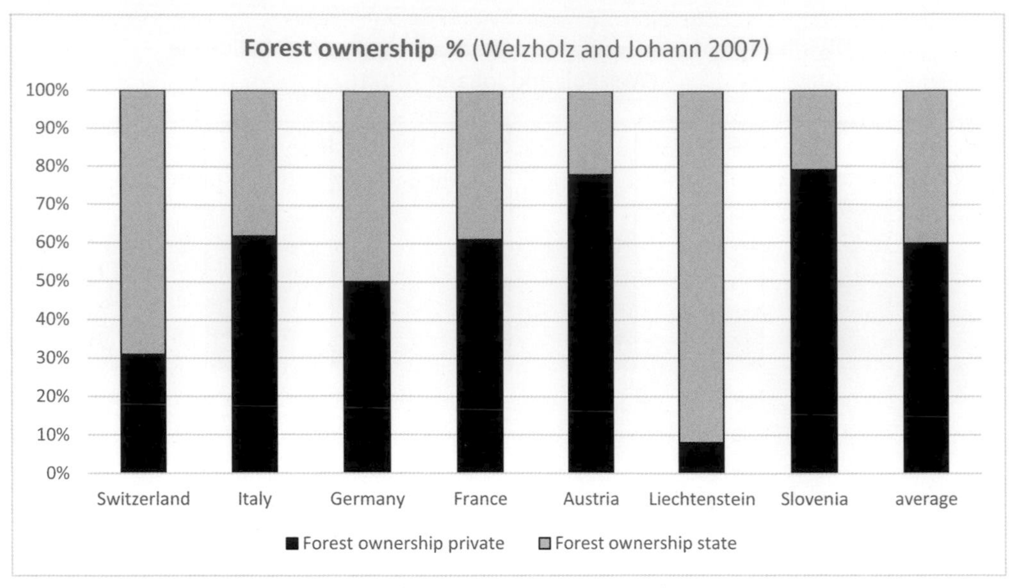

Figure 8.4 Graph of forest ownership. (Welzholz and Johann 2007)

the right to clear forest to cultivate grain for a certain time (three to five years). With the last sowing, tree seeds also had to be sown, so the forest could regrow. These seeds were collected in the vicinity. From the first colonization onwards, the cutting of single trees was the customary harvesting practice; this appeared to be the most viable system economically. A tree could provide thin ash poles for axles or wheels, thick larch trunks for ceiling beams, beech for the production of shafts for tools, spruce for roof trusses, larch shingles for roofing. Because of all these uses, different qualities and species of wood were in demand, and a diversity of rare tree species was maintained. These are now appreciated as noteworthy elements of the landscape (Wessely 1852: 392–395). Because of selective thinning, regeneration of the woods occurred spontaneously and sustainably. The outcome was an uneven forest that included different species of different ages and diameters characterized by spontaneous and permanent regeneration. Structural sustainability was thus guaranteed, even across a very small area (Schuler 1989; Küster 1995; Johann 2021).

Alpine areas with a long tradition of selective harvesting over many years are the French Jura, the Canton of Neuenburg (Swiss Jura), the Emmental, the Black Forest (Badensic section), forests in the Bregenz (Austria) and Allgäu (Germany) districts, the Bachern mountains and the Karawanken-Mountains in Northern Slovenia and the Dinaric high plateau (Karst) in Central Slovenia (former Carniola) (Schütz 2002).

Old-growth forests in protection areas

In the European Union, 23 per cent of the land is designated as protected areas. There are also large-scale protected zones: for instance, in the Alpine region, 24 per cent of the land is overlaid by legally more specific protected areas and further large parts by more generalized protection, for instance glaciers and high Alpine locations.

The Federal Office of Environment Bern (Switzerland) assumes that about 40–60 per cent of the Swiss forests helps to protect against natural hazards. According to forest law in the Bavarian Alps of Germany, about 60 per cent of the forests are designated as protection forest. In Austria, about 31 per cent of the forest area has an official protective function. In the Alpine region, the amount of protection forest is markedly higher – for instance, in the Tyrol, Austria, it is more than 66 per cent. In the Aosta Valley in Italy, about 80 per cent of the woods have a protective function. In spite of differences in data-collection methods, these figures clearly show the importance of protection forests in the Alpine regions. In recent decades, this importance has grown: for instance, regions that were formerly sparsely populated are now inhabited all the year round, and are accessible to tourists (Wehrli et al. 2007).

A direct protective function exists if protection directly depends on the existence of a forest in a particular place where there is a risk of potential damage. A typical example is a forest that protects against avalanches above a settlement. An indirect protective function can be found, for instance, in drainage areas for watercourses, where they can contribute to the reduction of erosion or flooding. There is certainly nothing new about the management of protective forests in the Alpine regions of Europe requiring special measures. In many countries, the multiple ecological functions of forests have been recognized. Already by the end of the Middle Ages, local farmers were striving to maintain forest cover. An example is the protection forests in Austria (1517 *Oberinntal*/Tyrol 1518 *Mölltal*/Carinthia), where the cutting of wood and litter harvesting was prohibited to avoid avalanches and gully erosion on the steep slopes above villages (Johann 2004), or in Switzerland, where the *Andermatt* – banning letter – from 1397 prohibited any utilization of wood or litter (Bürger-Arndt and Welzholz 2005).

An example from the High Tauern National Park (as it is called today), in Austria, is an illustration of management (Johann 2004). Farmers exclusively applied 'single-tree management' to the variably aged forests in order to cover the needs of each farmstead (*Hausholzbedarf*), adapting the size of trees selected for harvesting to required demands (i.e. sawn timber, fuel wood, fence wood, water pipes). Farmers estimated that trees matured for harvesting were between 90 and 150 years old at the time of felling. Each community regarded the forest directly above their farms as a protection forest that had to be carefully managed. They either belonged to communities or to individual farmers. In the seventeenth century, it was generally the case that the farm situated directly below a protection forest was responsible for its exploitation. Avalanche zones were specifically excluded from exploitation. In the eighteenth century, it was considered that the best protection against avalanches was a prohibition on cutting or chopping even a single tree without permission. In 1844, investigations by the authorities concerning the barely accessible mountain forests that grew on the steep slopes of what is today the National Park suggested that they were still well and evenly stocked with good wood. In these forest stands, branches that had been broken off by wind and snow lay unused on the ground in considerable quantity. The forester in charge of taxation felt that this quantity of deadwood inhibited the rejuvenation and growth of young trees. He also found that many forest stands were becoming 'overripe' or 'decaying'. He mentioned that in some stands so much deadwood was lying that it was scarcely possible to work through it. The lack of a market for this wood was seen as the main reason.

In Austria, from the Late Middle Ages until the mid-eighteenth century, forest ordinances and laws at local and regional level addressed the prohibition of clear cutting in protective forests throughout the whole country. The forestry law of 3 December 1852 also implied that on very steep sites and at high altitudes, forests should only be cut in narrow strips or by gradual single-tree felling, and that these areas should immediately be reforested with young trees. High forests growing at the upper timberline could only be harvested by the single-tree felling system (Schindler 1866: 11–12). Analytical researches of pollen in the High Tauern National Park indicate that it is mainly natural forest. At least at the timberline there is a natural mixture of tree species. In this case, in some protected areas, exploitation occurred 200 or more years ago.

Tree species composition, naturalness and structure of Alpine forests

Tree species and their regeneration, the naturalness of tree species distribution, the amount of deadwood and living, strong trees are important in the forest structure and in its perception (Hauk 2007).

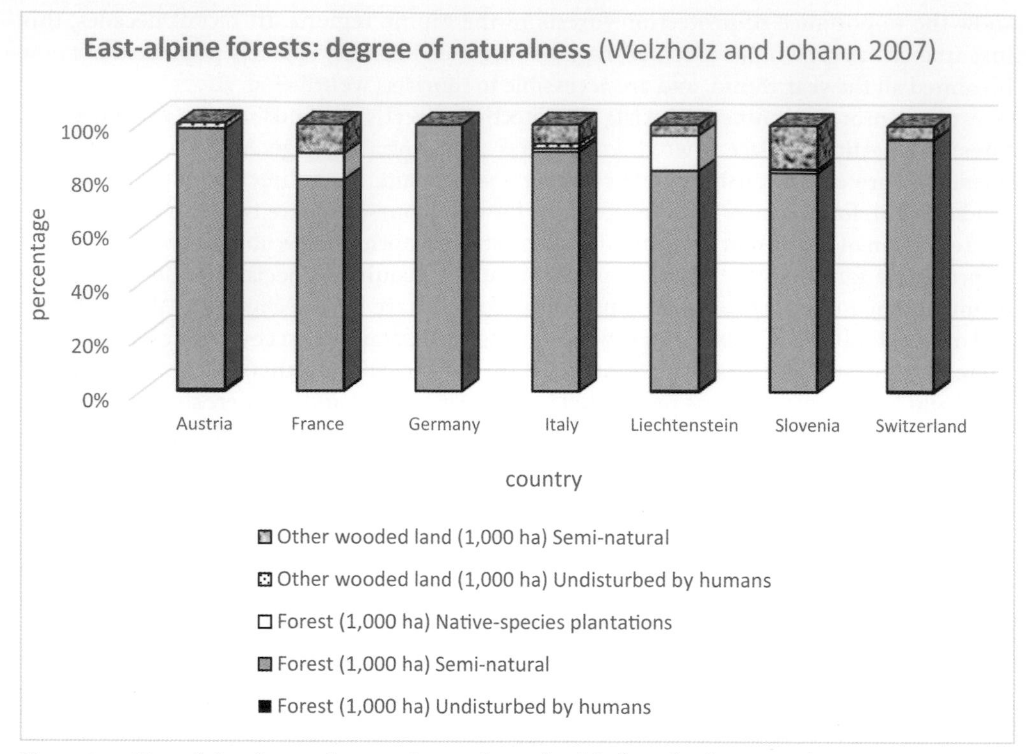

Figure 8.5 East-alpine forests degree of naturalness. (Welzholz and Johann 2007)

Only a small percentage of the forest - generally that at high altitudes - is undisturbed by people, but forests are still in a relatively natural state (ECEDG 2005). It is remarkable that the main part of the European Forest covered by the Alpine Convention is semi-natural; this comprises less-exploited forests that feature a natural blend of tree species with low amounts of ground vegetation (Fig. 8.5).

Owing to their geological history and glaciations, the Alps concentrate a very high number of different habitats and species in a relatively small area. The range in altitude and size of the Alps has allowed the development of a highly diverse flora. Extensive farming practices, transhumance and small-scale forestry have all contributed to a complex mosaic of different cultures and landscapes, and have considerably enhanced the already rich biodiversity of the region. Until recently, such activities have formed the mainstay of mountain economies across Europe. The Alps also act as a barrier between the Mediterranean climate in the south and the more temperate weather in the north, and this is reflected everywhere in the type of habitats and species. To the north, the lower slopes are dominated by deciduous trees, whereas in the south, they are mainly covered in evergreen woods. Coniferous forests prevail at higher altitudes and in the drier inland areas, where rainfall is considerably lower.

The main part of the area belongs to the high-forest climax communities. Further areas can be classified as long-term communities, where because of extreme soil conditions or natural constraints the development of vegetation stops at an earlier stage or is set back periodically by frequent external disturbances (e.g. regular avalanches). These communities are not impacted on by people, as they have always been excluded from exploitation. They are either areas without vegetation, or areas with vegetation but without forests or riparian forests. Some areas are stands of Green Alder in gulleys, shaped by periodic avalanches, and the steep slopes with Mountain Pine *Pinus mugo* (Permanent Secretariat of the Alpine Convention 2016).

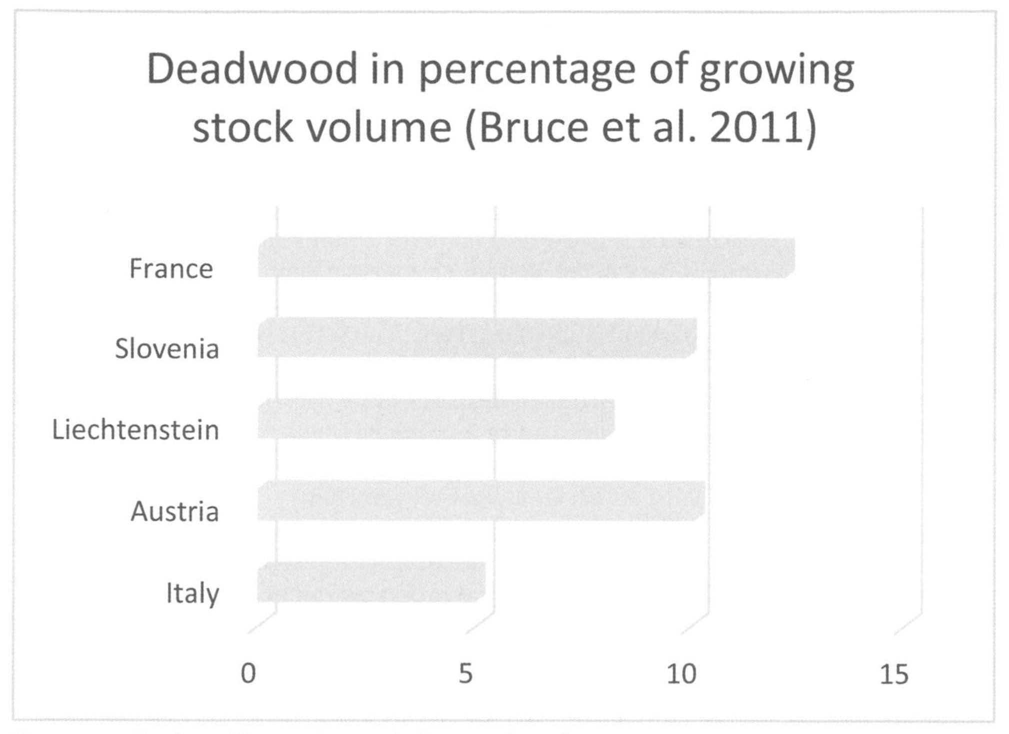

Figure 8.6 Deadwood in growing stock. (Brus et al. 2011)

As confirmed by Paolo Angelini, President of the Permanent Committee 2013–14 of the Alpine Convention (Angelini 2016: 3), Alpine forests overall display a good degree of naturalness; they are mostly mixed and contain an increasing diversity of species. Nowadays, Alpine forests mainly resemble the natural vegetation associated with predominantly native tree species. Regeneration is predominantly natural, and stand structures are moving away from the strictly even-aged and artificial. The predominant management-model mimics natural processes: clear-cuts on large areas are replaced by more structured cuttings and selection procedures. In relatively small areas, particularly on the southern slopes, many different types of tree species can occur (Brus et al. 2011). A survey conducted on the naturalness of Alpine-region hemeroby (a measure of the intensity of the culture of human use and in this sense a measure of the change in natural ecosystems), indicates the degree of human influence on the environment (Grabherr et al. 1998). In this study, each cover type was assigned a value within a range of 1 (unaffected) to 7 (artificial system). Unaffected cover types characterized by no anthropogenic influence include glaciers and virgin rocky areas. Forests were assigned values of around 2–3 points, while values between 3 (pastures) and 5 (permanent cultivation, arable areas) were assigned to agricultural areas (Permanent Secretariat of the Alpine Convention 2016: 18–19). The relatively high percentage of deadwood in the Alpine mountain forests is interesting (Fig. 8.6). Its medial volume lies near the optimal value of 10 per cent of the stock. The existence of very thick trees (veteran trees) of fir or beech points to rather extensive management, and so to a lower anthropogenic disturbance of the ecosystem.

Discussion

Old-growth forests in Europe are situated mainly at high altitudes and have to fulfil diverse and sometimes contradictory protection functions. They are responsible for safeguarding high biodiversity as well as protecting human beings from natural hazards.

Local communities make their living either from small-scale farming or increasingly from tourism (Wehrli et al. 2007).

A key measure to protect biodiversity is the setting up of nature reserves. Around 40 per cent of the Alpine region's surface area currently has some form of protection, with wide-ranging protection measures in force for nature reserves and the core areas of national parks. The list of Natura 2000 sites in the Alpine region (adopted in January 2004) indicates a total of 956 Sites of Community Importance, thus representing 37 per cent of the region. It was in 1911 when a general appeal to launch nature reserves in the Alpine region was made by several institutions. The first national park in the Western Alps was set up in Switzerland in 1914. In the Eastern Alps, an area of 4,000 ha was donated by a private forest owner to the German-Austrian Alpine Association to establish a National Park in 1918. This is the core of the present Hohe Tauern National Park (set up in 1984). In the Kalkalpen National Park, Austria, 26 per cent of the forests were classified as natural forest. The area's age, limited exploitation and difficult accessibility suggest that a large part of it might be labelled as virgin forest. In the core zone of the area, beeches with an age from 400 to 500 years can be found. A highlight is the oldest beech trees of the Alps, which are at least 525 years old.

From the perspective of nature conservation, these forests undisturbed by people are of great value, particularly if they are large and compact forest areas. Such forests can serve as reference areas, where natural ecological processes can be studied, and they can contribute to the development of close-to-nature forest management methods (MCPFE 2007). In the conditions of Central Europe, such forests occur only infrequently and hence are very precious. According to Moravčík et al. (2010), owing to fully functioning natural ecological processes, they should be left to self-regulating development without any human intervention.

In many forests with protective functions, the extreme environmental conditions, regular natural processes and formerly very intensive grazing combine with factors such as the eradication of large predators and an associated increase in herbivores to trigger instability. The breakdown and collapse of forest stands follows. One result of this is a failure of these forests' protective functions. While these stands contain a remarkably high number of old and very old trees, they lack variation in structure and regeneration is limited or absent. The stands are frequently 'gappy', and the amount of standing deadwood is high (Mörschel 2004).

Owing to the considerable amount of protection demanded by local communities, forests are increasingly important. However, they can only fulfil their protection roles if they are appropriately managed and the total prohibition of exploitation (as demanded by nature conservation bodies) is set aside. Sustainable utilization and conservation measures are of equal importance. This affects a range of sectors, including agriculture, forestry, hunting and fishing.

Regarding the protection forests, there is a dilemma. Being unexploited or only used very carefully by means of selective cutting, they can probably be considered to be old-growth forests. However, because they are unmanaged, they can constitute a danger for settlement zones. The need for protective management in individual protected forests is often difficult to estimate. Some recent studies indicate that under certain conditions, forests can provide good protection for a period of time, even without silvicultural measures. However, no general rules can be derived from them because the databases used are too small. As long as there is a lack of data documenting the development of protected forests without silvicultural interventions over longer periods, active protection forest care according to technically supported guidelines seems to be the best option. This approach is better than waiving protection forest care altogether. Examples from many countries in the Alpine region show similar assumptions, problems and perspectives concerning the development of management systems for mountain forests, and especially for protection forests (Wehrli et al. 2007).

Managing the ecosystems of protection forests aims to maintain them in a state in which effective protection is secured, with the ecosystem's functions reflecting its ability to provide human benefits. As the evolution of a dynamic forest ecosystem cannot be stopped, silvicultural measures are required to maintain both ecosystem integrity and protective functions. Ecosystem integrity is defined as the system's capacity to maintain its structure and functions using processes and elements that are characteristic for the ecoregion. This integrity implies that the forest is a stable environment, something that allows the ecosystem to function well in the long term. The main conditions that promote natural evolutionary processes and ecological stability in protection forests are diverse species, sufficient natural regeneration and an optimal forest structure (Dorren et al. 2004). In most Alpine countries, silvicultural measures in protective forests comprise the planting, tending and protecting of local tree species in areas where natural regeneration is absent. Protecting forests and protecting the human habitat are not mutually exclusive. Often, with minimal intervention, the protective function of forest stands can be maintained or even increased, provided interventions are made in a timely manner and in appropriate places (Nigsch 2009).

The principles and priorities of management are stipulated in the Mountain Forest and Soil Protection protocols of the Alpine Convention, which are also valid for mountainside farmed forests. According to these rules, the protective role of the mountain forests has a leading position, but this has to be handled in such a way that the targets of nature conservation and landscape maintenance are also considered (Schwarz et al. 2013). These regulations are largely in accord with the traditional single-tree felling method used by farmers, and with the close-to-nature silviculture recommended by Pro Silva (Pro Silva 2018). They also meet the recommendation by Oliver Rackham (2001) to leave regeneration to nature. Rackham pointed out that woodlands under traditional management have been cut down many times and have recovered without replanting. Management objectives should be achieved with the minimum necessary human intervention, being aimed at accelerating the processes that nature would accomplish by itself, albeit more slowly. This system works with natural populations of trees, ongoing processes and existing structures and using a cognitive approach. Both theory and practice take the forest to be a self-regulating ecosystem and manage it as such.

Conclusion

To sum up, it can be noted that in the countries covered by the Alpine Convention many forests that are managed by farmers and forests that have to deliver protective functions can to a great extent be regarded as 'natural'. They contain elements that can be assigned to old-growth forests. However, the climatic changes that can be expected will not only significantly change the risk of natural hazards in the Alps but will also destabilize the protection forests. This may be by damaging forests through heat stress, pests or storms. 'Risk dialogue' is a term that is being increasingly used in connection with natural hazards. With regard to the protection of old-growth forests and the role they play in the upkeep of mountain habitats, an open discourse is necessary. There need to be discussions about necessary provisions of active forest management and the prevention of logging in extensive forest areas, as demanded by some stakeholders. In the context of climate change and necessary active forest management, the decline of small-scale forest management in farmed forests through migration of the rural population must be considered and the issues addressed.

References

Angelini, P. (2015) *The Statement on the Value of Alpine Forests and the Alpine Convention's Protocol on Mountain Forests in the Framework of the International Forestry Policies beyond 2015*. Permanent Secretariat of the Alpine Convention, Innsbruck.

Bätzing, W. (2005) *Die Alpen. Geschichte und Zukunft einer Europäischen Kulturlandschaft*. C.H. Beck, Munich.

Brus, D.J., Hengeveld, G.M., Walvoort, D.J.J., Goedhart, P.W., Heidema, A.H., Nabuurs, G.J. and Gunia, K. (2011) Statistical mapping of tree species over

Europe. *European Journal of Forest Research*, 131(1), pp. 145–157. https://doi.org/10.1007/s10342-011-0513-5

Bürger-Arndt, R. and Welzholz, J.C. (2005) *The History of Protected Forest Areas in Europe – from Holy Groves to Natura 2000 Sites*. In: E. Johann (ed.) International IUFRO-Conference '*Woodlands – Cultural Heritage*'. News of Forest History Nr. III/(36/37)-1/2005, Vienna (Bundesministerium für Land- und Forstwirtschaft, Umwelt und Wasserwirtschaft), pp. 40–54.

Diaci, J. and Frank, G. (2001) Urwälder in den Alpen: Schützen und Beobachten, Lernen und Nachahmen. In: *CIPRA* (ed.) *Internationale Alpenschutzkommission CIPRA*, 2. Daten Fakten Probleme Lösungsansätze, Bern: Verlag Paul Haupt, Alpenreport, pp. 253–256.

Dorren, L.K.A., Berger, F., Imeson, A., Maier, B. and Rey, F. (2004) Integrity, stability and management of protection forests in the European Alps. *Forest Ecology and Management*, 195(1–2), pp. 165–176. https://doi.org/10.1016/j.foreco.2004.02.057

ECEDG (European Commission Environment Directorate General) (ed.) (2005) *Natura 2000 in the Alpine region*. Office for Official Publications of the European Communities, Luxembourg.

EEA (European Environmental Agency) (2002) *EEA. Europe's Biodiversity – Biogeographical Regions and Seas. The Alpine Region*. EEA report No 1/2002. www.alpconv.org. [accessed 20 August 2018]

FAO (Food and Agriculture Organization) (2002) *Proceedings: Second expert meeting on harmonizing forest-related definitions for use by various stakeholders*. FAO, Rome. http://www.fao.org/docrep/005/Y4171E/Y4171E34.htm. [accessed 22 August 2018]

Firm, D., Nagel, T.A. and Diaci, J. (2009) Disturbance history and dynamics of an old growth mixed species mountain forest in the Slovenian Alps. *Forest Ecology and Management*, 257, pp. 1893–1901. https://doi.org/10.1016/j.foreco.2008.09.034

Frank, G., Parviainen, J., Vandekerkhove, K., Latham, J., Schuck, W. and Little, D. (eds) (2007) *COST Action E27: Protected Forest Areas in Europe – Analysis and Harmonisation (PROFOR): Results, Conclusions and Recommendations*. Federal Research and Training Centre for Forests, Natural Hazards and Landscape, Vienna.

Grabherr, G., Koch, G., Kirchmeier H and Reiter K. (1998) *Hemerobie österreichischer Waldökosysteme*. Austrian Academy of Sciences. Wagner, Innsbruck

Guérin, J.P. (2008) The Alpine Nature – a Product of History. *CIPRA News*. https://www.cipra.org/de/news. [accessed 14 August 2018]

Hagedorn, F., Rigling, A. and Bebi, P. (2006) Wo Bäume nicht mehr wachsen. Die Waldgrenze. In: Eidgenössische Forschungsanstalt für Wald, Schnee und Landschaft (WSL) (ed.) *DIE ALPEN* 9/2006, pp. 52–55.

Hauk, E. (2011) *Biodiversität in Österreichs Wald*. In: BFW (ed.) *Waldinventur 2007/2009*. BFW-Praxisinformation Nr. 24, Vienna.

Johann, E. (2004) *Wald und Mensch. Die Nationalparkregion Hohe Tauern (Kärnten)*. Verlag des Kärntner Landesarchivs, Klagenfurt.

Johann, E., 2021. Wälder in den Alpen. Nutzung und Schutz. (Forests in the Alps. Utilizatio and Protection). In Karsten Berr und Corinna Jenal (Hg). 2022. *Wald in der Vielfalt möglicher Perspektiven. Von der Pluralität lebenswichtigere Bezüge und wissenschaftlicher Thematisierungen*. Springer VS: Wiesbaden, pp. 201–221

Koller, E. (1970) *Forstgeschichte des Salzkammergutes*. Österr Agrarverlag, Vienna.

Koller, E. (1975a) *Forstgeschichte des Landes Salzburg*. Salzburger Druckerei, Salzburg.

Koller, E. (1975b) *Forstgeschichte Oberösterreichs*. Landesverlag, Linz Oberösterr.

Kral, F. (1994) Waldgeschichte. In: *Österreichs Wald. Vom Urwald zur Waldwirtschaft*. Österr. Forstverein, Vienna.

Küster, H.G. *Geschichte der Landschaft in Mitteleuropa. Von der Eiszeit bis zur Gegenwart*. Verlag C. H. Beck, München.

Lund, H.G. (2002) Definitions of old growth, pristine, climax, ancient forests, and similar terms. (Online publication), Forest Information Services, Manassas, VA. http://old.grida.no/geo/GEO/Geo-2-408.htm. [accessed 18 August 2018]

Mathieu, J. (2015) *Die Alpen: Raum - Kultur – Geschichte*. Philipp Reclam jun. GmbH and Co.KG, Stuttgart.

MCPFE Liaison Unit Warsaw, UNECE and FAO (2007) *State of Europe´s Forests 2007*. The MCPFE Report on Sustainable Forest Management in Europe. Ministerial Conference on the Protection of Forests in Europe Liaison Unit, Warsaw.

Moravčík M., Merganič, J., Sarvasova, Z. and Schwarz, M. (2010) Forest Naturalness: Criterion for Decision Support in Designation and Management of Protected Forest Areas. *Environmental Management*, 46(6), pp. 908–919. https://doi.org/10.1007/s00267-010-9506-2

Mörschel, F. (2004) *Die Alpen: das einzigartige Naturerbe. Eine gemeinsame Vision für die Erhaltung ihrer biologischen Vielfalt*. WWF Deutschland, Frankfurt am Main.

Nigsch N. (2009) *Der Schutzwald in Liechtenstein. Konzept zur Erhaltung und Verbesserung der Schutzleistung des Waldes*. Amt für Wald, Natur und Landschaft Fürstentum Liechtenstein, Vaduz.

Norer, R. (2002) *Die Alpenkonvention. Völkerrechtliches Vertragswerk für den Alpenraum*. Institut für Wirtschaft, Politik und Recht der Universität für Bodenkultur Wien (ed.) Diskussionspapier Nr. 93-R-02 März 2002. Universität für Bodenkultur, Vienna.

Permanent Secretariat of the Alpine Convention (2016) *The Statement on the Value of Alpine Forests and the Alpine Convention's Protocol on Mountain Forests in the framework of the international forestry policies beyond 2015*. www.alpconv.org. [accessed 18 August 2018]

proHolz Austria (2013) Der Wald. *Zuschnitt* 51. September 2013, p. 15.

Pro Silva Europe (2018) Integrated forest management for resilience and sustainability across 25 countries. https://prosilvaeurope.wordpress.com/about-close-to-nature-forestry. [accessed 17 August 2018]

Rackham, O. (2001) *Trees and Woodland in the British Landscape*. Orion, London.

Schindler, K. (ed.) (1866) *Die Forst und Jagdgesetze der Österreichischen Monarchie*. Braumüller, Vienna.

Schuler, A. (1989) Changes of Forest Area and Forestry in the Swiss Prealps. In: Salbitano, F. (ed.), *Human Influence on Forest Ecosystems Development in Europe*. Proceedings of a workshop held in Trento, Italy, 26–29 September 1988. Pitagora, Bologna, pp. 121–127.

Schütz, J.-Ph. (2002) *Die Plenterung und ihre unterschiedlichen Formen*. Skript zu Vorlesung Waldbau II und Waldbau IV. Professur Waldbau ETH Zentrum, Zürich.

Schwarz, C., Marzelli, S., Lintzmeyer, F., Witty, S., Cuypers, S. and Brendt, I. (2011) *CIPRA. Leben mit alpinen Naturgefahren*. Ergebnisse aus dem Alpenraumprogramm der Europäischen Territorialen Zusammenarbeit 2007–2013. CIPRA Deutschland and ifuplan, Munich.

Straubinger, J. (2009) *Die Geburt einer Landschaft. Sehnsucht Natur*, Bd. 1. Books on Demand, Norderstedt.

Wehrli, A., Brang, P., Maier, B., Duc, P., Binder, F., Lingua, E., Ziegner, K., Kleemayr, K. and Dorren, L. (2007) Schutzwaldmanagement in den Alpen – eine Übersicht. *Schweiz. Z. Forstwes.*, 158(6), pp. 142–156. https://doi.org/10.3188/szf.2007.0142

Welzholz, J.C. and Johann, E. (2007) History of Protected Forest Areas in Europe. In: G. Frank, J. Parviainen, K. Vandekerhove, J. Latham, A. Schluck and D. Little (eds) *COST Action E 27 Protected Forest Areas in Europe. Analysis and Harmonisation (PROFOR) Results, Conclusions and Recommendations.* Vienna Federal Research and Training Centre for Forests, Natural Hazards and Landscape, Vienna, pp. 17–40.

Wessely, J. (1852) *Die österreichischen Alpenländer und ihre Forste*. Braumüller, Vienna.

CHAPTER 9

Biocultural Landscapes of Europe: A Journey with Oliver Rackham

Gloria Pungetti

Summary

Biocultural landscapes are embedded with high ecological and cultural value, and reveal the link between nature and culture. The understanding of this link is fundamental to studying the character of European landscapes, which endow the biocultural essence of our continent with its rich natural and cultural heritage. Oliver Rackham facilitated the understanding of this link in so many ways, especially through inspiring and helping to direct a number of pan-European initiatives on aspects of landscape history. This chapter provides a personal account of this journey, with Oliver Rackham at the helm.

Keywords: biocultural landscapes, European Landscape Convention, Florence Conference 2014, EUCALAND Programme, ECSLAND Programme

Introduction: the biocultural essence

Biocultural landscapes have been the subject of numerous global biocultural diversity initiatives. Among these is the UNESCO–Secretariat of the Convention on Biological Diversity (SCBD) Joint Programme on the Links between Biological and Cultural Diversity, which supported synergies between initiatives dealing with biological and cultural diversity around the world. Recognition by COP10, the Tenth Meeting of the Parties of the Convention on Biological Diversity (CBD), held in October 2010 in Nagoya, Japan, led to its dissemination in several places, including the Meeting held in Florence in April 2014, where European landscapes played a relevant role (SCBD-UNESCO 2014).

Florence is also where the European Landscape Convention (ELC) was adopted in October 2000 to promote the protection, management and planning of European landscapes, as well as the organization of European landscape cooperation. The ELC came into force on 1 March 2004 thanks to the Council of Europe and has been open for signature by pan-European states. It is not only the first international treaty concerned with European landscapes, but also an innovative initiative since it deals with all types of landscapes, whether outstanding or ordinary, rural or urban, natural or industrial, mountainous or coastal. A further novelty is that it puts people at its heart, stating that landscape means an area, as perceived by people,

Gloria Pungetti, 'Biocultural Landscapes of Europe: A Journey with Oliver Rackham' in: *Countryside History: The Life and Legacy of Oliver Rackham*. Pelagic Publishing (2024). © Gloria Pungetti. DOI: 10.53061/WDZI5893

whose character is the result of the action and interaction of natural and human factors (Council of Europe 2000). The biocultural character of landscape was thus evident from the moment the Convention came into force.

From a research point of view, the topic of biocultural landscape has been considered by several groups. Firstly, at European level a trio of projects was carried out since 2007 with the European Culture Expressed in Landscape (EUCEL) Initiative, linking the natural and cultural heritage of European landscapes, and demonstrating how these landscapes are perceived as common heritage that carries environmental and social values (Pungetti and Kruse 2010; Pungetti 2017).

Elsewhere, Asian scholars have developed biocultural concepts academically, with pioneering projects being discussed at conferences with scientists from other continents. The results on the diversity, functions and values of biocultural landscapes worldwide have been published with the aim of exploring biocultural diversity within the context of landscape ecology (Hong, Bogaert and Min 2014). Through different case studies, the biocultural understanding of Asian landscapes has been outlined, and methodologies for studying the links between their natural and cultural diversity have been proposed.

Thirdly, at the 2014 Florence Conference, 'Linking Biological and Cultural Diversity in Europe', arranged by the UNESCO-SCBD Joint Programme, it was stressed that European rural territory is predominantly a biocultural, multifunctional landscape that provides a crucial space for the integration of biological and cultural diversity (Agnoletti and Rotherham 2015; Agnoletti and Emanueli 2016).

The European context of biocultural landscapes is addressed in this chapter, which outlines the influence that Oliver Rackham had on the field. From now on, he will be referred to as OR.

Encounter: the inspiration

I met OR in Cambridge in 1991; he was introduced to me by my PhD Supervisor Alfred Thomas (Dick) Grove and his wife Jean, working at the Department of Geography of the University of Cambridge. Several field trips with them in Sardinia and later in Crete, enriched by the company of Jennifer Moody, revealed to me the relevance that landscape history would have in my studies. Field work with these outstanding scholars guided the direction of my PhD work at the Department of Geography, and consequently my first international articles (Pungetti 1993 1995).

At that time I was influenced, as were numerous others regardless of whether or not they were academics, by *The History of the Countryside* (Rackham 1986). I was impressed by the author's in-depth description of the countryside, and by his enquiry into and scholarly knowledge of diverse natural and cultural aspects of the British landscape, parallel to the story I was working on – in which the abiotic, biotic and cultural aspects of the Sardinian countryside contributed to delineate the landscape of that island. This work allowed me to see more clearly the picture, and finally to grasp the essence, of what I had already addressed in my first-degree thesis in Architecture, in 1986. With a multidisciplinary approach to territorial planning, my earliest research attempted to understand the value of historical analysis in environmental and landscape planning. That work was focused on the Etruscan sites of Emilia in Italy (Pungetti 1986).

Once in Cambridge, a few years later, I found myself fully immersed for the first time in reading *The History of the Countryside* for my PhD. This fascinating volume was for me like an orchestra, with the landscape elements – fields, hedges, trees, woods, ponds, marshes, paths, houses, animals and people – directed by the author in a harmonic composition that linked nature and culture, making everything I had studied before more comprehensible and enjoyable. My undergraduate research effort was beginning a new journey, from landscape history to the study of biocultural landscapes.

OR's work was truly inspirational for my PhD, which was connected to the MEDALUS Project funded by the European Commission Action Programme for the Environment of DG Research. The results were published in two volumes (Pungetti 1996; Makhzoumi and Pungetti 1999), while the MEDALUS research group of the University Cambridge, among the other partners involved, was coordinated by Dick Grove in cooperation with OR. The results of their pioneering research were published in *The Nature of Mediterranean Europe* (Grove and Rackham 2001) as profound reflections on all those years of work in the Mediterranean, encompassing many field trips, meetings with European scientists and interesting discussions on the topic.

OR and Dick Grove clearly illustrated in that volume the outcomes of our research, showing how a beautiful and fertile land, seen by most as a paradise, has changed over time to become degraded and desertified. More importantly, they pointed out that this has been caused by careless human management arising from later civilizations' lack of knowledge. The same framework was outlined in my PhD thesis, in 1996, where novel ideas were introduced more easily by working with these exceptional scholars.

Knowledge: the maestro

Being born in Italy and studying in Florence for my first degree, I have grown up with a strong sense of the 'Bottega–Maestro–Alumnus' system. The bottega (or workshop) was the European landscape, the Maestro was OR and I was the Alumna. Yet it was only after I had spent a decade in Cambridge that OR became my Maestro, and our discussion became deeper and more inspirational.

OR taught me many things, one of these being the importance of people in a landscape. As a consequence, in around 2005 we investigated the state of the art on this topic within the Cambridge environment, both in academia and in international non-governmental organizations on nature conservation and cultural heritage. It became clear that there was a gap in the research, and a request to develop more studies on people and landscape was made by the Cambridge scientific community. The need to research more deeply the cultural and spiritual values of landscape and nature, with particular attention being paid to local communities and their environments, was emphasized. After long discussions with several stakeholders in Cambridge, OR supported the development of a research group that could tackle the matter. In 2007, we finally set up the Cambridge Centre for Landscape and People (CCLP), with OR as the honorary director – guiding my role as chair and Jala Makhzoumi as project director.

CCLP is an interdisciplinary research centre of the University of Cambridge that studies and promotes dialogue on landscape and the people that live in it and shape it. Aiming to integrate the cultural and spiritual values of land and local communities into landscape and nature conservation, and socio-economic needs into sustainable development, it supports biocultural diversity and landscape.

OR was a pillar in our holistic research, facilitating the opening of new research avenues and pushing us to put forward original ideas. Together we gave lectures and courses on landscape and people, participated in landscape conferences, and organized biocultural landscape workshops, of which he was the master of ceremonies. In publishing the results of these events, OR passed on to us one of his incredible gifts: the ability to write books that are accessible to everyone. This was just one of his special insights to which I will always aspire.

Research and projects: collaboration

The deeper search for the link between landscape and people that was evoked by OR led me to develop the concept of culture expressed in landscape. This is the core of the EUCEL trilogy, which I coordinated with OR's involvement. EUCEL is a CCLP initiative made up of three programmes: European Culture expressed in Agricultural Landscapes (EUCALAND),

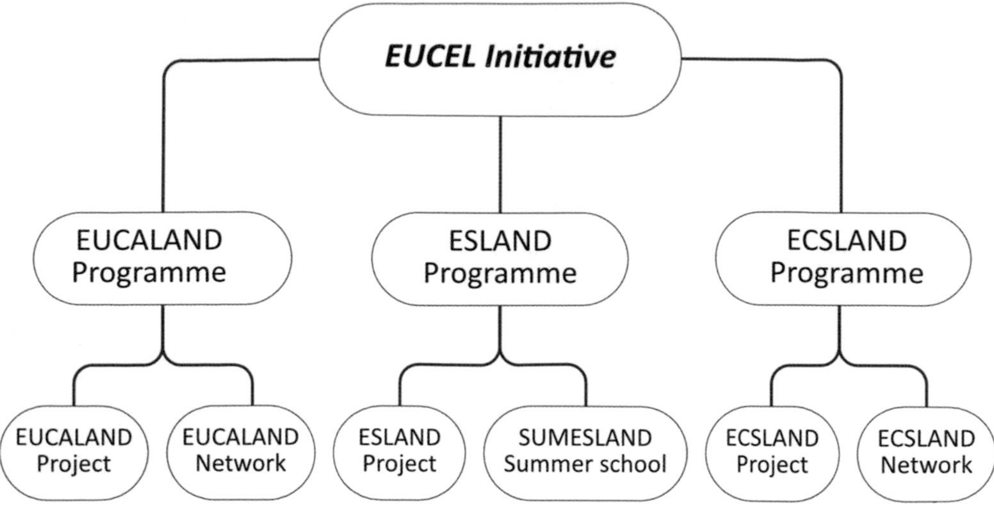

Figure 9.1 The EUCEL initiative by CCLP, University of Cambridge, UK. (G. Pungetti)

European Culture expressed in Island Landscapes (ESLAND) and European Culture expressed in Sacred Landscapes (ECSLAND) (Fig. 9.1).

The EUCALAND programme is divided into a project and a network. The project, developed in 2007–9 and co-funded by the Culture Programme of the European Commission, is a multi-disciplinary research project that is being carried out in 13 countries, with 15 partners and the involvement of 40 organizations (Fig. 9.2). It considers European agricultural landscapes as expressions of a common cultural heritage and shows the pressure that this is now under, owing to new developments and agricultural practices. Guidelines for future landscape planning and policies have been produced, together with a glossary, a conference, a travelling exhibition and a book (Pungetti and Kruse 2010). The EUCALAND network was developed to link the numerous partners. OR was extremely helpful in the production of the EUCALAND book, to which he also contributed (Rackham 2010), and was master of ceremonies for the EUCALAND conference that was organized at Corpus Christi College, Cambridge, in September 2009.

The ESLAND programme comprises a project and a summer school. Developed in 2011–13 and also co-funded by the Culture Programme of the European Commission, has been carried out in six countries, with eight partners and the involvement of 33 organizations (Fig. 9.3). Set up to improve the consideration of cultural heritage in European island landscapes, the project aims to promote an interdisciplinary approach for island landscape history, classification, identity and scenarios. ESLAND has been successful in raising awareness about the unique identity and values that European islands have for both islanders and mainland communities (see Pungetti 2013, 2017). OR's contribution greatly enriched the project, especially the book in which his last three chapters (Rackham 2017a, 2017b, 2017c) are published. He was again master of ceremonies at the ESLAND conference, which was held at the University of Sassari in October 2012, and at the SUMESLAND Summer School, which was organized in Sardinia in the summer of 2013.

The ECSLAND programme, comprising a project and a network, completes the EUCEL trilogy. It was also co-funded by the Culture Programme of the European Commission, being developed in 2013–15 (Fig. 9.4). The project, carried out in five countries, involved 12 partners and 20 organizations. With European landscape, cultural heritage and identity as the central issues of the EUCEL initiative, the ECSLAND project has been set out to explore the links between sacred landscape and cultural heritage, as well as to identify the unique values they have for European citizens.

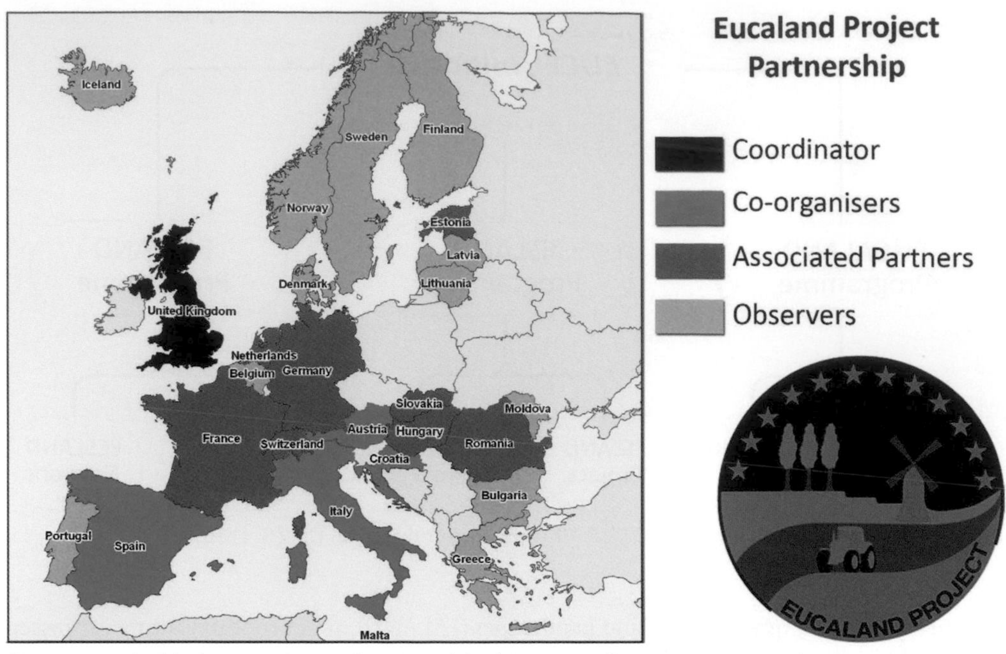

Figure 9.2 EUCALAND project and partnership. (G. Pungetti)

Figure 9.3 The ECSLAND context. (G. Pungetti)

Figure 9.4 The ECSLAND project. (G. Pungetti)

The conference organized at the University of Sassari in April 2015 was a tribute to OR, esteemed scholar and friend of the ECSLAND community (Fig. 9.5).

Research and writing: the books

CCLP was the beginning of the biocultural direction of our research, while EUCEL provided the possibility of applying the biocultural diversity concept to diverse types of European landscapes. The different research groups for the three projects, formed by OR, Jala Makhzoumi, me and the researchers involved in each project, worked constantly on the link between culture and nature, developing research material for the biocultural landscape approach.

We visited many European landscape types and met numerous people involved in the three projects. Although there were on occasion difficulties owing to different backgrounds, cultures and ideas, most of the time we enjoyed these differences, learning from the multicultural and interdisciplinary dialogue that the projects allowed. OR in particular enjoyed the different

Figure 9.5 Maguellonne Déjeant-Pons (ELC), Jala Makhzoumi (CCLP) and Gloria Pungetti (CCLP) at the ECSLAND conference in Sassari, paying a tribute to Oliver Rackham, 2015. (G. Pungetti)

cultures and the ecological and botanic diversity of the many landscapes that he visited in his red tones – socks, shirt, rucksack, notebooks and more.

During EUCALAND, the six research groups, made up by academics, practitioners and governmental officers from 13 European countries, worked on landscape description, history, classification, assessment, policy and planning (Fig. 9.6).

After exploring how agricultural landscapes are to be understood, not only in terms of farming and nature but also as a common heritage with environmental, social and cultural values, the characteristic components of agricultural landscapes were identified, their psychological and social influences on human well-being explored and possible developments for the future suggested. From this, a preliminary classification of agricultural landscapes was produced, intended to be debated across Europe, and moving beyond existing landscape classifications by considering landscape types viewed as products of history. At the same time, an attempt was made to compare the evolution of agricultural landscapes in different countries, and to consider the effect of similarities and differences among human societies. OR was particularly involved in this phase.

OR's contribution throughout the project was influential in terms of the concepts developed, and he also wrote a seminal paper, 'Landscape history of southern Europe' (Rackham 2010). In the introduction, the research honesty of a genuine scholar is well expressed: 'Southern Europe is complex: a much longer account than this could only summarize an imperfectly-known reality' (Rackham 2010: 95). This follows a narration of landscape changes that is written in a simple and succinct, yet comprehensive and accurate style: 'The story begins with mountain-building and climate change [...] To people of 9000 years ago the present mountains would generally be recognizable, but not the coast, and river deltas have utterly changed; most wild plants would be the same, but not the present animals' (Rackham 2010: 95).

Figure 9.6 Oliver Rackham with the EUCALAND research groups at Corpus Christi College, Cambridge, 2009. (G. Pungetti)

The story is completed by a description of past south European landscapes: from a time before people to prehistoric times; from Classical, Hellenistic and Roman times to Byzantine, medieval and post-medieval periods; and finally to modern times.

Later, with the ESLAND programme, we intended to lay down some principles for the understanding and appreciation of European islandscapes (Pungetti 2017). The initial goal was to improve consideration of their cultural heritage, describing them in terms of the past, present and future. Ultimately, it was intended to contribute to the implementation of European and global strategies, such as the European Landscape Convention (Council of Europe) and the World Heritage Convention (UNESCO). Therefore, six research groups on island landscape history, characterization, identity, scenario, toolkit and dissemination have worked together within the project, while linking with other experts outside it, in order to offer a methodology for researching European island landscapes in terms of nature and culture. The ESLAND research has tackled concepts and the character of European island landscapes, leading to publications, mapping and e-tools for their future development. We have worked on awareness-raising through publications, travelling exhibitions, presentations, websites and social media, all of which emphasize that these landscapes should be more culturally and sustainably oriented than at present.

ESLAND has studied Venice and Sardinia in Italy, Cyprus, Korčula in Croatia, the Isles of Scilly in the UK, Saaremaa in Estonia and Bornholm in Denmark. OR visited them all except the last. Thanks to his experience in island landscape changes (Rackham and Moody 1996; Grove and Rackham 2001), OR had a special project task: the coordination of the research group that focused on the landscape history of European islands. He met the many researchers involved, and we all remember him during our visits to the islands involved in the project, walking along the side of the road with his red notebook (Fig. 9.7), observing and recording everything from soil to sky, from trees to flowers, from fields to houses, from landscapes to people.

Figure 9.7 Oliver Rackham during an ESLAND project field trip in Sardinia, 2013. (G. Pungetti)

Coordinating the entire project, I had the privilege to meet with OR the many scholars who gathered together in Sassari at the ESLAND conference. The results of the ESLAND research community were published in *Island Landscapes* (Pungetti 2017), in which the last chapters on the topic by OR can be found.

In his first chapter, 'Island landscape history: the Isles of Scilly, UK' (Rackham 2017a), OR lays down the appropriate methodology for studying the history of island landscapes in Europe. He points out how the knowledge of island history is essential to understand islandscapes. His second chapter, 'Saaremaa: tackling landscape history in Estonia' (Rackham 2017b), employs this methodology to outline the main landscape changes that have taken place over the last centuries. In Saaremaa in particular, the landscape was radically changed and damaged under the Soviet regime, and new threats are now arising from the globalization of trees and plant diseases. In his third chapter, 'Landscape history of Cyprus: a preliminary account' (Rackham 2017c), OR points out that the current political division of the island seems to have had little effect on some landscape areas. In other areas, the land has been abandoned for decades and now reveals a separate type of landscape. These insights were also conveyed to the new researchers at the SUMESLAND Summer School of 2013 (Fig. 9.8), who enjoyed the lectures and stories that OR delivered.

Figure 9.8 Oliver Rackham with the teachers and students of the SUMESLAND Summer School in Sardinia, 2013. (G. Pungetti)

Just before OR died, we were working on the trilogy's third project: ECSLAND. The results of the discussion carried out during the years of the project's proposal development have been reported in some of our previous work (e.g. Pungetti, Hughes and Rackham 2012; Rackham 2015), but unfortunately, we have not been able to find any documents by OR on ECSLAND.

Research and teaching: education

Sardinia is a perfect location for the study of biocultural landscapes. Just before OR's death, after some time spent with him and Jala Makhzoumi in Sardinia and Cambridge, the three of us thought it was time to set up a new research organization: the Center for Biocultural Landscape and Seascape (CBLS). What better place than Sardinia, so beloved by OR? Set up first in Cambridge in 2014, with OR and Jala, the centre was officially instituted in 2018 by the University of Sassari (UNISS), and it has been dedicated to the memory of 'Professor Oliver Rackham, world leading expert in landscape'.

Within the CBLS framework, we are developing new courses and initiatives. We have been able to set up the first university course ever in Biocultural Landscape, which I teach to students from both sciences and humanities courses. This aims to provide a novel methodology for the reading of landscapes and their biocultural diversity, and to offer an understanding of the interrelationship between environmental and anthropic processes on land and sea.

At CBLS, we are also organizing workshops on biocultural landscapes, seascapes and islandscapes, evaluating their tangible and intangible values, and providing practical techniques for the appreciation of the links between local communities and landscape types. Furthermore, we are carrying out participatory seminars with students, local citizens, stakeholders and administrations to promote the understanding of biocultural diversity in landscape, culture and nature, and to allow strategic planning with the involvement of local people.

The development of all this has been made possible not only by my previous studies in particularly stimulating environments such as Florence and Cambridge, and the collaboration with my academic sister Jala Makhzoumi, but also and above all by the influence that OR has had on my research. My involvement until 1991 in the International Association of Landscape Ecology (IALE) has also been vital. My participation in an IALE conference in 1992 was thanks to Dick Grove and OR, who introduced me to the IALE community, and induced me to write my first article in *Landscape and Urban Planning* (Pungetti 1995). From that time, IALE has become an academic family, one where I have met many of my research companions.

Two decades later, in August 2011, the organizers of the IALE 8th World Congress in Beijing invited me to deliver a keynote speech: 'Biocultural diversity for sustainable cultural, sacred and ecological landscapes' (Pungetti 2013). OR joined me in Beijing, where we set up with Professor Sun-Kee Hong the worldwide IALE Working Group on Biocultural Landscape, to improve recognition and understanding of the cultural and spiritual values of species, sites, landscapes, seascapes and their ecosystems. This working group aims to support the link between cultural and ecological diversity, and to demonstrate ways in which biocultural landscape and seascape can contribute to the conservation of both nature and culture.

After the Congress, OR and I spent a marvellous week together with other IALE members, travelling in China across the biocultural landscapes between the Great Wall and the Yunnan Terraces. That was for OR the most enjoyable part of the Chinese journey.

Later, we chaired a few IALE symposia, where OR delivered astonishing talks such as 'Sacred groves in a Japanese satoyama cultural landscape: a scenario for conservation' in Beijing 2011 (Fukamachi, Rackham and Oku 2011), 'Significance of sacred trees and groves' (Rackham 2013) and 'Sacred groves and cultural landscape in Europe' (Rackham and Pungetti 2013) in Manchester in 2013.

Conclusions: the legacy

OR was a genuine maestro, a distinguished figure in any sphere. For me, this was twofold: he was an amazing teacher both in academia and in my personal life. He was able to convey complex thoughts in an accessible way, and to pass on to future scholars fundamental and traditional knowledge emerging from his research and experience. The ideas, debates and discoveries I shared with him were the additional ingredients for our constant Italian and English dinners, always enriched by his funny and rich stories of characters and places, set in our beloved English and Mediterranean landscapes. Whether alone or with our project companions, I always enjoyed his company and wisdom, now so missed by all but still enriching our academic lives.

We all remember enjoyable excursions with OR in many British woods and Mediterranean islands, his informative descriptions of places, archaeological sites and architectures, his language skills including Italian and Latin, his fun stories, red clothes and eccentric personality. Walking with him through the fields, absorbing his energy, passion and wisdom, brought a sense of clarity to the landscapes we were studying, even to the point that when we were dealing with ECSLAND some people said he was a 'sacred person' in our sacred landscapes, or 'a veteran tree, thus close to immortality', a 'landscape star' and 'the roots of an ancient oak living in us'. OR was able to enrich any field trip with exceptional passion, inspiration and wisdom.

For young researchers, it was reassuring to work with OR in our projects; we could trust his vast knowledge and his remarkable personality. He was a myth, and he is already a legend. We like to remember him as a gentleman in red, a renaissance man, an amazing fellow, a genius who never forgot the little things that make life meaningful. His legacy lives on in those who were inspired, and will be inspired, by him to appreciate the countryside, to understand people's impact on landscape and the essence of biocultural landscapes, and ultimately to learn the language of his beloved trees.

Wherever on the long and winding road you walked with your large hat, red socks and leather sandals, dearest Maestro, may your kindness and wisdom stand firm in our soul like the roots of an ancient oak.

References

Agnoletti, M. and Emanueli, F. (eds) (2016) *Biocultural Diversity in Europe*. Springer International Publishing, Switzerland. https://doi.org/10.1007/978-3-319-26315-1

Agnoletti, M. and Rotherham, I.D. (2015) Landscape and biocultural diversity. *Biodiversity and Conservation*, 24(13), pp. 3155–3165. https://doi.org/10.1007/s10531-015-1003-8

Council of Europe (2000) *European Landscape Convention*. Council of Europe, Strasbourg and Yale University Press, New Haven, CT.

Fukamachi, K., Rackham, O. and Oku, O. (2011) Sacred groves in a Japanese satoyama cultural landscape: a scenario for conservation. In: J. Bruce and F. Bojie (eds) *Landscape Ecology for Sustainable Environment and Culture, Proceedings of the 8th World Congress of IALE, Beijing, China, 18–23 August 2011*. IALE, Beijing, p. 150.

Grove, A.T. and Rackham, O. (2001) *The Nature of Mediterranean Europe: An ecological history*. Yale University Press, New Haven, CT.

Hong, S.K., Bogaert, J. and Min, Q. (eds) (2014) *Biocultural Landscapes: Diversity, functions and values*. Springer International Publishing, Switzerland. https://doi.org/10.1007/978-94-017-8941-7

Makhzoumi, J. and Pungetti, G. (1999) *Ecological Landscape Design and Planning: The Mediterranean context*. Spon-Routledge, London.

Pungetti, G. (1986) *Valore dell'analisi storica nella pianificazione territoriale: Il caso degli insediamenti Etruschi in Emilia*. Unpublished first degree thesis, Faculty of Architecture, University of Florence.

Pungetti, G. (1993) Landscape and environmental planning in the Po delta. *Landscape and Urban Planning*, 24, pp. 191–195. https://doi.org/10.1016/0169-2046(93)90098-X

Pungetti, G. (1995) Anthropological approach to agricultural landscape history in Sardinia. *Landscape and Urban Planning*, 31, pp. 41–56. https://doi.org/10.1016/0169-2046(94)01035-7

Pungetti, G. (1996) *Landscape in Sardinia: History, features, policies*. CUEC, Cagliari.

Pungetti, G. (2013) Biocultural diversity for sustainable ecological, cultural and sacred landscapes: the biocultural landscape approach. In: B. Fu and B. Jones (eds), *Landscape Ecology for Sustainable Environment and Culture*. Springer, Dordrecht, pp. 55–76. https://doi.org/10.1007/978-94-007-6530-6_4

Pungetti, G. (ed.) (2017) *Island Landscapes: An expression of European culture*. Routledge, London and New York. https://doi.org/10.4324/9781315590110

Pungetti, G., Hughes, P. and Rackham, O. (2012) Ecological and spiritual values of landscape: a reciprocal heritage and custody. In: G. Pungetti, G. Oviedo and D. Hooke (eds) *Sacred Species and Sites: Advances in biocultural conservation*. Cambridge University Press, Cambridge, pp. 65–82. https://doi.org/10.1017/CBO9781139030717.010

Pungetti, G. and Kruse, A. (eds) (2010) *European Culture Expressed in Agricultural Landscapes: Perspectives from the Eucaland project*. Palombi Editori, Rome.

Rackham, O. (1986) *The History of the Countryside*. J.M. Dent and Sons, London.

Rackham, O. (2010) Landscape history of southern Europe. In: G. Pungetti and A. Kruse (eds) *European Culture Expressed in Agricultural Landscapes: Perspectives from the Eucaland project*. Palombi Editori, Rome, pp. 95–101.

Rackham, O. (2013) Significance of sacred trees and groves. In: J. Porter (ed.) *Changing European Landscapes: Landscape ecology, local to global*. IALE European Congress, Manchester, 9–12 September 2013. Countryscape/IALE, Manchester, p. 4.

Rackham, O. (2015) Greek landscapes: profane and sacred. In: L. Käppel and V. Pothou (eds) *Human Development in Sacred Landscapes*. V&R Unipress, Göttingen, Germany, pp. 35–50. https://doi.org/10.14220/9783737002523.35

Rackham, O. (2017a) Island landscape history: the Isle of Scilly, UK. In: G. Pungetti (ed.) *Island Landscapes: An expression of European culture*. Routledge, London and New York, pp. 21–37.

Rackham, O. (2017b) Saaremaa: Tackling landscape history in Estonia. In: G. Pungetti (ed.) *Island Landscapes: An expression of European culture*. Routledge, London and New York, pp. 38–46.

Rackham, O. (2017c) Landscape history of Cyprus: a preliminary account. In: G. Pungetti (ed.) *Island Landscapes: An expression of European culture*. Routledge, London and New York, pp. 47–58.

Rackham, O. and Moody, J. (1996) *The Making of the Cretan Landscape*. Manchester University Press, Manchester.

Rackham, O. and Pungetti, G. (2013) Sacred groves and cultural landscape in Europe. In: J. Porter (ed.) *Changing European Landscapes: Landscape ecology, local to global*. IALE European Congress, Manchester, 9–12 September 2013. Countryscape/IALE, Manchester, p. 3.

SCBD-UNESCO (2014) *Progress Report on the Joint Programme of Work on the Links between Biological and Cultural Diversity*. UNESCO, Paris.

Part III

Mediterranean Studies

CHAPTER 10

Trees Grow Again: Greece and the Mediterranean in Oliver Rackham's Publications

J. Donald Hughes

Summary

Oliver Rackham is best known for his masterful works on the ecology of the British landscape, but also did thorough research and writing with collaborators on Crete, Greece and the European portion of the Mediterranean area. His interpretation of ecological change differs widely from the generally accepted view, which he calls the Ruined Landscape theory, or Lost Eden. His own view is that there was little or no deterioration between the end of the Bronze Age and the mid-twentieth century. He presents good evidence to back his assertions, but other interpretations are possible.

Keywords: Ruined Landscape theory, Lost Eden, Mediterranean Europe, ecological history

Preface

Comments in this chapter include a critical view of Rackham's writings on forests and deforestation, the causes and effects of erosion in the countryside, and the relationship between ancient and modern landscape processes. At the outset, I should reveal that I spent some time with Oliver in Australia. We travelled together to the coast and mountains, observing the environment and conversing pleasantly. He always wore red socks. One evening at John Dargavel's house, we made spaghetti by hand and enjoyed it with marinara sauce. On the other hand, Oliver was never appreciative of my work when referring to it in print. The reader may make allowances for these facts in what follows.

Introduction

Though deservedly best known for his monumental work on the British landscape and especially its forests (Rackham, 2006; reprint 2012), Oliver Rackham expanded his interests into the Mediterranean world, initially through a sweeping study of the archaeology, ecology and forests of Crete with Jennifer Moody and A.T. Grove, which resulted in the book *The Making of the Cretan Landscape* (Rackham and Moody 1996). Geographically more

J. Donald Hughes, 'Trees Grow Again: Greece and the Mediterranean in Oliver Rackham's Publications' in: *Countryside History: The Life and Legacy of Oliver Rackham*. Pelagic Publishing (2024). © J. Donald Hughes. DOI: 10.53061/HYHD2779

extensive work supported by the European Union on desertification in the European half of the Mediterranean Basin resulted in a tome entitled *The Nature of Mediterranean Europe: An Ecological History* (Grove and Rackham 2001). This book represents the results of work supported by the European Community under the Mediterranean Desertification and Land Use programme. It covers the ecological history of Mediterranean Europe, not including the African or Asian coastlands of the sea, and is a big book (9 in x 12 in), lavishly illustrated with striking photographs taken by the authors (mostly in colour), maps, drawings, charts and tables (Fig. 10.1). Those interested in the subject will also find an excellent summary of the sections dealing with Greece, also well illustrated, in the Twentieth J.L. Myres Memorial Lecture at Oxford (1999), *Trees, Wood, and Timber in Greek History* (Rackham 2001).[1] These works provided new opinions and countered the accepted view of the historical ecology of the northern Mediterranean. Rackham termed himself heretical because he rejected this view, which he called the Ruined Landscape theory or 'Lost Eden'.

The Ruined Landscape theory

This generally accepted idea is that from an originally rich and forested land, the Mediterranean area declined to a relatively depleted state from early times through classical Greek and Roman days, medieval and modern. Scholars versed in land-use management from as early as the mid-nineteenth century have investigated questions based on their observations. What forces have degraded the landscapes around the Mediterranean Sea, from the times when ancient empires rose and fell, through the so-called Dark Ages and the vicissitudes of medieval times to the Industrial Revolution and the world market economy? Why did the peoples of the Mediterranean, the heirs of ancient civilizations, inhabit a ruined landscape?

In the ancient world, then as now, change ruled, and human activities damaged nature, causing deforestation and overgrazing over wide areas, especially near cities and industrial centres. The results included erosion that bled soil down to the sea and impaired agriculture. The depletion of forests was not due to climatic changes alone, but to removal by human agency through clearing for agriculture, burning, for fuel in industrial processes and heating, timber and many other uses. Of course forests may regenerate over time, but use of land for other purposes such as grazing and agriculture, and severe erosion, can prevent growth of new trees. There was not total deforestation in the Mediterranean basin; in some districts dense forests survived, especially on inland mountain ranges. But many forests were thinned or were replaced by scrub. The demand for wood as fuel and for use in construction, including shipbuilding, repeatedly raised the price of timber and stimulated a search at ever-greater distances for sources accessible by sea and rivers. This represents the consensus among most of the scholars who have studied the subject.

It is not difficult to find ancient authors who observed environmental change, such as Cicero, who wrote: 'By means of our hands we endeavour to create as it were a second world within the world of nature' (Cicero, *De Natura Deorum*, 2.60). He, however, approved of the change as a benefit to humankind. Plato, on the other hand, said that the deforestation of Attica was long-lasting and caused disastrous erosion along with the drying of springs. Rackham dismisses this evidence because it occurs in a passage describing the war of Athens with the fabled continent of Atlantis (Grove and Rackham 2001: 288), but Plato is clearly making a comparison with Attica as it actually existed in his own day, pointing out that there were beams in buildings in Athens that came from mountains where only 'food for bees' (flowering plants and bushes) existed when he was writing (Plato, *Critias*, 111 B–D).

1 See also Rackham (1996), where the words 'Trees grow again' are used to counter the idea that use of forest resources produces deforestation.

Figure 10.1 Cover of Grove and Rackham (2001) *The Nature of Mediterranean Europe.* (J. Moody)

Rackham implies that the exponents of the Ruined Landscape view expected the Mediterranean to match conditions they were familiar with in Western Europe or eastern North America. But most of them actually lived and researched within the Mediterranean basin. Among the first was George Perkins Marsh, who served as United States ambassador to

Istanbul and then for 20 years to Italy. He observed in the Mediterranean area: 'the character and, approximately, the extent of the changes produced by human action in the physical condition of the globe we inhabit' (Marsh 1864: iii). And he warned that the result of 'man's ignorant disregard of the laws of nature' (Marsh 1864: 11) would be the deterioration of the land.

Another was Russell Meiggs, whom I visited several times near Oxford while he was writing *Trees and Timber in the Mediterranean World* (Meiggs 1982). He noted the rise in price of timber owing to scarcity. Similar comments were made by the ecologist Paul B. Sears, who wrote *Deserts on the March* (Sears 1980), and the soil conservationist Walter C. Lowdermilk, who surveyed the Mediterranean just before the Second World War (Lowdermilk 1944; 1953). The leading climate historian Henry Lamb believes that anthropogenic forest degradation began in about 250 BCE and has continued ever since. Clearance, shredding of trees for animal feed and grazing of sheep and goats continue to the present day. In fact, goats climb trees in Morocco to get at the Argan *Argania Spinosa* fruit, whose seeds pass through their digestive tracts and then are pressed for oil. They also climb trees to eat the foliage, not only in Morocco but also in Spain, Greece and elsewhere, including Montserrat and even Texas, for the latter purpose. Goats are extremely destructive of saplings and small trees. Papanastasis, a Food and Agriculture Organization (FAO) expert, avers: 'Domestic goats are blamed for much of the destruction of the Mediterranean forests ... It is not goats per se that are the real culprit but the continuous, uncontrolled overgrazing for which humans are responsible' (Papanastasis 1986: 44).

These astute men ascribed the decline of ancient civilizations to environmental problems such as deforestation, erosion and agricultural exhaustion. They concluded that the Greeks and Romans did degrade their environments, even if not to the extent of modern times, and it is not unreasonable to connect that process with the decline of classical civilization. But these modern authors did not simply ascribe environmental degradation to antiquity. They recognized that the Mediterranean basin has been subject to cycles of devastation and recovery, and that much of the devastation they saw was the result of medieval and modern mistreatment of the natural world. In passing, I should point out that I taught in Athens for a year, and visited Greece on five other occasions with the purpose of observing environmental conditions and changes.

Scientific Work

The consensus has only been underlined by contemporary studies of the landscape itself by scientists. One of these is Curtis Runnels, whose work in Greece has focused on the relation between human settlement and landscape through time, using archaeology and geological stratigraphy, and who says: 'Recent archaeological work is changing a long-standing view of the impact of agriculture on the land in Greece. The evidence mounts for episodes of deforestation and catastrophic soil erosion over the past 8,000 years. Many scholars believe they resulted from a long history of human land use and abuse' (Runnels 1995: 96).

Others have examined a variety of lines of scientific and written evidence in reconstructing regional ecology. For example, a study by Bruno Pinto and colleagues of northern Portugal over the long term during the Holocene concluded that in a district where the highlands had already suffered from tree removal, Roman agricultural technology along with economic and population relocation cleared the richer valley soils, with 'a major impact on the forests' (Pinto, Aguiar and Portidario 2010: 17).

Interpretations by Rackham and co-authors

Our authors Rackham and Grove have unexpected answers for the questions asked here. The landscape, they confidently assert, is not ruined. Greece in particular may look ruined to northerners not accustomed to its varying ecosystems. The state of the Mediterranean lands in, say, 1950, was no worse than it was at the end of the Bronze Age 3,000 years earlier. There is little evidence for deserts on the march, and if there were such evidence, it would not be a shame, because deserts are interesting landscapes full of biodiversity. Badlands, where gullies and other erosional features dominate, are stable landscapes that have remained in the same districts for centuries or millennia. See, for example, Figure 14.25 in Grove and Rackham (2001: 268), which portrays a landscape dominated by gullies with widely spaced almond trees. It may look like disastrous erosion, but the authors comment, 'In our opinion, although management is not ideal, this is a stable landscape.' Grazing by goats and sheep is not bad, since if the animals did not reduce the vegetation, it would be more vulnerable to catastrophic wildfires. Deforestation, they further assert, exists mainly in the imagination of artists and writers, who make the simplistic assumption that cutting down trees destroys forests, whereas trees grow again. In any case, the authors are not convinced that deforestation makes erosion worse. Everywhere throughout the book, whenever putative evidence arises for deforestation caused by human action, they exhaust themselves in searching for ways to explain it away. Conversely, they demand impossible levels of proof from those who judge that the evidence shows that anthropogenic forest removal can cause erosion. In fact, Rackham says: 'For the ancient Greeks it was normal that woods should be rare, remarkable, and often sacred to gods and nymphs' (Grove and Rackham 2001: 271).[2] These are surprising conclusions. Rackham is worth studying because he raises doubts about established ideas, which is usually a good thing, and he provides a wealth of evidence, even if that evidence is sometimes open to other interpretations.

Much of Rackham's argument is devoted to climate and its changes. Where alterations in the landscape such as erosion are undeniable, he tends to think they were caused by climatic episodes such as the Little Ice Age rather than by human agency. He is, however, sceptical both of the existence and the claimed effects of twentieth-century global warming. Other subjects he elucidates are geology and features such as badlands and limestone karst, and vegetation, including savanna ("Trees without Forests') (Grove and Rackham 2001: 190–216). Forests per se are discussed in a general chapter on plant life. Deltas and 'soft coasts' are of concern, as is the overuse of groundwater.

His comments on 'Aspects of Human History' are well supported by evidence. Since pollen studies indicate that more extensive forests existed in prehistory, he is willing to allow that humans did cause deforestation then and as late as the Bronze Age, but not afterwards: 'All the changes were complete by the end of the Bronze Age,' he avers (Grove and Rackham 2001: 166).

As to the Roman period, he baldly states that there was no degradation then (Grove and Rackham 2001: 80), which defies the evidence and strains credulity. How he can go on to claim that the plague under Justinian, late in the Roman Empire (AD 542), by reducing the population and therefore use of wood, may have caused an increase of trees and forests (Grove and Rackham 2001: 78), if there had not been a decrease before that, boggles the mind.

Contemporary technology and society

Coming down to modern times, Rackham and Grove indicate that the long period of stability of the resilient Mediterranean landscape may be at an end. They ascribe this partly to the use of technology, particularly the bulldozer (and they should add the chainsaw), which has scarred the slopes with dirt roads and unsupported 'false' terraces. No one can deny that fact, which is graphic in the landscape itself. They also condemn the electric pump, which enables the exhaustion of groundwater for ill-advised irrigation schemes that disrupt traditional agriculture and end up salinizing the soil. The authors think that the pernicious overuse of groundwater is about as close to desertification as Europe has come in recent decades. That is unobjectionable.

---

2 Pausanias, in his geography of Greece, gives the impression that sacred groves were rare bits of forest in a landscape that lacked trees for the most part.

Along with technology, the authors warn against the damage caused by land consolidators who buy up farms and remove their walls and hedges. Indeed, they distrust any plans of outsiders, governments and intergovernmental agencies, including the European Union that funds the authors. This is because they have a strong, understandable liking for local people, small farmers and the traditional Mediterranean way of living with the natural environment. Ignoring the fact that it is these people who actually operate the bulldozers and electric pumps, they say: 'Only continued occupation by people gaining their livelihoods locally can maintain the man-made diversity typical of Mediterranean Europe' (Grove and Rackham 2001: 365).

If this is more than nostalgia, the authors should look at the underlying social and economic forces that are wrenching local people from their former occupations, providing the technology and paying them to destroy their traditional settings. The book mentions some unfortunate results of the impact of the world market economy, but never uses the words capitalism or socialism in an analytical sense (or, indeed, any other).

Conclusions

In conclusion, *The Nature of Mediterranean Europe: An Ecological History* is a delight to the eye and a challenge to the explanatory faculty. It is far too technical, both in vocabulary and in approach, and too long, to appeal to the general reader. But it offers enough valuable information to make it an indispensable reference work for anyone seriously interested in the Mediterranean environment, and raises important issues that demand consideration. Rackham had a powerful mind and used it well.

References

Grove, A.T. and Rackham, O. (2001) *The Nature of Mediterranean Europe: An Ecological History*. Yale University Press, New Haven, CT.

Lowdermilk, W.C. (1944) Lessons from the Old World to the Americas in land use. *Annual Report of the Board of Regents of the Smithsonian Institution for 1943*. Government Printing Office, Washington, DC, pp. 413–427.

Lowdermilk, W.C. (1953) *Conquest of the Land through 7,000 Years*. Agriculture Information Bulletin No. 99. Government Printing Office, Washington, DC.

Marsh, G.P. (1864) *Man and Nature; or, Physical Geography as modified by Human Action*. Charles Scribners, New York. [reprinted 1965, D. Lowenthal (ed.), The Belknap Press of Harvard University Press, Cambridge, MA]. [accessed May 2021, https://publicdomainreview.org/collection/man-and-nature-1864].

Meiggs, R. (1982) *Trees and Timber in the Mediterranean World*. Clarendon Press, Oxford.

Papanastasis, V.P. (1986) Integrating goats into Mediterranean forests. FAO Corporate Document Repository. *Unasylva*, 154(38), pp. 44–52.

Pinto, B., Aguiar, C. and Portidario, M. (2010) Brief historical ecology of northern Portugal during the Holocene. *Environment and History*, 16(1), pp. 3–42. https://doi.org/10.3197/096734010X485283

Rackham, O. (1996) Ecology and pseudo-ecology: the example of Ancient Greece. In: G. Shipley and J. Salmon (eds) *Human Landscapes in Classical Antiquity: Environment and Culture*. Routledge, London, pp. 16–43.

Rackham, O. (2001) *Trees, Wood, and Timber in Greek History. The Twentieth J.L. Myres Memorial Lecture*. Leopard's Head Press, Oxford.

Rackham, O. (2006, reprinted 2012) *Woodlands*. (New Naturalist 100), Collins, London.

Rackham, O. and Moody, J. (1996) *The Making of the Cretan Landscape*. Manchester University Press, Manchester.

Runnels, C. (1995) Environmental degradation in Ancient Greece. *Scientific American*, 272(3), pp. 96–99. https://doi.org/10.1038/scientificamerican0395-96

Sears, P. (1980) *Deserts on the March*. University of Oklahoma Press, Tulsa.

CHAPTER 11

Friend or Foe? Oak Agroforestry Systems in the Mediterranean and the Role of Grazing

Thanasis Kizos

Summary

The importance and impact of grazing on Mediterranean treescapes is introduced and considered with evidence from Greece. The key role of Oliver Rackham and his collaborators in developing awareness and understanding the nature of long-term human impacts on the Mediterranean landscape is explained.

Keywords: Mediterranean treescapes, agroforestry, grazing, savanna

Introduction: Agroforestry systems in the Mediterranean

Oliver Rackham did not particularly like the use of 'agroforestry'. He preferred the term 'woodland' to refer to managed trees that are 'so close together that their canopies meet ... managed by coppicing or allowed to grow on into timber' (Rackham 2006: 21). He used 'wood-pasture' or 'savanna' for woodland where the understorey was also used for cultivation and/or grazing: 'trees are widely spaced and grassland, heather, etc. grow between them ... There are grazing animals (cattle, sheep, deer) and the trees are mostly a secondary land use' (Rackham 2006: 21). He used the term 'plantation' where 'trees have been put there ... [but] they have no relation to natural vegetation' (Rackham 2006: 28). He considered 'Mediterranean Savanna' ('trees without forest', Grove and Rackham 2001, Chapter 12) to be where: 'Usually the trees are wild, but orchards, olive-groves and 'agro-forestry' intercropping can be regarded as artificial savanna (Grove and Rackham 2001: 190).

Perhaps he did so because agroforestry seems to be a more 'policy-friendly' term. But one term or another has to be used to describe the diverse sets of management systems that include (a) wide variety of tree species; (b) varying degrees of cultivation of crops; (c) varying degrees of grazing; (d) different tree products. Let's take a look at these definitions. According to the World Agroforestry Center (ICRAF), agroforestry is defined as a

> collective name for land-use systems and technologies where woody perennials (trees, shrubs, palms, bamboos, *etc.*) are deliberately used on the same land management unit as agricultural crops and/or animals, either in some form of

Thanasis Kizos, 'Friend or Foe? Oak Agroforestry Systems in the Mediterranean and the Role of Grazing' in: *Countryside History: The Life and Legacy of Oliver Rackham*. Pelagic Publishing (2024). © Thanasis Kizos. DOI: 10.53061/GNCJ6247

> spatial arrangement or temporal sequence. In agroforestry systems, there are both ecological and economic interactions between the different components.
>
> Gordon, Newman and Williams 1997: 1

Further definitions are related to two differentiating qualities of these systems: the type of use of the understorey of the trees and the type of the plantation. Regarding the former, the type of land-use can be for: (a) arable land (in silvo-arable systems); (b) pastures and/or grazing of the understorey (in silvo-pastoral systems); and (c) arable crops and grazing (in agro-silvo-pastoral systems) (Nair 1993). For the latter, many different tree species have been used in agroforestry systems. In the Mediterranean, oaks, chestnuts and olives are among the landmarks, but walnuts, citrus trees, palm trees and hazelnuts have also been mentioned (Torralba et al. 2016).

Agroforestry plantations can be regular or irregular (Plieninger and Kizos 2011): in regular ones, trees are planted or grafted in ordinary rows and approximately equal spaces with tree densities varying according to the tree species and the location of the site (level or sloping field; type of soil; precipitation patterns). In irregular ones, trees are grafted upon wild ones or planted at irregular spaces in a savanna-style landscape with varying tree densities. Irregular plantations are usually linked with grazing.

In the Mediterranean, agroforestry systems date back to the Neolithic with a variety of trees and management systems (Blondel et al. 2010). Geographically, they are found in both the European, African and Asian parts of the basin, while they are encountered from the coasts of the sea to the tree limit on the mountains (Horden and Purcell 2000). Tree cultivation has been considered as a 'noble' act in the area (and in the Fertile Crescent and Persia), linked with notions of 'paradise' and aesthetically desirable landscapes. As management systems became more and more complicated and more knowledge was gained on the micro-ecology of different plant and animal species on one hand and more plants and animals were introduced on different localities on the other, trees, annual crops and grazing animals were mixed in systems that today are called agroforestry (Eichhorn et al. 2006). In this chapter, I examine the relationship between grazing and agroforestry systems, with some examples from oak grazing lands in Greece.

Oak agroforestry systems and grazing

Small ruminants such as sheep and goats are very well adapted to the vegetation and the climate of the Mediterranean. Sheep prefer annual plants but will eat tender leaves and branches of fresh growth of the evergreen Mediterranean shrubs and trees. Goats on the other hand, as Rackham famously noted in *the Making of the Cretan Landscape* (Rackham and Moody 1996) can climb anywhere and will eat shrubs and trees as well as annuals. Herders have managed different combinations of the two in varying pedoclimatic conditions: for example, goats dominate the highlands of Crete, parts of the High Atlas and the Middle East, while sheep are more abundant where more grass and fewer ravines are found. The vastly different qualities of meat and milk between the two (goat milk is the widely used animal milk with low fat content, while sheep milk is the one with the highest fat content among livestock) have made herders keep mixed flocks, with a few goats kept alongside sheep in most Eastern and Southern regions.

The role of grazing has been recognized as vital in maintaining soil fertility and complementing arable farming in the Basin (Kizos, Plieninger and Schaich 2013), but was also seen as a valuable resource that could utilize less fertile, less accessible and more marginal areas (Horden and Purcell 2000) in both transhumant and sedentary systems. Especially in oak grazing lands, oak savannas were ploughed for cereals, legumes (even tobacco) in fallow cycles of two or three years and grazed by sheep during fallow periods. Labour availability was a key component in all of these systems. The mechanization and industrialization of agriculture and animal husbandry affected in many ways all these systems: the population that sustained such systems has been reduced significantly, following social expectations and a shift of economy towards industry and services. Therefore, a tendency of abandoning or simplifying former complex management systems is recorded around the Mediterranean, and agroforestry systems have suffered from these developments as well as other management systems. Kizos, Plieninger and Schaich (2013) describe the process from integration (of animal husbandry, agriculture and forestry) to separation.

Two different processes are involved here: abandonment and intensification. In the first, grazing stops almost completely and the understorey is overgrown with shrubs and young trees, while the open spaces between the trees are gradually covered by annuals that are succeeded by shrubs and trees. The specific pedoclimatic conditions of the locality (as different micro-ecologies within the same area may follow different trajectories) determine if the landscape will eventually be transformed into maquis forest with older trees being slowly dying out and replaced by bunches and (if conditions allow) into dense forest (Fig. 11.1A). In intensification, grazing densities increase very much: Kizos, Plieninger and Schaich (2013) record densities that are ten times higher than the official policy guideline for the optimum grazing pressure. The understorey is kept almost bare, with dominance of unpalatable phrygana species, such as Thorny Burnet *Sarcopoterium spinosum*. In extreme cases, bare soil is exposed and rains can increase run-off erosion rates, resulting in badland formations and soil degradation. Grazing pressure puts the regeneration of dominant tree species at risk, with many older trees dying out without being replaced and very few young trees surviving grazing (Plieninger, Schaich and Kizos 2011). Figure 11.1C presents an idealization of the result of intensification.

Even if the results of intensification are very negative, abandonment seems to affect biodiversity and species richness as well. Psyllos et al. (2016) studied the dry mass produced in oak grazing lands and the species that animals graze (mostly annuals but also offshoots of phrygana and shrubs as well) in the village of Agra, in the environmentally marginal Western Lesvos. The results reveal a very high degree of diversity with the total number of identified species (or genera when species could not be identified owing to grazing) recorded for two consecutive years. Out of these species, only 70 per cent were identified in both years, indicating a very rich seedbank that climatic conditions and grazing pressure could turn on or off. The comparison of different grazing regimes revealed that although measurements of dry matter and number of species were higher on ungrazed plots than on all grazed ones, the differences were not so great, especially for regimes that left the area ungrazed for parts of the grazing season (September to June, see Table 11.1 for details). Abandonment and 'closing' of the open savanna oak landscape would gradually reduce these differences between ungrazed and grazed areas, as shrubs that establish after the first year of non-grazing reduce the number of annuals. Moderate grazing (Fig. 11.1B) appears therefore to play a major role in balancing between annuals and shrubs and retaining high levels of plant species.

Discussion: grazing as a friend or foe?

Grazing has been identified as a key driving force of change in Mediterranean landscapes, mostly as an agent of degradation and destruction, typically within the overall context of the 'ruined Mediterranean landscape' (see McNeill 1992, for a very characteristic example). Rackham, although he never denied the degrading effects of heavy grazing, was among the first to state that grazing was an integral part of the ecology of Mediterranean landscapes and Mediterranean forests. In fact, almost all of *The Nature of Mediterranean Europe: An Ecological History* is a long rebuttal to the Ruined Landscape theory. His legacy has helped us understand the finer grain and richer details of the resilience of Mediterranean landscapes and human impacts. For me, it has revealed that both the long and the short term have to be considered when landscape change and continuity is discussed. Oak agroforestry grazing

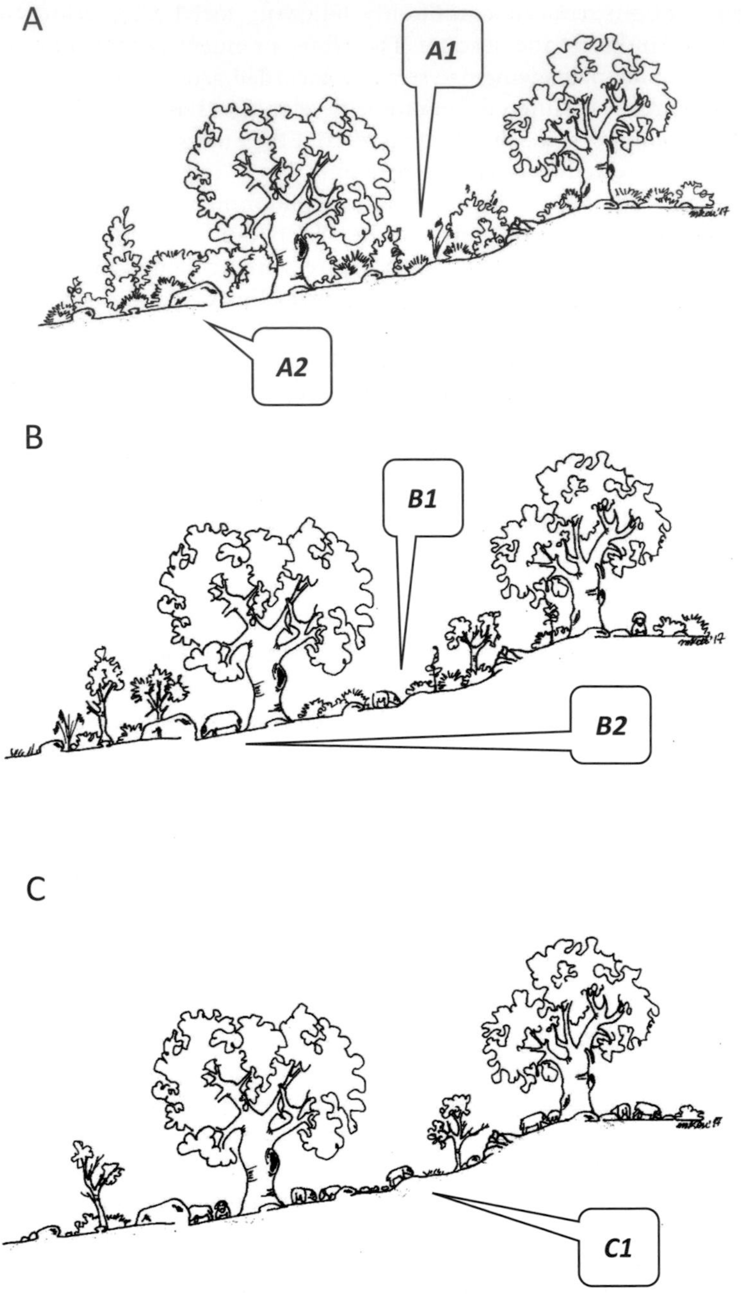

Figure 11.1 Agroforestry systems and grazing: different states of grazing pressure and agroforestry systems change (drawing by Maria Koulouri).

A: no grazing: understorey growth of shrubs and young trees (a1); spaces between trees (a2) are covered by annuals that are succeeded by shrubs and trees; the landscape turns into maquis forest and if the pedoclimatic conditions allow into dense forest.

B: moderate grazing: understorey is kept at low height with dominance of annuals and phrygana and/ or shrubs (b1); spaces between trees (b2) are covered by annuals and regeneration of dominant tree species continues with young trees surviving grazing.

C: heavy grazing: understorey is kept almost bare, with dominance of unpalatable phrygana (c1); regeneration of dominant tree species is at risk, with many older trees dying out without being replaced and very few young trees survive grazing.

lands particularly have to be treated with the nuanced view that grazing is neither a friend nor a foe. It is part of the system by definition and has to be treated as such. This raises a series of management issues, many of which are discussed by Torralba et al. (2016), which have to take into account societal needs, resilience and climate change, and whether to reintroduce and/or how to manage grazing.

Table 11.1 Species identified and dry mass produced in oak grazing lands in Agra, Western Lesvos, Greece in two consecutive years.

The total number of identified species or genera (when species could not be identified owing to grazing) was 93 for the two years, with 30 per cent found only in one of the two and 70 per cent in both years. (Source: Psyllos et al. 2016: all counts indicate five cages used to fence off grazing animals: Control cages were left ungrazed all season (September to June), grazed were continuously grazed, C1 was grazed until December and then fenced, C2 until April and then fenced and C3 until May and then fenced.)

		Mean Year 1	Mean Year 2	Minimum of 2 years	Maximum of 2 years
Species (including genus)	Control (N=19)	8.4	6.3	3	20
	Grazed (N=19)	5.7	6.1	2	9
	C1 (N=19)	7.7	5.7	2	10
	C2 (N=19)sss	8.1	5.2	2	14
	C3 (N=19)	1.4	6.1	0	10
	Total (N=95)	6.3	5.9	0	20
Dry mass (t/ha)	Control (N=19)	2,897.7	4,051.0	214	11,969
	Grazed (N=19)	881.4	1,352.5	31	6,151
	C1 (N=19)	2,150.8	3,202.4	172	17,382
	C2 (N=19)	1,816.7	1,296.5	47	12,402
	C3 (N=19)	257.0	2,033.1	0	6,852
	Total (N=95)	1,600.7	2,387.1	0	17,382

References

Blondel, J., Aronson, J., Bodiou, J.-Y. and Boeuf, G. (2010) *The Mediterranean Region: Biological Diversity in Space and Time.* Oxford University Press, Oxford.

Eichhorn, M.P., Paris, P., Herzog, F., Incoll, L. D., Liagre, F., Mantzanas, K., Mayus, M., Moreno, G., Papanastasis, V.P., Pilbeam, D.J., Pisanelli, A. and Dupraz, C. (2006) Silvoarable systems in Europe – past, present and future prospects. *Agroforestry Systems,* 67, pp. 29–50. https://doi.org/10.1007/s10457-005-1111-7.

Gordon, A.M., Newman, S.M. and Williams, P.A. (1997) Temperate agroforestry: an overview. In: A.M. Gordon and S.M. Newman (eds) *Temperate Agroforestry Systems.* CABI, Wallingford, pp. 1–8. https://doi.org/10.1079/9781780644851.0001

Grove, A.T. and Rackham, O. (2001) *The Nature of Mediterranean Europe: An Ecological History.* Yale University Press, New Haven, CT.

Horden, P. and Purcell, N. (2000) *The Corrupting Sea: A Study of Mediterranean History.* Blackwell, London.

Kizos, T., Plieninger, T. and Schaich, H. (2013) 'Instead of 40 sheep there are 400': Traditional grazing practices and landscape change in Western Lesvos, Greece, *Landscape Research,* 38(4), Special Issue: Animals and the landscape, pp. 476–498. https://doi.org/10.1080/01426397.2013.783905

McNeill, J.R. (1992) *The Mountains of the Mediterranean World.* (Studies in Environment and History), Cambridge University Press, Cambridge.

Plieninger, T. and Kizos, T. (2011) Assessing the sustainability impact of land management with the ecosystem services concept: towards a framework for Mediterranean agroforestry landscapes. In: Z. Roca, P. Claval and J. Agnew (eds) *Landscapes, Identities and Development.* Ashgate, Farnham, pp. 305–318.

Plieninger, T., Schaich, H. and Kizos, T. (2011) Land-use legacies in the forest structure of silvopastoral oak woodlands in the Eastern Mediterranean. *Regional Environmental Change,* 11, pp. 603–615. https://doi.org/10.1007/s10113-010-0192-7

Psyllos, G., Kizos, T., Hadjigeorgiou, I. and Dimitrakopoulos. P. (2016) Grazing land management and sheep farm viability in semi-arid areas: evidence from Western Lesvos, Greece. In: A.P. Kyriazopoulos, A. López-Francos, C. Porqueddu and P. Sklavou (eds)

Ecosystem services and socio-economic benefits of Mediterranean grasslands. *Options Méditerranéennes: Série A.* Séminaires Méditerranéens, 114, pp. 369–372.

Rackham, O. (2006) *Woodlands* (New Naturalist 100), Collins, London.

Rackham, O. and Moody, J. (1996) *The Making of the Cretan Landscape.* Manchester University Press, Manchester.

Torralba, M., Fagerholm, N., Burgess, P.J., Moreno, G. and Plieninger, T. (2016) Do European agroforestry systems enhance biodiversity and ecosystem services? A meta-analysis. *Agriculture, Ecosystems and Environment,* 230, pp. 150–161. https://doi.org/10.1016/j.agee.2016.06.002

CHAPTER 12

The Irreplaceable Trees of Crete

Jennifer A. Moody

Summary

Crete is home to many different kinds of ancient, historic and veteran trees. Dating them is often a challenge because many are opportunistic growers, have indistinct rings or are hollow. Almost all old trees in Crete are pollards or coppice stools. The old trees of Crete grow from the mountain tops to the coastal plains, and all show remarkable – though different – adaptations for surviving in the Cretan landscape. Today, many of Crete's great old trees are threatened by development and disease. Although some conservation efforts are under way on the island, they are unlikely to be enough.

Keywords: Crete, veteran and ancient trees, dendrochronology, conservation

Introduction

In the summer of 1968, a tall, lanky, rather shy young man walked into the Cretan landscape and fell in love with the trees. Meet Oliver Rackham, 29 years old and the expeditionary botanist for the Myrtos Fournoí Korifoí excavation in south-east Crete, a project directed by fellow 'Corpuscle' and Oliver's good friend Peter M. Warren.[1] A prodigious walker, Oliver hiked all over the region that summer, from Myrtos to Anatolí to Málles and up into the high Díkti mountains, where he encountered his first 'ancient' wild trees on the Great Island – the magnificent Prickly Oaks *Quercus coccifera* of Selákano (Fig. 12.1).[2] Fifty years later, one of the very trees he photographed on that fateful day was dedicated to his memory by the Cultural Association of Selákano with support from the Municipality of Ierápetra, the Natural History Museum of Crete, and the social cooperative partnership Eptástiktos (Fig. 12.2). In fact, three great Selákano Prickly Oaks and a magnificent Cretan Elm *Zelkova abelicea* in the White Mountains of west Crete (Figs 12.3 and 12.4), were dedicated to Oliver's memory in 2018 in celebration of his life and work on the island (Moody 2018).[3]

1 A member of Corpus Christi College, Cambridge.

2 Rackham, *RedNb* 143, pp. 762–763. References to 'Rackham, *RedNb* xxx' refer to Oliver's original red field notebooks, which were written from 1950 to 2015 and numbered by him from 1–698. They are archived in Corpus Christi College, Cambridge.

3 The tree dedication in the Omalos was sponsored by the Friends of Oliver Rackham, the Managing Committee of the Samaria Gorge, the Municipality of Plataniás, the Mediterranean

Jennifer A. Moody, 'The Irreplaceable Trees of Crete' in: *Countryside History: The Life and Legacy of Oliver Rackham*. Pelagic Publishing (2024). © Jennifer A. Moody. DOI: 10.53061/RWNP5374

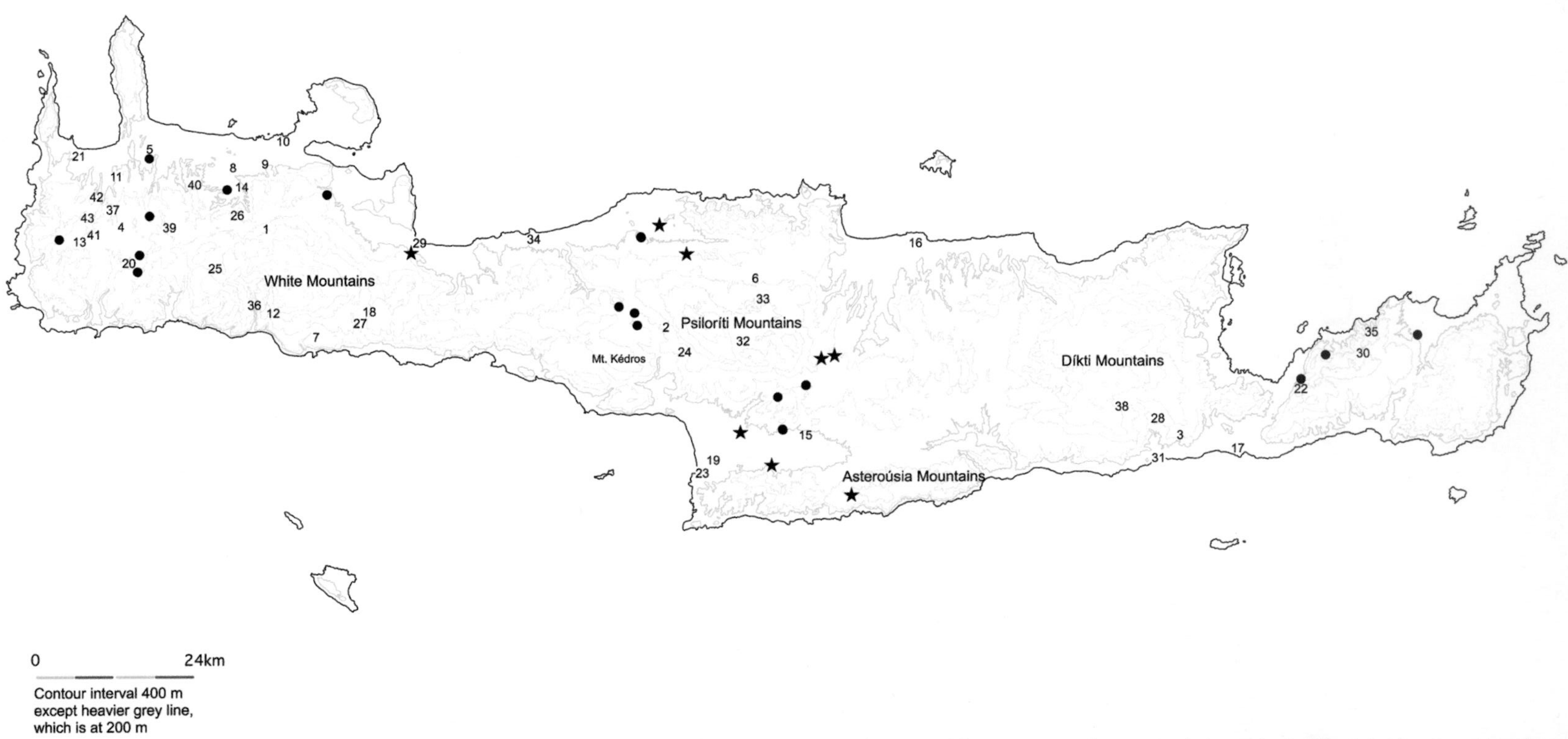

Figure 12.1 Locations mentioned: 1 Alíakes spring above Thérisso, 2 Amári valley, 3 Anatolí, 4 Ano Flória, 5 Ano Voúves, 6 Anógiea, 7 Anópolis, 8 Ayiá, 9 Boutsounária, 10 Chaniá, 11 Choudalianá, 12 Elighiás Gorge, 13 Elos, 14 Fournés, 15 Gortyn, 16 Heráklieon, 17 Ierápetras, 18 Imbros, 19 Kamilári, 20 Kándanos, 21 Kastélli Kíssamos, 22 Kavoúsi, 23 Kommós, 24 Kouroútes Amári, 25 Koutsopétra above the Omalós Plain, 26 Lákkoi Kydonías, 27 Lákkoi Sphakiá, 28 Málles, 29 Mathés, 30 Moulianá, 31 Myrtos, 32 Nídha Plain, 33 Páno Ambélia, 34 Réthymno, 35 Ríchtis Gorge, 36 Samariá Gorge, 37 Sássalo, 38 Selákano, 39 Sembronás, 40 Skonízo, 41 Strovlés, 42 Tsourounianá, 43 Vlátos (Source: J. Moody)

Stars are 'Miracle and healing' trees. Dots are Monumental Olives (J. Moody)

Figure 12.2 Great veteran Prickly Oak 'Selákano 1': (top) The tree as it was on 5 July 1968 when Oliver first visited Crete (Rackham, *RedNb 143*) (O. Rackham); (bottom) The tree 50 years later on 19 September 2018. Note the two large branches now missing from the tree. We estimate that this tree is between 850 and 1,000 years old. In the distance is 'Selákano 2', which we estimate is over 1,000 years old. (J. Moody)

Figure 12.3 The veteran Cretan Elm pollard dedicated to Oliver, August 2018. This tree has a circumference of *c.* 4.3 m and an estimated growth rate of 1 mm per year. We estimate it to be about 700 years old. The tree was noted and photographed by Rackham and Moody 36 years earlier in 1982, when they began to explore the landscapes of Crete. (See Rackham and Moody, 1996: 71 and Plate 7; Rackham, *RedNb* 291.) September 2019. (J. Moody)

Figure 12.4 Father Stylianos of Lákkoi blessing the Rackham *ambelitsiá* and chanting a memorial service for Oliver at 'Koutsopétrou', Crete, overlooking the Omalos Plain. Sheep and goat bells made up the choir, August 2018. (Ch. Orphanoudaki-Kapagioridou)

The remarkable trees of Crete divide into three somewhat overlapping categories: ancient, historic and veteran. Ancient and historic trees are trees with absolute ages. Trees older than 1,200 years are considered ancient, and trees between 100 and 1,200 years old are historic. Veteran trees are not simply old trees; they are trees with battle scars from lightning strikes, fires, hurricanes, tornados, floods, droughts, disease, axes and saws. Their gnarled and often hollow trunks are habitats for a multitude of organisms, such as fungi, lichens, mosses, insects, all manner of birds and small to medium-sized mammals (Rackham 1976, 1986, 2006; Siitonen and Ranius 2015; Stara and Vokou 2015). As habitats and companions, veteran trees are unique and irreplaceable. Oliver Rackham succinctly observed many years ago: "Ten thousand oaks of 100 years old are no substitute for one 500-year-old oak' (Rackham 1986: 152).

Dating old and veteran trees

Dendrochronology, the science of ageing trees by their annual growth rings, is uncommon in Crete; but recent activities by the Cretan Tree-ring Group and the Balkan Aegean Dendrochronology programme have had some success, particularly with cypress, fir, pine and deciduous oak (Christopoulou et al. 2019; Ważny et al. 2020).[4] Ageing trees by counting

Agronomic Institute of Chania, the Forestry Service of Chania, the Development committee of Lákkoi and the Mountaineering Club of Greece.

4 Both projects are directed by Professor Tomasz Ważny, Nicolaus Copernicus University, Toruń, Poland.

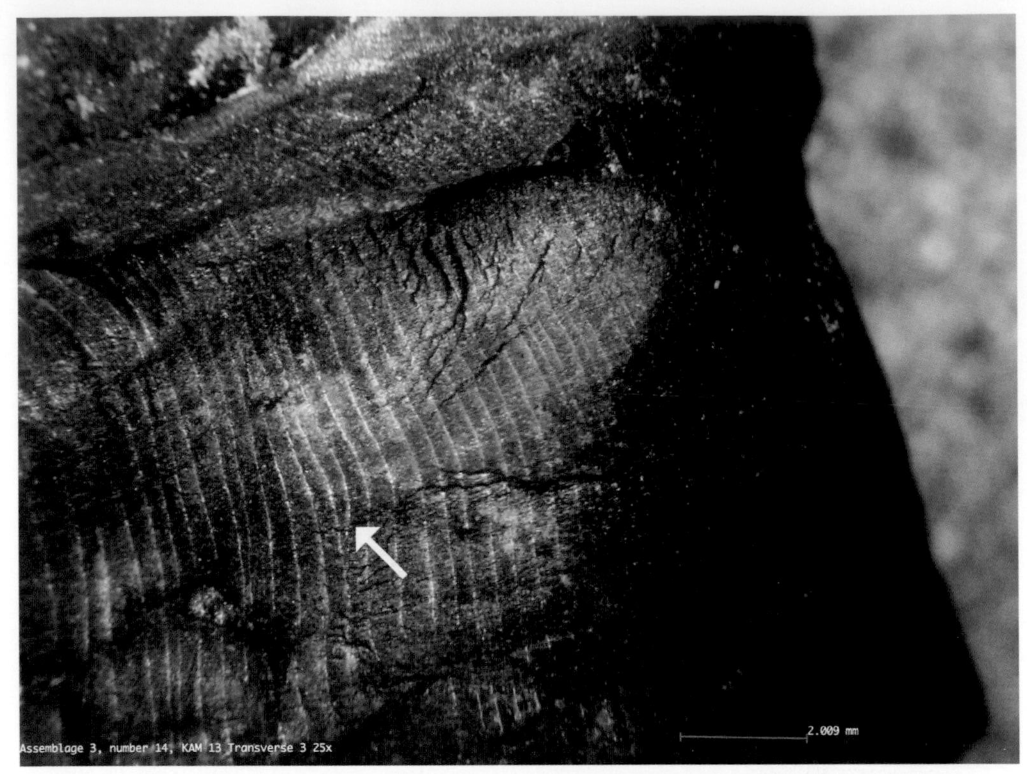

Figure 12.5 A Bronze Age fragment of cypress wood charcoal from archaeological excavations at Kamilári, Crete. The arrow points to a partial ring. (See Ntinou et al. 2019.) (J. Moody)

their annual rings is not entirely straightforward. Some trees such as cypress and olive are opportunistic growers. When conditions are harsh, they skip growing that year or generate only a partial ring. But when conditions are favourable, they might put on two or more rings. Such anomalous growth patterns are even preserved in ancient charcoal (Fig. 12.5), providing a rare window into ancient growing conditions.

Many old trees are hollow – especially planes, olives and chestnuts. They can also be multi-trunked stools – especially cypresses, prickly oaks and olives. Such trees require their missing rings to be estimated. For the time being, the best way to age these trees is to measure their diameter and circumference and then estimate a reasonable average growth rate based on the condition of the tree and its environmental setting. Radiocarbon dating is not sufficiently precise.

For olive trees, the growth rate ranges from about 0.5 to 3 mm per year, but it could be less or more. It is important to remember that 0.5 mm per year makes a dramatic difference to the age of a big tree. For example, the Ano Voúves Monumental Olive (which still fruits!) is claimed to be 2,000–4,000 years old (Fig. 12.6). Its diameter at 1 m above the ground is 3.7 m. A growth rate of 0.5 mm per year produces an age of about 3,700 years, and a rate of 1 mm per year produces an age of about 1,850 years. Given its favourable location in wet north-west Crete at an elevation of 270 m,[5] Oliver Rackham and I estimated its growth rate at 2 mm per year (at a minimum), resulting in an age of about 925 years. This is not an exact science!

5 This location falls within the 'Camelot zone' of the island, a frost-free belt between *approximately* 100 and 400 to 600 m where frost has been exceedingly rare and rainfall fairly consistent for thousands of years (Moody forthcoming).

Figure 12.6 The 'ancient' olive at Ano Voúves, Crete, July 2017. (J. Moody)

Characteristics of long-lived trees in Crete

Prickly Oaks, Cretan Elms and olives are not the only remarkable trees in Crete. There are also massive cypresses, pines, deciduous oaks, planes and chestnuts (Figs 12.7–12.13). The majority of these extraordinary trees are pollards (they have had their branches lopped off at 2 m or more above the ground) and coppice stools (they have been cut, browsed or burnt to ground level and have resprouted from their base). It is often observed that pollarding and coppicing prolong the life of a tree (Rackham 1976: 5–18; Genin 2018) but the reasons for this are not entirely understood. The most commonly cited reasons are: (1) it stimulates new growth, (2) it reduces the distance that water and nutrients have to travel to reach the new growth, and (3) it rebalances the relationship between the crown and the roots, when the latter have been limited in some way (Borchet 1976; Fortanier and Jonkers 1976; Del Tredici 1999–2000, 2001).

Cypress

The Mediterranean Cypress *Cupressus sempervirens* has been the iconic tree of Crete since antiquity, its wood admired for its durability and fragrance (Theophrastus, *Enquiry into Plants*). Perhaps the most famous examples are those in the grove of giant cypresses near the chapel of Aghios Nikólaos in the Samariá Gorge (Fig. 12.7a). In 1989, one of these trees stood about 35 m tall and had a circumference of 7.11 m. An old stump of one of these giants contained something over 500 rings (Rackham, *RedNb* 377).

The oldest cypresses on Crete are the gnarled, sun-bleached and wind-pruned specimens that dot the high mountains between 1,200 and 1,950 m, cracking the bare rock landscape with their powerful roots (Figs 12.7b and 12.7c). They are also the highest trees in Crete.

Figure 12.7 Ancient cypress: (a) One of the great cypresses near the chapel of Aghios Nikólaos in the Samariá Gorge with George Harrison, Jane Francis, Paula Perlman, Bailey Davis and Adrian Levine, June 2017. (J. Moody)

Figure 12.7 (*continued*) (b) Ancient cypress with over 1,000 rings measured by the Cretan Tree-ring Group with Oliver Rackham, Jenny Moody, Tomasz and Olga Ważny, Angathopí Sphakiá, August 2010 (Rackham, *RedNb 641*). (A. Helman-Ważny)

Figure 12.7 *(continued)* (c) Bleached and weathered great cypress tree, Angathopí Sphakiá, July 2014. (J. Moody)

Although primarily a mountain tree, relatively young cypresses are growing up in the foothills, especially on abandoned agricultural terraces (Rackham 2001: plate 18). They provide *terminus ante quem* dates for abandonment in the landscape. Some of these trees are several centuries old.

As noted earlier, cypress is an opportunistic grower, making it a challenge for dendrochronologists. However, the Cretan Tree-ring Group and the Balkan Aegean Dendrochronology programme have been able to reconstruct a reliable dating series based on living cypresses in the White Mountains that goes back to about AD 1100 (Ważny, Rackham and Moody, in preparation).

Cypress is fire resistant but may succumb if all of its foliage is burnt. All old cypresses have fire scars.

Where accessible along mule-tracks and roads, most sprawling old cypresses have been cut for firewood for centuries.

There is some evidence for Cypress Canker disease in Crete, caused by the fungus *Seiridium cardinale*, but it does not seem to be widespread (Xenopoulos and Diamandis 1985; Danti and Della Rocca 2017). Cretan cypress trees, however, are seriously plagued by Bark Beetles *Phloeosinus sp.* and have been for millennia. Their distinctive channels are well preserved on cypress wood-charcoal from the Bronze Age excavations at Chaniá.

Pine

Cretan Pine *Pinus brutia* is the main native pine of Crete. Its thick bark makes it more fire-resistant than its mainland cousin the Aleppo Pine *Pinus halepensis*, but they still love to burn their competition, which they do by draping the undergrowth with dry, combustible needles. Many of the great fires in the Cretan landscape are associated with pine, now and in eons gone by. Along the road from Ayiá to Fournés, a roadcut exposes red and yellow sediments that

Figure 12.8 Cretan pine pollard with Jenny Moody, Elighiás Gorge Sphakiá, July 2012 (Rackham, *RedNb 667*). (O. Rackham)

date to the Pleistocene. Low down in the roadcut, a gravelly layer contains small fragments of wood-charcoal, most if not all pine (Rackham, *RedNbs 418, 681*).

Pine is a good tree for dendrochronology with clear, readable rings. In Crete, however, very old trees are rare; the oldest pines found by the Cretan Tree-ring Group are just under 400 years old. Unusual for pines, the Cretan Pine pollards (Fig. 12.8).

In Crete, pines are sources of timber, resin and honey. Resining scars are common on older pines in the mountains but seem to have done little harm to the trees. The production of 'pine honey', however, is another story. Bees make this from the sugary secretions of the sap-sucking, scale-insect *Marchalina hellenica*, which has been in the Mediterranean for hundreds of years if not longer. When managed properly, the trees survive having their sap sucked out, but in the 1990s the introduction of too many insects by greedy beekeepers devastated pine forests in the White Mountains, especially around the mountain village of Anópolis (Rackham, *RedNbs 618, 635, 667* and *693*).

Prickly Oak

Prickly Oak is the most adaptable tree in Crete. It grows as an evergreen tree or shrub from the coastal plain to almost the tops of the White Mountains, which rise from sea level to over 2,400 m. They are tenacious of life and can survive indefinitely as shrubs, their height held down by some combination of browsing, fire and competition. Prickly Oak will even flower as a metre-high shrub and produce prodigious amounts of pollen that might lead a palynologist to think there was a towering oak forest. Once a Prickly Oak 'gets away', it can grow into a massive tree (see Fig. 12.2).

Prickly Oak is exceedingly fire tolerant. It sprouts readily after burning, making it sometimes difficult to tell if a giant stool is the result of repeated fires in the landscape (natural or otherwise), or coppicing (Fig. 12.9) (Rackham, *RedNb 597*).

Figure 12.9 Great Prickly Oak stool with Oliver Rackham, Lákkoi above Imbros Sphakiá, July 2005 (Rackham *RedNb* 597). (J. Moody)

Prickly Oak is unpopular with dendrochronologists because its wood is incredibly hard and has broken many an increment borer. Its annual rings are also indistinct.

Deciduous Oak

The main deciduous oak of Crete is *Quercus pubescens* var. *brachyphylla*. Some of these trees are huge, but few are older than 250 years. The oldest trees grow at higher elevations, such as the remarkable pollards at Páno Ambélia (880 m), between Anógiea and Nídha, the high mountain plain of the Psiloríti Mountains (Fig. 12.10 top) (Rackham, *RedNb 666*). Here the Cretan Tree-ring Group found trees with girths of 3.5 to 5.5 m and ages that ranged from 100 to 280 years. At low elevations, deciduous oaks are fast grown. We were surprised to find that the big oaks above Lákkoi Kydonías (550 m), with girths of 3 to 4 m, were only 52 years old (Fig. 12.10 bottom).

In Crete, deciduous oaks are invasive, and young trees can be seen colonizing abandoned agricultural terraces and hillsides where the soil is deep enough (Rackham 2001: plate 17).

Cretan Elm

The Cretan Elm is one of the world's rarest trees. Its genus is a Tertiary relict that survives today as six species scattered from the Mediterranean to Japan (Kozlowski et al. 2014). The Cretan Elm and its nearby cousin the Sicilian Elm *Zelkova sicula* are endemic to their island homes. Although not as close to extinction as its Sicilian cousin, the Cretan Elm (*'ambelitsiá'* in Greek) is strongly threatened by overgrazing, animal trampling, rural road construction and fire. 'Conservation of *Zelkova abelicea* in Crete', an initiative by the Mediterranean Agronomic Institute of Chania in partnership with the University of Fribourg and in collaboration with the four Forest Directorates of Crete (Chaniá, Réthymno, Heráklio and Lasíthi), aims to protect the trees and educate the public (see http://www.abelitsia.gr/en/).

Figure 12.10 Historic deciduous oaks: (top) Deciduous oak pollard with Oliver Rackham, Páno Ambélia above Anógeia, June 2012 (Rackham, *RedNb 666*). (bottom) Deciduous oak pollard with Olga Ważny, Lákkoi Kydonías, July 2013. (J. Moody)

Figure 12.11 Historic plane trees: (top) Veteran plane tree with Meryn Scott, Jenny Moody and friend, Alíakes spring above Thérisso Kydonías, September 1991 (Rackham, *RedNb 417*). (O. Rackham); (bottom) The same tree 26 years later with Adrian Newton, Lynn Davy and Jenny Moody, June 2017. (J. Francis)

Ambelitsiés grow mainly in the White Mountains of Chaniá between 900 and 1,800 m elevation. Small populations also survive on the other high mountains of Crete, including Mount Kédros – except for the Asteroúsia.

As with most elms, *ambelitsiés* have distinct annual rings good for dendrochronology. At high elevations, the species is slow grown and small individuals can be deceptively old (Fazan et al. 2012). Possibly the oldest Cretan Elms are stools that have been repeatedly cut (it is the preferred wood for making traditional shepherd's crooks in west Crete), burnt or browsed. Old pollarded *ambelitsiés* are uncommon; the grove that includes the grand specimen that was dedicated to Oliver in 2018 (Figs 12.3 and 12.4), probably contains more *ambelitsia* pollards than anywhere else in the world.

Plane

The plane trees of Crete can be enormous, but most do not compare to the giants of mainland Greece and elsewhere. They have distinct annual rings but hollow trunks, making estimating their age a challenge (Fig. 12.11). Their soft wood and tendency to self-pollard turns them into veterans at a relatively young age.

Crete is home to a variety of Oriental Plane that is evergreen, *Platanus orientalis* var. *Cretica*. It was recognized as an oddity even in Antiquity (Theophrastus, *Enquiry into Plants*: 1.9.5). Some consider this a Cretan endemic (Nikolakaki and Hajaje 2001), others a local mutation.

Canker Stain Disease (CSD) targets plane trees, especially the Oriental Plane, the dominant Mediterranean species. CSD is thought to have been introduced to Europe from the United States during the Second World War via military wood packaging (Tsopelos et al. 2017). CSD is devastating riverine woods and garden trees on the Greek mainland and elsewhere (Rackham, *RedNb* 676, 696). The disease is said to be unknown on Crete, but Oliver wondered if he saw it on the big planes at Boutsounária Kydonías (Rackham, *RedNb* 691). It would be an irreplaceable loss if the unique population of Cretan evergreen planes (fewer than 200 trees) succumbed to CSD.

Chestnut

The Sweet Chestnut *Castanea sativa* is likely an ancient introduction to Crete. *Castanea*-type wood charcoal was identified in Geometric to Archaic levels at Kommós (Shay et al. 1995: 158), and definite chestnut wood charcoal was identified in Roman levels at Kastélli Kíssamos.

Although some trees have been ravaged by disease and fire, today numerous impressive chestnut pollards and coppice stools grow in the moister parts of west Crete, especially around the villages of Elos, Strovlés, Sássalo, Flória, Kándanos, Sembronás and Skonízo, where the soils derive from the Phyllite-Quartzite series (Figs 12.12) (Rackham, *RedNbs* 280, 288, 363, 429, 485, 596). We estimate the oldest to be about 700 years.

Some Sweet Chestnuts lurk in the damper parts of central Crete, including the north slopes of the Psiloríti and Díkti Mountains. In the far east of the island, a few have been seen around Moulianá and in the Ríchtis gorge. Their stronghold, however, is in the west, where chestnut festivals are held at harvest time and the nut features in local cuisines.

Olive

The ancient and historic olives of Crete are not wild trees; they are cultivars (Figs 12.13 and 12.7m). Their origins, however, are unclear. Palynology and wood charcoal indicate that olive survived the Pleistocene on the island, but genetic studies have been unable to isolate populations of wild olives in the eastern Mediterranean because of cross-pollination with domestic varieties and the extremely long history of cultivation (Besnard and Berville 2000; Contento et al. 2001; Breton, Tersac and Berville 2006).[6]

6 No Cretan trees were sampled by any of these studies.

Figure 12.12 Historic chestnuts: (top) Great pollard chestnut with Oliver Rackham, 'Ntoulia' near Ano Flória Kíssamos, April 1988 (Rackham, *RedNb* 363). (J. Moody); (bottom) Old chestnut coppice stool with Jenny Moody, Vlátos Kíssamos, July 1981 (Rackham, *RedNb* 280). (O. Rackham)

Figure 12.13 Ancient olives: (top) Byzantine olive with Oliver Rackham, near Kouroútes Amári, April 1982 (Rackham, *RedNb* 290). (bottom) Monumental Olive at Kavoúsi Ierápetra, July 2017. (J. Moody)

Genetics suggests that olive cultivation may have first spread through ideas rather than actual trees. Places that already had wild olives, such as Crete, could have learned to tend their native trees, slowly selecting them for desirable features, perhaps through grafting, and developing their own distinctive varieties. Today, however, the common olive varieties found on Crete – Koronéiki, Mastoeídis-Tsounáti, Mastoeídis-Mouratoliá, Throumboliá and Chondroliá – grow throughout Greece. The age and origins of these varieties remain shrouded in mystery.

All of the Monumental Olives of Crete are grafted. It is also claimed that they are grafted onto wild Cretan olive stock. Although this is possible, even likely, it cannot be proven because no genetically wild olives have been identified on the island. The Ancients, however, would not be so bothered by scientific classification; they called any olive tree that was not tended (i.e. pruned and harvested) wild (Theophrastus, *Enquiry into Plants*: Book 3).

Olive wood frustrates most dendrochronologists because its growth rings are indistinct and hard to interpret. There has been some success in Spain (Terral and Arnold-Simard 1996; Arnan et al. 2012), but so far not in the eastern Mediterranean where attempts have been controversial (Cherubini et al. 2014; Erlich, Regev and Boaretto 2018).

Olives in situ are combustible but, unless the fire is exceedingly hot, will resprout from their base. They are frost intolerant and usually do not grow above 600 m on Crete. On the south slopes of the White Mountains, however, olives can grow up to 1000 m. Most of the great old olives of Crete grow within the 'Camelot Zone' (Moody forthcoming and footnote 5).

Olive trees in Italy, France, Spain and Portugal are being decimated by Olive Quick Decline Syndrome *Xylella fastidiosa*, a bacterium spread by various flying insects, including cicadas (Antonatos 2019; Schneider et al. 2020). Oliver voiced concerns about this disease in a lecture we gave on 'The Olive in Crete' in 2013 at the Laboratory of Tree-Ring Research (University of Arizona). Although it is not yet in Greece or Crete, the deadly pathogen is edging closer (DeAndris 2020). At the time of writing (2020), there is no cure; the only recourse is to cut down the infected trees.

Trees and meaning in the Cretan Landscape

In Oliver's iconic book *The History of the Countryside*, he wrote:

> There are four kinds of loss. There is the loss of beauty, especially that exquisite beauty of the small and complex and unexpected, of frog-orchids or sundews or dragonflies. There is the loss of freedom, of highways and open spaces, which results from the English attitude to landownership .... There is the loss of historic vegetation and wildlife, most of which once lost is gone forever. ... I am especially concerned with the loss of meaning. The landscape is a record of our roots and the growth of civilization. Each individual historic wood, heath etc. is uniquely different from every other, and each has something to tell us.
>
> Rackham 1986: 25–26

The stories that the remarkable trees of Crete tell range from the sacred to the profane and from today all the way back to the Bronze Age (3000–1200 BC). They demonstrate how meaningful trees have been and continue to be to the island's unique culture. Here, I focus on 'sacred' trees in the Cretan landscape, a topic Oliver had been exploring when he died in 2015 (Fukamachi and Rackham 2012; Pungetti, Hughes and Rackham 2012; Rackham 2015; Tsiakiris et al. *this volume*).

Some Cretan trees are attributed 'miraculous and healing' powers. Most – though not all – of these are associated with chapels or monasteries and are hung with votives of one kind or another (*támata*, clothing, locks of hair). In some cases branches are cut from the trees and taken home as blessings, or their fruits eaten as cures or prophylactics (Faure 1972; Psilakis 1992, 2005; Warren 1994; Goodison 2010).

Figure 12.14 Roots of the Holy Carob at Aghios Antónios covered with icons, Mathés Apokorónou, February 2017. (J. Moody)

The Carob of Aghios Antónios, Mathés, Crete

Near the village of Mathés in west Crete grew one such tree, a carob *Ceratonia siliqua*, which literally engulfed a small chapel. Several hundred years ago, a carob seed took root in the roof tiles of the fourteenth-century Venetian chapel of Aghios Antónios. In 1918, the tree was 15 m tall and had a girth of 2.5 m (Defner 1928: 85–88). Pruning the tree was frowned upon by the chapel's saint, and Khatzegake (1954: 7f.) reports that its leafy branches nearly touched the ground because neither man nor beast dare cut or eat them. Legend claims that once, when a man approached the tree with his axe, the saint blasted the axe from his hands to over 1 km away. It is also said that two squirrels took up residence in a hollow in the tree, as one might expect of a veteran tree, but this was surprising because squirrels have never existed on Crete. The squirrels vanished when the tree died during a fire in February 1982 (Loulakis, n.d.). The chapel was untouched by the fire, which was considered a miracle – one of the many associated with this tree. Today, the massive roots of the great carob survive inside the chapel and are hung with icons and votives (Fig. 12.14).

The Evergreen Plane of Gortyn

The most renowned healing tree in Crete not associated with a chapel or monastery is the evergreen plane at Gortyn. The origins of this tree's miraculous status are mythical and related to us by several ancient authors: Theophrastus (*Enquiry into Plants*: 1.9.5), Varro (*De Rustica*: 1.7), Pliny the Elder (*Natural History*: 12.1.5). According to them, it was under the shady boughs of a great plane tree at Gortyn that Zeus lay with Europa and she conceived their three sons, the legendary kings of Crete: Minos, Sarpedon and Radamanthys. This act caused the tree to become evergreen and never lose its leaves. Forever-after, Gortyn's evergreen plane tree

Figure 12.15 Hellenistic silver stater coins (330–270 BC) representing the city-state of Gortyn in south-central Crete. They illustrate the story of Europa, the plane tree and the bull. From Svoronos 1890, plate XIV (also see plates XIII and XV).

has been associated with fertility and in particular with the conception of sons. Even today, women hoping to conceive sons visit the tree there, which is said to be a descendant of the original, and they collect its leaves (Warren 1994: 262) – this is in spite of the surviving tree being rather small.

Testaments to this story are the coins minted at Gortyn in the Hellenistic period (Fig. 12.15). On one face they depict a woman, probably Europa, sitting in a big, pollarded tree. The other face shows a bull. In some versions of the coins the tree has leaves. Although the leaf shape is not diagnostic, the associations with Gortyn and the bull suggest it is a representation of the evergreen plane tree of legend.

Trees in Bronze Age Crete

The Bronze Age culture on Crete is known as Minoan; the term derived from the name of the legendary King Minos, who was conceived under the sacred plane at Gortyn. Although the stories associated with Minoan trees have been lost to the mists of time, their images have not. Nature features strongly in Minoan art and that includes trees.

Not all trees depicted in Minoan art are associated with obviously sacred or ritual themes. Some seem simply to be part of the landscape (Fig. 12.16), although it is hard to be sure (Krzyszkowska 2010). One indicator that a tree may be sacred is its association with built structures, some of which have been interpreted as shrines. The trees shown growing on top of shrines are all veterans, with gnarled and twisted trunks and withered branches, as is illustrated par excellence in the 'Ring of Minos' (Fig. 12.17).

The meaning and interactions between figures, trees, 'shrines' and other objects in these scenes is open to interpretation (Evans 1901; Marinatos 1989; Warren 1994; Younger 2009; Goodison 2010; Day 2012; Tully 2017). It does seem clear, however, that a story is being told and that trees were key characters in it.

Figure 12.16 One of two Minoan gold cups from Váphieo, showing a veteran olive tree. (J. Moody)

Figure 12.17 The 'Ring of Minos', showing two trees associated with structures, possibly shrines, and being attended to by nude figures. Both trees have withered branches and are growing from a bulbous base that could be interpreted as a 'lignotuber'. Lignotubers develop from repeated cutting. (Wikimedia Commons)

Conservation

Of all the ancient and historic trees on the island, it is Crete's gnarled old olives that have attracted the most attention. This is probably because, besides being remarkable trees, they are accessible and encountered daily as people go about their normal lives – unlike the ancient cypresses, Prickly Oaks and Cretan Elms that grow in remote corners of the mountains.

Changing fashions in agriculture, such as the preference for a different kind of olive or the planting of novel crops such as avocados and kiwis, combined with urban expansion have been deadly to many great old olive trees. They can be dug out, chopped up and turned into charcoal, or burnt on the spot (Fig. 12.18). Fortunately, several associations have organized on Crete to protect and record the island's old olives. An initiative by the Society of Cretan Olive Municipalities convinced the Prefecture of Crete to declare more than 14 trees and one ancient grove in the Amári Valley as National Heritage Monuments, referred to as Monumental Olive Trees.

Another initiative, called 'Saving the Ancient Olive Trees of Crete', is a collaboration between the Technological Education Institute of Crete and the Network of Cultural Associations of Crete. They have recorded over 107 'ancient' olive trees.

The internet has also done much to promote the protection of veteran olive trees. Some Monumental Olive Trees, such as the ones at Ano Voúves (Fig. 12.6) and Kavoúsi (Fig. 12.13 bottom), have become celebrities and have their own websites (go to www.cretanbeaches.com and search for Monumental Olive). They are even becoming tourist destinations. Such economic benefits will hopefully become incentives for registering and protecting more of these trees.

Individuals have also taken action. Stelios Stephanoudakis rescued a grand old olive from the charcoal-makers in 2005. He happened upon the tree, already pulled from the ground and sold for charcoal, near the mountain hamlet of Tsourounianá. Appalled, he bought it on

Figure 12.18 I was taken to see a huge, old olive that Shlomo Abramov (pictured) and Stelios Stephanoudakis had found near the chapel of Aghía Marína, Choudalinaná Kíssamos, and this burnt out hole is what greeted us, July 2012. The arrow points to the stump that is all that remains of this once historic tree. (J. Moody)

the spot, trucked it to his hotel grounds near Kastélli Kíssamos and replanted it. When I saw the tree there in 2012, it was thriving (Fig. 12.19). We estimate it is about 500–600 years old.

Losing these grand old trees diminishes all of us. They are homes to myriads of organisms. They are irreplaceable archives of landscape history (both the cultural landscape and the natural landscape), and they are remarkable beings and companions on our walks through the countryside. We can only hope that as more people become aware of the existence, the value and the meaning of these wonderful old trees in Crete – or elsewhere in the world – they will be inspired to protect and preserve them, and less likely to destroy them. Spread the word!

Acknowledgements

Oliver Rackham, who was himself likened to a great oak by Keith Kirby (2015), introduced me to the trees of Crete in the 1980s. Together we investigated their fascinating shapes and adaptations, their size and variety, and their many levels of meaning in the Cretan landscape (Rackham and Moody 1996). My past, present and future research into the Cretan landscape will always owe a profound debt to him.

I would also like to thank Erik and Birgitta Hallager for inviting Oliver and me to study the charcoal from the Greek-Swedish Excavations at Chaniá; Stavroula Markoulaki for allowing Oliver Rackham, Tomasz Ważny and myself to examine charcoal from her excavations at Kastélli Kíssamos; Alexandra Karetsou for allowing Oliver and me to make a preliminary study of the charcoal from the Alonaki and Juktas excavations; and Luca Girella and Maria Ntinou for inviting Oliver Rackham, Tomasz Ważny and myself to examine some of the charcoal from Kamilári.

Figure 12.19 The transplanted Tsourounianá Venetian olive tree, rescued from the charcoal-makers, happily growing at its new home, a beach hotel near Kastélli Kíssamos, July 2012. (J. Moody)

References

Antonatos, S., Papachristos, D.P., Kapantaidaki, D.Evr., Lytra, I.Ch., Varikou, K., Evangelo, V.I. and Milonas, P. (2019) Presence of Cicadomorpha in olive orchards of Greece with special reference to *Xylella fastidiosa* vectors. *Journal of Applied Entomology*. https://doi.org/10.1111/jen.12695

Arnan, X., López, B.C., Martínez-Vilalta, J., Estorach, M. and Poyatos, R. (2012) The age of monumental olive trees (*Olea europaea*) in northeastern Spain. *Dendrochronologia*, 30, pp. 11–14. https://doi.org/10.1016/j.dendro.2011.02.002

Besnard, G. and Berville, A. (2000) Multiple origins for Mediterranean olive (*Olea europaea* L. ssp. *europaea*) based upon mitochondrial DNA polymorphisms. *Life Sciences*, 323, pp. 173–181. https://doi.org/10.1016/S0764-4469(00)00118-9

Borchert, R. (1976) The concept of juvenility in woody plants. *Acta Horticulturae*, 56, pp. 21–36. https://doi.org/10.17660/ActaHortic.1976.56.1

Breton, C., Tersac, M. and Berville, A. (2006) Genetic diversity and gene flow between the wild olive (oleaster, *Olea europaea* L.) and the olive: several Plio-Pleistocene refuge zones in the Mediterranean basin suggested by simple sequence repeats analysis. *Journal of Biogeography*, 23, pp. 1916–1928. https://doi.org/10.1111/j.1365-2699.2006.01544.x

Cherubini, P., Humbel, T., Beeckman, H., Gärtner, H., Mannes, D., Pearson, C., Schoch, W., Tognetti, R. and Lev-Yadun, S. (2014) The olive-branch dating of the Santorini eruption. *Antiquity*, 88, pp. 267–273. https://doi.org/10.1017/S0003598X00050365

Christopoulou, A, Ważny, T., Moody, J., Tzigounaki, A., Giapitsoglou, K., Fraidhaki, A. and Fiolitaki, A. (2019) Dendrochronology of a scrapheap, or how the history of Preveli Monastery was reconstructed. *International Journal of Architectural Heritage*, 15(10), pp. 1424–1438. https://doi.org/10.1080/15583058.2019.1685023

Contento, A., Ceccarelli, M., Gelati, M.T., Maggini, F., Baldoni, L. and Cionini, P.G. (2001) Diversity of *Olea* genotypes and the origin of cultivated olives. *Theoretical Applied Genetics*, 104, pp. 1229–1238. https://doi.org/10.1007/s00122-001-0799-7

Danti, R. and Della Rocca, G. (2017) Epidemiological history of Cypress Canker Disease in source and invasion sites. *Forests*, 8(121). https://doi.org/10.3390/f8040121

Day, J. (2012) Caught in a web of a living world: tree–human interaction in Minoan Crete. *PAN: Philosophy Activism Nature*, 9, pp. 11–21.

DeAndreis, P. (2020) *Xylella* outbreak in Apulian buffer zone puts millenary trees at risk. *Olive Oil Times*, October 1, 2020. [https://www.oliveoiltimes.com/world/xylella-outbreak-apulia-buffer-zone-millenary-trees/86036].

Δέφνερ, M. (1928) *Οδοιπορικαί εντυπώσεις από την Δυτικήν Κρήτη*. Βιβλιοπωλείον Ιωάννου Ν. Σιδέρη, Athens.

Del Tredici, P. (1999–2000) Aging and rejuvenation in trees. *Arnoldia winter*, pp. 10–16.

Del Tredici, P. (2001) Sprouting in temperate trees: a morphological and ecological review. *The Botanical Review*, 67(2), pp. 127–140. https://doi.org/10.1007/BF02858075

Erlich, Y., Regev, L. and Boaretto, E. (2018) Radiocarbon analysis of modern olive wood raises doubts concerning a crucial piece of evidence in dating the Santorini eruption. *Nature. Scientific Reports*, 8: 11841. https://doi.org/10.1038/s41598-018-29392-9

Evans, A.J. (1901) Mycenaean tree and pillar cult and its Mediterranean relations. *The Journal of Hellenic Studies*, 21, pp. 99–204. https://doi.org/10.2307/623870

Fazan, L., Stoffel, M., Frey, D.J., Pirintsos, S. and Kozlowski, G. (2012) Small does not mean young: age estimation of severely browsed trees in anthropogenic Mediterranean landscapes. *Biological Conservation*, 153, pp. 97–100. https://doi.org/10.1016/j.biocon.2012.04.026

Faure, P. (1972) Cultes populaires dans la Crete antique. *Bulletin de Correspondance Hellenique*, 96, pp. 389–426. https://doi.org/10.3406/bch.1972.2142

Fontanier, E.J. and Jonkers, H. (1976) Juvenility and maturity of plants as influenced by their ontogenetical and physiological aging. *Acta Horticulturae*, 56, pp. 37–44. https://doi.org/10.17660/ActaHortic.1976.56.2

Fukamachi, K. and Rackham, O. (2012) Sacred groves in Japanese satoyama landscapes: a case study and prospects for conservation. In: G. Pungetti, G. Oviedo, and D. Hooke (eds) *Sacred Species and Sites: Advances in biocultural conservation*, Cambridge University Press, Cambridge, pp. 419–423.

Goodison, L. (2010) *Holy Trees and Other Ecological Surprises*. Just Press, Dorset, UK.

Genin, D. (2018) Chapter 13: Rejuvenating the elderly and aging the youngsters: ancient management practices in continuously renewed native ash tree forests in the High Atlas of Morocco. In: S. Paradis-Grenouillet, C. Aspe and S. Burri (eds) *Into the Woods: Overlapping Perspectives on the History of Ancient Forests*. Éditions Quæ, Versailles Cedex, France.

Khatzegake, A.K. (1954) *Traditional Churches of Crete*. Rethymnon, Crete. [in Greek: Χατζηγάκη, A.K. (1954) *Εκκλησίες Κρητικές Παραδοσιακές*.]

Kirby, K. (2015) 'Fall of a Great Oak – Oliver Rackham has died. BES Forest Ecology Group webpage. 16-Feb-2015. https://besfeg.wordpress.com/2015/02/16/fall-ofa-great-oak-oliver-rackham-has-died/.

Kozlowski, G., Frey, D., Fazan, L., Egli, B., Bétrisey, S., Gratzfeld, J., Garfi, G. and Pirintsos, S. (2014) The Tertiary relict tree *Zelkova abelicea* (Ulmaceae): distribution, population structure and conservation status on Crete. *Oryx*, 498, pp. 80–87. https://doi.org/10.1017/S0030605312001275

Krzyszkowska, O. (2010) Impressions of the natural world. landscape in Aegean Glyptic. In: O. Krzyszkowska (ed.) *Cretan Offerings. Studies in Honour of Peter Warren* (*BSA Studies 18*), British School at Athens, London, pp. 187–96.

Loulakis, K. no date. Περπατώντας την Κρητική παράδοση. Τα Δεντρα της Πιστης. *ΣΤΙΓΜΕΣ, το Κρητικό περιοδικό*. [*http://stigmes.gr*].

Marinatos, N. (1989) The Tree as a Focus of Ritual Action in Minoan Glyptic Art. In: W. Müller (ed.) *Fragen und Probleme der bronzezeitlichen Ägäischen Glyptik* (*Corpus der minoischen und mykenischen Siegel, Beiheft* 3, Gebr. Mann Verlag, Berlin, pp. 127–143.

Moody, J. forthcoming. Blowing in the Wind: The Seasonality of Foraging in Late Bronze Age Crete. Oxford Journal of Archaeology.

Moody, J. (2018) Oliver Rackham: his legacy and the ancient trees of Crete. *Kentro*, 21, pp. 16–19.

Nikolakaki, S. and Hajaje, †H. (2001) Phenology of flowering of the evergreen oriental planes (*Platanus orientalis* var. *Cretica*) endemic in the island of Crete. *Forest Genetics*, 8(3), pp. 233–236.

Ntinou, M., †Rackham, O., Moody, J. and Ważny, T. (2019) The wood charcoal remains from the Kamilari tholos tomb. In: L. Girella and I. Caloi (eds) *Kamilari: una necropoli di tombe a tholos nella Messará (Creta)*. [Monografie Scuola Archeologica Italiana di Atene e delle Missioni Italiane in Oriente 29] All'Insegna del Giglio, Athens, pp. 639–646.

Psilakes, N. (1992) *Μοναστηρια και ερημητηρια της Κρητης*, τομος A'. Καρμάνωρ, Herakleion.

Psilakes, N. (2005) *Λαικές τελετουργιες στην Κρήτη*. Καρμάνωρ, Herakleion.

Pungetti, G., Hughes, P. and Rackham, O. (2012) Ecological and spiritual values of landscape: a reciprocal heritage and custody. In: G. Pungetti, G. Oviedo and D. Hooke (eds) *Sacred Species and Sites: Advances in biocultural conservation*, Cambridge University Press, Cambridge, pp. 65–82. https://doi.org/10.1017/CBO9781139030717.010

Rackham, O. (1976) *Trees and Woodland in the British Landscape*. Archaeology in the Field Series. J.M. Dent and Sons, London.

Rackham, O. (1986) *The History of the Countryside: The full fascinating story of Britain's landscape*. J.M. Dent and Sons, London.

Rackham, O. (2001) *Trees, Wood and Timber in Greek History*. [The Twentieth J.L. Myres Memorial Lecture, New College, Oxford.] Leopard's Head Press, Oxford.

Rackham, O. (2006) *Woodlands*. (New Naturalist 100) Collins, London.

Rackham, O. (2015) Greek landscapes: profane and sacred. In: L. Käppel and V. Pothou (eds) *Human Development in Sacred Landscapes*, V&R Unipress, Göttingen, Germany, pp. 35–50. https://doi.org/10.14220/9783737002523.35

Rackham, O. and Moody, J. (1996) *The Making of the Cretan Landscape*. Manchester University Press, Manchester.

Schneider, K., van der Werf, W., Cendoya, M., Mourits, M., Navas-Cortés, J.A., Vicent, A. and Oude Lansink, A. (2020) Impact of *Xylella fastidiosa* subspecies *pauca* in European olives. *PNAS*, 117(17), pp. 9250–9259. https://doi.org/10.1073/pnas.1912206117

Shay, C.T. and Shay, J.M. (1995) The modern flora and plant remains from Bronze Age deposits at Kommos. In: J.W. Shaw and M.C. Shaw (eds) *Kommos I. The Kommos Region and Houses of the Minoan Town. Part 1: The Kommos Region, Ecology, and Minoan* Industries. Princeton University Press, Princeton, NJ, pp. 91–162. https://doi.org/10.1515/9781400852956.91

Siitonen, J. and Ranius, T. (2015) The importance of veteran trees for saproxylic insects. In: K.J. Kirby and C. Watkins (eds) *Europe's Changing Woods and Forests: From wildwood to managed landscapes.* CABI, Wallingford, UK, pp. 140–153. https://doi.org/10.1079/9781780643373.0140

Stara, K. and Vokou, D. (2015) *Τα μεγαλειώδη δέντρα του Ζαγορίου και της Κόνιτσας*. Πανεπιστήμιο Ιωαννίνων, Ιωάννινα. [Στάρα, Κ. and Δ. Βώκου]

Svoronos, J.-N. (1890) [republished 1972] *Numismatique de la Crete Ancienne.* R. Habelt, Berlin.

Terral, J.F. and Arnold-Simard, G. (1996) Beginnings of olive cultivation in eastern Spain in relation to Holocene bioclimatic changes. *Quaternary Research*, 46, pp. 176–185. https://doi.org/10.1006/qres.1996.0057

Tsopelas, P., Santini, A., Wingfield, M.J. and de Beer, Z.W. (2017) Canker Stain: a lethal disease destroying iconic plane trees. *Plant Disease*, 101(5), pp. 645–658. https://doi.org/10.1094/PDIS-09-16-1235-FE

Tully, C. (2017) Virtual reality: tree cult and epiphanic ritual in Aegean Glyptic iconography. *Journal of Prehistoric Religion*, 25, pp. 19–30.

Warren, P. (1994) Tree cult in Contemporary Crete. In: Λοιβη. εις Μνημην Ανδρεα Γ Καλοκαιρινου, Εταιρια Κρητικων Ιστορικων Μελετων, Herakleion, pp. 261–278.

Ważny, T., Tzigounaki, A., †Rackham, O., Moody, J., Helman-Ważny, A., Pearson, C., Giapitsoglou, K., Troulinos, M., Fraidhaki, A. and Apostolaki, N. (2020) Trees, timber and tree-rings in historic Crete, Byzantine to Ottoman. *Archaiologiki Erga Kritis*, 4(2016), pp. 297–307.

Ważny, T., †Rackham, O. and Moody, J. in preparation. Dendrochronology of the Cretan Cypress.

Younger, J.G. (2009) Tree tugging and omphalos hugging on Minoan gold rings. In: A.L. D'Agata and A. Van de Mortel (eds) *Archaeologies of Cult: Essays on ritual and cult in Crete in honor of Geraldine C. Gasell* (*Hesperia Supplement*, 42), Princeton University Press, Princeton, NJ, pp. 43–49.

Xenopoulos, S. and Diamandis, S. (1985) A distribution map for *Seiridium cardinale* causing the cypress canker disease in Greece. *European Journal of Forest Pathology*, 15, pp. 223–226. https://doi.org/10.1111/j.1439-0329.1985.tb00889.x

Zakharakes, S. (1961) Το Εκκλησάκι του Αγίου Αντωνίου εις Μάθε Αποκορώνου. *Kritiki Estia*, 9, pp. 384ff.

CHAPTER 13

Walking in Sacred Forests with Oliver Rackham: A Conversation about Relict Landscapes in Epirus, North-West Greece

Rigas Tsiakiris, Kalliopi Stara, Valentino Marini Govigli and Jennifer L.G. Wong

Summary

In his travels to the Greek mountains, Oliver Rackham frequently visited the sacred forests of Zagori and Konitsa and offered his own fundamental insights on how to interpret this complex and peculiar cultural landscape. In this chapter, we present Oliver Rackham's views and thoughts on the sacred forests of Epirus, north-west Greece, aiming at reconstructing the ecological history of these socio-ecological systems, through his own inimitable way. The manuscript is based mainly on Oliver's observations derived from our common visits to the forests enriched with extracts from our ongoing scientific work on the sites in an effort to answer his questions. Results attempt at unravelling the environmental history of Mediterranean mountains, providing valuable lessons on the sustainable use of natural resources, community management, and nature conservation.

We present his inspiring piece of work to the wider public to motivate scientists to study the delicate interlinks between natural and cultural processes in sacred natural sites across the Globe, integral part of our natural and cultural heritage.

Keywords: sacred groves, Epirus, relict landscapes, ecological history, shrubland, coppice-wood, grazing

Introduction

Oliver Rackham first visited the sacred forests of Epirus with Kalliopi Stara and Rigas Tsiakiris in September 2005 after the BioScene project conference held in Ioannina. This first, unscheduled visit was the starting point of a collaboration that took on an official form with Oliver's involvement in the THALIS-SAGE project (2012–15), as well as in the Konitsa Summer School lectures on Environmental History (2014). In his later visits, starting from Konitsa town we explored the villages of the Aoos valley and those of eastern Zagori, walking through the biggest sacred forests and searching for evidence of their ecological history (Fig. 13.1).

Rigas Tsiakiris, Kalliopi Stara, Valentino Marini Govigli and Jennifer L.G. Wong, 'Walking in Sacred Forests with Oliver Rackham: A Conversation about Relict Landscapes in Epirus, North-West Greece' in: *Countryside History: The Life and Legacy of Oliver Rackham*. Pelagic Publishing (2024). DOI: 10.53061/COCN2417

Figure 13.1 Photograph snapshot of Oliver's visits to Zagori and Konitsa, September 2005, February 2012 and July–August 2014. (K. Stara, R. Tsiakiris)

Oliver's untimely death in 2015 cut short his collaboration with colleagues in Epirus, where by then he was as a senior advisor to the 'Conservation through religion: the sacred groves of Epirus' THALIS-SAGE project in north-western Greece (2012–15). The subject of the THALIS project was the sacred forests of the Pindos mountain range, which have been protected for centuries through folk religious taboos and supernatural beliefs (Stara et al. 2016). Oliver's last contribution to the project is dated August 2014 and is a draft report titled 'Sacred Natural Sites: Sacred forests in Epirus'. This report is a first account of visits to a number of such forests that Oliver made with the authors in 2014 (Fig. 13.2). The draft report is marked as 'not for publication' as it was a work in progress. Oliver begins by stating:

> This is an account of visits by O.R., together with Kalliopi Stara, Rigas Tsiakiris, and Valentino Marini Govigli to a number of sacred forests in the districts of Konitsa and Zagóri in 2014, together with occasional visits to others in earlier years. This is O.R.'s summary and interpretation of what we saw. It would be impractical to identify which of us is responsible for each of the observations and interpretations.
>
> Rackham 2014: n.p.

In our research, we sought to

> reconstruct the ecological history of each of these forests since they were declared sacred, mostly in the 18th century. We have done this mainly on the evidence of existing living and dead trees and other plants, especially evidence of forest structure. Dates where possible are derived from annual rings, although sacred forests contain few stumps from which rings can be counted.
>
> Rackham 2014: n.p.

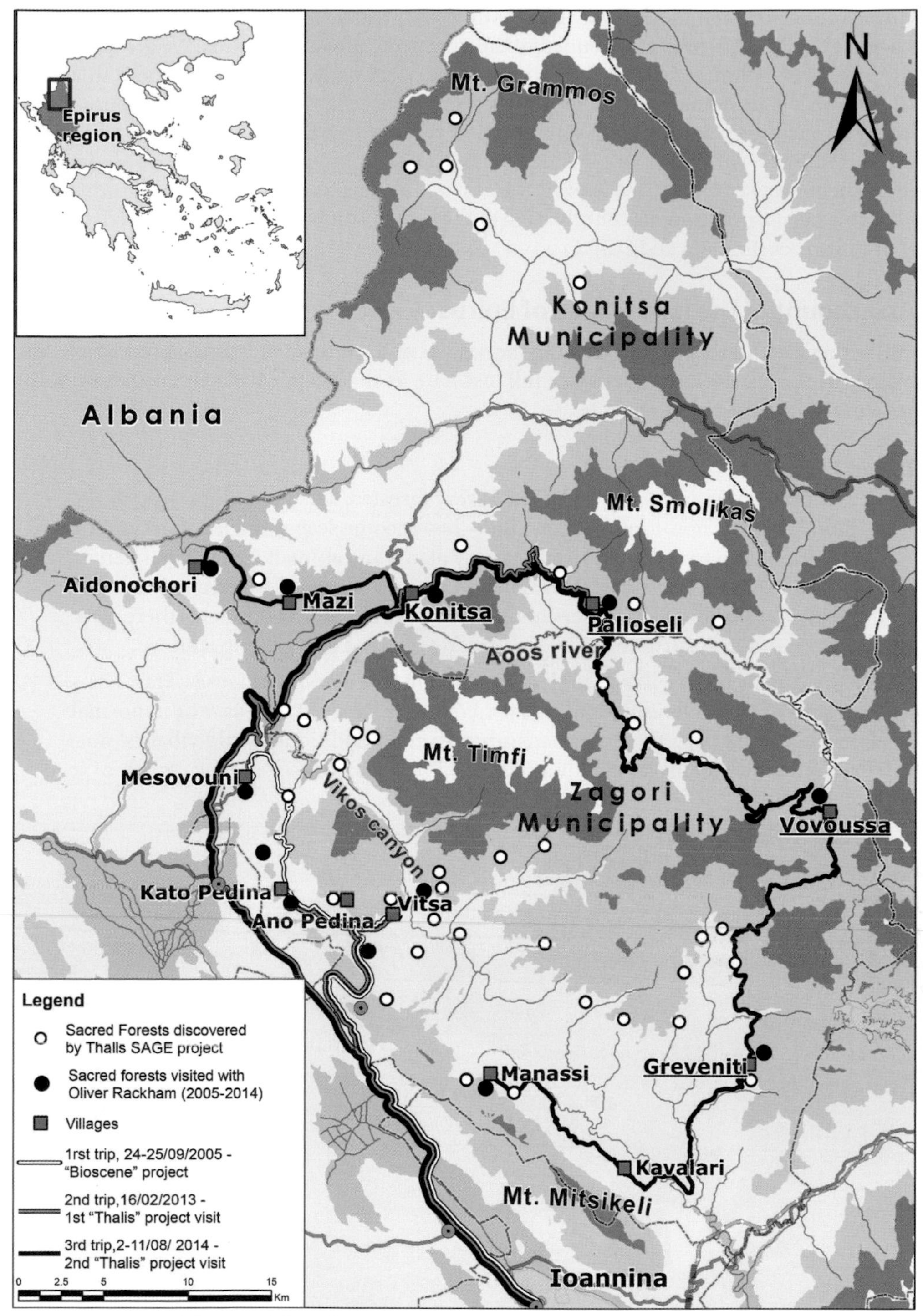

Figure 13.2 Sacred forests of the villages of Zagori and Konitsa municipalities (white circles) including those visited with Oliver Rackham (2005–14, black circles) and those reviewed in this chapter (underlined). (K. Stara, R. Tsiakiris)

Oliver's interpretation of what we saw on our visits greatly influenced and guided the later stages of the THALIS project and aided in the interpretation of the results. We greatly missed Oliver's input and we constantly had in our minds 'what would Oliver think about this?', but all we have left are memories and his draft report. In this chapter, we wish to honour Oliver's legacy to our project by revealing something of the way he worked in the field and to place the sacred forests of Epirus into his oeuvre as a work in progress. As an acknowledgement of the debt owed to Oliver by the project team, but also by us personally as his students, we offer this up as a posthumous continuation of the conversation on the nature of sacred forests started in 2005 (Figs 13.3 and 13.4).

An account of the landscapes of Epirus

Oliver prefaced his report by describing the natural and cultural landscapes, present use, and history of Epirus. We present Oliver's full text here as this remains a useful overview of the area and context of our later study.

Inland Epirus

> The mountainous interior of Epirus is very forested. Trees have increased to the point at which non-forest – open land – has become scarce and is attracting the attention of conservationists for the many plants and animals that will not persist in shade or need non-forest areas. Villages are now islands of habitation in a sea of forest and shrubland. A century and a half ago this was very different: there were islands of forest surrounded by well-tree'd farmland and grazing land.
>
> Epirus is a land of villages, settlements of a hundred or more houses. Hamlets – groups of tens of houses – and isolated houses are few, but may have been normal in the distant past; there are many unconfirmed traditions of 'old villages' now deserted. Every village had hundreds of hectares of cultivated land, much of it

Figure 13.3 Fieldwork with Oliver Rackham and Kalliopi Stara in Eastern Zagori, August 2014. (R. Tsiakiris)

Figure 13.4 Teaching by Aoos river in Konitsa Summer School, August 2014. (R. Tsiakiris)

terraced, growing cereals, pulses, and vines. There were hundreds of hectares of pasture land, either grassland or shrubland, grazed by the livestock (cattle, pigs, sheep, goats, horses, mules, and donkeys) of the villagers and by the flocks of itinerant Vlachs and Sarakatsani. Pasture occupied the steep slopes, high altitudes, and the badlands that in some areas cover more than half the territory. There was an infrastructure of *kaldirími* mule-roads for long-distance caravans of commercial transport, local footpaths, and irrigation canals. The limit of settlement is higher than elsewhere in Greece: many villages are at 1000–1200 m altitude. Consequently hay or tree-leaves need to be stored to keep the livestock alive in winter. The most productive grassland was meadow (*livádhi*), mown for hay. Other fodder came from leafy branches of trees (*kladhí*) dried and stored.

Many of the people were only partly dependent on agriculture for a livelihood: they were also traders and craftsmen. Often the men worked away from home (especially as stone-masons); agriculture and tree management was done mainly by women. Although there has always been some forest, it was nothing like as continuous as it is now, and there were many more non-forest trees. Savanna trees were scattered in pasture, and tens of thousands of 'working trees', pollards and shredded, were in hedges and terrace walls among the cultivated land.

In most European Mediterranean countries, mountains became depopulated in the twentieth century. Especially in the decades after World War II, agriculture became mechanized: agricultural engineers failed to respond to the needs of mountain farmers; new machines would not work on steep slopes, so that cultivation became restricted to plains. In Epirus, land abandonment has been especially severe. The region suffered severely from World War I, World War II, and the Greek Civil War. The establishment of the frontier between Greece and Albania in 1913 – where there had never been a frontier before – cut off trade routes

> and separated producers from consumers, especially in the decades when Albania was rigorously isolationist and cut itself off from other countries.
>
> Nearly all the villages are still inhabited, and give an impression of prosperity. Even in those that escaped damage in World War II, half or more of the houses are less than 40 years old. Gardening still flourishes – although there are also many abandoned gardens. For most of the day no domestic animal can be heard, not so much as a cock. Shepherding continues on a much reduced scale, not sufficient to hold down the increase of trees. A little hay is cut, probably on former cultivated land. In some areas there is still a regular programme of felling timber, though not enough to keep up with the growth of trees. An attempt at promoting mountain tourism seems now (2014) to have met with little success. Most of the inhabitants are either retired or have an outside source of income, such as an office job in Ioánnina or Athens. Restoring cultivation is likely to be frustrated by enormous numbers of wild pigs.
>
> Rackham 2014: n.p.

As described by Oliver, what we see today is a relict cultural landscape that echoes the prosperity of the years of the Ottoman occupation (1479–1913) giving way to population decline and degradation of farming extent and practices. This was especially so after 1945. As a consequence, the municipalities of Zagori and Konitsa are today the least-populated regions of Greece (National Population Census 2011). Historical information confirms that villages were formed from the consolidation of earlier hamlets during the sixteenth and seventeenth centuries (Papageorgiou 1995). Narratives ascribe the abandonment of old settlements with the appearance of snakes, demons or angels who made the place dangerous, while the location of extant villages are thought to be divinely mandated through epiphanies that identified the position for their central church. Often the church is the only building still standing in ruined settlements, functioning today as an outlying church (*xoklissi*) and accompanied by its ancient trees (Stara et al. 2016).

Most villages occur in the oak zone (around 1,000 m), based on a mixed system of agriculture and animal breeding, with the surrounding landscape characterized by terraces and *kladera* (groups or isolated shredded trees; see Halstead 1998). Occasionally, villages are set at higher altitudes, either inhabited all year round (e.g. Vradeto in Zagori at 1,340 m) or only during summer months (e.g. Aetomilitsa in Konitsa at 1,430 m, a Vlach village and one of the highest settlements in Greece).

The past wealth of Zagori and Konitsa is mirrored in the built environment and the historical infrastructure, which was not based exclusively on local production, as technical specialization (stone masons), mobility and even migration provided ways out of economic and demographic difficulties (Nitsiakos 2016). Thus, villages' wealth was augmented with repatriation resulting from male emigration to Europe, Africa and America (Stamatopoulou 1998). The distinctive architecture exhibited in most villages, especially in Zagori, is today protected as 'traditional' (Ministry of Environment and Energy, FEK 594/13-11-1978). However, there are also many new buildings (especially in the Aoos valley and eastern Zagori), which replaced houses burnt by German Nazis during the Second World War. In some villages of Eastern Zagori, all houses except the central church were burnt (e.g. Vovoussa, FEK 197/140/5-8-2008). Postwar abandonment further transformed the cultural landscape around villages, with severe consequences to present day biodiversity (Tsiakiris et al. 2009; Zakkak et al. 2014).

Environment

> The climate is mountain-Mediterranean, with cold winters as well as dry summers, but most summers have some rainfall. The growing season is in spring, extending into the summer if moisture is still available. Most trees (except the

> non-forest plane-tree) grow slowly because of the short season. The geology of most forested areas is either flysch or hard limestone. Flysch varies from thinly bedded micaceous sandstone – used for roofing slabs and paved kaldirími – to friable mudstone. It is very erodible and develops into badlands – landscapes of gullies that are bad for agriculture. Some of the badlands are still active, with trees and shrubs tumbling into the gullies, and slumps where modern roads intersect the unstable slopes. Much of the forest overlies former, probably Pleistocene, badlands. Limestone is karstic, weathering into cliffs and limestone towers.[1] Where limestone overlies flysch, it forms breakaways, where towers and boulders fall off and roll down the slope to menace villages below. Badlands and breakaways may have been more active in the Little Ice Age, when sacred forests were being declared.
>
> Rackham 2014: n.p.

This geological richness was recognized in 2010 with the inclusion of Vikos–Aoos in the UNESCO Global Geoparks; local traditional architecture is part of the geopark cultural heritage (Papaioannou and Kitsaki 2014). The inherent instability of the area has been amply demonstrated by historical and recent earthquakes, such as the one in 2016 in which falling rocks and damage to historic buildings and roads were triggered by a mild (5.3 on the Richter scale) event.

Place-names

> There is evidence of four rural languages: Greek, Vlach, Slavic, and Albanian – each from a different family of Indo-European languages. In the last 200 years different villages have spoken either Greek (in the west) or Vlach (in the east). Most villages have a name in Slavic (rarely Albanian or Vlach), even though none has recently spoken that language. Minor place-names tend to be Greek, Slavic, or Vlach. Thus Palioséli, a Vlach village with a Greek name, has a sacred forest with the Vlach name Miriáouwa. Kourí, the name of two sacred forests, is said to be the Turkish word for such forests. In the period of Hellenization in the 1920s and 1930s many places with names in other languages adopted Greek names (not usually a translation of the previous name). Here I give both the Greek name and its predecessor as recorded by earlier historians and travellers, for example *Eláti/Boults.*
>
> Rackham 2014: n.p.

The etymology of sacred forest toponyms was a key part of our research. Most sacred forests of the area are named *kourí*, indicating a piece of forest kept as a protected reserve for community needs (Grispos 1952). There are several alternatives proposed for the word etymology. It could derive from the Turkish verb *koru-mak* (take care of), or perhaps from the Latin *curare* with a similar meaning. Papanastasis et al. (2009) associates the word with the Greek noun *koura*, deriving from the words *kouros* (young) and *kourizo* (make young), and connected places named *kourí* to high grazing pressure areas and to the practice of tree shredding. However, in our study area shredding only occasionally occurs in sacred forests and is referred to in all cases as trespassing (Stara, Tsiakiris and Wong 2015). With the long influence of the Ottomans, we consider the name *kourí* in the Pindos mountains to have a

1 These are the original habitat of Horse Chestnut *Aesculus hippocastanum*, which as a wild tree is a rare cliff endemic (Rackham, original comment). Horse Chestnut is a very common ornamental tree all around Europe, native only to central Balkans and Greece where it is considered as endangered (Khela 2013). Zagori probably acted as glacial refugium for the species. A photograph taken by Oliver during our first visit in Zagori is included in his book *Woodlands* (2006: 277).

Turkish origin. In addition, the name *livádi* (literally meadow) is also used for several sacred protection forests above villages and these are generally open wood-pastures. However, the etymology of this word possibly derives from the ancient Greek *livás* (water drop) (Babiniotis 2002), because of the parallel forest function to regulate water along with its protective role (Stara and Tsiakiris 2010).

Savanna and problems of definition

> In Mediterranean countries it is often difficult to classify a particular hectare into forest or not forest. There are two intermediates: shrubland, where the trees are too small to count as forest, and savanna, where the trees are too wide apart. Shrubland contains more tree- and shrub-species per hectare than most forest, and has abundant low vegetation so as to be useful grazing land. Savanna produces a distinctive shape of tree; because the tree had no close neighbours, it has wide-spreading low branches, in contrast to the narrow crown and branchless stem typical of forest trees.
>
> These are the distinctions that we make. Whether local people (or officials) make the same distinctions, and at the same point on a continuum of structures, needs ethnographic investigation. The category of 'sacred forest' can include savannas, but appears not to include what we would call shrublands.
>
> A common change is for savanna to infill into forest as a new generation of trees fills the gaps between the original trees. The infill trees may be of the same or a different species. No obvious rules have emerged as to which tree infills which, except that fir seems not to be an original savanna tree. In this high-rainfall area it seems that grazing, rather than drought or fire, is the main determinant of savanna versus forest. Part of the recent increase of forest is at the expense of savanna.
>
> Rackham 2014: n.p.

Savanna and the problem of forest definition was one of Oliver's beloved conversation topics (Grove and Rackham 2001: 191–216, 185–87). In his original notes, Oliver used the word 'forest' (δάσος), which is a term also used in Greek legislation (Law 998/79) referring to areas covered fully or partially with native tree species regardless of size or structure.

In his visits, Oliver always carried old maps and old sketches showing scattered trees in bare landscapes. Additionally, he was especially interested in old photographs, drawings, etchings or postcards, particularly those related to Ioannina histories of the Ali Paşa Tepelenë era (1744–1822). He also used travellers' accounts (e.g. Leake 1835), with the intent to follow the same routes and compare landscape descriptions with the contemporary ones, locating old place-names, paths, plant species, monuments and so on. We were delighted to hear him reading these old texts, even seeing similarities in weather conditions, as in a case in Konitsa town in July 2014, when stormy days were similar and on the same dates as described in Baldacci (1899). Old pieces of timber such as the gigantic baulks of oak in the outer doors of Ioannina Castle were shown to be very helpful for the reconstruction of historical landscapes. The arrangements of veins showed that they had grown with horizontal branches similar to the trees found in open wood-pastures in Epirus. Similarly, oak lintels in one of the oldest houses in Kato Pedina, in Zagori (1759), yielded clues about the local scarcity and low quality timber available in the past, as they were crafted from malformed deciduous oaks grown in open landscapes.

In the THALIS project we used ethnographic research in combination with aerial photographs taken in 1945, to locate sacred forests and to see more clearly the dynamic of the savanna infilling process proposed by Oliver. Through these analyses, several sacred forests were shown to derive from savanna-like ecosystems, later infilled with various tree species (Pion 2014).

Tree species

'We have done this [the reconstruction of the ecological history] mainly on the evidence of existing living and dead trees and other plants, especially evidence of forest structure. (Rackham 2014: n.p.). Oliver always urged researchers to investigate carefully the dominant trees and shrubs for specific adaptations to grazing, sprouting and so on. As he noted in *The Making of the Cretan Landscape*, 'the present holds the key to the past' (Rackham and Moody 1996: 109). Therefore, he added in his THALIS report particular details for each species found in the area as important explanatory ecological information. The original text is stored by Cambridge Digital Library, which holds Oliver's digital archives.

Shrubland and coppice-woods

> Most local trees can exist in a browsed form as shrubs, growing up into trees if browsing ceases. Shrubland is the deciduous equivalent of the maquis of lower altitudes. It covers great expanses of hillside, for example on both the Albanian and Greek sides of the River Sarandapóros. Shrubland is typically of many trees and shrubs, including (roughly in descending order of abundance) hornbeam (*Carpinus orientalis*), junipers (*Juniperus oxycedrus* and *J. communis*), manna-ash (*Fraxinus ornus*), various deciduous oaks, hazel (*Corylus avellana*), maple (*Acer monspessulanum*), Cornelian cherry (*Cornus mas*). *Paliurus spina-christi* and *Cotinus coggygria* are abundant alongside roads. On grazed land these exist as bitten-down bushes, sometimes interspersed with pollarded or shredded savanna trees. When grazing ceases they grow up into an often impenetrable thicket of small trees, gradually sorting themselves out into hornbeam-dominated forest. Much shrubland had a previous existence as coppice-woods, periodically felled for firewood. The trees sometimes form stools 4 m or more in diameter, which are probably several centuries old; but how far these were produced by browsing or woodcutting is not yet determined.
>
> Rackham 2014: n.p.

Shrubland is the result of silvo-pastoralism and is the dominant vegetation type in much of Epirus, interspersed with wood-pastures and forest stands in less accessible areas. Shrub species differ between sites depending on geology and their grazing and fire history. In the case of limestone, Prickly (Kermes) Oak *Quercus coccifera* dominates the landscape, whereas Green Olive *Phillyrea latifolia* dominates areas in the badlands of Konitsa. Consecutive clear cuts and repeated use of those areas for charcoal production in the past (and still in Albania) might also have played a significant role for each species abundance and distribution.

Grazing history

> Most forests produce some pasturage. Dense or shady forests, especially beech, fir, or young pine, have little edible vegetation within reach of an animal, and are not much use as grazing land. Savanna with sparse trees may be nearly as productive of pasture as open land. Most of the forests that we saw are used as pasture, but the intensity of grazing and browsing vary from place to place with distance from the origin of the animals and variation in the density of trees. In Mázi, at the bottom of the forest, the trampling of a large flock of goats has destroyed almost all low vegetation, but upslope this extreme passes into a moderate degree of browsing. We are sometimes told that sheep and goats will not eat pine. This is not strictly true – young pine is sometimes browsed. Pine is evidently not preferred but will be eaten if there is not much else. In Kónitsa and Palaioséli many pines of the *c.* 250–300 year generation – in some areas of the forest – have gnarled and crooked bases, with many trees forked into two or three at the base. In

> Miriaouwa (Palaioséli) there are rare instances of pines that retain dead horizontal branches in the lower 2 m and in other instances torchwood cavities reveal the bases of lower branches that have long been occluded and overgrown. We infer that these forests passed through a stage in the 18th century as grazing land, in which animals browsed them into a bushy form, from which they got away into trees when grazing diminished.
>
> Rackham 2014: n.p.

Grazing in the sacred forest of Mazi has been mentioned by locals as one of this forest's past uses. However, in 1922, after a series of fires, the Forestry Service prohibited logging, grazing, and shredding for a fifteen-year period for goats and a ten-year period for sheep and cattle. These restrictions were reimposed at ten-year intervals with the agreement of the local community, though with conflicts with local shepherds concerning the impact of overgrazing in forest regeneration for protective and aesthetic reasons. This culminated in the complete exclusion of livestock from the sacred forest, removal of senescent old trees and the erection of a fence and gate around the protected area. The restriction on grazing promoted regeneration such that we recorded high densities of mainly Manna Ash *Fraxinus ornus* saplings under scattered large oak trees. Excessive grazing prevents successful oak regeneration, but absence of or low intensity grazing may also create conditions unfavourable to oak establishment. Disturbance and changes in management practices may also have affected oak populations. This is an example of how the infilling of open wood-pasture can be of markedly different tree species (Rackham 2006). Stand curves for the forests also had a story to tell. In Mazi, there were fewer than expected medium-sized oaks. It is known that until the end of the twentieth century, agro-pastoralists grazed their herds in the sacred forest. It was a shady area (*stalos*), where they could take their livestock in the hottest hours of the day. Grazing was intensive and mixed, with sheep, goats and cattle, while dairy livestock were milked in the forest. Cutting small trees and feeding them to animals in situ was allowed, as was grazing during the winter and foraging for acorns. These activities would have resulted in heavy trampling and soil compaction, as well as browsing of seedlings, which may have prevented effective seedling recruitment and hence the lack of medium-sized oaks.

Dead trees

> A natural forest in a steady state should contain a certain proportion of dead trees, depending on the ratio between the average longevity of a tree and the average time it takes to rot away and disappear once dead. Most of our forests appear to be short of dead trees: probable exceptions being Vovoúsa and Greveníti. This may be because dead or fallen trees might legitimately be taken away for firewood. However, in that case there should be visible stumps (or root-holes if the trees blew down), which we did not usually find. Most of the sacred forests have therefore not reached a steady state.
>
> Rackham 2014: n.p.

Deadwood is indeed a rare component in most of the sacred forests visited during the THALIS project, with the exception of a few sites that are often associated with stronger rules and stricter access restriction (e.g. Vovoussa and Greveniti). Deadwood, as remarked by Oliver, is a good indicator of low human disturbance. This is further proven by the low deadwood density in the agro-pastoralists' sites sampled in our later research, which, along with the evidence of old stumps, suggests that deadwood collection might have been allowed in certain cases.

Fire

> In most Mediterranean countries forest fires are a normal event at frequent or infrequent intervals. The low- and middle-altitude prickly-oak maquis of Epirus

> burns from time to time; where savanna has infilled with *Quercus coccifera*, a subsequent fire may kill the old savanna trees. The mountains of Epirus, including sacred forests, rarely burn, owing to the wet climate and lack of flammable trees. *Pinus nigra*, like all pines, is fire-adapted, but is among the least fire-promoting of pines. Single pines attract lightning, but are seldom set on fire, and if they are ignited the fire does not spread. In plantations in Spain, the average interval between fires for *Pinus nigra* was 217 years, compared to 22 years for *P. halepensis* (Grove & Rackham 2001: 235). In the wetter climate of Epirus, the fire return interval for *P. nigra* appears to be well over 300 years. An indicator of lack of fire is the abundance of the fire-sensitive *Juniperus oxycedrus* in many sacred forests. When individual pines are burnt at the base in connexion with gathering torchwood [resinous, flammable material suitable for making torches], such fires seldom spread to leaf-litter and never to other pines. Litter of black pines compacts into a non-flammable layer, especially where mixed with dead leaves of *Acer opalus* and other fire-excluding trees. We found traces of only small litter fires in sacred pine forest. Fir (*Abies*) does not easily burn but when it does is killed. An area of fir was burned on Mount Mitzikéli in the unusual fire year of 2007; most of the firs were killed but the intermingled broadleaved trees survived. Deciduous oaks sometimes have litter fires, but we found no evidence here. Beech, maples etc. are fire-excluding owing to their lack of flammable chemicals and their dense shade which excludes flammable grasses etc. We have encountered stories of forests being burned in World War II or the Civil War. This we have not verified. Burning the forests would be difficult now, but may have been possible in the past when the structure was more favourable to fire.
>
> Rackham 2014: n.p.

Fire is a basic element of the Mediterranean ecological history. In Epirus it was historically used by shepherds in all habitat types to create new pastureland or to remove understorey vegetation (still practised during winter even in the dense lowland oak forests near the Albanian border). The Black Pine *Pinus nigra* is a well-adapted species that can survive litter fires, owing to thick bark, self-pruning habits, and high crown. Evidence of several non-lethal ground-litter fires in a 815-year old forest stand (1196–2010) in Valia Calda (near Vovoussa sacred forest), confirms that fire has been a persistent factor in the development and ecology of current forest stands, likely affecting both forest structure and recruitment patterns (Touchan et al. 2012). Owing to land abandonment and the decline of goat grazing, dense, evergreen, homogenized, sclerophyllous maquis expands alongside young pine forest, invading former pastureland, oak wood-pastures and agricultural land, making them flammable. According to Forestry Service data from the 1930s, fires in the past were much more frequent but burnt relatively small areas, while recent fires have been more infrequent but have affected much larger areas – something that is encouraged by climate change.

Such wildfires threaten sacred forests that are no longer isolated patches of woodland (e.g. a rare stand of old Montpelier Maples *Acer monspessulanum* in the monastery forest of Agia Paraskevi in Ano Pedina village was burned in 2000). Old fire evidence has been found in few sacred forests (e.g. Kato Pedina, Mazi, Palioseli); however, its influence in present-day vegetation remains an open question for further research.

Woody ground vegetation[2]

> Tree seedlings are locally abundant but generally absent – a consequence of shade combined with browsing. Where they do occur they often do not match the canopy

2 Originally 'Herbaceous plants and woody ground vegetation'. As Oliver noted, 'Our visit was in summer, so these remarks are very incomplete.' Therefore, the original text is not included.

> trees. For instance, in Mázi Kourí, an oak forest with very few ash trees, the young trees are predominantly ash. Oak seedlings generally do not persist beyond one or two years: oak regeneration may be mainly via ground-oak [dwarf, scrubby oaks often encountered in the Mediterranean].
>
> Rackham 2014: n.p.

Owing to the heterogeneity of habitats within even a single sacred grove, seedling number as noted by Oliver fluctuates across patches, more abundant in open forest areas and scarcer in dense shrubby forest patches (Pion 2014). The THALIS forest inventory proved that browsing in grazed sacred forests has a direct effect on the abundance of seedlings and saplings of deciduous oaks and their morphology (Marini Govigli et al. 2020). On the contrary, sacred forests with stronger prohibition rules, as the excommunicated forest of Greveniti, are closer to natural forests and exhibit gap dynamics of regeneration (Cullen 2015).

Sacred forests of Epirus: relict landscapes of the past

> Sacred forests are relict landscapes. Usually they are not 'foresters' forests' managed for timber production, consisting of uniform straight young or middle-aged trees felled according to a regular programme. Most of them have been used, but not in this way: they contain evidence of grazing and of harvesting products other than timber. They often contain trees more than 250 years old and veteran trees which are of importance as historical antiquities and as habitats for animals, birds, and lichens.
>
> Rackham 2014: n.p.

Our research on the sacred forests in Epirus dates back to 2000 when we first appreciated that local taboos played an important role in the protection of specific wooded areas, mainly wood belts above mountain villages (Stara 2000). After almost 20 years of research, we discovered more than 80 such forest and groves (Avtzis et al. 2018) and many more sacred natural sites (Stara et al. 2015). Oliver was fascinated by this peculiar element of the cultural landscape of the area, and this formed the core of our long-lasting friendship and collaboration across the Pindos Mountains and their environmental history.

Sacred forests[3]

> Most villages contain more than one church. Other churches are scattered round the landscape, sometimes associated with springs or hilltops or with (rumoured) deserted villages. Many churches have one or a few trees of wild species, especially oaks, associated with them. These are usually not working [pollards or coppice] trees – except (as at Pigí) where a new churchyard has included pre-existing trees.
>
> Sacred forests are areas of forest, hectares in extent. They are said to have been declared by a definite liturgical procedure – called 'cursing' or 'excommunicating' – involving five or preferably seven priests and a procession to delimit the sacred area. Banned activities normally included felling timber, woodcutting, and sometimes grazing, on pain of death or other temporal or spiritual penalty. Sometimes the community could authorize felling timber for some public purpose such as building a bridge. Sacred forests are now often inconspicuous in the landscape, hidden among much more extensive non-sacred forest, but before land abandonment would often have been much of the total forest.
>
> Our informants considered them often to have been declared sacred because they fulfilled some public function. They were expected to protect the village from events happening above: 'badland' erosion, breakaways (trees were intended to

3 The subchapter 'Sacred forests' was originally placed before 'Tree species'.

> catch falling rocks before they reached the houses), avalanches etc. or to protect the village water supply. These are the interpretations put on the forests by the present inhabitants, and may or may not be the reasons why they were originally declared sacred. Many sacred forests contain an isolated church, which is usually difficult to date. Churches tend to be immediately surrounded by their own sacred trees, which may be different from those of the sacred forest, for example the great oaks around the chapel in the sacred forest of Vovoussa.
>
> Rackham 2014: n.p.

Solitary trees or groves accompanying outlying churches are a common element of the Greek landscape. Dr A. Stengel, an Austrian forest expert invited as a consultant by the Greek government at the beginning of the twentieth century, said that 'the only way of reforesting Greece is to condense the network of outlying churches' (Grispos 1973: 144). A survey of 824 individual trees of 231 sacred natural sites in 23 villages of Zagori revealed 51 different tree species, the majority of which were evergreen/deciduous oaks and maple species (Stara et al. 2015). Dedications to a church and most often to the church of the village patron saint or excommunication were used as (de)sanctification practices during the Ottoman occupation. After the establishment of the modern Greek state, special forestry regulations replaced past practices. However, locals continued to honour taboos associated with the supernatural guardians of the sacred forests (Stara 2012b; Marini-Govigli et al. 2021). In his footnotes, Oliver added: 'to what extent excommunication of forest had a basis in Orthodox formal theology remains to be investigated'. Excommunication constitutes the heaviest sentence that can be imposed on Orthodox Christians by exclusion from the society of the Church, being deprived of its mysteries, condemned to a cursed life as trespassers, which applies also to their descendants, and the damnation of the soul to eternal hell. From the later Byzantine period, and particularly during the Ottoman occupation, excommunication was employed as a common practice for economic or social reasons (Mihailaris 2004). In essence, it imposed an abstract threat to protect trees, forests and other natural resources from potential trespassers (Stara 2012a). In our visit to the archives of the Ecumenical Patriarchate of Constantinople (2014), we found applications for excommunication in private cases of an economic or social nature, but none specifically related to forest protection.

Even less current today, the term 'sacred' originally included an aspect of fear (Verschuuren et al. 2010), while supernatural punishments related to trespassing have been perceived as the result of a saint's anger or curses. These could range from warnings or accidents during tree felling to misfortune, economic catastrophe, sudden deaths or tragedies as severe as the annihilation of whole families. The many deaths ascribed to trespassing in the forest of Agia Paraskevi in Vovoussa induced such fear that people avoided using the land in the vicinity of the sacred grove, preferring to recognize it as a saint's property (Stara et al. 2016). Oliver quoted a similar story in his manuscript from a nearby area in Epirus *Eleftherokhóri* (near Paramythiá), *Makrokhórafo*, visited in 1988:

> In 1988 Jennifer Moody and O.R. noted the isolated oak forest of Eleftherokhóri and asked the local policeman about it. 'This forest belongs to the Church, and nobody is allowed to cut down a tree in it.' 'But we found a stump and counted the annual rings.' 'I know who did that. He was killed in a motorcycle accident shortly after.' [Local people said that the forest] was given to the church of Ay. Varvára to stop the Turks felling the trees. 'To fell any tree brings death.' However, the oaks are all pollarded.
>
> Rackham 2014: n.p.

Ethnographic research indicated that sacred forests in Epirus were a sophisticated and locally adapted management system of forest resources. According to local oral history, the unstable geology of those mountains is one of the reasons for the creation of sacred forests, functioning in parallel as a last resort in times of need or preserving ecosystem services in common lands. Sacred forests might exceptionally even give their timber for the benefit of the community in order to build churches, schools or bridges, as in the case of Konitsa town during the Second World War, as stated earlier by Oliver. Our subsequent field research additionally confirmed the protective function of sacred forests against natural hazards and especially against rock falls in several villages (e.g. Aristi, Dikorfo, Koukouli, Manassi, Mikro Papingo, Pades, Palioseli). Taking into account several variables such as precipitation, geology, inclination, orientation, spring presence, vegetation type and distance from village infrastructures, we statistically tested several landscape models that successfully proved the non-random location of sacred forests (Tsiakiris et al. 2017). Furthermore, according to Kalantzi (2008), who studied the history of landslides in the wider area of Ioannina, the majority of catastrophic events since 1960 (88 per cent) were due to extreme weather (rain/snowfall and ice; note that our research area has one of the highest levels of precipitation in Greece, with some places getting over 2,400 mm/year). Here, as elsewhere, protection forests help safeguard against landslides.

Walking in Pindos mountain's sacred forests with Oliver Rackham

Oliver's report contains notes on 12 sites in Epirus dating back to 1988. However, we present in this chapter only five sites, in which we conducted further detailed ethnographic work and tree surveys (Table 13.1). Along with Oliver's original text, we add a short concluding paragraph that summarizes some of the final findings of the THALIS project. Our intention is to introduce the reader to some of the sites on which Oliver worked and to show how our later work supported Oliver's findings and sometimes questioned them.

Kourí forest, Mazi village

> On the hill above the village is a sacred forest, abruptly differentiated in structure and composition from the grazed shrubland around it. Slope is unstable, with limestone boulders perched precariously on a steep flysch slope. In the forest are three churches: 1. Modern hilltop church of Prophétes Elías, with a few recently planted trees. 2. Ruined church beneath it. Two phases of building: earlier has

Table 13.1 Summarized data on five sacred forests visited with Oliver Rackham.

Town / village name	1. Mazi	2. Konitsa	3. Palioseli	4. Vovoussa	5. Greveniti
Sacred forest name	Kouri	Kouri	Mereao	Agia Paraskevi	Toufa
Mean altitude (m)	480	600	1,080	1,000	980
Area (ha)	10.40	115.70	22.40	6.80	43.29
Dominant Vegetation	Broadleaved oaks	Black pine	Black pine	Black pine	Beech
Reason of Protection	Protective wood belt, aquifers	Protective wood belt, aesthetic	Protective wood belt	Worship	Protective wood belt
Population (1–3: year 1895, 4–5: 1870)	188	2,759	1,044	210	1,500
Population (year 2011)*	187	2,492	45	102	193
Villagers occupation (present)	Agro pastoralists	Farmers, retailers, craftsmen, public servants	Retired, woodcutters	Woodcutters, tourism	Retired, agriculturalists, woodcutters

* Recent data of the National Population Census (2021) is still not available on the village level but only for municipalities. Zagori municipality shows a further ~10% decrease in its population, as the year 2021 has in sum 3,384 inhabitants while in the year 2011 it had 3,724 inhabitants

> bonding timbers of round juniper; later of sawn oak. Rafters are from pole-sized oaks, smaller than any oaks in the forest now. 3. Biggish church of Evangelístria, dated 1755. Around it are the biggest oaks. Big, rather closely set though crooked deciduous oaks; hornbeams. Evidence of shredding (hornbeam) and woodcutting (oak – could be of dead trees). The most sharply defined of the woods that we have seen. A modern barbed-wire fence, with a functional gate, follows approximately the limit of the oaks. Grazing continues inside, but much less than outside. One of the few woods with a new generation of trees – which does not match the existing trees. These are all ash, whose seedlings are abundant and get away wherever there is a gap in the oaks. Oak seedlings, though frequent, never get away.
>
> Rackham 2014: n.p.

The sacred forest at Mazi is named Kouri and dedicated to the village patron Virgin Mary. According to locals it was protected mainly as a protective wood-belt. Big detached boulders suggest rockfalls and erosion from shallow gulleys that can be seen above the village (Fig. 13.5). In the past, the forest was valued for its role in replenishing village aquifers, while areas under the bigger trees were used as *stálos* (literally livestock shelters that protected from the sun in the heat of the summer). From 1975 to 1977, the community was allowed by the Forestry Service to collect and selectively cut firewood for villagers' needs. Then in 1982, the ecclesiastical and community councils agreed with the Forestry Service to fence the forest to restrict grazing, ignoring the opposition of some local families. In 1995, several veteran trees were cut, and part of the resulting income was used for site enhancement works. The forest is now composed of a mixture of dense forest with an impenetrable understorey and more open forest with large deciduous oak trees. After performing the tree inventory, we observed how the Mazi sacred forest comprises a mix of five species of oaks (Macedonian Oak *Quercus trojana*, Turkey Oak *Q. cerris*, Downy Oak *Q. pubescens*, Hungarian Oak *Q. frainetto* and Sessile

Figure 13.5 Kouri Forest at Mazi, 'On the hill above the village is a sacred forest, abruptly differentiated in structure and composition from the grazed shrubland around it', 2 August 2014. (K. Stara)

Oak *Q. petraea*) together with their hybrids with a sclerophyllous understorey that differs in composition from the dominant canopy. The southern area is mainly Green Olive (probably former patches of shrubland), while the northern area and the edges are mainly dominated by Oriental Hornbeam *Carpinus orientalis* and Manna Ash *Fraxinus ornus*, which will probably become the future canopy as ash saplings outnumber young oaks. The oldest and largest trees were not located next to the churches, but near the southern gate to the forest. This cluster of trees was found to be 220–240 years old, some with old carved crosses in their bark. The oaks lean towards the south-east, but uprooted trees have fallen in a north-north-west direction (Marini Govigli et al. 2015).

Kourí forest, Kónitsa town

> On the talus slope of Mount Æminádhia, which towers nearly 1000 m above the town of Kónitsa. Geology is obscure, with soft rock and a tendency to badland. The forest might protect the town from avalanches; there are a few boulders but little scope for rockfalls. The lower slope is hornbeam (*Carpinus*), and could have been either pasture or coppice. The forest is adjacent to the cave-chapel of Panayía, with an important spring. Pine (nearly all nigra) is dominant, with a savanna structure; usually 2 ages of pines. Older trees (one 2.67 m girth) probably *c.*200 years (younger than in other pinewoods). Many older pines have multiple trunks or gnarled bases, suggesting period of growing up out of pasture in early 19th cent. Abundant torchwood scars, usually charred; frequent evidence of litter fires. A few lightning scars. Dead trees remarkably rare. Gaps between older pines suggest a phase of felling, using slide-ways for taking logs down the steep slope. This was long enough ago (*c.*100 years?) for stumps to have disappeared. A generation of fir has filled gaps between remaining pines. Local fires may have helped the establishment of fir. Fir continues to increase, invading the hornbeam zone. Remains of trenches and rifle-pits from World War II. Pines should be cored to establish demography.
>
> Rackham 2014: n.p.

Konitsa is a small town in the municipality of the same name. On the south-west edge but outside the forest, on a ridge overlooking the Aoos gorge, is a medieval archaeological site called *Kasrti*. The forest is named Kouri, and it is only this name that suggests it maybe a sacred forest. Locals describe it only as 'protective' (Fig. 13.6). According to the Forestry Service data, even the name Kouri referred to only part of the forest in the past. From 1928, the forest has belonged to the municipality, which manages it as a protected site. Part of the forest, in the vicinity of the houses of the upper town, was burned in 1940. Afterwards it was logged for poles and fuel wood, and the lower hornbeams were shredded and grazed during the Second World War, the civil war (1941–50) and until the 1960s. In 1953, the municipality decided to manage the forest, removing mature trees in order to use the money gained for technical works aiming to enforce the forest's protective character in its most degraded parts (Konitsa Forestry Service archives). Dendrochronology revealed the largest Black Pines were no older than 160 years and firs no older than 120 years (Kyparissis et al. 2015). Aerial photographs from 1945 reveal the recent expansion of the young forest towards the town, covering the gullies. The broadleaves had their roots suppressed by grazing, but with reduced grazing are now covering the area with a dense scrubland. This has been invaded by fir, which is well established on the edges of the pine forest as well as within it.

Miriáouwa forest, Palioseli village

> Sacred forest is immediately above the large village, on flysch capped with hard limestone. Unstable slope dissected by dormant badland gullies: principal church of the village is distorted by a landslip. The forest is mainly of pines higher on

Figure 13.6 Kouri Forest in Konitsa, 'The forest might protect the town from avalanches; there are a few boulders but little scope for rockfalls', 1 February 2013. (R. Tsiakiris)

the slope and oaks below. It is divided vertically by a main gulley containing the springs that provide a water supply and drive a mill. To the W of the gulley the village is menaced by a breakaway from which limestone boulders (5 m or more across) tumble from time to time; here the oak belt is broader. E of the gulley the instability takes the form of badlands, some of which have been active within the lifetime of the present pines; the oak belt is narrow and the forest is mainly of pines. The western forest contains an ikonostasi dated 1868.[4] Oaks comprise at least three deciduous species (*Q. pubescens, cerris, trojana*) as well as the evergreen *Q. coccifera*, here at the exceptional altitude of 1,200 m. They are smallish, partly dead at the top, very slow-grown, and appear to suffer from drought on this south aspect. One stump, only 94 cm girth, dated from *c.*1734; we found another oak 2.76 m girth. In part they are mingled with hornbeam shrubs. Pines are partly blackened by a moderately fierce fire (fuelled by bracken?); in places they infill savanna oaks. Pines are sometimes multi-stemmed; occasionally torchwooded; sometimes bear small incisions probably for marking paths. Oak may infill pine, or pine may infill oak. Unusually, pine and oak both have advance regeneration: pine as abundant seedlings, oak as areas of ground-oak. We heard the story that the oaks of the western forest were planted to protect the village. We cannot verify the story – the planting would have to have been early 18th cent. – but we find that the oaks tend to occur in pairs. The eastern forest lies above a chapel of Ay.

4 *Iconostasi* (literally icon stand): a shrine comprising a stove or metal box containing icons and an oil lamp that remains lit most evenings.

> Demétrios and a graveyard. Around the chapel is a grove of oaks of 4 ages, some of them big hollow veteran trees. The younger oaks have evidently grown up recently out of ground-oak. Pines in the eastern forest are of three generations: original spreading savanna trees, an early infill, and a much younger generation. Lightning scars are frequent. Original trees and early infill both show torchwood scars, especially near pines that lightning has struck. We found local charring but no general evidence of fire. One 'original' pine, 2.10 m girth, we estimated to date from c.1766. Trees of the early infill often show stumps of low branches – occasionally whole dead branches, or remains of branches displayed in torchwood cavities – that show they passed through a bushy phase, probably through being bitten down, before they got away and grew up into trees. Juniper is an abundant understorey, indicating former grazing and lack of fire. At the top of the forest is a small patch of remaining grassland in the midst of the surrounding recent forests.
>
> Rackham 2014: n.p.

The Palioseli sacred forest, named Mereao, literally meaning commonland, is referred to as excommunicated. It is maintained mainly for its protective character. Huge pines in the western part are leaning over owing to unstable substratum near the location of a network of springs as well as a small landslide. Some in the eastern part have been uprooted. In addition, some pine tree-roots on the steeper eastern slope, where there are many small gullies, are exposed by more than 80 cm, suggesting significant erosion events. Rockfalls are reported by locals, and we observed huge rocks and small boulders stuck in the bases of broadleaf trees. Locals relate that some of this is deliberate as small stones were placed in bases of the trees to stabilize them. Hornbeams, oaks and pines show quite different size distributions. Hornbeam is only present in smaller size classes suggesting that it has only recently started to infill the available open space. Oaks on the other hand have hardly any young saplings (though a large quantity of seedlings), with a good proportion of trees around 45 cm. in diameter, reinforcing the idea that they may have been planted. This is an open question to be investigated. Pines, on the other hand, are present across the grove at all life stages and sizes. The age analysis suggests that the forest might date back to around 1750, matching Oliver's initial estimations. The oldest oak (Turkey Oak) cored gave an age of 345 years, while the oldest pine was 278 years old (Kyparissis et al. 2015). Deadwood in the forest is rare and it seems to have been collected, even though locals claim to be opposed to cutting or collection of wood, saying that this practice will destroy the village (Fig. 13.7).

Ay. Paraskevé/La Stevinieri, Vovoúsa/Voiása village

> Vovoúsa is a large village that still practises woodcutting and a little cultivation. The sacred forest lies to the SE, coming down almost to the river Aóos, and does not lie above the village. The very steep (but not rocky) flysch and ophiolite slopes have some tendency to form badlands. On the top is a chapel of Ay. Paraskevé, with a date-stone (not in situ) that appears to read 1755. Its bonding timbers are sawn oak. The forest has exceptionally big pines, mostly of savanna shape, probably of two generations, some over 30 m high, one 3.66 m girth. They appear not to have gnarled bases. These are infilled with oak (also *Ostrya* and *Carpinus* near bottom of slope). Torchwood scars are frequent; one has been enlarged to communicate with the hollow interior either for an ikonostasi or to get at a nest of bees. Rare evidence of fire. This forest has unusually many standing dead and fallen pines. (Whether the fallen trees were already dead is not clear.) Some of these have been sectioned and yield the following approximate dates of origin: 1690 for a tree 1.33 m circumference at 20 m high 1690 for a tree 2.3 m

Figure 13.7 Mereao forest in Palioseli, 'Pines are sometimes multi-stemmed; occasionally torchwooded', 7 August 2014. (K. Stara)

> in circumference at about 15 m high 1770. The last few decades are exceptionally slow-grown. Around the chapel are a few very big oaks, including an ancient hollow *Quercus dalechampii*. There is a dense stand of young hornbeam with a few pollard and other *Ostrya*.
>
> Rackham 2014: n.p.

Vovoussa's sacred forest hosts the church of the village's miraculous patron Agia Paraskevi (Stavinere in Vlach) and it is named after her. There are apparently no other reasons than worship to explain the strictly protected character of the forest, except perhaps the protection of an ancient mule road and past infrastructure from landslides, as the river cuts the bottom of the hill (Fig. 13.8). Locals visit the forest frequently to pray and hold a service to celebrate the patron saint every 26 July. This mixed broadleaved and pine forest, as Oliver pointed out, contains an unusually high number of dead trees with several gigantic uprooted pines. The average basal area of the forest is 6.8 m² per ha, which is much higher than any of the other sampled sacred forests. This reinforces the suspicion that the trees in this forest have been left alone and the site was mainly used for religious purposes. Dating and inventory of the trees have revealed that black pines (oldest individual sampled aged 351 years) once probably dominated the forest, only recently being infilled by hornbeams and oaks (mean age about 60–80 years), excluding large, old oaks near the church (Kyparissis et al. 2015). At least five periods of forest protection and expansion in well-defined concentric belts can be identified. These correlate with local stories that the protected area expanded as villagers decided to dedicate fields adjacent to the forest to the saint as a memorial or as a bequest. Cutting of trees and removal of wood incurs a supernatural punishment as severe as death, and this local cult is very much alive. Forestry operations to fell some trees according to an approved management plan by the Forestry Service in the 1970s and later on have been cancelled owing to strong (and sometimes violent) local objections and demonstrations.

Figure 13.8 Agia Paraskevi forest in Vovoussa. 'The forest has exceptionally big pines, mostly of savanna shape', 8 August 2014. (R. Tsiakiris)

Toúfa forest, Greveniti village

> This forest lies on the upper slopes above the large village of Grevenı́ti, which is unstable and prone to landslips. The lower slopes are young to middle-aged pine and beech forest, coppice-woods, and overgrown shrubland. At the top is a large surviving open area with a chapel of the Prophet Elias. The slope is on flysch, at nearly its maximum angle of rest, and somewhat unstable with landslips. Near the bottom are loose boulders of unknown origin. There are two distinct types of forest, dominated by pine and beech. Pines are of two generations, probably not widely different in age. The original trees (one 2.54 m circumference) kept their lower branches alive but upturned into a candelabrum shape. Occasional torchwood scars. Some have remains of a bushy base indicating severe browsing when young. An understorey of beech is developing. In a shallow dry valley the forest changes to beech. Here are some of the biggest beeches we have seen, one 5.31 m diameter and many 40 m or more high. Some of the beeches appear to be pollarded. There are also big *Ostrya* and an ivy 57 cm girth. The junction of the two types is rather abrupt, with a narrow zone of intermingling: scattered old torchwooded pines appear to have been invaded by expanding beech. In this zone are some uprooted dead beech. A few *Acer opalus* are as tall as the beeches. There are also some long-dead remains of gnarled oaks, some of which are directly adjacent to living beeches. One of these oaks had been cut with an axe at least 50 years ago. Ground vegetation includes sanicle, Epipactis, Viola. A mule-path, still occasionally used, leads up through the beech forest in the direction of Métzovo.
>
> Rackham 2014: n.p.

Greveniti village sacred forest is said to stabilize the slope above the village. Greveniti has a long history of landslides (also within the forest) and several buildings have been distorted by a landslip in the last decade. Locals blame the increase in number and severity of landslips on land abandonment and the subsequent deterioration of the sophisticated network of canals used to irrigate the vegetable fields, as well as the cementing of small roads and alleys within the village, all of which has changed water flow patterns. Springs are located within the forest and between the protected forest and village. People recount narratives referring to the excommunication ritual. This in combination with place-names inside the forest that refer to open land give us evidence for a more open past landscape. Our additional analyses revealed how Greveniti is now a relatively homogeneous European Beech *Fagus sylvatica* forest, probably a shift from a former pine-dominated community (still present in its north-east boundaries) and a large-scale infill of areas with scattered trees of a maple-deciduous oak pastoral woodland community (a few remaining dying veteran individuals can be found in the western part), (Fig. 13.9). This is substantiated by the age analysis, which shows that the oldest beech trees sampled are only about 255–266 years old (Kyparissis et al. 2015). Evidence of occasional grazing can be identified around the village and at the entrance of the sacred forest, where broadleaves (some with the characteristic shape of shredded trees) are found along the lower boundary of the tall beech forest. Regeneration of beech is lush both in forest gaps created by naturally uprooted and dead beech and on the boundaries of the forest. Deadwood is abundant, although after 1982 a council decision changed Forest Service rules allowing villagers to collect dead trees for firewood (Greveniti Community Council, 60/1982), and few torchwoods can be found only near the upper boundary of the pine forest. There is some embankment in the bases of trees, which indicates local events of erosion. Gigantic individual beech trees (some pollards and likely the parents of the younger trees) within the sacred forest probably marked the ancient mule path from the villages of Zagori to the nearby town of Metsovo in the time when Greveniti used to be a prosperous community and the capital of the wider area.

Figure 13.9 Toufa forest in Greveniti, 'In a shallow dry valley the forest changes to beech. Here are some of the biggest beeches we have seen', 9 August 2014. (K. Stara)

Oliver Rackham's legacy

Oliver's ideas and thinking still spur us on to new research in several areas of special interest. In light of this, we quote his last five subchapters without any additional comment with the aim to further inspire new scientists and researchers.

The idea of sacred forests

> The principle of banning [placing restrictions on] forests on slopes to protect habitants against upslope hazards goes back in the Alps at least to the middle ages, as shown by the place-name Bannwald. Originally it meant forests that were supposed to intercept avalanches or falling rocks, but the idea was extended in modern times to forests valued for protecting water supplies or for nature conservation. ['Bannwald' can also mean a hunting district under ban, i.e., forest preserve]. These Alpine forests, however, are protected by the secular rather than the spiritual authorities. Invoking the authority of the Church rather than the State appears to be a Balkan phenomenon.
>
> As far as we know, most of the sacred forests are not medieval, but were declared in the eighteenth or nineteenth century. Why this was so we do not yet know. Zagóri then was not so isolated as it became in the 20th century. Reasons for protecting forests could have been derived from the Alps, or from the theories of foreign scientists at the time (in the same way that British imperialists reserved areas of forest on tropical islands in the belief that they would prevent climate change (Grove 1995).
>
> Rackham 2014: n.p.

Uses of sacred forests

> In the majority of these forests there has been little or no woodcutting within the length of time needed for stumps to rot away. If trees were felled for ecclesiastical or communal purposes, most of the direct evidence has disappeared. Working trees – pollarded or shredded – exist within sacred forests, but are less common than outside. Usually there was no agreement on the type of use of working trees. Thus in Aidhonokhóri we found pollard, shredded, and maiden individuals of a number of species mixed apparently at random.
>
> Most of the sacred forests have been savanna, typically rather dense savanna with 40–60% tree cover, mostly now infilled. They have been through a period as grazing land, as shown by gnarled tree-bases, residual juniper, or ground-oak. Reduction of grazing – but not elimination – is evidently one of the effects of being declared sacred. Most sacred forests, however, are sharply differentiated from the shrublands and bushy grasslands in which grazing still continues or has recently declined: these have a wider range of trees and shrubs including *Fraxinus*, *Cornus mas*, *Cratægus*, and (near roads) *Paliurus* and *Cotinus*. Sacredness did not eliminate torchwood-gathering, which was evidently regarded as not significantly damaging the tree.
>
> Rackham 2014: n.p.

Ecological and historical significance of sacred forests

> Sacred forests should not be thought of as 'wilderness' or 'virgin forest' untouched by human intervention: this concept is proving elusive as evidence grows of the pervasiveness and early date of human activities. Rather, they are rare examples

of long-established sites of minimal intervention, withdrawn from most human activities centuries ago. Since then they have followed their own trajectories of development. They are historical features that record natural developments and particular types of human activity during, and sometimes before, they were declared sacred. They are concentrations of what once were working trees, now gradually diminishing in the rest of the landscape through neglect and casual woodcutting.

Sacred forests are especially significant for their large numbers of veteran trees. These have not yet been much studied in Greece, but elsewhere they are known to be the specific habitats of animals and plants for which young and middle-aged trees are of no use. These include birds that nest in holes in trees, bats that use hollow trees or loose bark, the multitude of insects that live on rotten wood in particular parts of a veteran tree, the spiders that prey on those insects, and the lichens that live on rain-tracks or old dry bark under overhangs.

Rackham 2014: n.p.

Conservation of sacred forests

Sacred forests seem to be little known outside their immediate communities. Their prohibitions are still active but gradually fade as communities decline and disperse. They are not widely appreciated in the outside world. They are, in practice, open to the public, but we doubt whether any tourist ever finds them.

Most sacred forests are outside the official forestry organization. They do not call for active management and are not threatened by neglect. We hear, however, of communities who consider felling the trees to raise a little money. The profit from such a sale is likely to be minimal, given the poor quality of the timber and the high cost of felling and transport on what are mostly steep and inaccessible sites. (In non-sacred forests the amount of felling now going on is not enough to keep up with the growth-rate of the trees.)

Rackham 2014: n.p.

Conclusions

The sacred forests of Epirus are among the world's oldest protected areas: some of them have been protected for nearly twice as long as Yellowstone National Park (United States) or Epping Forest (England). Sacred forests are extremely varied in structure and composition, from forest to savanna and from nearly pure pine to pure oak to beech, making it difficult to generalize.

Rackham 2014: n.p.

Final remarks

During our collaboration, Oliver raised many research questions encouraging us to study the demography of trees (from cores and sections) and the effect of grazing ('what species do cattle, sheep, goats eat?'). He wondered 'how long does a stump take to disappear, how many years does a dead tree take to fall and how many years does a fallen dead tree take to disappear?', and 'how effective are sacred forests as protection against falling rocks and other upslope hazard?'. Thanks to the THALIS project, many of the questions raised by Oliver have been successfully investigated (see Kyparissis et al. 2015; Stara et al. 2015, 2016; Tsiakiris et al. 2017; Avtzis et al. 2018; Muggia et al. 2018; Marini Govigli et al. 2020, 2021; Benedetti et al. 2021; Diamandis et al. 2021; Stara 2021, 2022; Zannini et al. 2021; Moudopoulos-Athanasiou 2022; Roux et al. 2022); yet others remain to be further explored.

Figure 13.10 An ancient beech in Greveniti forest, 9 August 2014. We dedicated this remarkable tree to Oliver who was astonished by its size. The tree has a characteristic candelabrum shape that symbolizes the long interaction between humanity and nature, a beloved topic of Oliver's ecological research history. (K. Stara)

The sacred forests of the villages of Zagori and Konitsa from 2015 are part of UNESCO's National (Hellas) Intangible Cultural Heritage Index as a sophisticated locally-adapted management system in the small scale of the community. Our proposal has been dedicated to the memory of Oliver Rackham. Moreover, in collaboration with the municipality of Zagori, the local community and the cultural association of Greveniti, we dedicated to his memory the giant monumental beech tree that impressed him so much (Fig. 13.10). In *The Ancient Trees of Zagori and Konitsa* (Stara and Vokou 2015), there are instructions on how to find his tree inside the forest. Moreover, our team study proved that sacred forests, even small in size and designated for other services in the past, have nowadays acquired a new role to play in biodiversity conservation (Avtzis et al. 2018), and this accompanies their intangible qualities. Lastly the Hellenic Ministry of Culture and Sports in 2022 submitted a nomination file for the inscription of Zagori as a Cultural Landscape on the UNESCO's World Heritage List, including these sacred forests among the most important element of the area, due to their universal value – as had been noted by Oliver: 'the sacred forests of Epirus are among the world's oldest protected areas'.

Acknowledgements

This research was co-financed by the European Union (European Social Fund) and Greek national funds through the Research Funding Programme THALIS – University of Ioannina. We would like to thank the scientific coordinator of THALIS–SAGE project, John M. Halley, our colleagues Aris Kyparissis, Theofilos Vanikiotis, Nikos Markos and Stavros Stagakis (University of Ioannina, Greece) for providing us with their unpublished data on dendrochronology, MSc students Nathalie Pion and Robert Cullen (Bangor University, UK) for their data on specific forests of the area and Alkis Mpetsis for his help in the preparation of the map.

We would like to extend particular thanks to the Border Crossings Network and especially Professor Vassilis Nitsiakos for inviting Oliver to Konitsa for the ninth Konitsa Summer School in 2014 to contribute to the course 'Environmental history and cultural ecology of the Mediterranean and the Balkans. The case of Pindus and the adjacent borderlands'. Moreover, we would like to thank the Forestry Services of Ioannina, Konitsa and Metsovo for access to their archives and all local participants to the study for their time and hospitality during field work and especially Charissis Karakoglou (Konitsa), Giorgos and Maria Grentziou and Giannis Milionnis (Palioseli), Michalis Kontogiannis (Iliochori) Antonis Stagogiannis and his family (Vovoussa), Giannis Stergiou and Vassilis Nolas (Doliani), Christos and Glykeria Nikolaki (Greveniti), Miltos Bukas and Evi Vaimaki (Dikorfo) and Nikolaos and Rodokleia Stara (Kato Pedina).

References

Avtzis, D., Stara, K., Sgardeli, V. Betsis, A. Diamandis, S., Healey J.R., Kapsalis, E., Kati, V., Korakis, G., Marini Govigli, V., Monokrousos, N., Muggia, L., Nitsiakos, V., Papadatou, E., Papaioannou, H., Rohrer, A., Tsiakiris, R., Van Houtan, K.S., Vokou, D., Wong, J.L.G. and Halley, J.M. (2018) Quantifying the conservation value of sacred natural sites. *Biological Conservation*, 222, pp. 95–103. https://doi.org/10.1016/j.biocon.2018.03.035

Babiniotis, G. (2002) *Dictionary of Modern Greek Language*, 2nd edition. Centre of Lexikology, Athens. (in Greek)

Baldacci, A. (1899) *From my travel to Albania. Pindos and Smolikas.* Parnassos Literary Association, Athens, pp. 152–174. (in Greek)

Benedetti, Y., Kapsalis, E., Morelli, F. and Kati, V. (2021) Sacred oak woods increase bird diversity and specialization: Links with the European Biodiversity Strategy for 2030. *Journal of Environmental Management*, 294: 112982. https://doi.org/10.1016/j.jenvman.2021.112982

Cullen, R. (2015) Forest dynamics in a Fagus sylvatica dominated forest in North West Greece. Unpublished MSc thesis, Bangor University, UK.

Diamandis, S., Topalidou, E., Avtzis, D., Stara, K., Tsiakiris, R. and Halley, J.M. (2021) Fungal diversity in sacred groves vs. managed forests in Epirus, NW Greece. *Journal of Microbiology and Experimentation*, 9(5), pp. 142–154. https://doi.org/10.15406/jmen.2021.09.00335

Gripsos, P. (1952) Kouri in Epirus. *Hpeirotiki Estia*, 18, pp. 296–299. (in Greek)

Grispos, P. (1973) *Forest History of Modern Greece.* Editions of the Forest Service Applications and Training 25, Athens. (in Greek)

Grove, R. (1995) *Green Imperialism: Colonial expansion, tropical island Edens and the origins of environmentalism, 1600–1860.* Cambridge University Press, Cambridge.

Grove, A.T. and Rackham, O. (2001) *The Nature of Mediterranean Europe: An ecological history.* Yale University Press, New Haven, CT and London.

Halstead, P. (1998) Ask the fellows who lop the hay: leaf-fodder in the mountains of Northwest Greece. *Rural History*, 9(2), pp. 211–234. https://doi.org/10.1017/S0956793300001588

Kalantzi, F. (2008) Meleti katolisthisseon apo istorika, vivliofica dedomena kai xartografisi stin periochi Ioanninon. [Study of landslides from historical data, bibliography and mapping in Ioannina region]. Unpublished MSc thesis, University of Patras. (in Greek)

Khela, S. (2013) *Aesculus hippocastanum.* The IUCN Red List of Threatened Species 2013: e.T202914A2757985.

Kyparissis, A., Stagakis, S., Markos, N. and Vanikiotis, V. (2015) *Study of climatic and cultural effects by dendrochronology of selected trees.* Report to Thalis-SAGE project, University of Ioannina, Greece.

Leake, W.M. (1835) *Travels in Northern Greece*, Vol. I. J. Robwell, London.

Marini Govigli, V., Healey, J.R., Wong, J., Stara, K., Tsiakiris, R. and Halley, J.M. (2020). When nature meets the divine: effect of prohibition regimes on the structure and tree-species composition of sacred forests in Northern Greece. *Web Ecology*, 20, pp. 53–86. https://doi.org/10.5194/we-20-53-2020

Marini Govigli, V., Efthymiou, A. and Stara, K. (2021) From religion to conservation: unfolding 300 years of collective action in a Greek sacred forest. *Forest Policy and Economics*, 131, 102575. https://doi.org/10.1016/j.forpol.2021.102575

Marini Govigli V., Pion, N., Cullen, R., Healy, J.R. and Wong, J. (2015) *Aging trees in six sacred forests in Zagori.* Report to Thalis-SAGE project, University of Ioannina, Greece.

Mihailaris, P.D. (2004) *Excommunication: Adaptation of a punishment to the needs of the Ottoman Occupation*, 2nd edition. National Research Foundation – Center of Modern Greek Research 60, Athens. (in Greek)

Moudopoulos-Athanasiou, F. (2022) The Celebration of St. Viniri in Băeasă (Vovousa): Approaching the Archaeology of the Sacred Forests in Northwest Greece. *International Journal of Historical Archaeology.* https://doi.org/10.1007/s10761-022-00664-5

Muggia, L., Kati, V., Rohrer, A., Halley, J., and Mayrhofer, H. (2018). Species diversity of lichens in the sacred groves of Epirus (Greece). *Herzogia*, 31, pp. 231–244. https://doi.org/10.13158/099.031.0119

Nitsiakos, V. (2016) *Peklari. Social Economy in a Greek Village.* LIT Verlag, Münster.

Papageorgiou, G. (1995) *Economic and Social Mechanisms in Mountainous Areas. Zagori (middle 18th–20th centuries.* Pizareios foundation editions, Ioannina. (in Greek)

Papaioannou, H. and Kitsaki, G. (2014) *Vikos-Aoos Geopark: Nature narrates its history. Development Agency 'Epirus S.A.'*. Region of Epirus, Ioannina.

Papanastasis, V.P., Mantzanas, K., Dini-Papanastasi, O. and Ispikoudis, I. (2009) Traditional agroforestry systems and their evolution in Greece. In: A. Rigueiro-Rodróguez, J. McAdam and M.R. Mosquera-Losada (eds) *Agroforestry in Europe: Advances in agroforestry*, Vol. 6. Springer, Dordrecht, pp. 89–109. https://doi.org/10.1007/978-1-4020-8272-6_5

Pion, N. (2014) Effect of grazing management practices on the structure of sacred groves in Epirus, Greece. Unpublished MSc thesis, Bangor University, UK.

Rackham, O. (2006) *Woodlands.* (New Naturalist 100). Collins, London.

Rackham, O. (2014) *Sacred Natural Sites: Sacred Forests in Epirus.* University of Ioannina, Ioannina, Greece. [Unpublished manuscript for THALIS -SAGE project 'Conservation through religion: the sacred groves of Epirus' (2012–2015)]

Rackham, O. and Moody, J. (1996) *The Making of the Cretan Landscape.* Manchester University Press, Manchester and New York.

Roux, J.L., Konczal, A.A., Bernasconi, A., Bhagwat, S.A., De Vreese, R., Doimo, I., Marini Govigli, V., Kašpar, J., Kohsaka, R., Pettenella, D., Plieninger, T., Shakeri, Z., Shibata, S., Stara, K., Takahashi, T., Torralba, M., Tyrväinen, L., Weiss, G. and Winkel, G. (2022) Exploring evolving spiritual values of forests in Europe and Asia: a transition hypothesis toward re-spiritualizing forests. *Ecology and Society*, 27(4): 20. https://doi.org/10.5751/ES-13509-270420

Stamatopoulou, Ch. (1998) *Greek Traditional Architecture: Zagori.* Melissa, Athens.

Stara, K. (2000) The impact of perceived land-use values on biodiversity conservation in the Vikos – Aoos National Park, Papingo, Greece. Unpublished MSc thesis, Bangor University, UK.

Stara, K. (2012a) Northern Pindos National Park excommunicated forests. In: J.M. Mallarach (ed.) *Spiritual Values of Protected Areas of Europe Workshop Proceedings.* Federal Agency for Nature Conservation, Bonn, pp. 69–74.

Stara, K. (2012b) Trees and the sacred in modern Greece. *Landscape*, 2(11), pp. 60–63.

Stara, K. (2021) The lonely guardians of history and time. In: F. Moudopouloa-Athanansiou (ed.) *In Artsisti: a mountain community between past and future.* Society of Contemporary History, Athens, pp. 141–163. (in Greek)

Stara, K. (2022) When matter becomes immaterial. The sacred forest as a local management system of the 'commons' of the past and an element of contemporary Intangible Cultural Heritage. In: Nitsiakos V, Drinis, I.N. and Potiropoulos, P. (eds). *Cultural Heritage: New readings - Critical approaches.* Ars Nova, Athens, pp. 113–132. (in Greek)

Stara, K. and Tsiakiris, R. (2010) The sacred woods called 'meadows' in Zagori, Pindos mountains, Greece. In: A. Sidiropoulou, K. Mantzanas and I. Ispikoudis (eds) *Range Science and Life Quality – Proceedings of the 7th Panhellenic Rangeland Congress* [Ministry of Environment, Energy and Climate Change], Thessaloniki, pp. 57–62. (in Greek)

Stara, K. and Vokou, D. (eds) (2015) *The Ancient Trees of Zagori and Konitsa.* University of Ioannina, Ioannina. (in Greek)

Stara, K., Tsiakiris, R., Nitsiakos, V. and Halley, J.M. (2016) Religion and the management of the commons: the sacred forests of Epirus. In: M. Agnoletti and F. Emanueli (eds) *Biocultural Diversity in Europe* (Environmental History 5). Springer-Verlag, Berlin, pp. 283–302. https://doi.org/10.1007/978-3-319-26315-1_15

Stara, K., Tsiakiris, R. and Wong, J.L.G. (2015) The trees of the sacred natural sites of Zagori, NW Greece. *Landscape Research*, 40(7), pp. 884–904. https://doi.org/10.1080/01426397.2014.911266

Touchan, R., Baisan, Ch., Mitsopoulos, I.D. and Dimitrakopoulos, A.P. (2012) Fire history in European black pine (*Pinus nigra* Arn.) Forests of the Valia Kalda, Pindus Mountains, Greece. *Tree-Ring Research*, 68, pp. 45–50. https://doi.org/10.3959/2011-12.1

Tsiakiris, R., Gribilakou, L., Betsis, A., Rubas, E., Stara, K. and Halley, J.M. (2017) Can inhabitants' testimonies about Zagori and Konitsa sacred forests protective function be confirmed by spatial predictive models? *18th Hellenic Forestry Congress proceedings.* Hellenic Forestry Association, Thessaloniki, pp. 1080–1087.

Tsiakiris, R., Stara, K., Pantis, I. and Sgardelis, S. (2009) Microhabitat selection by three common bird species of montane farmlands in northern Greece. *Journal of Environmental Management*, 44, pp. 874–887. https://doi.org/10.1007/s00267-009-9359-8

Verschuuren, B., Wild, R., McNeely, J.A. and Oviedo, G. (2010) *Sacred Natural Sites: Conserving nature and culture.* Earthscan, London.

Zakkak, S., Kakalis, E., Radovic, A., Halley, J.M. and Kati, V. (2014) The impact of forest encroachment after agricultural land abandonment on passerine bird communities: the case of Greece. *Journal for Nature Conservation*, 22(2), pp. 157–165. https://doi.org/10.1016/j.jnc.2013.11.001

Zannini, P., Frascaroli, F., Nascimbene, J., Persico, A., Halley, J.M., Stara, K., Midolo, G. and Chiarucci, A. (2021) Sacred natural sites and biodiversity conservation: a systematic review. *Biodiversity and Conservation.* https://doi.org/10.1007/s10531-021-02296-3

CHAPTER 14

Historical Ecology and the History of 'Individual Landscapes': Oliver Rackham's Field Visits to Liguria (North-West Italy)

Roberta Cevasco, Diego Moreno and Charles Watkins

Summary

In this chapter, we outline the importance and influence of Oliver Rackham's field visits to Liguria. We focus in particular on a visit he made in September 1984. Part of the trip was planned to study the use of historical photographs in documenting on the spot, using repeat photography, the historical dynamics of vegetation cover. But the visit also touched on the problem of the wildwood, the definition of ancient woodland in Mediterranean mountains and the significance of indicator species (here Saw-wort *Serratula tinctoria*). We also consider some of the subsequent developments in the use of historical ecology for the history of individual landscapes at particular sites.

Keywords: Liguria, degradation and environmental crisis, Mediterranean woodland degradation hypothesis, photo-monitoring, indicator species

Two Ligurian Field Visits

Oliver Rackham made two field visits to the Ligurian Apennines in the 1980s and 1990s. The first was a ten-day field trip in 1984 with Diego Moreno, together with some colleagues and students from the University of Genoa studying historical geography. This visit followed on from several years of scientific collaboration with Oliver shared since 1979 with Piero Piussi and Diego Moreno. The latter also attended the Flatford Mill Course and had the chance to do fieldwork and visit many ancient woodland sites in Cambridgeshire with Oliver in 1980. This collaboration produced the first collection of essays in Italian that presented the historical approach nurtured by Oliver when an edition of the historical journal *Quaderni Storici* devoted to woodland history was published (Moreno, Piussi and Rackham 1982).

The second field visit was spent in the Upper Vara Valley in 1996. Here Oliver joined the annual University of Nottingham undergraduate field visit in Ligurian Eastern Apennines, which included 30 undergraduate geographers and historians, the staff including Charles Watkins and Ross Balzaretti from Nottingham, Diego Moreno and Roberta Cevasco from Genoa and Don Sandro Lagomarsini from the local village of Cassego (Balzaretti, Pearce

Roberta Cevasco, Diego Moreno and Charles Watkins, 'Historical Ecology and the History of 'Individual Landscapes': Oliver Rackham's Field Visits to Liguria (North-West Italy)' in: *Countryside History: The Life and Legacy of Oliver Rackham*. Pelagic Publishing (2024). DOI: 10.53061/KTIQ6027

and Watkins 2004). Some of Oliver's field observations collected in the Val di Vara in 1996 were published in *The Nature of Mediterranean Europe* (Grove and Rackham 2001). During the field visit some customary but illegal fires had been lit during the night by commoners to improve the quality of their pastureland. This provided Oliver with a number of opportunities to observe the good and bad effects of fire on the vegetation and soils. He made use of fieldwork on the slopes of Monte Porcile to study the relationship between different tree and shrub species and occupational burning, noting that 'fire was very sensitive to topography and vegetation' (Grove and Rackham 2001: 227). During the field-course, a local newspaper reported that the Forest Service was hunting the mountains for the 'pyromaniacs', to which Oliver joked: 'In order to give them a medal?'. He also worked with Roberta Cevasco on a profile sketch through terraces with pasture and chestnuts in the Valle Lagorara. Elements of the historical cyclical processes used to improve the production of nuts and grazing were identified (Cevasco and Poggi 1997; Cevasco et al. 1999; Cevasco 2004: 166). We also studied the Bertignana terraces at Varese, which Oliver noted 'are the earliest known to us with a documented date' (Grove and Rackham 2001: 110). He fully engaged with student projects such as one examining the destructive effects of 'galaverna', the encasing of tree branches with thick ice, on trees at Valletti, with much evidence of shredding of oaks in neighbouring villages (Figs 14.1 and 14.2). Oliver was, as always, a lively and entertaining companion on the field visit, and his observations and ideas provided valuable evidence of the value of historical ecology for the students and staff who accompanied him.

These two visits were not the only ones Oliver made to study the Italian countryside. He worked on an archaeological project at Cinque Finestre, where the dig of a Roman villa was directed by Andrea Carandini – subsequently one of the leading Italian classical archaeologists – and also visited Sardinia, Venice, Florence and other Italian regions, contacting many Italian scholars and researchers. A complete account of his Italian research experience – including his Ligurian visits – must be drawn together, on the basis of his own field notebooks. In this chapter, we focus on his 1984 Ligurian visit and on his valuable contribution concerning the use of historical photography and fieldwork to examine precise changes in vegetation (Sheail 1980; Métailié 1986; 1988; Moreno and Montanari 1988; Piana et al. 2012). This included the sketching of vegetation during fieldwork on a transparent film to identify key differences on the spot.

Oliver Rackham's Ligurian fieldwork in 1984

Oliver's field visit of 8–17 September 1984, reconstructed using Diego Moreno's field notes and personal recollections, was characteristically both intense and fruitful. It started with a visit to the historian Edoardo Grendi (Raggio 2004) – one of the founders of the Italian micro-historical approach to social history – in his home at Bersano near Tortona. Discussions were wide ranging and included historical ecology, hedges, documentary evidence for ancient woodland and vernacular architecture. Oliver was particularly impressed by the huge wooden frame of the well-conserved eighteenth-century roof of the *cascina* (courtyard farmhouse) at Bersano. At Genoa, he visited the Archivio di Stato map collection and the medieval city with the help of Ennio Poleggi. On 11 September, we visited the terraced vineyard landscape of the Polcevera Valley, the terroir of Coronata white wine, and the Costa di Begato. We tasted the products and discussed with the owners local practices used to fertilize the terrace soils including using Tree Heather *Erica arborea* (in Genoese dialect *brugo*) as a green manure.

Two days, 12–13 September, were spent at Cassego (Upper Vara Valley), were Oliver met Don Sandro Lagomarsini and visited the School and Museo Contadino (Peasant Museum). He visited the Lago di Bargone, an important site for the development of pollen studies in

Figure 14.1 Students studying the impact of 'galaverna' at Valletti, near Varese Ligure, March 1996. (C. Watkins)

Figure 14.2 A recently shredded Turkey Oak tree at Teviggio, near Varese Ligure, August 1996. (C. Watkins)

the nearby Val Petronio, and also the remains of an ancient wood named Selva di Pessino on the watershed of the Vara and Taro valleys near the Passo di Cento Croci. Considerable attention was given to documenting the local methods used to shred Turkey Oak *Quercus cerris* and the use of firewood in local cooking systems. In these few days, Oliver finished (and possibly sent home) a paper devoted to the concept of wildwood and its limits. He typed up a paper copy at Cassego – during his two-day stay at Case Bottini – which is now in Diego Moreno's personal archive. We are not sure whether this paper has been published, but it was circulated and was of great importance in developing in subsequent years Sandro Lagomarsini's approach to historical ecology and environmental conservation issues (Lagomarsini 2017).

On 14–15 September, returning to Genoa, we visited the Fontanabuona Valley and spent a day studying the ancient grassland of the Montagna di Fascia (Moreno 1990; 2018). Here the use of historical photographs at two or three identifiable places provided evidence of the historical dynamics of shrubs and pines connected with the abandonment of local haymaking practices. The visual documentation produced by Oliver was published some years later in an essay on the historical ecology of these grasslands in a rescue archaeology report (Moreno, Croce and Montanari 1992; 1993). Finally, on 16 September, Oliver visited several sites in the Foresta Demaniale del Lerone in the mountain facing the Ligurian Sea to the west of Genoa. This is an area well known in the geological literature as Gruppo di Voltri, because the Alpine and Apennine belts come into contact (Capponi et al. 2016). During this visit Oliver observed the location of an indicator species (a small isolated population of the herbaceous plant Saw-wort *Serratula tinctoria*) growing along the ancient path that crosses the northern slope of Monte Tardia near the Passo della Gava at the head of Lerone Valley. In the next section, we consider the implications of the unpublished field documents made in 1984 and relate them to subsequent research.

Repeat photography at Ponte Negrone

Sarasini (1984, 1986) studied the historical photography collection of the Regional Forestry Inspectorate for Liguria of the Corpo Forestale delle Stato. His research showed that the archive included 2,173 original prints and negatives dating from 1925 to 1968. It included a nucleus of 927 photographs and negatives received from the former Archivio Fotografico Comando Coorte di Genova della Milizia Nazionale Forestale, mainly from around Genoa. Sarasini identified the precise location of over 500 different sites from the detailed original captions on the photographs. One group documented the upper part of the Lerone Valley, which extends from sea level up to 1,122 m at Monte Reisa in the Gruppo di Voltri. (Fig. 14.3)

One negative plate (ng 223) had a precise original caption 'Bacino del Rio Lerrone vicino a Ponte Negrone', dated 1927 (Fig. 14.4). Following established practice, the photograph was taken before some planned afforestation and slope consolidation (Greco 2017). The administration was inspired by the Mediterranean woodland degradation hypothesis, and the photographs were intended to document degradation reaching an irreversible state with *suoli scheletrici* (stone-rich soils) and *roccia madre nuda* (exposed rock under layer) (see, for example, Di Tella 1940). The slopes appear covered by *cotica secca con graminacee* (a dry and poor turf), which would have been interpreted as an immediate consequence of pastoral use of the land and especially a pastoral fire regime.

It was possible in 1984 to identify the viewpoint precisely, but, as Oliver noted on his drawing, after 57 years it was 'impossible to repeat from the exact viewpoint because now obstructed by pines'. Oliver made a drawing (Fig. 14.5), and a photograph (Fig. 14.6) was taken

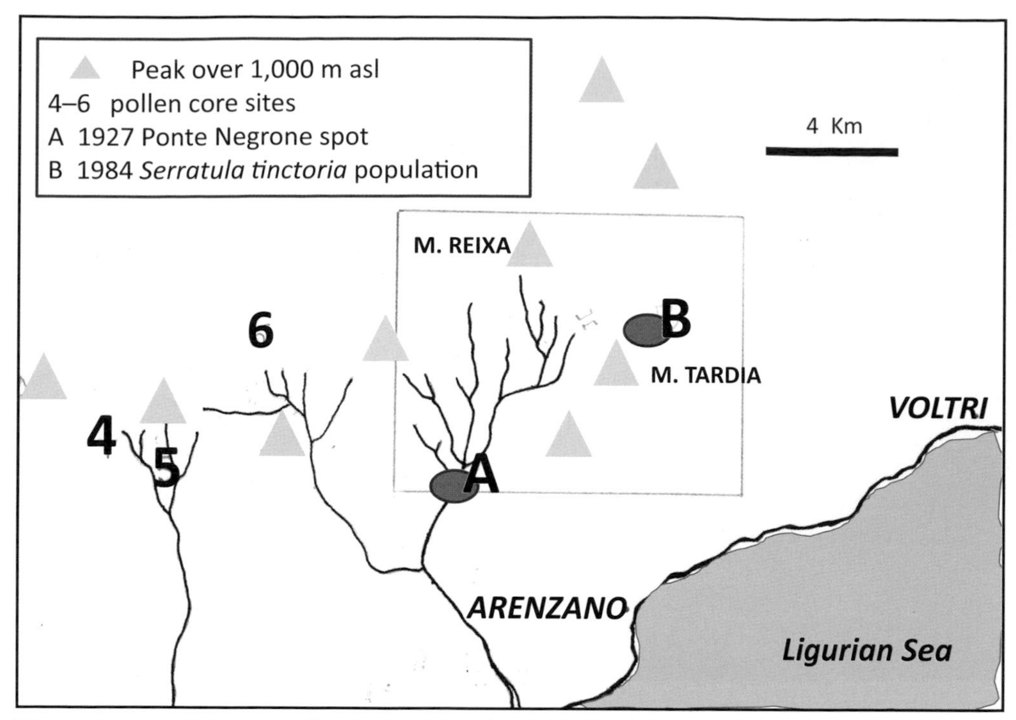

Figure 14.3 Upper Lerone Valley Location Map (central square) and sites visited in 1984. (the authors)

to replicate as closely as possible the same view. Oliver's drawing shows features from the 1927 photograph in red and the 1984 fieldwork in black. Comparison of the 1927 photograph with the drawing and the 1984 photograph allows us to identify significant changes in the populations of trees, shrubs and herbs. One example is given by the scree slope at the centre of the drawing, which remains a similar size and shape to its form in 1927. In the geobotanical analysis of the Lerone Valley, these habitats where considered highly unstable and subject to the continuous movement of the stones (Guido and Petroni 1975). However, the fieldwork suggested that on the contrary the scree slopes appear a relatively stable feature in the dynamic of this 'individual landscape' with the same living shrubs. Common Hazel *Corylus avellana* and Turkey Oak appear to be continuously 'coppiced' by the micro movement of the stones caused by cyclical temperature changes, in combination with fire events, and remained with similar dimensions (Fig. 14.7).

The pines have grown following afforestation after 1927. Some may have derived from seeds from the sparse population of Maritime Pine *Pinus pinaster* found nearby, following a change in the forestry regime and the creation of the Beigua Regional Park in 1985, which adopted what was termed 're-naturalization', in other words the abandonment of productive forestry. We photographed the same spot in August 2019 (Fig. 14.8). The policy of re-naturalization continues, and this has allowed the continued growth of pines, both Maritime Pine and Austrian Pine *Pinus nigra*, together with Cedar of Lebanon *Cedrus libani* introduced as part of the afforestation scheme. Other species include Manna Ash *Fraxinus ornus*, Holm Oak *Quercus ilex*, European Hop Hornbeam *Ostrya carpinifolia*, Common Alder *Alnus glutinosa* and Tree Heather. There is field evidence that fire has affected some of the area to the right of the photograph, south of the River Lerone.

Figure 14.4 'Bacino del Rio Lerrone vicino a Ponte Negrone', 1927. (Regional Forestry Inspectorate, Liguria, Corpo Forestale dello Stato)

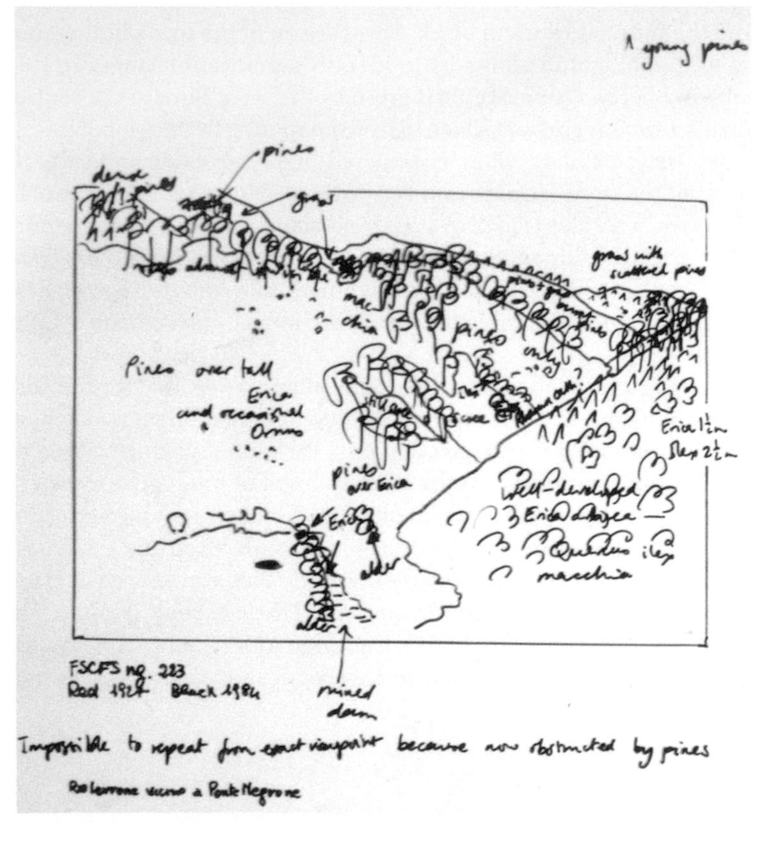

Figure 14.5 Oliver Rackham's tracing over the 1927 photograph. (Private collection)

Figure 14.6 Photograph of the 1927 spot, September 1984. (D. Moreno)

Figure 14.7 Oliver in the scree near a coppiced Common Hazel *Corylus avellana*, 1984. (D. Moreno)

There are no Maritime Pine in the 1927 photograph, but in the lower part of the valley some groups of this pine are well documented in the 1822/3 'Consegne dei Boschi' of the commune di Arenzano. On 7 May 1822, a local landowner in the Lerone valley Antonio Maria Ghigliotti noted that:

> woodland belonging to me located in the municipality of Arenzano ... Canale di Lerone ... place called Piano D'Anave [*sic*] ... mountainous woodland consisting of wild pine plants with low trunks and coppiced in the branches and of the age in assortment from a year up to fifteen and ordinarily they are cut of any age to use them for firewood. Also included in said woodland are some parcels of alder trees ('verna' dial.) and some other willow trees ('gabba' dial.) and quite a few small cypress trees, all of which are used for firewood ... with a declaration ... there are also saplings (sic) of low-stemmed oak aged fifteen to twenty approximately never been cut and used for firewood.[1]
>
> Archivio di Stato di Genova 1822: 215

This description shows that pines of autochthonous origin were very well integrated into the local agro-silvo-pastoral system in the early nineteenth century. They were intensively managed, and leaves were collected to allow the development of a richer layer of herbaceous and shrubby plants for grazing and hay making. The trees were regularly shredded to produce

1 bosco di mia spettanza situato nella commune di Arenzano ... Canale di Lerone ... luogo ditto Piano D'Anave [*sic*] ... bosco montuoso consistente in Piante di Pini selvatici di basso fusto e cedui nelli rami e dell'età in assortimento da un anno sino a quindeci ed ordinariamente soglionsi tagliare di qualunque età per servirsene di legna da fuoco v'è compresa anche in detto bosco qualche particella d'alberi di verna e qualch'altra di gabba ed alquanti piccoli alberi di cipresso tutti che servono per legna da fuoco ... con dichiarazione ... vi sono compresi anco degli arboscili (sic) di Rovere di basso fusto dell'età di anni quindeci in venti circa non stati mai tagliati e datti per legna da fuoco. (our translation).

Figure 14.8 Photograph of the 1927 spot, September 2019. (C. Watkins)

firewood for limekilns, the lime being used in local industries in nearby Cogoleto and Voltri, and for green manure for local vineyards (Moreno et al. 1993).

From at least the late medieval period until 1855, the area was common land (*comunaglie*) freely used by the inhabitants of Arenzano and nearby communities for regulated grazing by sheep and cattle. From 1855 until 1910, the access regime of these lands changed and they became communal land, owned by the commune but let out to individuals for grazing and hay production. Between 1910 and the 1970s, the area was a state-owned area (the Lerone Forest State Domain) managed by the Special Forestry Agency to protect what was identified as a highly degraded area (Moneta 2016). This degradation and environmental crisis had been caused, according to a contemporary technical description, by the transition from regulated common sheep grazing to individual access in the mid-nineteenth century (Carretto 1910).

Interestingly, similar variation is inscribed in the local pollen rain. Six pollen-fertile sites along the main ridge and along the maritime slope of the Voltri mountain group have been cored and analysed (Moneta 2016). Three of these sites are of particular interest for the recent history of the Upper Lerone Valley: 4-Prariondo, 5-Canei and 6-Fretto (Fig. 14.3). In each case, there are two clear vegetation phases. The first, radiocarbon dated to before the nineteenth century (dated 220 ± 45 BP), has a complex vegetation structure characterized by local agropastoral fire practices. The second phase is less biodiverse and dominated by Gramineae. The spore curves of coprophilic fungi and those of some pollen taxa indicate that these two phases are linked to the changes in the management of the common lands in the mid-nineteenth century.

Saw-wort

About 4 km to the north-east of the Ponte Negrone site on the grassy slopes of Monte Tardia, on the 'ancient' common lands of Crevari, Oliver Rackham and Diego Moreno noted clumps of Saw-wort (Jefferson and Walker 2017). Diego recounts at the beginning of his book *Dal documento al terreno* [From the document to the field] (Moreno 1990: 7, 2018: 5) how Oliver pointed out that the 'remarkable flowering of herbaceous plants distributed in homogeneous and contiguous clumps' in England could have been considered 'archaeological' evidence of a 'woodland margin' (Fig. 14.9). This hypothesis was supported by research on the topographical history of the slope: a document of 1612 indicates that these 'ancient common lands' formerly had more tree cover and were converted to pasture between 1500 and 1600 (Moreno 1990).

Figure 14.9 Saw-wort *Serratula tinctoria* site on the north slope of Monte Tardia, 1984. (D. Moreno)

Many years later, populations of Saw-wort were observed to increase rather than decrease following controlled burning at the Pian Brogione site in the common lands of Casanova (Val Trebbia, western Apennines). This was an unexpected result of an experiment that had been undertaken to increase the biodiversity of the heathland SIC (Cevasco et al. 2015; Cevasco and Moreno 2015). Oliver's 1984 observation on the Crevari mountains associated with the detail found in the Casanova mountains led us to look for other traces of this forgotten plant. On a European scale, the distribution of Saw-wort appears anomalous and relict. It can only be understood by going down to the topographic scale of an individual site and to its particular ecology considered as a historical product. This has already been noticed in Liguria for Round-Leafed Birthwort *Aristolochia rotunda* and Lily of the Valley *Convallaria majalis*, whose respective distributions are not linked to climatic or geological factors but to local land-management practices (Cevasco 2007; Bruzzone 2015).

Observational and sedimentary sources produced for the Pian Brogione site and the wider Trebbia-Aveto watershed area highlight several interesting issues:

1. The increase in the population of Saw-wort after the controlled fire experiment for the 2010 SIC project.
2. The pollen and anthracological (charcoals) diagram produced by paleobotanists (Menozzi and Montanari 2010; Cevasco 2018) shows a (relative) peak of Asteraceae, to which belongs Saw-wort, datable to the sixteenth century. The peak follows the greatest episode of fire recorded by the diagram starting from the twelfth century (to be precise, the quantity of microcarbons is equal to that recorded in the twelfth century when the fire precedes the planting of chestnut trees).
3. In the mid-sixteenth century, when the seigneurial jurisdiction changed from the Fieschi to the Doria family, the ancient mule track known as the Strada del Cifalco running along the watershed Trebbia-Aveto just above Pian Brogione was chosen by Tuscan muleteers as their preferred route to save time and taxes on the journey to France.
4. The pollen from Common Sainfoin *Onobrychis vicifolia* – a forage plant known in Val d'Aveto as lupinella, introduced in France in the fifteenth century and 'adopted' in Val d'Aveto only in the second half of the nineteenth century – has been documented in traces in the pollen diagram of Lake Rezzo (upper Val d'Aveto) around the sixteenth century. It appears that it had been tried out as a forage plant much earlier than was previously thought.

These are four small clues indicating that in the sixteenth century the routes of muleteers and merchants between Tuscany and France left sedimentary and documentary traces in the Trebbia-Aveto area (Tigrino 2007). Moreover, the intersection of the data on the increase in Saw-wort at Pian Brogione following the controlled fire and the pollen and anthracological diagram shows that Saw-wort is favoured by controlled fire practices and is an indicator of such activity.

If we remove the historical context from observation, Saw-wort is not a particularly useful plant to characterize a specific ecological area, because it is found in variable habitats (from woods to flooded meadows) and with strong differences in pH. However, if we adopt the micro-view of historical ecology, the same plant can take on the value of an 'indicator species'. This might be, for instance, as an indicator of fire as at Pian Brogione, or of a woodland margin, as Oliver noted at Monte Tardia. It may also be an indicator of historical gathering and collection activities. In nearby northern Tuscany, for example the practice of 'making Cerretta in Cerbaia', the practice of collecting Saw-wort (called Serretta or Cerretta) to make a yellow dye for woollen cloth, is recorded in the Cerbaie Hills from the second half of the fourteenth century up to the mid-sixteenth century (Piussi and Stiavelli 1995; Zagli 2004; Malvolti 2014). The plant grew in common lands, including wetlands and wooded pastures of oak and chestnut trees. Cerbaie, between the two lakes of Bientina and Fucecchio that were reclaimed at the end of the eighteenth century, was an important transit area between the Lucchesia and the Valdarno, used by shepherds, sheep, merchants and muleteers travelling along the Via Francigena to France and the Maremma (Municipality of Fucecchio; Malvolti 2014). Oliver's teaching remains important in exploring the topographical history of those particular Saw-wort populations, and not the history of generalizations about Saw-wort.

Conclusions

Oliver Rackham's field visits to Liguria provided a very significant stimulus to scholars and researchers working on environmental and cultural heritage. They gave valuable insights into the benefits to be derived from the careful application of historical ecology. They emphasized the importance of studying precise, locally placed histories in understanding why particular species grow in particular places.

His historical approach has been developed at the University of Genoa's Centre of Interdepartmental Research: the Laboratorio di archeologia e storia ambientale (LASA). It has been a crucial context for many research projects and has been nurtured by PhD theses of the special course in Historical Geography applied to understanding the cultural and environmental heritage and a multiple sources approach to studying sites. LASA applies a microhistorical and topographical approach devoted to the planning and conservation of rural heritage and landscapes. This multiple-source field approach allows historical and pre-historical environmental processes to be connected. The study of palaeoecological sources (such as pollen and charcoal) in conjunction with historical sources permits the detailed study of lost multiple land-use systems. This method has been used to discover several local agro-silvo-pastoral systems that have generated many 'individual landscapes' (Pescini, Montanari and Moreno 2018; Moreno et al. 2019). This approach has strong connections with an historical geography that bridges natural sciences and humanities. 'Individual landscapes' are studied using the perspective of historical ecology. The materiality of the landscape is a starting point from which can be derived knowledge of immaterial components such as practices of production, use of land, use of vegetation and the associated local knowledge.

However, a lot of work remains to be done to ensure that the intellectual legacy of Oliver Rackham's scholarship and research is not lost in Italy. For instance, the Italian text of the European Landscape Convention (ELC) signed in Florence in 2000 adopted various definitions that emphasize 'perceptions' and 'representations' of landscape over the materiality of landscape. This perception approach to the landscape often presents a false 'natural' dimension that disguises and overshadows the living features of the landscape itself. These features, often unrecognized and ignored by planners, hold precisely located historical evidence that can be vital for understanding present environmental processes (Montanari and Moreno 2008).

References

Archivio di Stato di Genova, Prefettura Sarda, Boschi e Foreste, *Consegne dei Boschi e Selve*, 1822, busta 215.

Balzaretti, R., Pearce, M. and Watkins, C. (eds) (2004) *Ligurian Landscapes: Studies in archaeology, geography & history*, Vol. 10. Accordia Research Institute, University of London, London.

Bruzzone, R. (2015) *Dalla foglia al folio. Un erbario figurato del XVI secolo e il suo contesto*. Sagep, Genoa.

Capponi, G., Crispini, L.,Federico, L. and Malatesta, C. (2016) Geology of the Eastern Ligurian Alps: a review of the tectonic units. *Italian Journal of Geosciences*, 135(1), pp. 157–169. https://doi.org/10.3301/IJG.2015.06

Carretto, G. (1910) *Gli usi civici nelle provincie di Cuneo, Genova e Porto Maurizio*. Tipografia Nazionale di G. Bertero e C, Rome.

Cevasco, R. (2004) Multiple use of tree-land in the Northern Apennines during post medieval times. Linking clues of evidence. In: R. Balzaretti, C. Watkins and M. Pearce (eds) *Ligurian Landscapes: Studies in archaeology, geography & history*, Vol. 10. Accordia Research Institute, University of London, London, pp. 155–177.

Cevasco, R. (2007) *Memoria verde. Nuovi spazi per la geografia*, Diabasis, Reggio Emilia.

Cevasco, R. (2018) *Serratula tinctoria* L. Tracce nascoste in una zolla di brughiera. In: D. Moreno, *Dal documento al terreno. Storia e archeologia dei sistemi agro-silvo-pastorali.* [reprinted in C. Montanari and M.A. Guido (eds) *Attualità di una proposta storica*] Genoa University Press, Genoa, pp. 337–356.

Cevasco, R. and Moreno, D. (2015) Historical ecology in modern conservation in Italy. In: K. Kirby and C. Watkins (eds) *Europe's Changing Woods and Forests: From wildwood to managed landscapes.* CABI, Wallingford, pp. 227–242. https://doi.org/10.1079/9781780643373.0227

Cevasco, R., Moreno, D., Balzaretti, R. and Watkins, C. (2015) Historical chestnut cultures, climate and rural landscapes in the Apennines. In: D. Harvey and J. Perry (eds) *The Future of Heritage as Climates Change.* Routledge, London, pp. 130–147.

Cevasco, R., Moreno, D., Poggi, G. and Rackham, O. (1999) Archeologia e storia della copertura vegetale: esempi dall'Alta Val di Vara. *Memorie della Accademia Lunigianese di Scienze 'Giovanni Capellini'*, LXVII, LXVIII, LXIX (1997, 1998, 1999), pp. 244–256.

Cevasco, R. and Poggi, G. (1997) Pratiche storiche, castagneti terrazzati e controllo della biodiversità: note di ecologia storica nella valle di Lagorara (sec. XIX–XX). In: *Conservazione e biodiversità nella progettazione ambientale. Atti del 1° congresso, Perugia, 28–30 Novembre 1996*, Quaderno 8. IAED, Rome, pp. 102–112.

Di Tella, G. (1940) *Il bosco contro il torrente*, Touring Club Italiano, Milan.

Greco, S. (2017) *Una foresta di carte. Materiali per una guida agli archivi dell'Amministrazione Forestale* Ministero della Difesa, Rome.

Grove A.T. and Rackham, O. (2001) *The Nature of Mediterranean Europe: An ecological history.* Yale University Press, New Haven, CT and London.

Guido M.A. and Petroni, P. (1975) Flora e vegetazione della valle del torrente Lerone (Appennino Ligure occidentale). *Webbia*, 29, pp. 643–716. https://doi.org/10.1080/00837792.1975.10670033

Jefferson, R.G. and Walker, K.J. (2017) Biological flora of the British Isles: *Serratula tinctoria. Journal of Ecology*, 105, pp. 1438–1458. https://doi.org/10.1111/1365-2745.12824

Lagomarsini, S. (2017) *Coltivare e custodire. Per una ecologia senza miti.* Libreria Editrice Fiorentina, Florence.

Malvolti, A. (2014) *La comunità di Fucecchio nel Medioevo. Boschi, acque, campagne.* Ricerche sul territorio fucecchiese tra Medioevo ed età moderna, studi 1976-2013, Studi Fucecchiesi – 2, Tipografi a Monteserra Vicopisano.

Menozzi, B. (2010) Analisi biostratigrafiche a Pian Brogione. In: Cevasco, R., Cevasco, A., Menozzi, B., Montanari, C., Parola C. and Stagno, A. *Progetto Interventi di valorizzazione degli habitat prioritari e delle Zone Umide all'interno del SIC IT331012 Lago Marcotto-Roccabruna-Gifarco-Lago della Nave, Final Report, Provincia di Genova, Area 11 Sviluppo Territoriale, Sviluppo Sostenibile e Risorse Naturali*, Università degli Studi di Genova, LASA Laboratorio di Archeologia e Storia Ambientale sezione geografico storica DISMEC e sezione botanica DIPTERIS, Genoa, pp. 1–68.

Métailié, J.P. (1986) Photographie et histoire du paysage: un exemple dans les Pyrénées luchonnaises. *Revue géographique des Pyrénées et du sud-ouest*, 57(2), pp. 179–208. https://doi.org/10.3406/rgpso.1986.3050

Métailié, J.P. (1988) Une vision de l'aménagement des montagnes au XIXe siècle: les photographies dela RTM . *Revue Géographique des Pyrénées et du Sud-Ouest.* 59(1), pp. 35–52. https://doi.org/10.3406/rgpso.1988.3105

Moneta, V. (2016) *Fonti biostratigrafiche per la storia dei paesaggi rurali del gruppo montuoso del Beigua (Liguria, Italia).* PhD thesis. Università degli studi di Genova Scuola di Dottorato di Ricerca Società, culture, territorio corso geografia storica per la valorizzazione del patrimonio storico-ambientale Ciclo XXVII.

Montanari, C. and Moreno, D. (2008) Beyond perception: towards a historical ecology of rural landscape in Italy. *Cuadernos Geográficos*, 43(2008-2), pp. 29–49.

Moreno, D. (1990) *Dal documento al terreno. Storia e archeologia dei sistemi agro-silvo-pastorali.* Il Mulino-Ricerche, Bologna.

Moreno, D. (2018) *Dal documento al terreno. Storia e archeologia dei sistemi agro-silvo-pastorali.* [reprinted in C. Montanari and M.A. Guido (eds) *Attualità di una proposta storica*] Genoa University Press, Genoa.

Moreno, D., Cevasco, R., Pescini, V. and Gabellieri, N. (2019) The archeology of woodland ecology: reconstructing past woodmanship practices of wooded pasture systems in Italy. In: F. Allende Álvarez, G. Gomez-Mediavilla and N. López-Estébanez (eds) *Silvicultures – Management and Conservation.* IntechOpen, London, pp. 66–85. https://doi.org/10.5772/intechopen.86101

Moreno, D., Croce, G.F. and Montanari C. (1992) Antiche praterie appenniniche. In: R. Maggi (ed.) *Archeologia preventiva lungo il percorso di un metanodotto* (Quaderni della Soprintendenza Archeologica della Liguria, No. 4) Colombo, Chiavari, pp. 159–176.

Moreno, D., Croce, G.F., Guido, M.A. and Montanari, C. (1993) Pine plantations on ancient grassland: ecological changes in Mediterranean mountains of Liguria, Italy during the 19th and 20th centuries. In: C. Watkins (ed.) *Ecological Effects of Afforestation.* CABI, Wallingford, pp. 93–110.

Moreno, D. and Montanari, C. (1988) The use of historical photographs as source in the study of dynamics of vegetational groups and woodland landscape. In: Salbitano, F. (ed.) *Human Influence on Forest Ecosystems Development in Europe: Proceedings of a workshop held in Trento, Italy, 26–29 September 1988.* Pitagora Editrice, Bologna, pp. 317–373.

Moreno, D., Piussi, P. and Rackham, O. (eds) (1982) Boschi storia e archeologia. *Quaderni Storici*, 49, pp. 7–163.

Pescini, V., Montanari, C. and Moreno, D. (2018) Multi-proxy record of environmental changes and past land use practices in a Mediterranean landscape: the Punta Mesco Cape (Liguria- Italy) between the 15th and 20th century. *Quaternary International Journal*, 463, pp. 376–390. https://doi.org/10.1016/j.quaint.2017.03.060

Piana, P., Balzaretti, R., Moreno, D. and Watkins, C. (2012) Topographical art and landscape history: Elizabeth Fanshawe (1779 – 1856) in early nineteenth-century

Liguria. *Landscape History*, 33, pp. 65–82. https://doi.org/10.1080/01433768.2012.739397

Piussi, P. and Stiavelli, S. (1995) Storia dei boschi delle Cerbaie. In: A. Prosperi (ed.) *Il Padule di Fucecchio: la lunga storia di un ambiente naturale*. Edizioni di Storia e Letteratura, Rome, pp. 123–136.

Raggio, O. (2004) Michrohistorical approaches to the history of Liguria: from microanalysis to local history: Edoardo Grendi's achievements. In: R. Balzaretti, M. Pearce and C. Watkins (eds) *Ligurian Landscapes*. Accordia Research Institute, University of London, pp. 97–103.

Sarasini, G. (1984) *Fonti archivistiche per la storia forestale della Liguria (XIX–XX secolo), Il fondo fotografico storica dell'Ispettorato Regionale delle Foreste pe la Liguria (1925–1965)*. Tesi di Laurea. facoltà di Lettere e Filosofia, Università degli studi di Genova.

Sarasini, G. (1986) La fotografia forestale: un fondo di archivio a Genova. In: *Studi in memoria di Teofilo Ossian de Negri*, Vol. III. Cassa di Risparmio di Genova e Imperia, Genoa, pp. 160–173.

Sheail, J. (1980) *Historical Ecology: The documentary evidence*. Institute of Terrestrial Ecology, Cambridge.

Tigrino, V. (2007) Giurisdizione e transiti nei 'feudi di Montagna' dei Doria-Pamphilj alla fine dell'Antico Regime. In: A. Torre (ed.) *Per vie di terra. Movimenti di uomini e di cose nella società di Antico Regime*. Franco Angeli, Milan, pp. 161–183.

Zagli, A. (2004) *Fra boschi e acque. Comunità e risorse nelle Cerbaie in età moderna*. Istituto Storico Lucchese, sezione Valdarno, Le Cerbaie. La Natura e la Storia, Pacini Editore, Pisa, p. 109.

Part IV

Approaches to Countryside Research

CHAPTER 15

Oliver Rackham, Archives and Ancient Woodland Research

Melvyn Jones

Summary

The remarkable legacy of Oliver Rackham, in terms of countryside history and especially with regard to woodlands and wooded landscapes, owes much to his meticulous delving and research into the archives. This chapter takes a similar exploratory journey to examine woodland histories discovered in part through the archives in Sheffield and South Yorkshire, England.

Keywords: history, documentations, archives, coppice, wood-pasture, Domesday

Introduction

Oliver Rackham, from his earliest research and publications to his last published work, placed great emphasis on using documentary evidence. In the very first volume of the journal *Landscape History* Oliver writes on the subject of historical information and notes that without such information it is impossible adequately to assess the importance of individual woods and woodland types, or to draw up rational management plans (Rackham 1979).

He put this assertion into practice whether researching and writing about the history of trees and woodland in general or about specific woods or specific trees. An impressive list of information can emerge from the documentary study of a wood: its status (i.e. the date when it is first mentioned as a named wood), past management practices, changing markets for wood and timber, other past uses of the site (e.g., charcoal making, mining and quarrying), the significance and in some cases the dating of boundary and internal earthworks, and woodland clearances and planted extensions. Moreover, it can confirm or overturn inferences from field surveys. Although Oliver emphasized that a study of a wood or woods based on documentary evidence alone would be weak, he also stressed that a history based on vegetation alone without the support of written evidence and earthworks would be, to use his words, rash, incomplete and insecure. Documentary evidence can also invalidate factoids and pseudo-history, matters Oliver was continually banging his fist on the table about. Moreover, arguments about the ecological and heritage value of sites, as he also pointed out, are greatly strengthened by combining field and documentary evidence. His

Melvyn Jones, 'Oliver Rackham, Archives and Ancient Woodland Research' in: *Countryside History: The Life and Legacy of Oliver Rackham*. Pelagic Publishing (2024). © Melvyn Jones. DOI: 10.53061/ILKY7287

outlook is beautifully exemplified in one of his lesser-known books, *The Woods of South-East Essex* (Rochford District Council 1986b).

Needless to say, Oliver Rackham has inspired me to undertake documentary research for a period of nearly 40 years. In the rest of this chapter, I would like to share with readers some of my most memorable findings from documents, mainly those stored in Sheffield Archives. The vast majority of the woodlands referred to in what follows were first documented before 1700, many of them as early as the late medieval period, and therefore can be defined as ancient woodland (see, for instance, Eccles 1986; Jones 1986a; Carey 2009).

Specifically, I shall discuss Domesday woodland types and distribution in South Yorkshire; a range of documentary evidence recording the practice of coppice-with-standards management including an annotated list of 49 South Yorkshire woods (most of which survive) from the late sixteenth century belonging to the Earl of Shrewsbury and a woodland management plan for the 30 woods on the Marquis of Rockingham's South Yorkshire estate extending from 1749 to 1770 ending 'and so begin the circle again'; a study of changing woodland management and markets for wood and timber in Ecclesall Woods, a 300-acre wood, using the annual woodland accounts over a period of a century and a half from 1751 to 1900; and finally, again utilizing estate accounts, management practices and markets for wood and timber on Thomas Watson-Wentworth's Irish estate in the early eighteenth century.

Domesday woodland types and distribution in South Yorkshire

If we take the results of William the Conqueror's great national survey of 1086 at face value, then woodland cover had been drastically reduced by the late eleventh century and the countryside was not covered by the boundless woodland of people's imagination. Discovering the details of Domesday woodland cover in South Yorkshire became a necessity for me in 1986 after I read Rackham's section on Domesday woodland cover nationally in *Ancient Woodland: Its history, vegetation and uses in England* (1980: 11–127). There, Rackham calculated that the Domesday survey of 1086 covered 27 million acres of land of which 4.1 million were wooded, that is 15 per cent of the surveyed area. His figure for the West Riding of Yorkshire is 16 per cent (Rackham 1980: 111).

So, I turned to the two-volume translation of the Domesday Book for Yorkshire (Faull and Stinson 1986) and began my calculations and mapping. My own calculation for woodland cover at Domesday for South Yorkshire is just under 13 per cent. By way of comparison, woods today, including plantations, cover just over 6 per cent of the region. What this means is that in the eleventh century, South Yorkshire was relatively sparsely wooded even by today's standards.

The Domesday surveyors in South Yorkshire in 1086 gave woodland measurements for each manor in almost every case in leagues (12 furlongs or 1.5 miles) and furlongs (220 yards or one-eighth of a mile) and in most cases recorded how woods were utilized. When the data are mapped (Fig. 15.1), noticeable variations in the distribution and types of woodland are clearly discernible. In the Dark Peak, the Pennine Fringe and in the Exposed Coalfield zones, that is in the western three-fifths of the region, woodland was relatively extensive, with a substantial number of communities having more than 1,000 acres of woodland. In contrast, in the Magnesian Limestone belt and in the Humberhead Levels further east, the picture was different. In those areas, woodland was more scattered, and amounts in individual communities were generally smaller than to the west. Additionally, the Magnesian Limestone belt, although only covering about one-eighth of the land area of the region, contained nearly a third (10 out of 33) of the places in which woodland was not recorded at all. Significantly there is almost a total absence of Old English and Old Norse woodland clearance place-name elements such as -ley, -royd, -field and -thwaite in the Magnesian Limestone belt, suggesting very early clearance of woodland there.

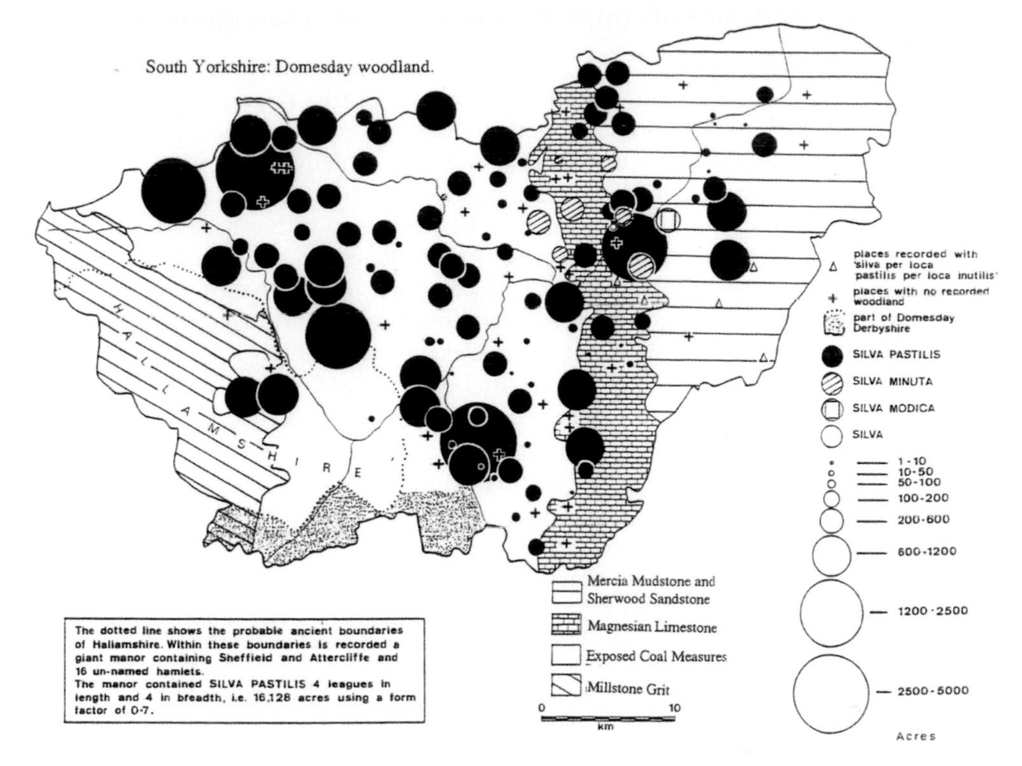

Figure 15.1 South Yorkshire: Domesday woodland. (M. Jones, 2009)

The types of woodland recorded in South Yorkshire at Domesday also suggest a shortage of woodland in some places in the east of the region, particularly in the Magnesian Limestone belt, and a relative abundance further west. When woods were relatively abundant and populations relatively small, they would have been able to be exploited for timber and underwood *and* as common pastures for cattle, sheep and pigs; that is, as wood-pastures. Domesday woodland in South Yorkshire was described in four main ways: as *silva*, *silva modica*, *silva minuta* and *silva pastilis*. *Silva* is simply 'woodland', the meaning of *silva modica* is not clear, *silva minuta* is 'coppice' and *silva pastilis* is 'wood-pasture'. Of the 111 manors in which woodland was recorded, 102 had wood-pastures and seven had coppices. All seven occurrences of coppice woods were in the eastern half of the region, two in the eastern part of the Coal Measures and five on the Magnesian Limestone. Although wood-pastures were found throughout South Yorkshire, they were very extensive, and the only type of woodland found in the western half of the region in the Millstone Grit country and throughout most of the Coal Measures.

Wooded commons, the *silva pastilis* recorded at Domesday and its descendants, were unfenced woods in which underwood and timber were harvested but in which the animals of commoners were allowed to graze freely. Commoners usually also had the right of the underwood and deadwood, but the timber trees usually belonged to the lord of the manor and permission had to be granted by the lord of the manor to the tenant. This was known as the right of estovers (*estoveir* is Old French for 'be necessary' which is a translation from the Latin phrase *est opus*: it is necessary). These manorial rights consisted of separate rights called botes. These rights allowed tenants to take timber, underwood, deadwood and bracken from wooded commons to build and repair their houses (housebote), to make and repair their hedges (hedgebote or haybote), to repair their farm equipment (ploughbote and cartbote) and for firewood (firebote).

Documentary evidence of coppice-with-standards management in South Yorkshire

The development and practice of coppice management in this region is remarkably well documented. The archival records include not only estate accounts but also contracts, diaries, leases, estate management plans, maps and the records of religious houses and surveys. These are all sources into which Oliver Rackham delved in his quest to uncover the history of woodland management, and whose example I have followed with relish.

Throughout South Yorkshire, the form of coppice management called coppice-with-standards, which combined the production of underwood with that of single-stemmed timber trees, emerged as the most important form of woodland management, in economic terms, during the medieval period, and continued to be so until at least the second half of the nineteenth century (Fig. 15.2). In this kind of woodland management, most of the trees were periodically (every 15 to 30 years) cut down to the ground to a stool, and from the stool grew multiple stems, the coppice or underwood. Some of the trees were not coppiced but were allowed to grow on to become mature single-stemmed trees, and these were the standards. They would normally grow through a number of coppice cycles, as the coppice rotations were known, and therefore were of various ages. Those that had grown through only one coppice cycle, less than 30 years old at most, were called wavers (usually written as weavers in South Yorkshire); those that had grown through two coppice cycles and were therefore between 40 and 60 years old were called black barks in South Yorkshire; and those that had grown through three or more coppice cycles and were therefore at least 60 years old were called lordings.

The earliest-known surviving documentary record of coppice-with-standards management in South Yorkshire is a lease written in Latin at the relatively late date of 1421 (Hall 1914: 122, 325–326). The lease refers to unnamed woods on a farm at Norton (then part of Derbyshire), and contains a number of clauses concerning the right to cut underwood and timber, charcoal burning and keeping animals out of the woods for three or four years after cutting. The woods on the farm are referred to as *le Spryng bosci*, an interesting mixture of French, English and Latin, 'spring' being the usual name for a coppice-with-standards in South Yorkshire from the fifteenth to the nineteenth century. Timber in the document is referred to by the Latin word *maerimium*.

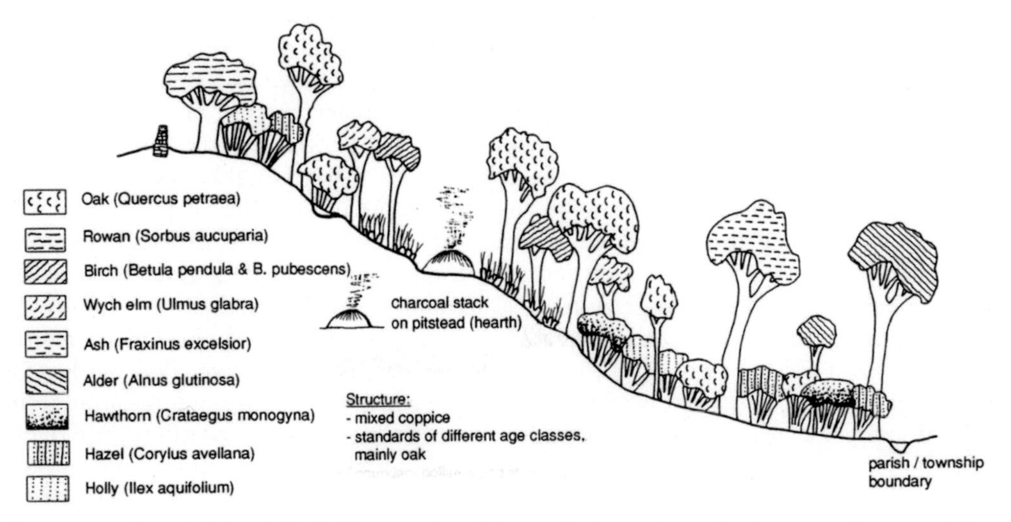

Figure 15.2 A typical South Yorkshire coppice-with-standards wood on sloping ground on a parish edge. (M. Jones, 2009)

Two other late medieval records of coppice-with-standards have also survived, both in the same geographical area as the 1421 record. The first is a lease dated 1462 and written in English, which refers to a number of woods in Norton parish including 'herdyng wood', the old name for the present Rollestone Wood. The lease noted that the lessees, John Cotes and John Parker, had been granted permission by the lord of the manor, William Chaworth, 'to fell downe cole (to make into charcoal) and carry the said Woddes', preserving for the owner 'sufficiaunt Wayvers after the custom of the contre' (Hall 1914: 123). Wavers, as already noted, were the young timber trees left to grow among the felled underwood, indicating that the woods concerned were coppices-with-standards. Wavers were also mentioned in the second document, also a lease, which was written in 1496, also in English, which refers to two woods in the Sheaf valley in Sheffield, one of which was the still surviving Hutcliff Wood. The two woods were the property of Beauchief Abbey, and the lease records that the 'abbot of Beacheff' had granted permission to the lessees 'to cooll [to make into charcoal] ii certen wodds', the woods to be left 'weyverd workmonlyke' (Beauchief Muniments in Sheffield Archives, BM994). Significantly, the document also refers to a bloom hearth (a primitive furnace) and a dam (the local name for a pond at a water-powered industrial site). Undoubtedly, the increasing dominance of coppice management, at least in the western half of South Yorkshire, was closely related to the expansion of metal smelting and related trades such as nail-making and edge tool manufacture.

Nearly a century later, in 1574, William Dickenson, bailiff of Sheffield, recorded in his diary that he went to four spring woods in Totley called Fraunces Fields, the Carre, Husters and Long Spring, which were compartments of the present-day Gillfield Wood (Fig. 15.3) and marked 968 spyres or spyeres and '1 ashe' (Miscellaneous Documents in Sheffield Archives, MD192, folio 72r). A 'spyre' was a spear or spire; terms which were generally used to refer to young oaks of about 20 years' growth that had been selected for growing on to become timber trees.

Figure 15.3 Aerial view of Gillfield Wood on the southern edge of Sheffield today showing its irregular shape, typical of ancient woods, and thick hedges surrounding fields created by woodland clearance; all the surrounding fields have Old English or Old Norse woodland clearance names. (Sheffield City Council)

But coppice management was also a feature of the eastern part of the county by the sixteenth century at least. At the dissolution of Roche Abbey in the Magnesian Limestone belt in 1546, a grant to Henry Tyrrell included 60 acres of coppice in four separate woods, two of which (Norwood and Hell Wood) have survived to the present day. Two other woods belonging to the abbey, including a spring wood of 15 acres, are also listed. There were also 800 oaks and ashes of 60 and 80 years' growth in the abbey's coppice woods and in other places in the abbey demesnes 'parte tymbre and part usually cropped and shred' (Aveling 1870: 130–31). This reference to shredding—the cutting off of side branches to produce a crop of poles and leaf fodder for animals—is one of only two instances of this practice that I have found in documents relating to woodland management in South Yorkshire. Coppice woods on the Magnesian Limestone were also felled for charcoal making. In 1553, the 69-acre (28-ha) wood called Anston Stones, which still survives, was leased to two woodmen described respectively as a 'blacksmith' and a 'smetheman'. They were to 'wayve the said wood with all kind of oake', leave the wood 'wayved woodman lyke' and have 'sufficient turfe & hylling for the colinge of the said wood' (Yorkshire Archaeological Society Archive, YAS DDS 4/46).

By the early seventeenth century, spring wood management appears to have been general. In an undated document written for the seventh Earl of Shrewsbury, the major landowner in south-west Yorkshire, who succeeded to the title in 1590 and died in 1616, 49 spring woods were listed (Lambeth Palace Library, London, Shr PLPL/Ms698, folio 1). Significantly, they were listed as belonging to the earl's forges and contained references to charcoal making such as 'Granowe Spring—20 years ould redie to cole—100 ac' and 'Thorncliff Spring one half about 9 years old tother halfe coalable—30 ac' (Fig. 15.4). In 1637, John Harrison, in his

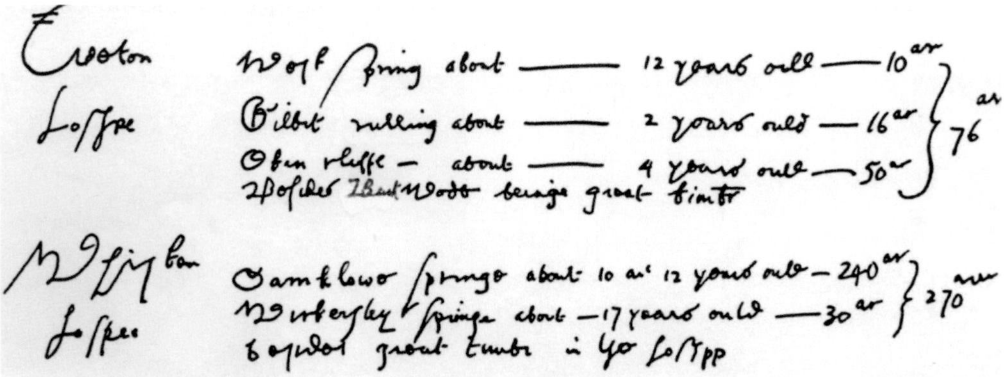

Treeton Lordship	West Spring about 12 years old -10 acres	
	Guilthwaite Ridding about 2 years old – 16 acres	76 acres
	Oaken Cliff about 9 years old - about 50 acres	
	Besides Burnt Wood being great timber	
Whiston Lordship	Canklow Spring about 10 and 12 years old – 240 acres	
	Wickersley Spring about 17 years old – 30 acres	270 acres
	Besides great timber in the lordship	

Figure 15.4 Extract from and transcription of an undated document written between 1590 and 1616 listing the seventh Earl of Shrewsbury's South Yorkshire coppice woods. (Shrewsbury Papers, Lambeth Palace Library, SHrPL PL (Ms698, folio 1))

survey of the manor of Sheffield, listed 36 spring woods in which the underwood varied in age from four years to forty.

Another aspect of coppice management becomes clear during the seventeenth century. This was the practice of subdividing large coppice woods into compartments and felling them at different times. Ecclesall Woods, for example, were made up in the seventeenth century of more than 20 different woods varying in size from 8 acres (3.2 ha) to 45 acres (18.2 ha) (Wentworth Woodhouse Muniments in Sheffield Archives, MP46). And a map drawn up in 1810 (Fig. 15.5) shows that the 243 acres (98.3 ha) of Canklow Wood had been divided since at least 1797 into 11 compartments, which had been coppiced between that year and 1807 (Arundel Castle Muniments in Sheffield Archives, ACM She 169). The original map also shows for most of the compartments the number of black barks and lordings (collectively called reserves) and wavers left standing in each coppiced compartment.

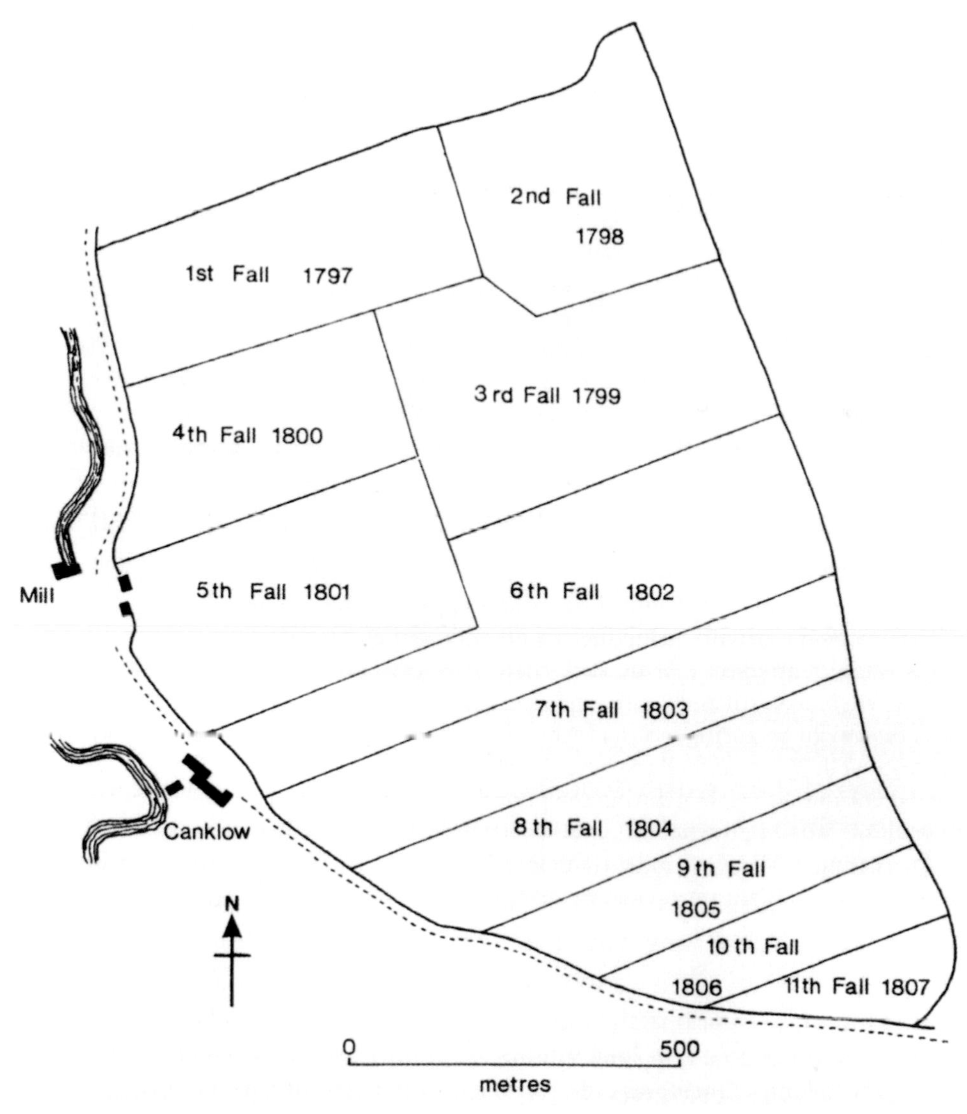

Figure 15.5 Redrawing of a map of the compartmented Canklow Wood in 1810; the original map also gives the number of 'weavers' and reserves left standing in seven of the compartments. (Arundel Castle Muniments, ACM She 169)

In the seventeenth century, the iron industry in south-west Yorkshire had achieved a high degree of sophistication. By the 1650s, the most powerful ironmaster in the region was Lionel Copley (Goodchild 1996), who entered into a succession of agreements with local landowners to fell and coal their spring woods. The surviving leases illustrate contemporary coppice practice. For example, in 1657 Copley entered into a ten-year agreement with the second Earl of Strafford of Wentworth Woodhouse to fell the underwood and selected timber trees in 13 of the earl's woods. Under the contract Copley was to cut 1,000 cords of wood (in South Yorkshire this was a pile of wood 4 ft wide, 8 ft long and 4 ft high) each year for charcoal making. He was allowed to cut 'young timber trees, Lordings, Black Barks, powles, coppices and Springwoode' together with 'the Bark thereof'. The lessee was also instructed to make sure that 'all the said Springwoode [is] well and sufficiently weavered' and that the coppice was 'workmanlike cutt downe ... and the stowens [stools] thereof neare to the roote so as best preserve for future growth and next springing thereof'. He was asked to burn the 'Ramell' in places that would be 'least prejudiciall to the weavers and Springwood which shall be left to grow' (Wentworth Woodhouse Muniments in Sheffield Archives, WWM D778). 'Ramell' was the small brushwood.

Various aspects of eighteenth-century spring wood management in South Yorkshire are well illustrated in two schemes devised by Thomas Wentworth, first Marquis of Rockingham, who inherited the Wentworth estates in 1723. In 1727, he devised what he called 'A Scheme for making a yearly considerable Profit of Spring Woods in Yorkshire' and in 1749 what he described as 'A Scheme for a Regular Fall of Wood for 21 years...' (Wentworth Woodhouse Muniments in Sheffield Archives, WWM A 1273). In the 1749 scheme, a 21-year rotation was used so that the woods coppiced in 1749 would be cut again in 1771, or as the Marquis put it, '& so begin the Circle again' (Fig. 15.6). This meant that the marquis's 876 acres (355 ha) of coppice-with-standards woodland in South Yorkshire would produce a regular crop of 40 acres (16 ha) of underwood a year. The marquis stipulated that there were to be five black barks (mature timber trees, 40–60 years old) and 70 wavers (sapling timber trees) left in every acre of felled coppice. The Marquis of Rockingham was fortunate that on his estate, besides hundreds of acres of coppice woods, he also had deposits of ironstone, and he linked the charcoaling of the former with the mining of the latter. In 1749 he wrote that: 'whereas it is the Iron Men that keep up the Price of the Wood, especiall care must be taken that the Iron Stone be never let for a longer time than the Woods are agreed for' (Wentworth 1749, WWM A 1273).

In the eastern parts of South Yorkshire, coppice-with-standards management also continued to be important during the seventeenth and eighteenth centuries. On the Duke of Leeds' estate centred on Kiveton Park, there were 17 woods in the early 1700s in Harthill, Thorpe Salvin and North and South Anston, five of them described in 1739 as 'timber woods' and another eight as spring woods, including the surviving Anston Stones Wood, Hawks Wood, Lob Wells Wood and Old Spring Wood. Among the products made from the timber and wood cut in these woods during the period in question were hop poles, scaffold poles, cordwood, pit wood ('puncheons'), heft wood, hazel hoops, hedge bindings and, perhaps most interesting of all, 562 'straite oaken trees', which were taken by 'land and water to his majestys yard at Chatham' in 1701 and yielded £473 in income (Yorkshire Archaeological Society Archive, YAS DD5/35).

In South Yorkshire in the post-medieval period, when a coppice wood or a compartment within a coppice wood was going to be felled and sold on the open market, the sale was preceded by a valuation of the wood, which involved marking the timber trees that were to be left standing and marking and valuing the timber trees that were to be felled. The valuation of the compartment was called setting out, and this involved marking (with paint and/or numbering with a scribe iron), valuing the timber trees to be felled and computing an overall value for the underwood. 'Top and lop' (branches of the timber trees) was often valued separately. If it were an oak wood, the value of the oak bark would also be calculated.

A Scheme for a Regular Fall of Wood for 21 Years to come from the Year 1749 of about 42 Acres a Year the Coppice Woods in Yorkshire amounting to about nine Hundred Acres, in which Calculation Scholes Wood, all timber Trees & Woods in the Park at Wentworth are not included, nor the Woods in Northamptonshire, particularly Withmail Park of 100 Acres nor a Wood of about 30 Acres at Yesthrop Park which was felled Anno 1726 – Reserves to be at least 5 Black Barks & 70 Wavers – Philip Wood 2 Acres 2 Roods being Holted is not included nor the Wood by the Pyramide.

			Acres
Tinsley Park at Nine Falls for the Years 1749, 1750, 1751, 1752 1753, 1754, 1755, 1756 & 1757			350
Anno 1758	Bassingthorp Spring		37
Anno 1759	Great Thorncliffe		37
Anno 1760	Harley Spring 18^A Luke Spring 8 – 2 Goss Wood 8 Acres Bank Spring 4 Acres	totall	38 – 2
Anno 1761	Giles Wood 24 Acres King's Wood 11 Normandale Springs 5 Bolderfall 2A 2R	totall	42 – 2
Anno 1762	Tindle Brig Spring 4A 2R Wadsworth Spring 1 – 2R Westfield Ing Spring 3 Acres Simon Wood 25 – 2R Birkfield Spring $7^A – 2^R$ Coney Garth Spring $2^A – 2^R$	totall	43 – 2
1763	Law Wood – 40 Acres		40 – 0
1764	Street Wood – 13 Rowing Spring $4^A – 2^R$ Littlewood Ing Spg 5 Acres 2 Roods Thorncliff Bottoms 5^A Blackmoor Bottoms 1^A Longland Spring & Longley Bottoms 5^A Little Thorncliffe 3	totall	37 – 0
1765, 1766, 1767 being three Years Fall	upper Linthwait $5^A – 2^R$ Golden Smithys 4 Rainbergh 115	totall	124 – 2
38 acres Per Annum			
Anno 1768, 1769 & 1770	Westwood exclusive of all Wasts		126 – 0
being a Fall of 42 Acres Per Year. & so begin the Circle again excluding all wast to fall			876

N.B. no Spinneys or Plantations in Wentworth Park & Demesnes are Included – There are also some little Reins & Spinneys up and down not taken notice of – also hedge Rows A Wood at Yesthorpe Park 30 Acres

1749

have now bought 400 Acres more called Edlington Wood 1750 bought Ld Gallway's Estate in Hoyland on which is 60 acres of wood so now the woodland is encreased near one half of that which may be brought into Regular Falls

Figure 15.6 The first Marquis of Rockingham's woodland management scheme for the period. (Wentworth Woodhouse Muniments, WWM A 257–485 and WWM A700–744)

It was not unusual for the timber, the underwood together with the top and lop, and the bark to be sold separately.

The next stage was to advertise the sale, and this was done until the mid-19th century through handbills (Fig. 15.7). At the sale, normally held at a local inn, the woodward would write the valuation(s) on a ticket and put it folded on the table in front of him. Within a specified time, all those bidding for the timber, underwood and other products included in the sale had to put their bids on separate tickets on the table. This was sometimes done three times, with the woodward announcing the highest bidder on each occasion. The highest bidder became the purchaser provided the bid equalled or exceeded the estate valuation. The surviving contract for a fall drawn up and dated 14 January 1823 shows that the purchaser was given until the end of September 1824 to fell, saw and remove the timber, peel the bark and make charcoal. The contract specifically instructs the purchaser to cut down the trees in a 'workmanlike manner ... so as to encourage the future growth of the roots'. There were strict

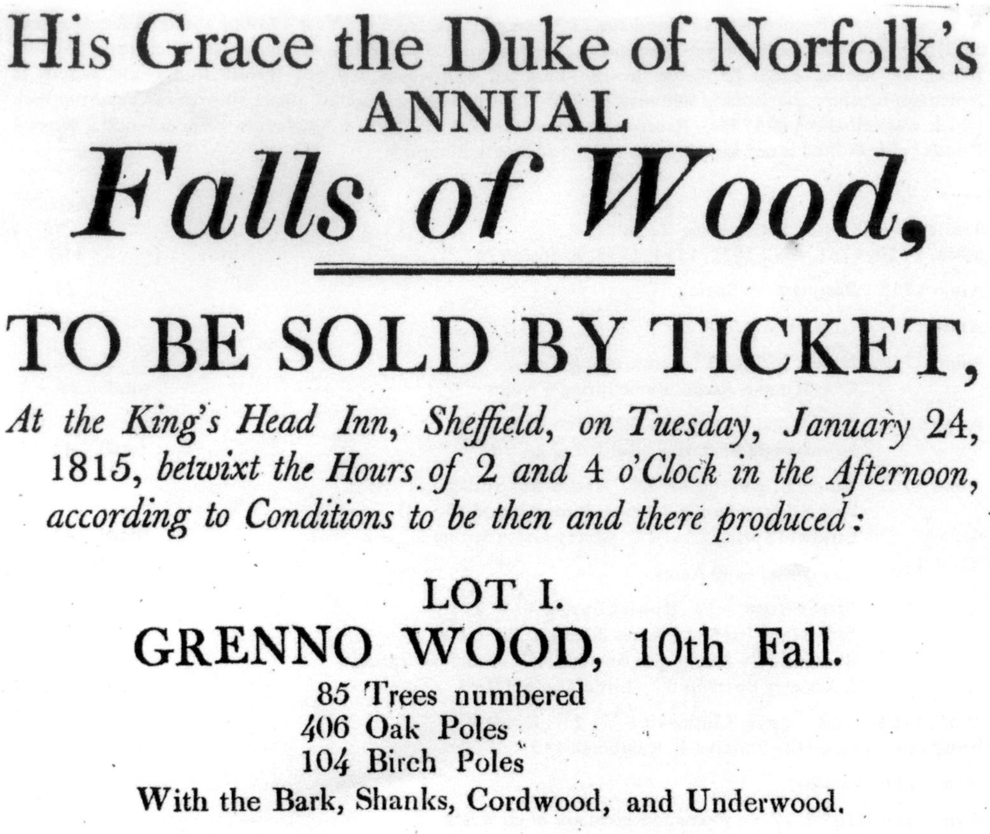

His Grace the Duke of Norfolk's
ANNUAL
Falls of Wood,

TO BE SOLD BY TICKET,
At the King's Head Inn, Sheffield, on Tuesday, January 24, 1815, *betwixt the Hours of* 2 *and* 4 *o'Clock in the Afternoon, according to Conditions to be then and there produced:*

LOT I.
GRENNO WOOD, 10th Fall.
85 Trees numbered
406 Oak Poles
104 Birch Poles
With the Bark, Shanks, Cordwood, and Underwood.

Figure 15.7 Extract from a handbill announcing the sale of wood and timber, 1815. (Arundel Castle Muniments, ACM She 303)

rules about making sawpits in the wood and gathering turf to cover charcoal stacks (called 'covering sods'). (Arundel Castle Manuscripts in Sheffield Archives, ACM She 303).

Changing woodland management and markets for wood and timber in Ecclesall Woods using the annual woodland accounts over a period of nearly two centuries from 1715 to 1900

The two case studies that follow make use of the detailed woodland accounts kept by the Wentworth-Fitzwilliam estate, whose headquarters were at Wentworth Woodhouse. The estate's woodland accounts for both its South Yorkshire and Irish estates covering almost three centuries survive in Sheffield Archives. It cannot be emphasized enough that accounts are not as dull as most people imagine, and they allow details of woodland management to be reconstructed with surprising precision.

When Oliver Rackham wrote about managed woodlands, he always emphasized at the beginning the difference between the two products of a coppice-with-standards: wood from the coppiced underwood and timber from the standard trees (e.g. 1986a: 67; 2006: 7; 1986b: 5–6). And these differences became very obvious when the documentary records for Ecclesall Woods, a 300-acre area of surviving woodlands on the southern edge of Sheffield, were analysed (Jones and Walker 1997).

Changes in the surrounding region impacted directly on the way that Ecclesall Woods were managed and on the markets that were found for the wood and timber felled in the woods.

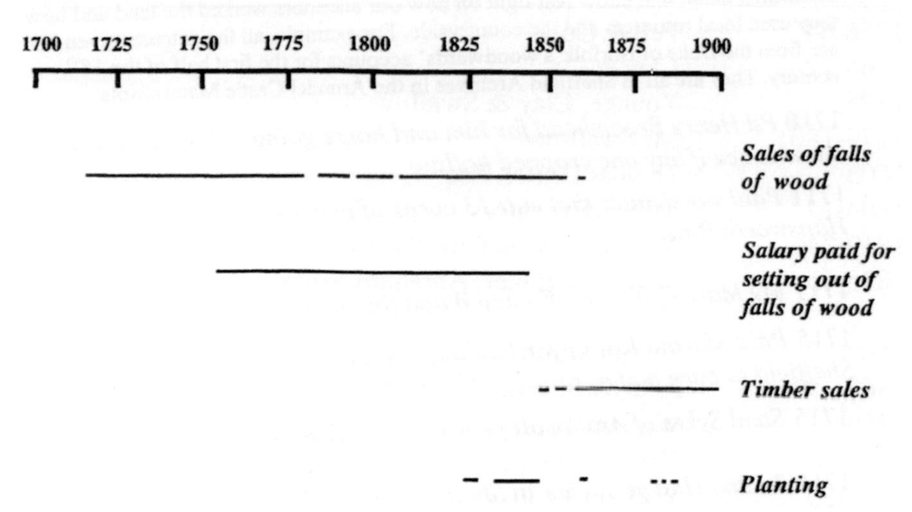

Figure 15.8 Indicators of changing management practices in Ecclesall Woods, 1715–1901. (Wentworth Woodhouse Muniments, WWM A 257–485 and WWM A700–744)

And these changes were meticulously recorded in the Bright estate and their successor, the Wentworth Woodhouse estate, account books between 1752 and 1901 (Wentworth Woodhouse Muniments in Sheffield Archives, WWM A 257-485 and WWM A700-744). The woods were gradually changed from coppice-with-standards into high forest or canopy woods (Fig. 15.8). The reason for this was the decline in the market for charcoal from the coppice wood and an expansion in the market for timber for the coal mining industry (pit props) and sleepers for the railways. This did not necessarily mean the extinction of charcoal making, which could use branch wood and roots.

The first sign that things were changing appeared in 1824, when £68 7s 6d was paid to a Mr Proctor for 27,500 larches and ashes for 'filling up the falls in Ecclesall Wood and Tinsley Park'. In the following year, a Mr Oldham was paid £99 for planting 'Forest trees' in Ecclesall Woods, Edlington Wood and Tinsley Park. Further planting took place in 1826. There was then a gap until 1830 when planting started again, continuing until 1845. After that, planting was only recorded specifically in Ecclesall Woods in 1885, 1886, 1888 and 1889, although it was recorded on the Wentworth estate throughout the second half of the nineteenth century and probably included more planting in Ecclesall Woods. Numbers and species of trees planted were sometimes recorded. Following the large planting of 1824, 13,000 trees were planted in Ecclesall Woods in 1831 and 21,000 trees were planted in Ecclesall Woods and Edlington Wood in 1833. Larch and ash, as noted earlier, were recorded as the species planted in 1824, and oaks and elms are also specifically recorded.

By the mid-nineteenth century, the process of change from coppice-with-standards to high forest was complete. This is reflected in the sale arrangements and the buyers of wood and timber. Only falls of wood or coppice or sales of falls of wood were recorded in the account books from 1715 to 1847. Falls of wood were also recorded from 1848 to 1852 and from 1856 to 1859. After 1859, sales of falls of wood were not recorded again in the 42 years up to 1901. In 1848, a timber sale was recorded for the first time. A further timber sale took place in 1851 and then continuously from 1856, until the records end in 1901.

The annual account books that have survived from 1756 to 1901 also record meticulously the purchasers of wood and timber from Ecclesall Woods, and these too reflect the changing structure of the woods (Fig. 15.9). John Fell and Richard Swallow of Chapeltown furnace, which

1756-64	Fell & Co
1766-72	Clay & Co
1775	Younge, Clay & Swallow
1779-1801	Mr Swallow
1803	Thomas Fenton Esq
1804-17	Milnes Executors
1819-34	Messrs Newton Chambers & Co
1835-65	Windle & Baker, Abraham Windle, John Goodwin, Joseph Smith
1866-67	Joseph Smith
1868	John Swinscow; Joseph Smith; William Toplis
1869	William Toplis
1870	George Rawlins
1871-80	William Topliss
1881-82	George Rawlins
1883-1901	William Toplis

Figure 15.9 Purchasers of wood and timber from Ecclesall Woods, 1756–1901. (Wentworth Woodhouse Muniments, WWM A 257–485 and WWM A700–744)

was charcoal-fuelled, were the buyers of falls of wood at Ecclesall from the 1750s until the beginning of the nineteenth century. From 1819 to 1834, the buyers were Newton Chambers of Thorncliffe Ironworks, which had coal mines, ironstone pits and an extensive railway network. By the mid-1830s, industrial buyers had disappeared, to be replaced by timber merchants. Abraham Windle or Windle and Baker, timber merchants, with premises in Sheffield, at Wortley and at Deepcar, bought the falls of wood or falls of timber from 1835 until 1864. John Goodwin, timber merchant, bought a fall of timber in 1865; Joseph Smith, also a timber merchant, bought the timber in 1866 and 1867; and three buyers, John Swinscoe, Joseph Smith and William Toplis, again all timber merchants, bought the timber in 1868. Then, with the exception of the years 1870 and 1881 and 1882, when George Rawlins, another timber merchant from Gleadless, was the buyer, William Toplis was the only buyer until the last surviving account book in 1901. Joseph Smith and Sons had three sawing and planing mills in Sheffield. John Swinscoe had Norfolk saw mills and three other premises in Sheffield. William Toplis was a Chesterfield timber merchant.

Woodland management for constructional timber on the Watson-Wentworth Irish estates in the first half of the eighteenth century

The headquarters of the Watson-Wentworth estate in South Yorkshire was at Wentworth Woodhouse where, by the end of the eighteenth century, it was surrounded by an estate covering 18,000 acres. The Watson-Wentworth family of Wentworth Woodhouse had on the South Yorkshire estate more than 30 coppice-with-standards woods that were managed with great sophistication, and they exported their approach to woodland management to their Irish estates. These lay in six blocks, five in County Wicklow and one in County Kildare (Jones 1986b; 1998). This estate totalled an astonishing 91,000 acres or 37,000 ha. By 1750, the estate contained 2,356 acres (954 ha) of coppice and scrub woods, of which 30 woods, comprising 1,989 acres (805 ha), were coppice woods proper (Fig. 15.10). They

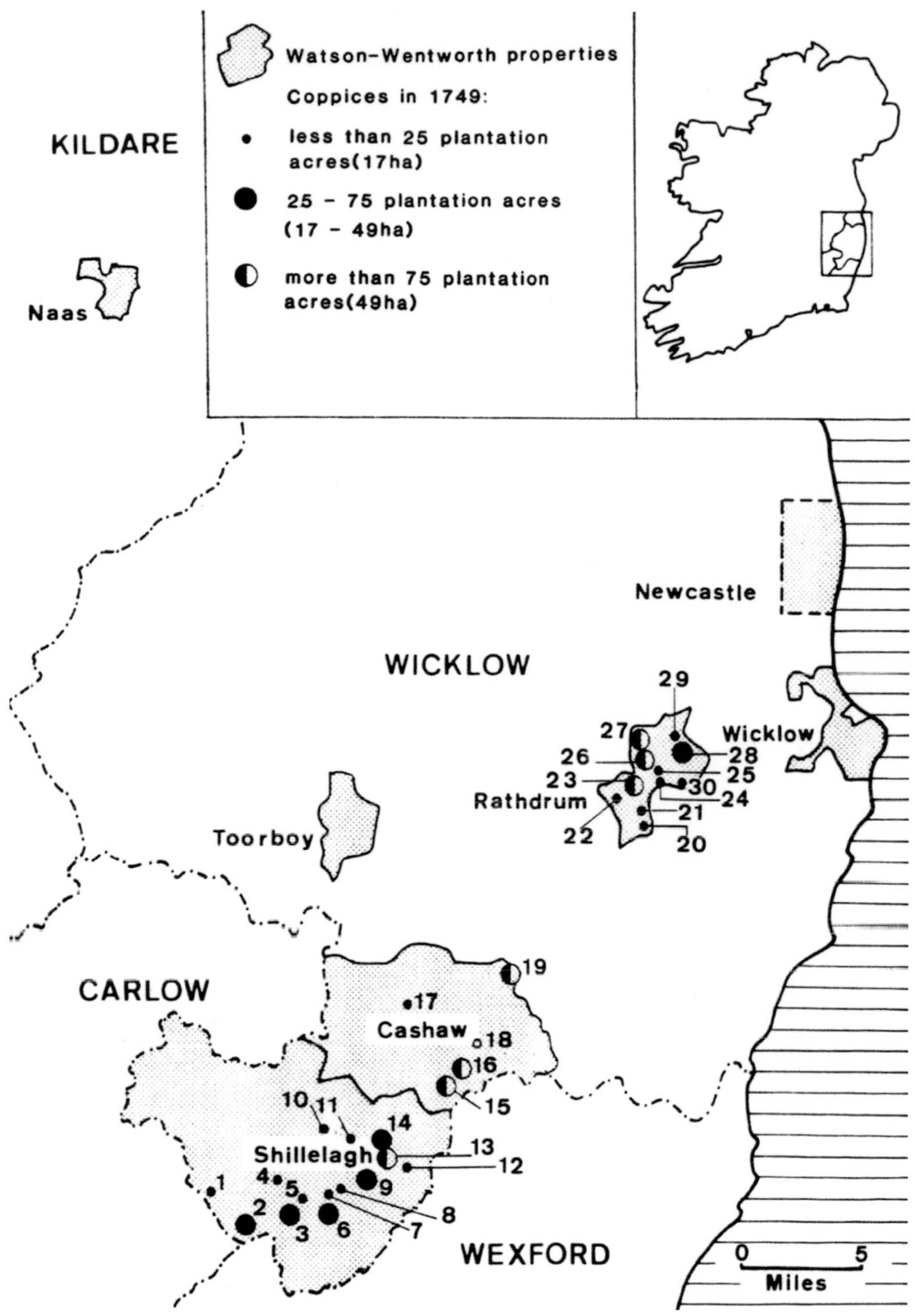

Figure 15.10 Coppice woods on the Watson-Wentworth Irish estate in County Wicklow, *c.* 1750. (Wentworth Woodhouse Muniments, WWM A767)

1. Moylisha; 2. Raheengraney; 3. Balisland; 4. Ballard & Minmore; 5. Ballyknockers; 6. Cronyhorn; 7. Carrig; 8. Coolattin Scrub Wood; 9. Coolattin Wood; 10. Conelea; 11. Nickson's Brow; 12. Paulbeg; 13. Tomnifinnogue & Ballykelly; 14. Balyraheen; 15. Killaveny; 16. Coolalug; 17. Corndog; 18. Tomcoyle; 19. Ruddenagh; 20. Upper Corballis; 21. Lower Corballis; 22. Round Coppice; 23. Ballygannon; 24. Glasnarget; 25. Key's Coppice; 26. Stump; 27. Cronybyrne; 28. Ballynakill; 29. Barnbawn; 30. Bahana

varied in size from 202 to 3 acres. Detailed accounts of their management have survived for the period 1707 to 1749. And like the documentary records for Ecclesall Woods, the archival records of these Irish woods allow the researcher to discover in great detail the management practices employed and the markets for the coppiced underwood and the timber trees. Emphasis here will be placed on the markets for the timber trees (Wentworth Woodhouse Muniments in Sheffield Archives, WWM A764, WWM A766, WWM A767 and WWM A 770).

The Irish coppice woods were managed as coppice-with-standards like their South Yorkshire counterparts. Between 1707 and 1749 coppice cycles varied from 16 to 33 years with a mean of 25 years. A scheme for all the woods in 1749 assumed a 22-year cycle. The standards were overwhelmingly oak, but ash and alder were also mentioned. The 1749 coppicing scheme laid down that at each fall 60 standards per plantation acre (37 per statute acre) should be left standing. The standards (generally referred to, as in South Yorkshire, as reserves) were not even-aged. The 1748 scheme stipulated that at each fall ten black barks (that had grown through at least two coppice cycles) and 50 wavers (that had grown through no more than one coppice cycle) should be left.

Besides underwood for charcoal making, building timber was a very important product of these Irish woods in the first half of the eighteenth century. It was sold by the named piece and in undifferentiated lots. Named pieces included unworked wood described as poles and saplings and semi-finished and finished articles such as 'riberrys' (cleft spars), principals, purlins, beams, collar beams, hammer beams, rafters, laths, shingles, lintels, doorcases, clapboard and 'window stuff'. Named industrial items included helves, millshafts and timber for waterwheels. Timber was also provided for a substantial number of named building projects including Dublin Barracks, for which £1,423 was received in 1708–9 (this must have been the famous Collins Barracks, now a museum, which was built between 1704 and 1710) and Dublin 'Colledge' in 1719.

Timber was also supplied for the construction of courthouses at Athy, Carlow and Wicklow, repairs to market houses at Blessington and Newtown Mount Kennedy, new churches at Coolkenna, Donard, Inch and Kilcullen, church repairs at Ballymore, Baltinglass, Carnew, Clonegall, Donaghmoor, Hackettstown, Hollywood, Kilcommon, Lymrick and Tullow, a new gaol at Carlow, five bark mills and a fulling mill. Timber, along with carpenters and joiners, was also provided for building four 'school houses' between 1713 and 1718. Fifteen pieces of timber were also sold for making a bridge near Donard. Ship timber, like general building timber, was also an important product of the standard trees of the Irish woods. Ship timber was sold squared, sawn and in the round. It was sold in the woods, at timber yards and delivered, sometimes at the estate's expense, sometime at the buyer's. It was generally sold direct to shipbuilders whose buyers came to the estate, but some went to dealers and some was carried to Wicklow 'to be laid on the Murrow for Sale'. The Murrough is the shingle beach that stretches northwards from the town.

All types of ship timber were sold: boat boards, bowsprits, deck beans, futtocks, gunwales, keels, keelsons, knees, masts (fore masts, main masts and mizzen masts), rabbet bends (a rabbet was a notch or groove into which another timber could be inserted), rudders, ship frames, ship planks, skegs and stems, besides many 'bend trees' for unspecified uses. Scaffolding poles and bilgeways (the cradles in which the ship was built and from which it was launched) were also sold to shipbuilders, and treenails, described in the accounts as 'trunnils or 'sipp pins', were sold by the thousand.

Most of the ship timber went to Dublin and Wicklow, although a substantial proportion of that carried to Wicklow was then shipped to England. Between 1707 and 1720, 21 shipbuilders and ship timber dealers were mentioned by name. Of the 15 for whom a location was given, two, apparently dealers, were from within the estate itself: (one from Arklow, one from Wexford), four were from Wicklow, four were from Dublin and five were from Whitehaven in Cumbria.

Conclusions

I hope readers agree that any study of the woods referred to in this chapter would be incomplete without searching for, evaluating and including the documentary evidence cited. It was Oliver Rackham who started me and many other woodland researchers on this adventure into documented woodland history.

In 2006, in his Foreword to the New Naturalist volume *Woodlands*, he claimed that he was now a rather old-fashioned botanist and says that 'for good or ill, I haven't a particular theory to promote'. Don't you believe it: the book was as thought-provoking as ever. While again stressing that to undertake a thorough and complete woodland study required ecological and archaeological investigations to accompany documentary research, he devoted a whole chapter to documentary sources and how to use them.

Last but not least, it must be recorded that Oliver had a wicked sense of humour. Discussing the uses of willow in *The History of the Countryside* (1986a: 216), he described a use that has thankfully fallen into disuse – for making canes. He cited an inquest in 1301 that was held on a schoolmaster from Oxford who had fallen off a willow pollard into the River Cherwell while cutting rods with which to beat his pupils!

References

Aveling, J.W. (1870) *The History of Roche Abbey*. Robert White, Worksop.

Carey, M. (2009) *If Trees Could Talk: Wicklow's Trees and Woodlands over Four Centuries*. National Council for Forest Research and Development, Dublin.

Eccles, C. (1986) *South Yorkshire: Inventory of Ancient Woodland*. Nature Conservancy Council, Peterborough.

Faull, M.L. and Stinson, M. (1986) *Domesday Book: Yorkshire*, 2 volumes. Phillimore, Chichester.

Goodchild, J. (1996) Lionel Copley: a seventeenth century capitalist. In: M. Jones. (ed.) *Aspects of Rotherham: Discovering Local History*, Volume 2. Wharncliffe Publishing, Barnsley.

Hall, T.W. (1914) *Descriptive Catalogue of the Jackson Collection*. J.W. Northend, Sheffield.

Jones, M. (1986a) Ancient woods in the Sheffield area: the documentary evidence. *Sorby Record*, 24, pp. 7–18.

Jones, M. (1986b) Coppice wood management in the eighteenth century: an example from County Wicklow. *Irish Forestry*, 43, pp. 15–31.

Jones, M. (2000) The absentee landlord's landscape: the Watson-Wentworth Estate in eighteenth century Ireland. *Landscapes*, 1(2), pp. 33–52. https://doi.org/10.1179/lan.2000.1.2.33

Jones, M. (2009) *Sheffield's Woodland Heritage*. (4th edition) Wildtrack Publishing, Sheffield.

Jones, M. and Walker P. (1997) From coppice-with-standards to high forest: the management of Ecclesall Woods 1715–1901. In: I.D. Rotherham and M. Jones (eds) *The Natural History of Ecclesall Woods, Part 1, Peak District Journal of Natural History and Archaeology*. Special Publication No. 1, July 1997, pp. 11–20.

Jones, M. (1998) The rise, decline and extinction of spring wood management in south-west Yorkshire. In: C. Watkins (ed.) *European Woods and Forests: Studies in Cultural History*, CABI, Wallingford, pp. 55–72.

Rackham, O. (1979) Documentary evidence for the historical ecologist, *Landscape History*, 1, pp. 29–33. https://doi.org/10.1080/01433768.1979.10594337

Rackham, O. (1980) *Ancient Woodland: Its History, Vegetation and Uses in England*. Edward Arnold, London.

Rackham, O. (1986a) *The History of the Countryside*. J.M. Dent and Sons, London.

Rackham, O. (1986b) *The Woods of South-East Essex*. Rochford District Council, Rochford.

Rackham, O. (2006) *Woodlands*. (New Naturalist 100), Collins, London.

CHAPTER 16

From Household Equipment to Countryside in Eleventh-Century Bavaria

Richard Hoffmann

Summary

Considering Oliver Rackham's powers of explanation and investigation, this chapter provides examples of his approach to the subject and the power of critical interpretation from documentary evidence. Rackham's craft was honed in the English countryside but transferred easily to, say, Europe or to North America. Furthermore, Oliver's research and consequent writing was a model to stimulate and guide others.

Keywords: environmental history, culture, nature, microhistory, monastic records

Introduction

In publications and in person, Oliver Rackham taught ways to see landscapes of the past in the present and in dusty relics of the past, even those overlain by multiple intervening layers of meaning. As my own career evolved from agrarian to environmental history, Rackham's successive writings showed possibilities I might envision for my own work and, perhaps more importantly, provided clear and telling exempla for students, be they fourth year history undergraduates or post-graduate MA and PhD candidates. One key lesson came at the level of microhistory, not the social incidents or personal vignettes commonly featured by that approach, but the story of a place. We environmental historians have come to like the term 'socio-natural site' to refer to a locale where intellectual constructs of culture and nature are precipitated into life by actions of people in a particular place, resulting in a hybrid of the natural and the cultural (Hoffmann 2014: 2–16, 378–79). But without indulging in such stereotypical academic theorizing, Rackham's talk and sketches of specific English woodlands or later of Cretan landscapes showed the way. Although neither I nor my own advanced students ever worked in the neighbourhood of Oliver's own explorations in England or elsewhere, and he and I shared a space and audiences on only three occasions, I thought of him as both model and (unwitting) mentor. This chapter is both memoir and exercise in landscape microhistory to honour his memory.

Encounters with Oliver

At our first face-to-face meeting in spring 1996, Oliver and I shared an impromptu demonstration of his approach by together reconstructing a specific subalpine Bavarian landscape

Richard Hoffmann, 'From Household Equipment to Countryside in Eleventh-Century Bavaria' in: *Countryside History: The Life and Legacy of Oliver Rackham*. Pelagic Publishing (2024). © Richard Hoffmann. DOI: 10.53061/HGRC3840

from one scrap of eleventh-century parchment. My late York colleague Elinor Melville and I had organized an interdisciplinary colloquium of scholars and students from North America, Latin America, Europe and India at York University entitled 'Humans and Ecosystems before Global Development: Problems, Paradigms, and Prospects for Early Environmental History'. We wanted to show the North American pioneers of the self-conscious subdiscipline how we could practise it on situations long before industrialization, modern state bureaucracies, fossil fuels and other characteristics of modernity. We invited Oliver (and some other contributors to this present volume) as established practitioners of the craft both to present and to comment. On the last afternoon, we did small mentored workshops on sources and I offered one on this textual object, Tegernsee Abbey cellarer Gotahelm's inventory from 1023. To my surprise and perhaps trepidation, Oliver joined the dozen or so mostly students in the room. The workshop soon became a dialogue on tools and equipment, their uses and what they could tell about the community and its landscape. Engrossed students watched the show. No, there were and are no recordings, transcripts or notes. I cannot replicate what either Oliver or I then said, but I came away with admiration for his erudition (and I thought he was a botanist!) and a full enough grasp of the potential in this text to offer a summary, sadly less lively, re-enactment of what it conveys of a community, a valley, a lake at the edge of the Bavarian Alps nearly a millennium ago. Voluminous abbey archives from the ensuing eight centuries can then trace the evolution of human uses and natural landscape in this specific socio-natural site, but that is not my purpose here (Fig. 16.1).

Tegernsee Abbey

Beside a deep lake fed by streams from high Alpine valleys some 50 km south of Munich,[1] mid-eighth-century members of a noble Bavarian family established a monastery (for what follows see Wessinger 1885; Redlich 1931; Hartig 1946; Angerer 1968; Hemmerle 1970). Some later legends say they chose the Tegernsee for its rich fishing, others for its relative seclusion from political tensions as the Frankish Carolingians suppressed Bavarian autonomy. Following generations considered the house a Carolingian royal abbey, but with that dynasty's decay the monastery's rich endowment of estates on the Bavarian plain downstream of the lake tempted local strongmen, leaving it moribund from the 880s and victim of a destructive fire in 970. In 978, Emperor Otto II declared Tegernsee an imperial abbey and sponsored refoundation by monks brought from Trier who followed the Benedictine rule. A generation later, Abbot Ellinger (1017–26 and 1031–41) was actively reviving other old houses of the region, Benediktbeuern among them.[2] Well into the thirteenth century, more than 100 choir monks and as many novices, students and other personnel sustained Tegernsee's reputation for political and spiritual leadership and cultural achievement. Its scriptorium (writing office) produced both religious and secular works. Subsequent almost stereotypical monastic cycles of decline and revival matter here only to explain the thirteenth- and fourteenth-century loss of most earlier administrative records, and also a building boom after 1450 and transformation into a gem of baroque architecture from the 1670s. All ended in 1803 when a modernizing Bavarian monarchy secularized all its monasteries. Tegernsee's rich library and extensive late medieval and early modern archives were transferred to Bavarian state institutions in Munich and the buildings put to various secular uses, including as a country residence for the queen mother. Munich elites discovered areas around the lake to be ideal locations for their own summer homes, gradually turning the landscape into a fashionable resort. Wooded cover

1 At an altitude of 725 m the Tegernsee covers 9 km^2 to a depth of 72 m. Peaks on the crest of its watershed range between 1,300 and 1,800 m.

2 Benedictine reform in Germany was contemporary with that associated with Cluny in France and St. Dunstan in England.

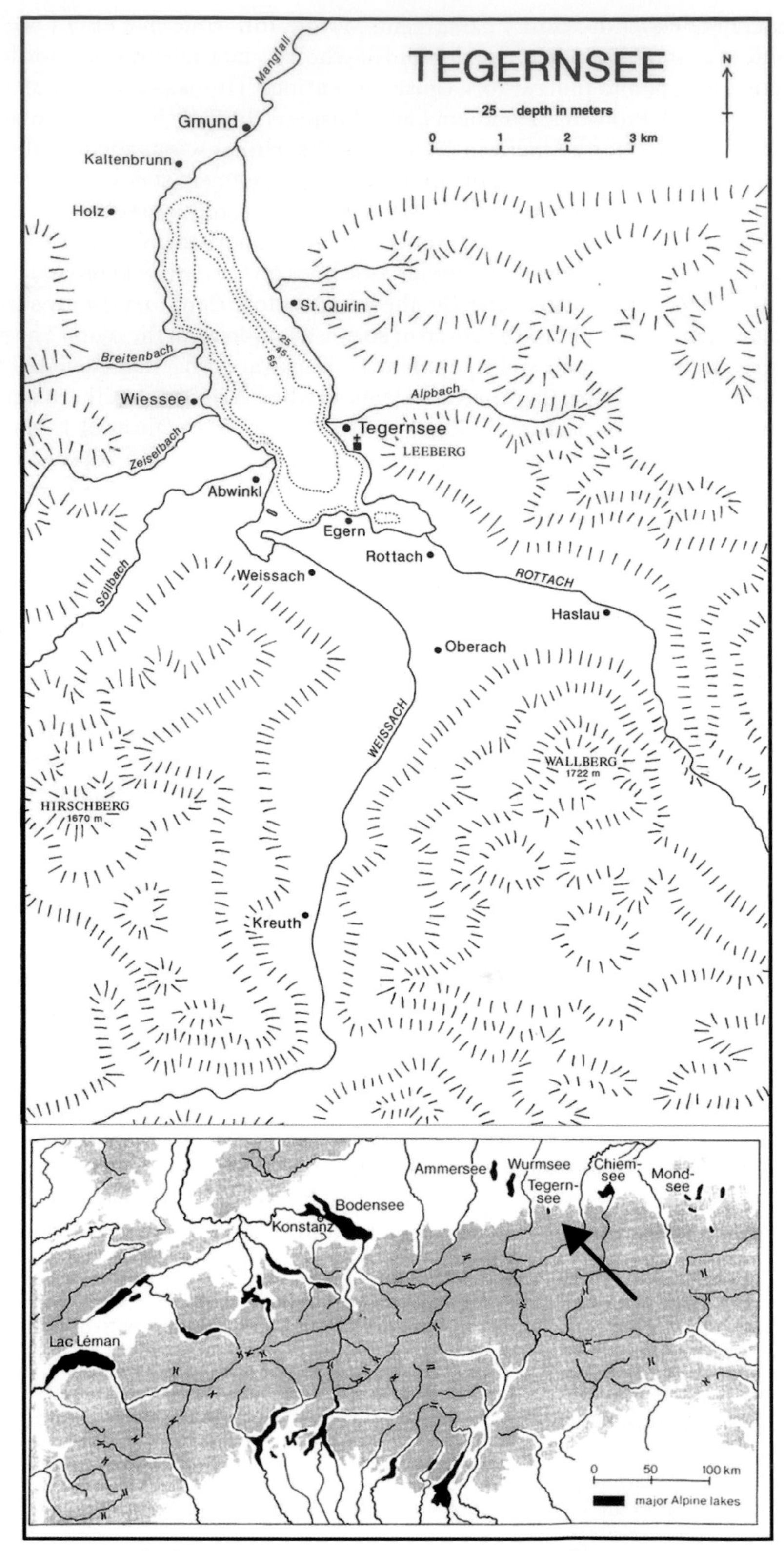

Figure 16.1 Tegernsee: (top) landscape detail showing the location of Tegernsee Abbey; (bottom) Central Alps showing the major lakes. Arrow points to the Tegernsee. (R. Hoffmann)

now starts 1 km or so back from the lake's east and west shores, and the cleared areas of the two larger southern valleys are about the same width. The lake remains cold and relatively poor in plant nutrients.

An eleventh-century cellarer's inventory

Bayerische Staatsbibliothek [henceforth BSB], Clm 18181, fol. 118v, is a single parchment sheet put to use at Tegernsee in the eleventh century as the inside back cover or back flyleaf binding a codex of grammatical and rhetorical works. But already in the nineteenth century this leaf was recognized as a reuse of a sheet with slightly earlier writing, an inventory of the cellarer's office. The text is in Latin, but with added interlinear glosses in early Bavarian vernacular. While text and glosses are in contemporary Tegernsee script, they are not by the same hand and the glosses use a paler and browner ink.[3] The following transcript marks the original lines with /, expands abbreviations and also indicates all breaks in the running text.

> Haec sunt instrumenta piscalia siue etiam quaeque ad usum cellarii pertinentia . quae / Gotahelmus ł inuenta. siue plura . . . . *[2 words erased]* acquisita inibi relinqui officium / cellarii restituens XVI kal. februarii Anno dominice incarnationis Mxxiii: / Incubas uinarias xxxvi. quindecim et duas[4] ex eis cum vino episcopi[5] quinque / cum ceruisa . et quatuor magnas . quatuorque paruulas . duas / cum vino et metone . Lagoenas saugmaritias . xl / et decem ferro circumdatas . Sellam cum hultia et filtro atq,[6] / freno . Massam unam . Bipennes . iiii . Secures viii et vnam paruvlam[7] / Dolatorias viii . Falces foenales . xii . Falciculas . iii . / Aratrum . ligonem . I . Vomerem . I . Iuga . iiii . Ascias . ii . / Caldarios . vi . Sartagines . vi . Fuscinulam . chrouvil[8] . ii . / Cramuculas . hala[9]. iii . Endire prantreita[10] . I . Sergam . I . Graticulas rostisarn[11] .ii. / Terebellos . nabagera[12] . vii . Roscios[13]. nuila[14] . ii . Malleolos . hamara[15] . iii . Incudes . anapoz[16]. iii . / planatoria . poumscapun[17]. iii . Runcina . rita[18]. I . Furca . kartgapala[19]. I . Funes . seili[20]. viii . / Vuangas .

3 Dating and relation of the hands in this codex follow Eder (1972: 87). Glosses and base text were separately published in Steinmeyer and Sievers (1879–1922). Their glosses and their reference words are given in vol. 3 (1882, *Sachlich geordnete Glossare*, item MCXL, p. 657). The entire Latin text (without glosses) is in vol. 4 (1898, *Alphabetische geordnete Glossare ... Handschriftenverzeichniss*, as Hs 430, pp. 562–63). I have closely examined the original manuscript and a good photocopy but not the facsimile reportedly in Chroust (1931: Lieferung 1, Tafel 8a). Note that codicological and linguistic studies have paid no attention to what the text has to say.

4 *et duas* inserted above line.

5 *episcopi* inserted above line.

6 *atq* hard to read.

7 *et vnam paruvlam* inserted above line.

8 *chrouvil* above line.

9 *hala* above line.

10 *prantreita* above line.

11 *rostisarn* above line.

12 *nabagera* above line.

13 Steinmeyer and Sievers (1879–1922, vol. 3: 657) read as a scribal error for *Roscinos*.

14 *nuila* above line.

15 *hamara* above line.

16 *anapoz* above line.

17 *poumscapun* above line.

18 *rita* above line.

19 *kartgapala* above line.

20 *seili* above line.

> hovvn[21]. ii . Duo dolia . potegun[22] cum holeribus . duo cum leg/uminibus . Carros . vvagans[23]. iii . Leugas . slitun[24]. iii . Sedatios . hasiper[25]. v . / Sagenas . segina[26] iii . Retia strumalia . chelnezzi[27] . xxxiiii . Retia / lacunaria . rinanchera[28] . vii . Retia stagnaria . senezzi[29] . vi . Vnum tripnezi . / Funes satis . Rivaream . I . Naues . vi . Caseos . lxxii . / Lini septem scoti . et cannaui nouem . Manuale . I . Inuaccaritio Liutperti . Taurum unum . Vaccas . xvii . / Vitulas . iii . Inarmentario Ruotperti . Vaccas . xii . / Vitulas . iiii . et Taurum. Inarmentario / *[short blank space]* Engilmari Tauros duos . Vaccas . xxv . / Vitulos . iiii . et quinque[30] *[erasure and a blank space]* / In caprificio Gotahelmi . Capras . xxiiii[31] . Hedos . x . / Itemque inofficio Rihham . Hircos . ii . Capras . xxx.iiii . / Hedos . x . In uaccaritio Adal . Tauros . ii . / iuvencos . ii . Vaccas . xxiiii . Vitulos . vi
>
> *Drawing of 2 interlocking heart-shaped leaves*
>
> Tegernsee attinet *[check mark]*

A reasonable English translation would be:[32]

> These are the fishing gear and other items for the use of the cellarer which Gotahelm either found or ... acquired [and] left there in the cellarer's office 17 January in the year of the Lord's Incarnation 1023.
>
> Wine casks, 36. 15 and 2 of them with wine of the bishop, 5 with beer, and 4 large and 4 small. 2 with wine and mead. Jugs for use with pack saddles, 40, and 10 covered in iron. A saddle with a saddle-cloth and felt pad and bridle. Club, 1. Double axes, 4. Axes, 8, and 1 small. Adzes, 8. Hay sickles, 12. Small sickles, 3. Plough. Mallet, 1. Ploughshare, 1. Yokes, 4. Trowels, 2. Kettles, 6. Frying pans, 6. Small tridents, 2. Pothooks, 3. Firerake, 1. Grill, 1. Gridiron, 2. Augers, 7. Planes [?], 2. Hammers, 3. Anvils, 3. Tree scrapers [?], 3. Small scythe, 1. Fork, 1. Ropes, 8. Spades, 2. 2 jars with herbs; 2 with pulses. Carts, 3. Sleds, 3. Winding reels, 5. Seines, 3. Basket-traps, 34. Panel nets for whitefish, 7. Stillwater or lake nets, 6. 1 drive net.[33] Enough ropes. Dock, 1. Boats, 6. Cheeses, 72. Lines, 7 of 'scot' and 9 of hemp.[34] Manual,[35] 1. In Liutpert's cattle farm: bull, 1; cows, 17; heifers, 3.

21 *hovvn* above line.

22 *potegun* above line.

23 Steinmeyer and Sievers (1879–1922, vol. 3: 657; vol. 4: 563) read as *Garros*, which is not required by the mss. *vvagans* above line.

24 *slitun* above line.

25 *hasiper* above line.

26 *segina* above line.

27 *chelnezzi* above line.

28 *rinanchera* above line.

29 *senezzi* above line.

30 *et quinque* erased, then another illegible erasure.

31 The first I looks either crossed out or halved.

32 By the author.

33 The German *tripnezzi* is here in the main text, not the gloss.

34 In this context dictionaries of medieval Latin offer no plausible meaning for *scoti* (a tax, Scots, a coin). Could this be a scribal error for *scorti*, 'hide' or 'skin'?

35 Medieval Latin *manuale* can refer to a handbook, a book of specifically liturgical procedures, a towel, a specific article of liturgical clothing or a head of cattle. In the inventory's context, the first and last seem more plausible. I have translated as 'manual' because four and five centuries later the Tegernsee cellarer's office possessed three successive codices that laid out its own standard operating procedures and those for the monastic establishment: Bayerische Hauptstaatsarchiv [henceforth BHSA] KL Teg 185¼ 'Talmud der Hofkellerei', BHSA KL Teg 185½ 'Encheiridion seu liber manualis' and BSB Clm 1469 'Manualis oeconomicum monasterii ... 1583'. All treat such

In Ruotpert's dairy farm: cows, 12; heifers, 4; and a bull. And in Engilmar's dairy farm: bulls, 2; cows, 25; heifers 4; and 5 [left blank]. In Gotahelm's goat farm: she-goats, 24; kids, 10. And the same at the Reitham estate:[36] he-goats, 2; she-goats, 34; kids, 10. In Adal…'s cattle farm: bulls, 2; steers, 2; cows, 24; heifers, 6.

Two entwined water lilies[37]

Belongs to Tegernsee

Observations on an eleventh century cellarer's inventory

In a monastery properly run under the Rule of St Benedict, the abbot was the 'father' of the community responsible for the monks' spiritual and physical well-being. The cellarer (*cellarius, Kellner*) served as the abbot's leading subordinate for the material life of the convent, and was therefore the economic manager. Good cellarers could make good abbots. At Tegernsee, most of the known tenth- and eleventh-century cellarers who survived their abbots also succeeded them. Gotahelm served under Abbot Ellinger and, after the latter went to reform Benediktbeuern in 1026, also under Albin (1026–31). Ellinger resumed his Tegernsee office in 1031, and Gotahelm followed his old mentor as abbot of Benediktbeuern (1032–62). The document from 1023 suggests a well-run establishment.

The equipment is inventoried in mostly functional groups for lumbering, arable agriculture, livestock, metal work, carpentry, gardening, fishing, the kitchen and storage or overland transport. Collectively these indicate how the monastery made use of its diverse landscape.

Start with the woodland, exploitable with more than a dozen axes, large and small. From the valley lands around the lake to the upper slopes, workers for the monastery had access to a diverse species mix, with oak, beech and birch at lower levels and fir or pine higher up. Despite a constant need for fuel, the inventory holds no clear sign of coppice. Double axes are commonly tools for felling trees and single or small ones for processing the trunk and limbs. Adzes start the shaping process, making timbers as small as feasible for the laborious job of getting them from the woods to the point of use. Streams feeding the lake are not as large as the nearby rivers Isar or Inn, both used to float logs in the Middle Ages, but perhaps the monks and their subjects could use spring run-off to get their timbers down to the lake and then tow them to the abbey site.

Most of Tegernsee's cereal-growing estates lay downstream of the lake's outlet along the river Mangfall or further out on the Bavarian plain. Later cellarers certainly received grain rents from peasant farms there and used these revenues in part to supply subject servants from the mountains who made cheeses. Gotahelm's office still did some arable farming in the vicinity of the abbey, possessing a plough with an [iron] share and the mallet needed to set that in place, as well as four yokes to harness oxen. Small sickles harvested grain but more numerous scythes (literally hay sickles) suggest the upland location was better for producing fodder. Likewise among the garden implements appear another scythe and the mildly puzzling *planatoria*, glossed as *boumscapun*, literally tree scrapers, suggesting a bill hook or other implement to prune or cut leafy hay. Were the monastic garden plots large enough also to use the plough, or were the spades and fork enough to prepare the garden soil for herbs and vegetables?

---

matters as when the cloister gates are opened and closed each day, seasonal practices for work and like material matters under the purview of the cellarer. The eleventh-century monastery had need of a similar authoritative reference tool. Were, however, the inventory's word *manuale* not followed by the number *I* [1], it could refer to the succeeding passages, which do count livestock.

36 The modern hamlet Reitham is 7 km north of Gmund. The Tegernsee *Amt* here was later named for Warngau, 1 km away.

37 The *Seerose* or water lily often served as a symbol for the abbey.

Mown hay fed livestock, but surely so did rough or wood-pasture; some output from a pastoral sector was held in 72 cheeses. Named individuals (monks, servants, other subjects?) were responsible for six specific livestock farms holding 187 head.[38] Two 'cattle farms' and two explicitly 'dairy farms' had 108 bovines, mostly cows and heifers but each also with one or two bulls. Likewise on two goat farms she-goats predominated; although both also held juveniles, one lacked a male sire. Dairy products were certainly important in monastic diets, but one wonders who in an institution that foreswore meat may have eaten the beef or veal from dry cows or male calves. The very object on which the inventory is inscribed becomes an important reminder that calves and young goats provided another resource for this institution: raw materials for the vellum or parchment of its busy scriptorium.

The lake had an important role in the monastic resource mix.[39] Six boats provided platforms to deploy dozens of basket traps and four specialized kinds of nets. One design targeted the whitefish (here probably *Coregonus renke*),[40] for which the Tegernsee and other lakes on the northern slope of the Alps were long famous. Other gear was meant for different fishes in the shallows. There were ropes and reels to wind them, a dock or pier and what may be fishing lines, though hooks are conspicuously absent. Successful operation of this equipment demanded the detailed local environmental knowledge, skills and teamwork shown in fifteenth-century and later records to have been provided by a half-dozen full-time abbey servants. The stubbornly abstinent monks of Tegernsee then delighted in their own *Renken*, preferably smoked.

Gotahelm and his subordinates also possessed tools to store, transport and modify the primary products of lake, arable, meadow, pasture and woodland. Their kitchen had a grill, gridiron, frying pans and kettles, with hooks to hang the pots, utensils to stir or spear and a rake to handle the coals. Carpenters could work the timbers with augers, planes and hammers. Or were the latter to help forge on the three anvils the whole catalogue of iron tools? Craft skills were needed indoors as well as out. Of the 36 casks (work of a cooper?), 15 held wine, two more wine for the bishop, both likely imported. Five more contained surely local beer and two wine and mead, leaving four large and four small unidentified. Forty (ceramic?) jugs could be transported on pack saddles and another ten were bound in iron. There was only one set of riding tack, but three carts or wagons for ground suitable for wheeled vehicles and three sledges for other surfaces. Missing are the horses or draft oxen, though the latter could have been commandeered from peasant tenants. But those people were not objects of surviving eleventh-century records from Tegernsee.

The cellarer's inventory, like all historical documents, omits much. In the absence of other such administrative records from the eleventh-century abbey, the lacunae cannot really be filled. Still, the text from 1023 establishes that a generation after the community's revival it was well equipped to fulfil the Rule of St Benedict by exploiting a diverse natural and anthropogenic environment. Estate managers had learned, surely from long-resident peasants, to adapt to a landscape quite different from that of Trier. The monastic household was prepared to work in the waters of the lake at their doorstep, meadows and fields along the valley floor, and the wooded slopes and pastures that rose toward the peaks. Eleventh-century sources

38 The only easily located livestock operation was at Reitham, the hamlet some 7 km north of the lake's outlet at Gmund. The inventory provides no evidence regarding use of high alpine pastures, which occur in some thirteenth-century records.

39 Gotahelm's inventory provides the first firm evidence of the Tegernsee fishery, for which technical, organizational, cultural and environmental particulars become richly available at the end of the Middle Ages (see Hoffmann 1994, 1995, 1997).

40 The taxonomy of European whitefishes (genus *Coregonus*) is famously convoluted and disputed. For one current view, see Kottelat and Freyhof (2007: 364, 368).

can establish a base line for the sorely lacking environmental history of the Alps' northern slopes. A deep scholarly concern for the particulars of such a medieval landscape is a valued legacy from Oliver Rackham.

References

Angerer, A. (1968) *Die Bräuche der Abtei Tegernsee unter Abt Kaspar Ayndorffer 1426–1461 verbunden mit einer textkritischen Edition der Consuetudines Tegernseenses.* Studien und Mitteilungen zur Geschichte der Benediktiner-Ordens, 18. Erganzungsband (Ottobeuren: Kommissionsverlag Winfried Werk).

Chroust, A. (1931) *Monumenta palaeographica, Denkmäler der Schreib-kunst des Mittelalters, Schrifttafeln in lateinischer und deutscher Sprache.* II. Série, Bd. 1, O. Harrasowitz, Leipzig.

Eder, C.E. (1972) *Die Schule des Klosters Tegernsee im frühen Mittelalter im Spiegel der Tegernseer Handschriften.* Münchener Beiträge zur Mediävistik und Renaissance-Forschungen, Beiheft, Arbeo Gesellschaft, Munich.

Hartig, M. (1946) *Die Benediktinerabtei Tegernsee 746–1803.* Schnell and Steiner, Munich.

Hoffmann, R.C. (1994) The craft of fishing Alpine lakes, ca. A.D. 1500. In: D. Heinrich (ed.) *Archaeo-Ichthyological Studies. Papers presented at the 6th Meeting of the I.C.A.Z. Fish Remains Working Group, Schleswig.* Beiheft zur Offa, 51, Wacholz Verlag, Neumünster, pp. 308–313.

Hoffmann, R.C. (1995) Fishers in late medieval rural society around Tegernsee, Bavaria – a preliminary sketch. In: E.B. DeWindt (ed.) *The Salt of Common Life: Individuality and choice in the medieval town, countryside and church. Essays Presented to J. Ambrose Raftis.* Studies in Medieval Culture, vol. 36, Medieval Institute Publications, Kalamazoo, MI, pp. 371–408.

Hoffmann, R.C. (1997) *Fishers' Craft and Lettered Art: Tracts on fishing from the end of the Middle Ages.* University of Toronto Press, Toronto, pp. 73–190. https://doi.org/10.3138/9781442674929

Hoffmann, R.C. (2014) *Environmental History of Medieval Europe.* Cambridge University Press, Cambridge.

Kottelat, M. and Freyhof, J. (2007) *Handbook of European Freshwater Fishes.* Published by the authors, Cornol, Switzerland and Berlin.

Redlich, V. (1931) *Tegernsee und die deutsche Geistesgeschichte im 15. Jahrhundert,* Schriftenreihe zur bayerischen Landesgeschichte, Bd. 9, Biederstsein, Munich. (Reprinted 1974, Scientia, Aalen)

Steinmeyer, E. and Sievers, E. (eds) (1879–1922) *Die althochdeutschen Glossen,* 5 vols. Weidmann, Berlin.

Wessinger, A. (1885) Kaspar Aindorffer, Abt in Tegernsee 1426–1461. *Oberbayerisches Archiv für vaterländischen Geschichte,* 42, pp. 196–260.

CHAPTER 17

It's a Fair Coppice: Methodological Considerations of the History of Woodland Management

Péter Szabó

Summary

This chapter presents historical evidence in relation to the uses of European woodlands for the production of coppice, especially for firewood. This account develops ideas initiated by Oliver Rackham to examine the management of woods for fuelwood coppice – a dominant use of many landscapes prior to the emergence of economic 'scientific' forestry.

Keywords: woodland management, coppice, fuelwood, 'scientific' forestry

Introduction

Coppicing is the repeated cutting on short rotation of stems growing from permanent stumps (called stools) in broadleaved forests. The ability of trees to grow back when cut down (or broken) is in all likelihood older than humanity and could be an evolutionary reaction to an outside force, such as trampling by large animals, breakage by windstorms or tree-felling by beavers (Rackham 2006: 72–73). Most (but not all) conifers lack this property, and it is an open question why certain tree species coppice and others do not. This already puzzled the Greek philosopher Theophrastus in the fourth century BC (Theophrastus, *Enquiry into Plants*, Book II. I–II.4). There are two main reasons why people coppice: first, the freshly growing stems are relatively thin and can be turned into firewood with minimum effort; and second, coppicing ensures that trees regenerate reliably and produce the same quantity and quality of wood in each cycle. Oliver Rackham spent a lifetime studying the history and current ecology of coppice woods. To a large extent through his works, coppicing is now generally acknowledged to have formed the basis for firewood production in practically all parts of broadleaved Europe (and even outside Europe) until the breakthrough of 'scientific' forestry in the nineteenth century (Rackham 1976, 1980, 2006).

In spite of its former widespread use, coppicing was all but eradicated in Europe by the 1950s (Fig. 17.1).[1] 'Scientific' forestry, with its focus on high forests preferably of conifers,

1 Except in the Mediterranean, where it was practised for longer. Nonetheless, many existing coppices in the Mediterranean are also out of use (Unrau et al. 2018).

Péter Szabó, 'It's a Fair Coppice: Methodological Considerations of the History of Woodland Management' in: *Countryside History: The Life and Legacy of Oliver Rackham*. Pelagic Publishing (2024).
DOI: 10.53061/QLUX8152

Figure 17.1 An ancient coppice wood in the south-eastern Czech Republic. Note that coppicing is no longer active and the shoots are much older than they used to be allowed to grow, November 2013. (R. Hédl)

strongly opposed coppicing. This opposition sometimes took the form of direct repression: the Czech forest law of 1995 banned the regular cutting of stands younger than 80 years. For two centuries, coppicing was looked down upon and snubbed by those in positions of power in forestry administration. It was presented as a backward method that produced scrubby growth instead of the orderly stands of tall trees that foresters preferred to see. In an unexpected turn of events, coppicing was rediscovered by nature conservationists as well as foresters in the past half-century (Buckley and Mills 2015a). Plant conservationists are interested in coppicing because many woodland plants seem to be adapted to quick successions of light and dark periods in forests, which result from coppicing. In the absence of light, such plants – many of which are rare and endangered – disappear (Kopecký, Hédl and Szabó 2013). Coppicing is a management tool that can also promote important invertebrates and birds (Beneš et al. 2006; Fartmann, Müller and Poniatowski 2013; Fuller 1992). By contrast, foresters have rediscovered coppicing as a cheap and reliable way to produce firewood. When fossil fuel supplies seemed inexhaustible and nuclear power stations promised to solve energy problems once and for all, firewood appeared to be a thing of the past. By now, however, people are much more aware of the problems connected with fossil fuels and nuclear energy. Firewood has found itself back in the spotlight as one of the desired renewable sources of energy. With growing prices for wood, coppicing can be a viable economic option in certain regions (Kneifl, Kadavý and Knott 2011; Buckley and Mills 2015b).

Coppicing in Europe goes back to prehistory; that is, it has been practised for millennia. Although its basic outlines have remained the same, it has been neither spatially nor technologically static. Untangling the history of coppicing is important if we are to better understand the consequences of its abandonment and the effects of its potential reintroduction for conservation purposes. The history of coppicing can be approached through several types of sources, ranging from biological to historical. With his extraordinary skills across disciplines, Oliver Rackham was instrumental in discovering, contextualizing and using many of these sources. His research also clearly showed that conclusions about past (and, by implication, about future) woodland management are most solid when based on several types of sources. In this chapter, I try to follow in Rackham's footsteps to systematize and critically examine the ways coppicing can be detected from prehistory until the twenty-first century and give special attention to recent developments in each field. Given the limits of this contribution, I cannot claim to provide an exhaustive review, and my treatment especially of better-researched topics – such as tree-rings or written sources – must inevitably remain cursory.

Evidence to point to the existence of coppicing in the past can be grouped into three basic categories: written historical sources, pollen and wood.[2] In the following, I present these source types describing their properties, advantages and pitfalls.

Written evidence

Some written sources offer straightforward descriptions of coppicing. Here is the classic example of Hayley Wood (Cambridgeshire, England) from 1356, discovered and quoted by Rackham: 'a certain Wood called Heylewode which contains 80 acres by estimate. Of the underwood of which there can be sold every year, without causing waste or destruction, 11 acres of underwood, which are worth 55s. at 5s. an acre' (Rackham 1975: 26).[3] This text describes a seven-year coppice cycle. Just how widespread this practice was in Europe is demonstrated by the description of a wood called Mulschachen in today's south-eastern Czech Republic 58 years later, in 1414: 'When it becomes seven years old, one estimates it to be worth about 10 lb. den' (Bretholz 1930: 184).

Antique authors were also aware of coppicing. Pliny the Elder, for example, suggested an eight-year coppice cycle for chestnut trees to be used as vineyard stakes (Pliny the Elder, *Naturalis Historiae*, book 17, ch. 34). Descriptions of coppicing became common in the Middle Ages, and countless examples are available from the thirteenth to the twentieth centuries. Until the eighteenth century, the most explicit mentions of coppicing can be found in estate conscriptions of various kinds, such as the examples given here. After that, especially in Central and Western Europe, forest management plans contained very detailed descriptions of coppice management, including maps of coppice compartments (e.g. Bürgi 1999a; Müllerová, Szabó and Hédl 2014) (Fig. 17.2). Coppicing also found its way into prescriptive legislation in the Middle Ages. For example, the 1476 forestry law of Venice ordered a ten-year coppice cycle for all community forests (Appuhn 2009: 112). Legislation usually turned out to be too rigid to reflect realities on the ground: the monks of Montbenoît in 1719 must have felt frustrated explaining that they could not stick to Colbert's famous 1669 *Ordonnance des eaux et forêts*, which prescribed a ten-year coppice cycle, because the pine trees in their forests simply would not grow back from the stump (Matteson 2015: 60). The general trend was for the coppice cycle to get longer throughout the centuries, culminating in about 40 years – although there were many exceptions, and coppice cycles of around 15 years were common even in the nineteenth century. There are only guesses at why the cycle lengthened. Rackham (1975: 31) was among the first to address the issue, which he attributed to the gradual slowing down of growth rates resulting from soil phosphate exhaustion through coppicing. Later, he added the spread of

2 The presence of certain plants, which Rackham called 'coppicing plants', in woodland can also indicate a coppicing history for the given wood (Rackham 2003: 413–38). Being rather indirect evidence, I do not cover these here. Nor do I include archaeobotanical investigations (mostly based on animal dung) of animal fodder.

3 Translation in original. In this chapter, all translations are mine except when indicated otherwise.

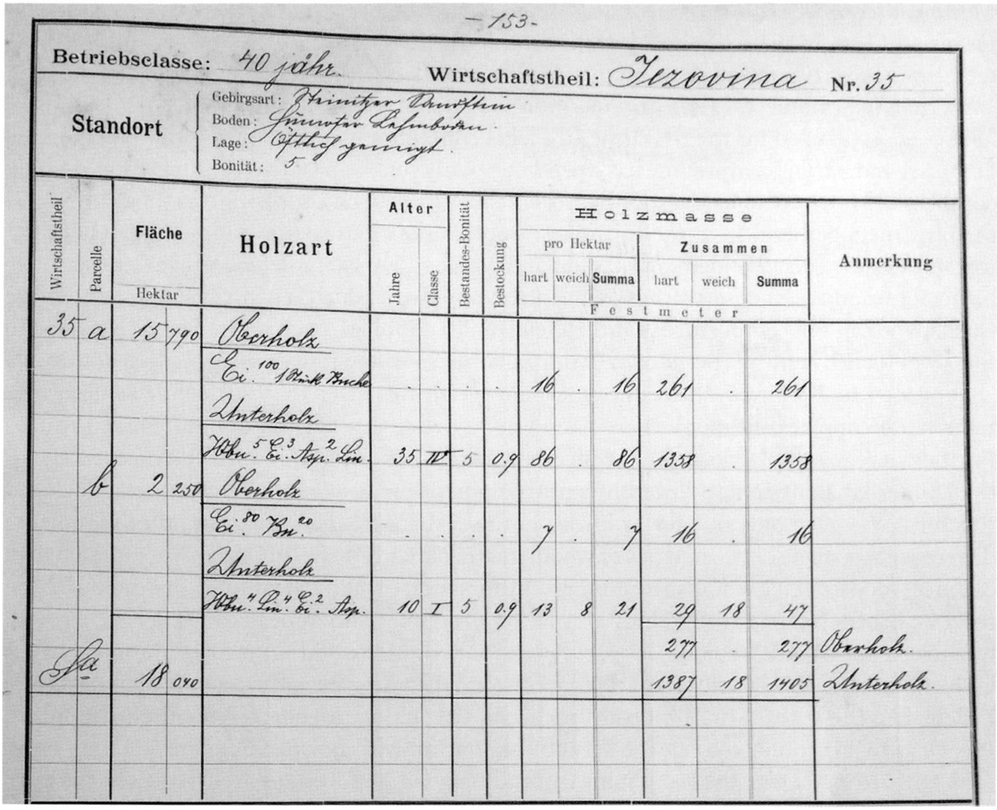

Figure 17.2 Detail from the forest management plan of the Bohuslavice forest district in the south-eastern Czech Republic from 1899. The document specifies the coppice cycle (40 years) as well as the tree species for standards (*Oberholz*) and underwood (*Unterholz*). (Original kept at the Moravian Provincial Archive (Moravský zemský archiv) fond F 31 kniha 388)

chimney building as another possible explanation (Rackham 2003: 140). Others suggested an increased market for larger logs and a shift towards less labour-intensive wood-pasture systems after the Black Death (Galloway, Keene and Murphy 1996).

Most written sources are not quite as explicit as those noted here. When working with these, there is no certainty that our interpretation is correct. When researching single forests, it is best if several types of evidence corroborate each other. One type of such evidence is the names of forests, either geographic names or general woodland typology – although the two are often difficult to tell apart. In many European languages, there are specific words to denote coppices (e.g. French *taillis*, German *Niederwald*, Italian *ceduo*, Hungarian *sarjerdő*) (for more examples, see Unrau et al. 2018). Sometimes these are used to describe forests; sometimes they are (parts of) names of forests. By and large, they denote that the wood in question was coppiced at some point, although they neither establish how permanent coppicing was nor are much help in dating it. In addition to simply faulty interpretation, several factors make the usage of these terms challenging. First, words change through time. The same word can mean different things in different periods, and the same phenomenon can be referred to by different words in different centuries. Some older words that arguably denoted coppices have gradually disappeared from common usage, such as the medieval Hungarian *eresztvény* (Szabó 2005). New words either developed organically or were created to purpose. The two processes could freely overlap. The German *Niederwald*, for instance, appears to be a medieval word, but was not used as a technical term to denote coppices until the nineteenth century.

Mittelwald (coppice-with-standards) is claimed to have been created by the renowned forester Heinrich Cotta in the early nineteenth century. In the Swiss case, the promotion of these terms even contributed to shaping the structure of forests that more strictly matched ideal cases than their predecessors (Bürgi 1999b). In Spain, today's most common term for coppices (*monte bajo*) was used for the first time in a legal context only in 1574 (Ortego and de Lomana 2017). Second, strict terminology has been in use only for the past two centuries. Before that, coppices did not have to have a specific name at all, especially if coppicing was the commonest form of management. In some countries, woodland structure (i.e. coppicing versus high forests) was not a decisive element in woodland typology. For example, in Denmark, descriptive terms for woodland focused rather on resources and ownership (Fritzbøger 2004). With some words, one can only conjecture that they referred to coppices, such as the Swedish *surskog* in the seventeenth century (Bergendorff and Emanuelsson 1996). It is possible that *Buschwald* (as opposed to *Hochwald*) in western German territories after the fourteenth century also referred to coppices, although there is no way to prove this beyond doubt (Hausrath 1928). Furthermore, even if a word for coppices existed, it was often not applied consistently. In the Domesday Book in late eleventh-century England, *silva minuta* referred to coppices, but this term was used only in a few counties in the eastern Midlands, and in other regions no difference was made between types of woodland (Rackham 2003). Similarly, *silva caedua* meant coppice woods when used by some Roman authors (notably Pliny the Elder), but did not when used by others (Harris 2018). Many words were regional and appear somewhat obscure in the given context today, such as *Maß*, which meant coppice wood in southern Moravia in the early modern period. Third, terms can have several meanings in addition to coppicing. While I identified the Latin term *rubetum* in medieval Czech documents as referring to coppices, it is clear that it could also refer to scrubby vegetation with no specific management (Szabó et al. 2015). Nonetheless, the geographical distribution of hundreds of occurrences of the term seemed to confirm that it was mostly used to describe coppices – even keeping in mind that such argumentation is to a certain extent circular.

The existence of coppicing can be deduced from various other types of sources as well. Richard Keyser brought scholarly attention to the 'Book of sales of the woods of Champagne' and similar documents in thirteenth-century France (Keyser 2009). These documents recorded leases of woodland for cutting for a given period. Although never mentioning coppicing explicitly, lease periods of 6 to 20 years, accompanied by the condition that the merchants can 'cut each tree only once so that it grows back quickly' (Keyser 2009: 372),[4] make it clear that the lease period in fact equalled the coppice cycle. As opposed to the explicit mentioning of coppice cycles, Galloway, Keene and Murphy (1996: 454) termed such information 'implicit coppice cycles'. Furthermore, bans on pasturing were often introduced for a few years after cutting, usually a reliable indicator of coppicing because young coppice shoots are a favourite for browsing animals and need to be protected until they grow tall enough. Keyser's French account books also mentioned the presence of (or, more precisely, the obligation of merchants to leave behind a certain number of) standard trees (*bailivaux*). Coppice woods sometimes (but not always) contained a set of straight and tall trees grown from seed to be used for construction purposes. Just like coppice cycles, the number of standards also made it to medieval legislation, most famously to the 1376 forest code of Charles V of France, which ordered that eight to ten standards be left on every *arpent* (roughly equivalent to an acre) (Isambert, Decrusy and Jourdan 1824: 462). Account books in the strict sense can also reveal the existence of coppicing. In their classic study, Galloway, Keene and Murphy (1996) employed hundreds of manorial accounts in a wide circle around London to assess the fuelwood consumption of the English capital in the fourteenth century. They

4 Translation in original.

convincingly argued that fuelwood came from coppices. Similar research was carried out for early modern Madrid and Paris (Boissière 1990; Bernardos et al. 2011; Ortego and de Lomana 2017). I tried to demonstrate the use of account books to detect coppicing with the example of the Choustník (southern Bohemia) woodland account book of 1447 (Szabó 2018). This book made no mention of cutting cycles, but its closer examination leaves little doubt that it described a fully functioning coppice system. Wood in this region was sold either by the territory or by the piece. Because an average single oak cost one sixth of the average value of all the wood cut on approximately 900 m^2, the system described was clearly coppice underwood with standards. High forests were so unusual that sometimes they were referred to as such. When – rarely – high forests were sold by the territory (rather than selling individual trees in them), they brought in three to four times more money than the average coppice compartment of the same size. Comparing the area cut in 1447 with what we know about woodland sizes on the estate, I even ventured to estimate the coppice cycle, which appeared to be somewhere between 6 and 18 years: a plausible number for the region and period.

Pollen

Pollen is the substance plants use to transfer their reproductive material. Pollen grains of individual species or genera are highly characteristic. In favourable conditions (mostly when not exposed to oxygen), grains are extremely resistant. They can survive many thousands of years and still be recognizable under the microscope. In lake and mire sediments, pollen forms temporal layers that can be studied to untangle vegetation history (Faegri, Kaland and Krzywinski 1989). However, the connection between fossil pollen and past vegetation is anything but straightforward. Individual tree species vary tremendously in their production of pollen. The ability of pollen grains to fly also differs, and grains have different taphonomic characteristics. Furthermore, every pollen site is unique: some reflect local vegetation while others gather their pollen from a larger area, depending on size and local geography. Another important issue is the calculation and presentation of pollen counts. Usually 500 grains per sample are counted and the results are presented as percentages in the overall sum. This means that an increase in the pollen percentage of a tree species need not reflect any change in the presence of that tree in the given landscape. Alternative techniques (such as measuring pollen influx) exist to overcome this challenge, but these are not routinely used everywhere (Birks and Berglund 2018). A number of researchers looked at temporal changes in pollen percentages to infer past management practices. For example, Gardner (2002) interpreted a series of fluctuations (over 72–228 years) in oak, hazel and hornbeam pollen as indicators of coppice or pollard management (the latter is similar to coppicing except that trees are cut higher, out of the reach of browsing animals) in Neolithic and Copper Age Hungary. In spite of his conclusions, he admitted that the temporal resolution of the pollen core was too coarse to register actual coppice cycles. He conjectured that 'there may be a threshold level at which the signal of woodland management is apparent in the pollen record' (Gardner 2002: 551).

More examples of similar research exist, including the explanation of the mid-Holocene elm decline as the consequence of large-scale pollarding for leaf-fodder, which would have drastically reduced flowering (and therefore pollen production) in elm (Troels Smith 1960; Parker et al. 2002). Oliver Rackham (1980) pointed out that it would have been virtually impossible for early Neolithic populations to achieve this effect by themselves: there were too many elms and too few people. Rasmussen (1990) rightly remarked that pollen cannot provide 'definite direct evidence' for coppicing (or pollarding) in prehistory.

Other than subjective interpretations of pollen profiles, the only research known to me that tried to empirically test the connection between coppicing and pollen was published by Waller, Grant and Bunting (2012) for three British woods. They applied a two-stage strategy: first, they followed the number of flowers and the quantity of pollen produced by lime *Tilia* spp.,

Common Hazel *Corylus avellana* and Common Alder *Alnus glutinosa* in active coppice woods throughout the coppice cycle. This was combined with vegetation surveys to establish whether different stages in the coppice cycle produce distinct pollen signals. These data were then used to produce model simulations of the effect of coppicing on the pollen record, and finally these simulations were compared to actual pollen diagrams. Unsurprisingly, the results were not straightforward. In general, coppicing seemed to push woodland composition from canopy species to understorey species and to make more sensitive species altogether invisible in the pollen record. This was consistent with existing pollen cores, especially in East Anglia. However, similar changes in pollen diagrams can be caused by multiple factors; therefore, conclusions were stronger when corroborated by other evidence, such as fossilized wood from prehistoric trackways (of which see more in the next section, below) in the Thames estuary. Waller, Grant and Bunting also pointed out two important limitations of the interpretation of pollen diagrams in search of past coppicing. First, once the wood is not coppiced entirely, the potential signs of coppicing can be completely masked (also mentioned in Rasmussen 1990). Second, size matters: the interaction between the size of coppice compartments and that of the catchment area of pollen sites is an major source of reconstruction error (Bunting, Grant and Waller 2016a). The same authors also examined non-arboreal pollen as a potential tool to identify coppicing and arrived at the same inconclusive results: the pollen of plants other than trees can reflect coppicing, but the interpretation of these signals is ambiguous (Bunting, Grant and Waller 2016b). At present, it seems that pollen diagrams have clear potential to detect coppicing in past woodlands, but more methodological work is needed to realize this potential. As Rackham put it in his usual succinct manner, 'Palynological criteria are wanted to differentiate between the following: forest; savanna [...]; coppice wood; hedges and non-woodland trees; and (in Mediterranean countries) maquis' (Rackham 2006: 80).

A promising direction of future research could be the search for multiple scenarios, of which coppicing can be one, rather than for the elusive 'correct' reconstruction (Bunting, Grant and Waller 2016a).

Wood

Wood comes in many forms, from fossilized pieces in palaeoecological cores to charcoal fragments in archaeological excavations to living tissues. As far as the history of coppicing is concerned, wood has the advantage over pollen that it can provide more direct evidence of management. Regardless of the form in which the wood gets to the researcher, it has four characteristics that can be used to detect coppicing in the past: anatomy, cambial age (see below), diameter and ring width.

Rackham did pioneering work in wood anatomy, more precisely in the analysis of external morphology. In his 1977 paper on three prehistoric wooden tracks (raised wooden walkways in marshy terrain) in the Somerset Levels, he used his practical knowledge of hazel coppicing to examine archaeological wood remains (Rackham 1977). His conclusions still fascinate today, for he claimed to have worked out the actual method hazel stems were cut from the stools some four millennia ago. He also established that rather than harvesting larger contiguous areas at once, the makers of the tracks preferred to choose individual shoots. The main aim of coppicing, he contended, may have been the production of leaf fodder for animals. Similar work on prehistoric trackways was done in other European countries, for example in Denmark or in the Netherlands as early as 1986 (Malmros 1986; Casperie 1986). In the latter case, the author deduced the existence of willow coppices (including their approximate location!) in the Iron Age. Because at this point metal tools were already available, willow shoots were cut differently from hazel rods in Neolithic Somerset, which were partly torn away.

Living coppice stools of greater antiquity also prove that coppicing existed in the past. Rackham pointed out many times in the 1970s that stools expand with each successive cutting (because shoots usually appear on the outside), which gives an opportunity to (at least tentatively) date them and establish a *terminus post quem* for coppicing in the given wood (Rackham 1975; 1976). He inferred that many large coppice stools were from the Middle Ages, if not earlier. Pigott's (1989) calculations on lime stools confirmed this view. Systematic efforts at dating stools are rare. A prominent example is Vrška et al. (2016). Through a great amount of manual work, Vrška and colleagues uncovered the root system of several large oak coppice stools in a national park with a known coppicing history in the southern Czech Republic, thereby proving that stems not apparently connected above ground were indeed parts of a single tree. They estimated the largest stools to be more than 800 years old. However, not all multistemmed trees are necessarily coppice stools. Dutch researchers showed that similar structures can emerge through the effects of grazing (Copini et al. 2005).

Wood reacts to the physical impact of cutting as well as to the specific environmental conditions created by coppicing – for instance by changes in the size and distribution of cell types, pore size or by the formation of false rings. These were described in detail based on examples by Schweingruber (2007), and the specific effects of coppice management were studied by Girardclos et al. (2018). Distinguishing between the signals of various types of management (pollarding, coppicing, leaf removal, browsing by animals) is evidently very difficult, if possible at all. Equally challenging is to recognize such signals in archaeological rather than recent wood. This area of research is perhaps the least developed among those discussed in this chapter, and therefore offers the greatest potential. So far, wood anatomical research on archaeological wood appears to focus on pollarding rather than coppicing (Haas and Schweingruber 1993). For example, Bernard, Renaudin and Marguerie (2006) compared living shredded trees with archaeological material. In addition to ring width, similarities in the mean porosity of the rings between recent and archaeological material led the authors to assume the existence of shredding in medieval France (see also Thiebault 2006). In a similar fashion, Bleicher (2014) combined wood anatomical evidence with ring width to draw conclusions about pollarding in the Neolithic in Switzerland.

In contrast to wood anatomy, annual tree-ring width, wood diameter and cambial age have been more often used to make inferences about coppicing. Trees in temperate climates develop a new annual ring each year, because the speed of growth slows down during the vegetation season. The structure of earlywood (spring growth) differs from that of latewood (summer growth), which makes a pattern recognizable to the naked eye. The width of an annual ring is determined by a host of factors, and tremendous amounts of literature exist in dendrochronology and forestry that describe and interpret these factors. The sum of individual tree-rings – plus bast (the inner bark/vascular material) and bark, if present – make up wood diameter. Cambial age is the number of annual rings from the pith (the centre of the tree) in a wood sample. In other words, cambial age tells the researcher how long the tree lived rather when it lived (which is absolute age). The difference between the birth and death dates of a tree equals its cambial age only for the trunks of standard, seed-grown trees. This is especially significant for coppicing: a coppice shoot can have the cambial age of ten years, while the entire tree (the stool) can be hundreds of years old.

That the width of annual tree-rings reflects growing condition around trees has been known for a long time in Europe, and early observations on this are attributed to none other than Leonardo da Vinci (Schweingruber 1988: 256). Later on, it was realized that mechanical damage to the tree can also leave behind recognizable signals in ring width. Measurable reactions in tree-rings can be of two types: a sudden increase ('release') or decline in ring width depending on the type of tree (standard or stool) as well as the part of the tree (trunk or shoot). As ever, Oliver Rackham was among the first to utilize this source type to study the management history of woodland, and his research aptly illustrates the opportunities offered by tree-ring width for the study of historical coppicing. In 1975, he analysed the annual rings of an ash coppice stool in Hayley Wood, which showed sudden declines at about every 14 years followed each time by prolonged recovery. Rackham connected this to coppicing, when after harvest the stool's ability to grow is drastically reduced (Rackham 1975: 32). Conversely, standard trees can show sudden increases when they get more light either through the cutting of standard trees close to them, as demonstrated by Rackham for Hayley oaks, or by the cutting of underwood (Altman et al. 2013) (Fig. 17.3). Because tree-ring width is influenced by a large number of factors, connecting its changes to management events remains challenging. As also suggested by Rackham, coppice shoots themselves have their particular growth dynamics. Especially in the first few years of their life, they grow very fast. This makes it possible to recognize coppice shoots (and by implication the existence of coppicing) among archaeological wood samples (Rackham 1975: 113; see Bleicher 2014 for a recent overview of tree-ring patterns in the context of prehistoric management). This assumption was tested by Haneca, Van Acker and Beeckman (2005), who compared ring width growth trends in current coppices with archaeological material in Belgium. They observed remarkable similarities, which they tentatively interpreted as signs of coppice management in the Roman and medieval periods. Naturally this method raises its own questions, and Copini and colleagues repeatedly criticized it (e.g. Copini, Sass-Klaassen and den Ouden 2010), arguing that their own oak samples from Dutch coppice forests showed that initial strong growth in itself was not sufficient to distinguish coppice shoots from young seedlings. However, they mentioned that the comparison of earlywood patterns may be a better key to identify the fingerprint of coppicing. Deforce and Haneca (2015) extended such investigations to well-preserved charcoal fragments from fourteenth-century Belgium, reiterating earlier conclusions about the clearly recognizable growth pattern of coppice shoots. Even more recently, Girardclos et al. (2018) countered these

Figure 17.3 An ancient coppice wood south-east of Brno in the Czech Republic in 1953. This was the period when coppicing was being rapidly abandoned. Note the various stages of coppicing and the standards. (kontaminace.cenia.cz)

arguments based on evidence from a current oak coppice in northern France. Their statistical analysis did not show significant differences in growth rates between coppice underwood and standards. Whether these contradicting conclusions have to do with differences between species (oak appears to produce less clear a signal than other common tree species) or with variations in local conditions and management remains to be seen.

Cambial age and wood diameter have been commonly used in the reconstruction of past woodland management. This is arguably because these properties – as opposed to ring width, the proper measurement of which requires a laboratory – can be recorded and studied by non-specialists. The basic assumption underlying this type of research is that populations of wood samples from managed stands show a pattern in their cambial age and diameter distribution distinct from samples from unmanaged (or differently managed) stands. Strong clustering especially in age distribution can be interpreted as the fingerprint of management. Following up on Rackham's observation in the Somerset Levels, Anne Crone (1987) took samples from existing coppices as comparative material, and came to the conclusion that the age and diameter structure of Somerset wood remains does not support the idea of formal coppicing in the Neolithic. It is clear that because this kind of investigation is based on statistical inferences about population samples, larger amounts of wood are needed to draw meaningful conclusions. Probably the first to bring more rigorous methodology into such research was André Billamboz from the 1980s onwards (Billamboz 1992). He termed his investigations based on thousands of prehistoric wood remains from southern Germany 'dendrotypology'. Welmoed Out and her colleagues recently provided further methodological sophistication in trying to resolve the most important question (already addressed by Crone): whether wood samples from managed and unmanaged stands can be distinguished by cambial age and diameter alone (Out, Vermeeren and Hänninen 2013; 2018). They constructed theoretical models for distribution patterns, which were compared with data from existing coppices and archaeological material. Their conclusions were cautious but promising, detecting management based on cambial age and diameter analysis is possible, but that the 'strength of the conclusions will … depend on the diameter of the investigated branches, the size and context of samples and the availability of additional indications such as the physical characteristics of the wood' (Out, Vermeeren and Hänninen 2013: 4095).

They recommended using at least 50 wood remains larger than 2 cm in diameter. A related issue, which is especially relevant for archaeological remains, is the reconstruction of original wood diameter sizes from charcoal fragments. Several competing methods were developed, which either compared archaeological finds with contemporary charcoal burning sites, or used mathematical calculations that modelled the way wood breaks into smaller pieces of charcoal (Dufraisse 2008; Ludemann 2008). Once a reconstructed size distribution is available, it can be analysed with the same methods as non-carbonized archaeological finds (Paradis-Grenouillet et al. 2015).

Rackham's legacy

Many authors quoted in this chapter pointed out that the best way to approach the history of coppicing is using multiple sources and lines of investigation. This brings us full circle to Oliver Rackham's work. He was unique in his ability to analyse written sources with the same confidence as tree-rings or plant distribution. While such universal knowledge appears unattainable for most of us, we may take comfort in noting that the individual disciplines involved in research on coppice history have been moving forward by leaps and bounds in recent decades. With much more research produced and with methods getting increasingly sophisticated and difficult to comprehend but for a handful of specialists, nowadays it is hardly possible to be up to date and active in more than one of the disciplines described here. The way forward is through interdisciplinary teamwork. In addition to building an atmosphere of mutual trust and scientific respect, the biggest challenge in such teamwork is always to unite the individual researches and results in a common interpretative framework. This is where Oliver Rackham's legacy is most relevant today.

Acknowledgement

The research leading to this chapter was supported by long-term research development project RVO 67985939 from the Czech Academy of Sciences, and grant GA17-09283S from the Czech Science Foundation.

References

Altman, J., Hédl, R., Szabó, P., Mazůrek, P., Riedl, V., Müllerová, J., Kopecký, M. and Doležal, J. (2013) Tree-rings mirror management legacy: dramatic response of standard oaks to past coppicing in Central Europe. *PLoS ONE* 8, 2: e55770. https://doi.org/10.1371/journal.pone.0055770

Appuhn, K. (2009) *A Forest on the Sea. Environmental expertise in Renaissance Venice.* The Johns Hopkins University Press, Baltimore, MD.

Beneš, J., Cizek, O., Dovala, J. and Konvicka, M. (2006) Intensive game keeping, coppicing and butterflies: the story of Milovicky Wood, Czech Republic. *Forest Ecology and Management*, 237(1–3), pp. 353–365. https://doi.org/10.1016/j.foreco.2006.09.058

Bergendorff, C. and Emanuelsson, U. (1996) History and traces of coppicing and pollarding in Scania, south Sweden. In: H. Slotte and H. Göransson (eds) *Lövtäkt och stubbskottsbruk.* Kungl. Skogs-och Lantbruksakademien, Stockholm, pp. 235–304.

Bernard, V., Renaudin, S. and Marguerie, D. (2006) Evidence of trimmed oaks (*Quercus* sp.) in north western France during the early Middle Ages (9th–11th centuries AD). In: A. Dufraisse (ed.) *Charcoal Analysis: New Analytical Tools and Methods for Archaeology.* Archaeopress, Oxford, pp. 103–108.

Bernardos, J., Hernando, J., Madrazo, G. and Nieto, J. (2011) Energy consumption in Madrid, 1561 to *c.*1860). In: G. Massard-Guilbaud and S. Mosley (eds) *Common Ground: Integrating the social and environmental in history.* Cambridge Scholars, Newcastle upon Tyne, pp. 316–339.

Billamboz, A. (1992) Tree-ring analysis in archaeodendrological perspective. The structural timber from the South West German lake dwellings. In: T.S. Bartholin, B.E. Berglund, D. Eckstein, F.H. Schweingruber and O. Eggertsson (eds) *Tree-Ring and Environment.* Lund University, Lund, pp. 34–40.

Birks, H.J.B. and Berglund, B.E. (2018) One hundred years of Quaternary pollen analysis 1916–2016. *Vegetation History and Archaeobotany*, 27, pp. 271–309. https://doi.org/10.1007/s00334-017-0630-2

Bleicher, N. (2014) Four levels of patterns in tree-rings: an archaeological approach to dendroecology. *Vegetation History and Archaeobotany*, 23, pp. 615–627. https://doi.org/10.1007/s00334-013-0410-6

Boissière, J. (1990) La consommation parisienne de bois et les sidérurgies périphériques: essai de mise en parallèle (milieu XVe–milieu XIXe siècles). In: D. Woronoff (ed.) *Forges et forêts. Recherches sur la consommation proto-industrielle de bois.* Éditions de l'École des Hautes Études en Sciences Sociales, Paris, pp. 29–56.

Bretholz, B. (ed.) (1930) *Das Urbar der Liechtensteinischen Herrschaften Nikolsburg, Dürnholz, Lundenburg, Falkenstein, Feldsberg, Rabensburg, Mistelbach, Hagenberg und Gnadendorf aus dem Jahre 1414.* Anstalt für Sudetendeutsche Heimatforschung, Reichenberg and Komotau.

Buckley P. and Mills, J. (2015a) The flora and fauna of coppice woods: winners and losers of active management or neglect? In: K.J. Kirby and C. Watkins (eds) *Europe's Changing Woods and Forests.* CABI, Wallingford, pp. 129–139. https://doi.org/10.1079/9781780643373.0129

Buckley P. and Mills, J. (2015b) Coppice silviculture: from the Mesolithic to the 21st century. In: K.J. Kirby and C. Watkins (eds) *Europe's Changing Woods and Forests.* CABI, Wallingford, pp. 77–92. https://doi.org/10.1079/9781780643373.0077

Bunting, M. J., Grant, M. J. and Waller, M. (2016a) Approaches to quantitative reconstruction of woody vegetation in managed woodlands from pollen records. *Review of Palaeobotany and Palynology*, 225, pp. 53–66. https://doi.org/10.1016/j.revpalbo.2015.10.012

Bunting, M. J., Grant, M. J. and Waller, M. (2016b) Pollen signals of ground flora in managed woodlands. *Review of Palaeobotany and Palynology*, 224, pp. 121–133. https://doi.org/10.1016/j.revpalbo.2015.10.001

Bürgi, M. (1999a) A case study of forest change in the Swiss lowlands. *Landscape Ecology*, 14, pp. 567–76. https://doi.org/10.1023/A:1008168209725

Bürgi, M. (1999b) How terms shape forests: 'Niederwald', 'Mittelwald' and 'Hochwald', and their interaction with forest development in the canton of Zurich, Switzerland. *Environment and History*, 5, pp. 325–344. https://doi.org/10.3197/096734099779568263

Casperie, W.A. (1986) The two Iron Age wooden trackways XIV(Bou) and XV(Bou) in the raised bog of southeast Drenthe (the Netherlands). *Palaeohistoria*, 28, pp. 169–210.

Copini, P., Buiteveld, J., den Ouden, J. and Sass-Klaassen, U.G.W. (2005) *Clusters of Quercus robur and Q. petraea at the Veluwe (the Netherlands).* CGN/DLO Foundation, Wageningen.

Copini, P., Sass-Klaassen, U. and den Ouden, J. (2010) Coppice fingerprints in growth patterns of pedunculate oak (*Quercus robur*). In: T. Levanic, J. Gricar, P. Hafner, R. Krajnc, S. Jagodic, H. Gärtner, I. Heinrich and G. Helle (eds) *TRACE - Tree Rings in Archaeology, Climatology and Ecology*, Vol. 8. GFZ Potsdam, Potsdam, pp. 54–60.

Crone, A. (1987) Tree-ring studies and the reconstruction of woodland management practices in antiquity. In: G.C.J. Jacoby and J.W. Hornbech (eds) *Proceedings of the International Symposium on Ecological Aspects of Tree Ring Analysis*. US Department of Energy, Washington, DC, pp. 327–336.

Deforce, K. and Haneca, K. (2015) Tree-ring analysis of archaeological charcoal as a tool to identify past woodland management: the case from a 14th century site from Oudenaarde (Belgium). *Quaternary International*, 366, pp. 70–80. https://doi.org/10.1016/j.quaint.2014.05.056

Dufraisse, A. (2008) Firewood management and woodland exploitation during the late Neolithic at Lac de Chalain (Jura, France). *Vegetation History and Archaeobotany*, 17, pp. 199–210. https://doi.org/10.1007/s00334-007-0098-6

Faegri, K., Kaland, P.E. and Krzywinski, K. (1989) *Textbook of Pollen Analysis*. John Wiley and Sons, Chichester.

Fartmann, T., Müller, C. and Poniatowski, D. (2013) Effects of coppicing on butterfly communities of woodlands. *Biological Conservation*, 159, pp. 396–404. https://doi.org/10.1016/j.biocon.2012.11.024

Fritzbøger, B. (2004) *A Windfall for the Magnates: The development of woodland ownership in Denmark c. 1150–1830*. University Press of Southern Denmark, Odense.

Fuller, R.J. (1992) Effects of coppice management on woodland breeding birds. In: G.P. Buckley (ed.) *The Ecology and Management of Coppice Woodlands*. Chapman and Hall, London, pp. 169–92. https://doi.org/10.1007/978-94-011-2362-4_9

Galloway, J.A., Keene, D. and Murphy, M. (1996) Fuelling the city: production and distribution of firewood and fuel in London's region, 1290–1400. *Economic History Review*, 49, pp. 447–472. https://doi.org/10.1111/j.1468-0289.1996.tb00577.x

Gardner, A.R. (2002) Neolithic to Copper Age woodland impacts in northeast Hungary? Evidence from the pollen and sediment chemistry records. *Holocene*, 12, pp. 541–553. https://doi.org/10.1191/0959683602hl561rp

Girardclos, O., Dufraisse, A., Dupouey, J.L., Coubray, S., Ruelle, J. and Rathgeber, C.B. (2018) Improving identification of coppiced and seeded trees in past woodland management by comparing growth and wood anatomy of living sessile oaks (*Quercus petraea*). *Quaternary International*, 463, pp. 219–31. https://doi.org/10.1016/j.quaint.2017.04.015

Haas, J.N. and Schweingruber, F.H. (1993) Wood-anatomical evidence of pollarding in ash stems from the Valais, Switzerland. *Dendrochronologia*, 11, pp. 35–43.

Haneca, K., Van Acker, J. and Beeckman, H. (2005) Growth trends reveal the forest structure during Roman and medieval times in Western Europe: a comparison between archaeological and actual oak ring series (*Quercus robur* and *Quercus petraea*). *Annals of Forest Science*, 62, pp. 797–805. https://doi.org/10.1051/forest:2005085

Harris, W.V. (2018) The indispensable commodity: Notes on the economy of wood in the Roman Mediterranean. In: A. Wilson and A. Bowman (eds) *Trade, Commerce and the State in the Roman World*. Oxford University Press, Oxford, pp. 211–236.

Hausrath H. (1928) Beiträge zur Geschichte des Nieder- und Mittelwaldes in Deutschland. *Allgemeine Forst- und Jagd-Zeitung*, 104, pp. 345–348.

Isambert, F.A., Decrusy, M. and Jourdan, A.J.L. (1824) *Recueil général des anciennes lois françaises, depuis l'an 420 jusqu'à la révolution de 1789*. Vol. 5. Paris.

Keyser, R. (2009) The transformation of traditional woodland management: commercial sylviculture in medieval Champagne. *French Historical Studies*, 32, pp. 353–384. https://doi.org/10.1215/00161071-2009-002

Kneifl, M., Kadavý, J. and Knott, R. (2011) Gross value yield potential of coppice, high forest and model conversion of high forest to coppice on best sites. *Journal of Forest Science*, 57, pp. 536–546. https://doi.org/10.17221/32/2011-JFS

Kopecký, M., Hédl, R. and Szabó, P. (2013) Non-random extinctions dominate plant community changes in abandoned coppices. *Journal of Applied Ecology*, 50, pp. 79–87. https://doi.org/10.1111/1365-2664.12010

Ludemann, T. (2008) Experimental charcoal-burning with special regard to charcoal wood diameter analysis. In: G. Fiorentino and D. Magri (eds) *Charcoals from the Past: Cultural and Palaeoenvironmental Implications*. Archaeopress, Oxford, pp. 147–158.

Malmros, C. (1986) A Neolithic road built of wood at Tibirke, Zealand, Denmark. Contribution to the history of coppice management in the Sub-Boreal period. *Striae*, 24, pp. 153–156.

Matteson, K. (2015) *Forests in Revolutionary France: Conservation, Community, and Conflict, 1669–1848*. Cambridge University Press, New York. https://doi.org/10.1017/CBO9781107338197

Müllerová, J., Szabó, P. and Hédl, R. (2014) The rise and fall of traditional forest management in southern Moravia: a history of the past 700 years. *Forest Ecology and Management*, 331, pp. 104–115. https://doi.org/10.1016/j.foreco.2014.07.032

Ortego, J.H. and de Lomana, G.M.G. (2017) Firewood and charcoal consumption in Madrid during the eighteenth century and its effects on forest landscapes. In: E. Vaz, C.J. de Melo and L.M.C. Pinto (eds) *Environmental History in the Making, Volume I: Explaining*. n.p., Springer, pp. 321–340. https://doi.org/10.1007/978-3-319-41085-2_18

Out, W.A., Vermeeren, C. and Hänninen K. (2013) Branch age and diameter: useful criteria for recognising woodland management in the present and past? *Journal of Archaeological Science*, 40, pp. 4083–4097. https://doi.org/10.1016/j.jas.2013.05.004

Out, W.A., Hänninen, K. and Vermeeren, C. (2018) Using branch age and diameter to identify woodland management: new developments. *Environmental Archaeology*, 23, pp. 254–66. https://doi.org/10.1080/14614103.2017.1309805

Paradis-Grenouillet, S., Allée, P., Vives, G.S. and Ploquin, A. (2015) Sustainable management of metallurgical forest on Mont Lozère (France) during the Early Middle Ages. *Environmental Archaeology*, 20, pp. 168–183. https://doi.org/10.1179/1749631414Y.0000000050

Parker, A.G., Goudie, A.S., Anderson, D.E., Robinson, M.A. and Bonsall, C. (2002) A review of the

mid-Holocene elm decline in the British Isles. *Progress in Physical Geography*, 26, pp. 1–45. https://doi.org/10.1191/0309133302pp323ra
Pigott, C.D. (1989) Factors controlling the distribution of *Tilia cordata* Mill at the northern limits of its geographical range: IV. Estimated ages of the trees. *New Phytologist*, 112, pp. 117–121. https://doi.org/10.1111/j.1469-8137.1989.tb00316.x
Pliny the Elder. *Naturalis Historiae Libri XXXVII*. Vol. III. *Libb. XVI–XXII*. K. Mayhoff (ed.). B.G. Teubner, Leipzig. 1892.
Rackham, O. (1975) *Hayley Wood: Its History and Ecology*. Cambridgeshire and Isle of Ely Naturalists' Trust, Cambridge.
Rackham, O. (1976) *Trees and Woodland in the British Landscape*. J.M. Dent and Sons, London.
Rackham, O. (1977) Neolithic woodland management in the Somerset Levels: Garvin's, Walton Heath and Rowland's tracks. *Somerset Levels Papers*, 3, pp. 65–71.
Rackham, O. (1980) *Ancient Woodland: Its history, vegetation and uses in England*. Arnold, London.
Rackham, O. (2003) *Ancient Woodland: Its history, vegetation and uses in England*. 2nd edition. Castlepoint Press, Dalbeattie.
Rackham, O. (2006) *Woodlands*. (New Naturalist 100) Collins, London.
Rasmussen, P. (1990) Pollarding of trees in the Neolithic: Often presumed – difficult to prove. In: D.E. Robinson (ed.) *Experimentation and Reconstruction in Environmental Archaeology*. Oxbow Books, Oxford, pp. 77–99. https://doi.org/10.2307/j.ctvh1dp6m.12
Schweingruber, F.H. (1988) *Tree Rings: Basics and applications of dendrochronology*. Kluwer Academic Publishers, Dordrecht. https://doi.org/10.1007/978-94-009-1273-1_5
Schweingruber, F.H. (2007) *Wood Structure and Environment*. Springer, Berlin and Heidelberg.
Szabó P. (2005) *Woodland and Forests in Medieval Hungary*. Archaeopress, Oxford. https://doi.org/10.30861/9781841716947
Szabó P. (2018) Középkori cseh erdőgazdálkodás a choustníki uradalom erdőszámadásainak tükrében. In: D. Mérai, Á. Drosztmér, K. Lyublyanovics, J. Rasson, Zs. Papp Reed, A. Vadas and Cs. Zatykó (eds) *Genius loci: Laszlovszky 60*. Archaeolingua, Budapest, pp. 113–116.
Szabó P., Müllerová J., Suchánková S. and Kotačka M. (2015) Intensive woodland management in the Middle Ages: spatial modelling based on archival data. *Journal of Historical Geography*, 48, pp. 1–10. https://doi.org/10.1016/j.jhg.2015.01.005
Theophrastus. *Enquiry into Plants*. Translated by A.F. Hort (Loeb Classical Library). William Heinemann, London. 1916.
Thiebault, S. (2006) Wood-anatomical evidence of pollarding in ring porous species: a study to develop? In: A. Dufraisse (ed.) *Charcoal Analysis: New Analytical Tools and Methods for Archaeology*. Archaeopress, Oxford, pp. 95–102.
Troels-Smith, J. (1960) Ivy, mistletoe and elm. Climatic indicators–fodder plants: a contribution to the interpretation of the pollen zone border VII–VIII. *Danmarks Geologiske Undersøgelse II*, 4(4), pp. 1–32. https://doi.org/10.34194/raekke4.v4.7000
Unrau, A., Becker, G., Spinelli, R., Lazdina, D., Magagnotti, N., Nicolescu, V.N., Buckley, P., Bartlett, D. and Kofman, P.D. (eds) (2018) *Coppice Forests in Europe*. Albert Ludwig University of Freiburg, Freiburg.
Vrška, T., Janík, D., Pálková, M., Adam, D. and Trochta, J. (2016) Below- and above-ground biomass, structure and patterns in ancient lowland coppices. *Forest-Biogeosciences and Forestry*, 10, pp. 23–31.
Waller, M., Grant, M.J. and Bunting, E. (2012) Modern pollen studies from coppiced woodlands and their implications for the detection of woodland management in Holocene pollen records. *Review of Palaeobotany and Palynology*, 187, pp. 11–28. https://doi.org/10.1016/j.revpalbo.2012.08.008

CHAPTER 18

Oliver Rackham and Shadow Woods

Ian D. Rotherham

Summary

This chapter discusses the emergence of the 'shadow woods' concept and the pivotal role and influence of Oliver Rackham in the awareness of smaller veteran trees and in the historic lineage of anciently wooded sites. This vision of shadow woods and the fluxing of biodiversity through space and time relates to the long-term discussions between Oliver and for example George Peterken and Frans Vera, regarding the nature of European primeval landscapes. These debates were framed particularly by opinions on woodland origins and the interactions between woods and wood-pastures.

Keywords: Shadow woods, ghost wood, lost wood, ancient wood, woodland, wood-pasture, wooded common

Introduction and overview

Like Frans Vera, I came across Oliver's seminal 1980 book on ancient woodland when it first came out. The volume arrived at Sheffield University's library and was an absolute revelation, transformatory in terms of how I viewed the world and what I wanted to study. His 1986 book on the history of the countryside, although an altogether lighter read, was equally mind-blowing. However, in 1980 I was an impoverished PhD student at Sheffield University and the book was out of my league in terms of purchasing it. Nevertheless, like Frans at a similar time, I took the book, and one dark night illegally photocopied the lot on the departmental copier. By the late 1980s I was collaborating with a dear friend, the late Oliver Gilbert (see Gilbert, 1989), who pioneered interest in the ecology of previously neglected urban woodlands. Indeed, it was he that declared how Oliver Rackham was remarkable in being one of the few academics in ecology and landscape studies whose ideas had changed the way researchers around the world thought.

I had studied under Donald Pigott as my professor at the University of Lancaster, and his influence had whetted my appetite for woodland ecology. However, at that time I was more of a pure ecologist than a historian, but Rackham's writing opened a whole new world. In terms of research, this was very much forward-thinking history in order to help an understanding of ecology, but it also triggered an interest in countryside history from a previous generation, especially Hoskins (*The Making of the English Landscape* 1955) and Darby's Domesday Geography series (e.g., Darby 1955).

Ian D. Rotherham, 'Oliver Rackham and Shadow Woods' in: *Countryside History: The Life and Legacy of Oliver Rackham*. Pelagic Publishing (2024). © Ian D. Rotherham. DOI: 10.53061/ZHEH8716

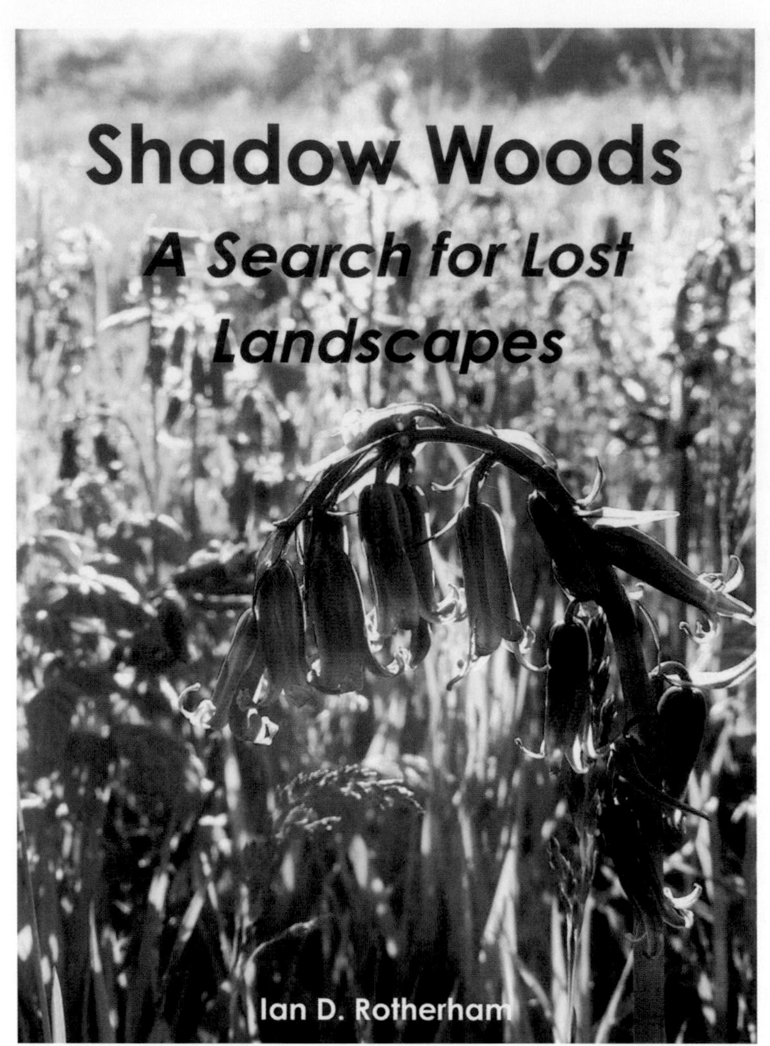

Figure 18.1 Cover to *Shadow Woods* (Rotherham 2017) showing shadow wood bluebells, *Hyacinthoides non-scripta*, Owler Bar, Peak District, 2016. (I. Rotherham)

A further step in the process for me was the work of a scholar at the then Sheffield Polytechnic, now Sheffield Hallam University, Melvyn Jones. A social geographer by training, Mel had become fascinated by the history and management of woodlands across Sheffield and South Yorkshire. Again stimulated by Rackham's work, Mel was actively researching local archives for historical materials and discovered a treasure trove of materials relating to the region's woods (see, for example, Jones 2009 and in *this volume*). The combination of all these wonderful sources of inspiration triggered a gradual transition into landscape studies, environmental history and countryside history. The ability to 'read' landscape history from the humps and bumps on the ground, from the soil and from the trees and vegetation became a fascination. Furthermore, at this time none of the region's woods was considered by their owners and managers to be 'ancient'; and it was suggested that the oldest was planted perhaps 200 years ago. Yet emerging from Mel's work was evidence in archives of exactly how these woods were being managed as far back as the sixteenth century, for example, and they were clearly 'ancient' as defined by Rackham and by Peterken (Peterken 1981).

By now I was working as the head of Sheffield City Council's ecological advisory unit, and was collaborating with local historians at the city's museum and with the regional archaeology services. We soon established a small study group with the Peak District National Park Authority, which we called the Landscape Conservation Forum (LCF): our remit was to promote joint interdisciplinary working and exchanges of ideas. Local amateur natural historians and what in the past would have been described as 'antiquaries' were also interested in the region's woods, and were pressing me to help find answers about their heritage and to unravel the complex woodland histories. In particular, a local businessman Richard Doncaster was fascinated by pit-like structures called Q-pits found in local woods. This was one of the factors that led us in 1992 to hold a first LCF conference on Ancient Woods ('Ancient Woodlands – their archaeology and ecology – a coincidence of interest' (Fig. 18.2)), and we invited Oliver to that meeting. Sadly, owing to a prior commitment, he was not able to attend, and indeed I recently refound his letter very graciously declining. My former undergraduate mentor Donald Pigott stepped in as our keynote speaker. In 2003, however, Oliver did attend our major 'Working and Walking in the Footsteps of Ghosts' conference, also held in Sheffield. Frans Vera, George Peterken, Donald Pigott, Chris Smout and Oliver Rackham were among a tremendous line-up of authorities from around the world, and the conference included the formal launch of the second, updated edition of Oliver's *Ancient Woodland: Its history, vegetation and uses in England* (Rackham 2003) (Fig. 18.3).

Ancient woods and ancient woodland

When setting out to investigate anciently wooded landscapes, our objective was to understand their ecology, to help place the trees in their human and landscape context, and through this to connect the approaches of ecology, history and archaeology. These were at that time largely disparate disciplines. At the time, we knew very little about the archaeology in woods, but then when we went on site with archaeologists and asked them for answers, it was clear that neither did they. What soon emerged was that archaeologists often recognized 'archaeology *in* the woods', for example a medieval smelting site or a Romano-British villa. However, they clearly did not recognize or understand the 'archaeology *of* the woods', in other words the evidence left of past woodland management. Furthermore, it soon became apparent that local historians and local history groups actively avoided woodlands because 'they are woods and there's nothing there'. Following the 2003 conference and events with colleagues across Europe, we began to undertake exchanges and develop shared approaches. Increasingly, we were also running woodland projects with the Heritage Lottery Fund, and undertook a major national review of the use of ancient woodland indicator plants for the Woodland Trust, English Nature, the Forestry Commission and the British Ecological Society. This produced several published outputs on indicators and landscapes (e.g., Glaves et al. 2009a, b, c; Rotherham 2011; 2013a). The most significant activity was the work with the Woodland Trust, English Heritage and the Forestry Commission, supported by the Heritage Lottery Fund and on 'woodland heritage' with a major output being *The Woodland Heritage Manual: A guide to investigating wooded landscapes* (Rotherham et al. 2008). This was a first attempt to bring together the skills, insights and approaches of ecologists, archaeologists, historians, and foresters to share ideas on ancient woodlands and their heritage. There followed smaller regional projects on industrial woodlands, and on woodland heritage for local community groups.

The idea of being able to use worked trees and indicator plants to inform our understanding of ancient woods had taken hold and was proving to be a potentially powerful tool. Furthermore, the problems of trying to understand woodlands with single-discipline approaches were becoming apparent. For example, a major archaeology survey of South Yorkshire's ancient woodlands took place on the back of a project called 'Fuelling the Revolution'. This was based on Melvyn Jones's remarkable studies that showed how the ancient woods of the region had produced wood, fuelwood, and charcoal to power the Industrial Revolution. However, at the presentation of the final report, when the consultant archaeologists were asked what they made of the trees, the answer was 'Oh, we didn't look at those – we don't do trees.' That really rang alarm bells.

Nevertheless, shared approaches and useful toolkits were emerging and being tested by colleagues and collaborators nationally and internationally. However, with the national review of the use of botanical indicator species, some questions were beginning to arise. One key driver for the work was that the Woodland Trust were increasingly being challenged over ancient woodland status of sites by developers and their consultants. Essentially if a site was shown to be ancient woodland, then there was a modicum of planning protection from destructive development and a presumption against this. However, the Ancient Woodland Inventory (e.g., Bevan 1992) by this time used by local authorities and agencies, was not established to be tested in a court of law or at a public enquiry. Consultants were thus able to challenge the status of a woodland designation. A series of reviews (including one of all British regional lists of woodland indicator plants then in use) was undertaken in parallel with a programme of expert seminars. The project was supported by the Woodland Trust, the Forestry Commission, the British Ecological Society and other stakeholders. An output from the work was a protocol (Rotherham 2011) that could be developed and which took information from a range of disciplines to assess ancient woodland status. This was later

Figure 18.2 Proceedings of the 1992 Sheffield conference. (I. Rotherham)

used (successfully) to defend woodland sites at public and planning enquiries. In terms of ancient woodland indicators, we were confident that a given suite of plants evaluated in a local and regional context could provide evidence for a site being a medieval 'wood'. Intriguingly, though, the evidence took you back to enclosure and naming as a 'wood' in medieval times, but not necessarily before that. A question that then emerged from the studies concerned the nature of a locale 'before' it became a 'wood'; that is, when it was unenclosed. A popular assumption once the idea was established of 'ancient woodlands' in the English landscape was that these were not only ancient but also primeval: the ancient woods were links to a so-called wildwood. This misconception lingers even today – of a countryside covered by densely packed trees in primeval woodland of which our ancient woods are tiny remnants. However, the evidence had long been accumulating to dispel this myth. Work by scholars such as Pigott (1993) and Day (1993) clearly established that most if not all ancient woods (at least in lowland England) held evidence of human settlement and activities and of periods without tree cover. In other words, even the known ancient woods had times historically when they had not been wooded. This ends the wildwood myth, but also triggers the question of what this unwooded but treed countryside was like.

Figure 18.3 Oliver Rackham's 2003 book launched at the 2003 Sheffield conference. (I. Rotherham)

Figure 18.4 Veteran Common Hawthorn at Longshaw in the Peak District, 2009. (I. Rotherham)

A further context to this research journey was our involvement in what I call the 'Vera debate': the questions raised by Frans Vera (Vera 2000) about the nature of primeval Europe. Oliver Rackham and George Peterken also joined in a discussion on this theme at the 2003 Sheffield woodland conference. This fascinating debate raised the issue of where our 'ancient coppice woods' sat in the longer timeline of European treescapes and particularly in relation to the vision emerging of a European savanna landscape as opposed to wall-to-wall dense forest. Ultimately, there seemed to be broad agreement that there were open savanna-like wood-pastures and closed-canopy forest, but the questions then arose of how big the patches were, where they were, and how far apart. Other questions related to the ecological drivers of the closed-canopy areas and of the open grasslands and savanna; and the debates still go on today in relation to rewilding. It was from these discussions that the idea of 'shadow woods' began to develop. However, a further key was a field visit to Chatsworth House and Park in Derbyshire for the conference dinner and a pre-dining field visit. On arrival at Chatsworth Park, Oliver took centre-stage in a landscape of great trees. However, as the audience gathered under a large oak-tree Oliver headed off up the hill to a far more modest and rather wizened Common Hawthorn *Crataegus monogyna* (Fig. 18.4). He declared this to be far more interesting, in all probability very old and mostly overlooked by researchers. In retrospect, this was probably where our interest in smaller veteran trees began in earnest.

The Act of Commons

In seeking to connect an understanding of landscape history to ideas of countryside origins and of future rewilding, it has become obvious that many of those discussing or researching fail to consider basic historical sources. The most obvious documentary source on the English countryside in medieval times is the Domesday account of 1086. While this is neither comprehensive nor uniform in its coverage, and it is not always spatially reliable, it does provide a remarkable view into the pre-feudal landscape. It often tells you how much woodland you have in a manorial estate and, moreover, whether it is coppice (i.e., a wood) or wood-pasture (e.g. wooded common, chase, park, forest). In most cases, it is the latter (wood-pasture) that predominates and the former (coppice) that is generally rare. This, I argue, provides an insight into the pre-feudal, pre-Conquest English countryside and there are clear echoes of Vera's European primeval wood-pasture savanna. So, with reference to the debates noted earlier and to the writings of Della Hooke on trees in the Saxon landscape (1986, 2011), we have a vision of the countryside leading up to the Norman Conquest. There are coppice woods in this landscape, and we know, for example, that the Romans practised formal coppice management too, but it is predominantly an open landscape of unenclosed treescapes – wood-pastures of various sorts. In many cases, the trees were widely spaced in open, fluid landscapes that mixed grazing and trees as savanna or wood-pasture. There is also emerging archaeological evidence of massive, tall, straight oaks in built structures in both pre- and post-Roman Britain, and this can be taken to indicate the areas of close-grown, closed-canopy forest from which these timbers were cut in the wider landscape of savanna and wetland.

Yet if we fast-forward a few centuries to say the fourteenth or fifteenth century, then in England we are in a feudal English landscape of manors controlled in the main by Norman overlords. Moreover, this countryside is largely fixed, mapped, documented and accounted for, and with rising human populations the wood-pasture tradition is being displaced by coppice woods. Wood-pasture is an effective system of wood and timber supply with small human populations in expansive countryside. Coppice-with-standards systems provides for sustainable 'cut-and-come-again' resources of timber and wood. These woods were enclosed, named, and protected from both grazing animals and the local peasantry. This is a landscape transformed from pre-Domesday and begins to be recognizable in our modern world. The change from the pre-feudal to the Norman medieval countryside is also reflected in legislation of the time, namely that 'Magna Carta of the landscape', the Act of Commons or Statute of Merton of 1235. Again, correspondence with Oliver suggested that this was a hugely significant watershed in England countryside history and something to be considered further. The idea of trees outside woods was debated at our Sheffield conference 'Trees Beyond the Wood' in 2013 (Rotherham et al. 2013). Sadly, Oliver could not attend as he had taken up a 'once in a lifetime' invitation to visit Ethiopia and to examine the country's unique treescapes. However, the concept of trees outside formally enclosed woods was gathering momentum, and the Woodland Trust was soon to launch a successful Heritage Lottery Project on this theme.

A big challenge is to recognize the dynamic nature of the countryside over time as it and the associated ecology flux and change throughout history. The consequence in landscape ecology, is a fluid mix of succession, of continuity, and of both temporal and spatial connection. Above all, the systems are not static, and yet they have a dynamic resonance and stability throughout the centuries. Nor are these systems 'natural', but they result from people and nature interacting to generate eco-cultural landscapes. The question is, then, where do the trees fit into this mosaic of fluidity and stability or succession? Furthermore, with our research having identified and established ancient medieval woods in the English countryside, we were faced by a mix of apparently contrary evidence and information. So-called woodland indicator plants occur in situations where there is clearly no sign of enclosed woodland, and there is also plentiful evidence of past woods in areas without woodland today. In seeking to unravel this tangled history of our countryside, it is important to address the nature of woods, wood-pastures, lost woods, ghost woods and shadow woods (Rotherham 2017, 2018). Some definitions are necessary.

Lost woods, ghost woods and shadow woods

Woods

These are treed sites enclosed from the wider countryside, named, and managed over subsequent centuries. Protected from large grazing herbivores at least during the early years of the coppice cycle or after major tree-felling, this approach to tree management originated with the Romans. Some woods occurred in the Saxon landscape (Hooke 1986, 2011), but most originated in the two to three centuries following the Norman Conquest (Rackham 1980, 1986).

Wood-pastures

These are treed landscapes but mix tree-cover with large grazing herbivores, both wild and domesticated. The most widely recognized forms of wood-pastures are parks, chases, forests, and wooded commons (Rackham 1980, 1986; Rotherham 2007).

Lost woods

Enclosed woods often suffered damage, loss or destruction over the centuries following their establishment. So lost woods are sites that were enclosed and named but were subsequently removed or lost by conversion to farmland, by urban development, under infrastructure such as roads or simply by opening-up and reversion to a grazed landscape (e.g., Lewis 2019). Some ancient woods, such as Gardom's Coppice in the Peak District, still exist but have been lost from memory and even from maps. This ancient wood, now shrouded by secondary birch growth, has over 1,000 veteran coppice trees that were overlooked by contemporary surveyors.

Figure 18.5 Veteran shadow wood hawthorn below White Edge in the Peak District, 2018. (I. Rotherham)

Ghost woods

When woods are lost, then destruction may be total – with a site completely removed. However, in many cases the ghost of a lost wood can be seen in the landscape, with physical features such as woodbanks, walls, charcoal hearths, and lanes. In many cases, veteran trees and ancient woodland botanical indicators mark the area of a past wood now etched into the modern landscape. Even in intensively farmed or highly urbanized areas, indicators survive alongside field-names, and place-names associated with woodland use (Rotherham 2017).

Shadow woods

The concept of shadow woods as lost Domesday landscapes is one of the exciting outputs from this ongoing research (Rotherham 2017). These are remnants of once extensive Domesday wood-pastures that persisted unenclosed into the medieval as wooded commons. In other words, it was from these landscapes that our woods originated, and remarkably, having survived periods of medieval enclosure and especially the intensive parliamentary enclosures, some areas remain (Fig. 18.5). These were initially identified by detailed field surveys of flowering plant indicators, smaller veteran trees, waxcap fungi, soils, and other evidence. Not being woods, these sites often have limited documentary evidence associated with them, though some Peak District moors were described in nineteenth-century maps as having plentiful smaller trees and being 'wood-pastures'.

Lost woods and shadows

Perhaps the biggest issue in terms of shadows and ghosts is that of being able to see them, to recognize them and then to understand them. First, we need to see them in the landscape. It has been an interesting experience to go in search of lost landscapes, of hidden ecologies,

Figure 18.6 Veteran shadow wood of rowan, hawthorn and Sycamore Maple *Acer pseudoplatanus* below White Edge in the Peak District, 2018. (I. Rotherham)

and forgotten histories. Yet once you begin to see the evidence and to read the signs, a whole new landscape comes to life. It is there all the time, but we simply fail to see it (Rotherham 2013b, 2017, 2018).

The importance and significance of the smaller trees

A key to finding the shadow woods is through their smaller veteran trees (Fig. 18.6). These may be either diminutive but ancient oaks, for example, or species such as Common Hawthorn, Common Holly *Ilex aquifolium*, European Rowan *Sorbus aucuparia* and even birch *Betula* spp., which are inherently small in stature. It was such trees that Oliver pointed out as exciting back at the 2003 Sheffield conference. These specimens are to be found around the upland fringe zones and in lowland areas that survived the parliamentary enclosures. Some may be survivors of managed wooded commons and others occur in former royal forests or in ancient parks. In terms of shadow woods, there appear to be three main categories of relict treescapes: 1) open wooded common with oak *Quercus* spp., birch, rowan, holly and hawthorn and often with woodland indicator plants; 2) ancient willow *Salix* spp. and Common Alder *Alnus glutinosa* carr woods on wet areas and occurring from the uplands down to lowland zones but especially in the western parts of Britain; 3) ancient hanging oakwoods on boulder, scree and talus slopes below crags, edges and other rocky outcrops.

Number 2, the ancient willow and alder sites, have multi-stemmed alders and huge multi-stemmed willows growing horizontally as clones through wet landscapes. They are often species-rich in terms of their botany and presumably so for their invertebrate interest too, though so far this has not been researched.

Number 3, the hanging oakwoods have massive 'medusoid' (i.e., multi-stemmed) oaks that appear to result from human management and extreme climatic impacts. These are accompanied by holly, rowan and birch, but are often botanically poor and associated with recent histories of intensive sheep-grazing.

In considering these trees, despite their small stature, they are very old. Oak, holly, and willow are capable of being long-lived as multi-stemmed specimens, and birch is also capable of being long-lived in upland situations especially if coppiced. However, the rowans and hawthorns are particularly intriguing, and this was something that Oliver Rackham was especially excited about. Many trees such as the European Beech *Fagus sylvatica*, Common Hawthorn and European Rowan, for instance, can put down adventitious roots inside a stem-break or rot-pocket, and this is widely recognized. These roots often grow vertically downwards into and through a mat of rotting deadwood and presumably are deriving (rescuing?) nutrients from this material. At the 1992 Sheffield conference, Ted Green pointed out the significance of his observations at Windsor Great Park with massive, hollow oaks surviving the 1987 storm (Green 1993). They were, he asserted, recycling their own dead innards so that nutrients from the long-dead and now rotted xylem material were taken back into the living tree. This might occur by absorption from the tree hollow or else, as nutrients leached down to ground level, by uptake via the extensive mycorrhizal root-plate of the great tree.

This same process appears to apply in the upland shadow woods, and I suspect its inception is triggered by environmental stress such as drought in these nutrient-poor, extreme situations. Some trees such as rowans and birches on exposed upland crags may be broken by high winds, and hollowing then ensues, but in other cases with rowan and hawthorn, hollowing is probably triggered by stress. This process raises some critical issues and challenges in the aging and describing of these trees. As they hollow out, then branches emanating from the main stem send down endocormic, adventitious roots as described earlier. This potentially strengthens the stem and facilitates the capture of nutrients released from the rotting wood back into the living trees. As with the beech trees and the oaks, any nutrients leached down to ground level will be rapidly reabsorbed by the dense, mycorrhizal root-plate. Scavenging nitrogen, phosphorus, and potassium in these nutrient-stressed soils is a means to survival; and seen in this light, the hollow, rotten core of the tree is not a sign of poor health and decline but an effective adaptation to stress.

There are further complications too, since it is clear that both hawthorn and rowan send down roots inside the trunk and over time these morph into new stems. This means that younger bark and stem materials occur *within* the older trunk rather than merely accumulating on the outside. When tree-coring to attempt to age these trees, we can get an estimate for the age of the outer layers until hitting rotten wood, and these are typically anywhere between around 50 years to 100 years plus for rowan and hawthorn. However, beyond this solid layer we break into the hollowed core, and in some cases multi-layered younger stem growth. Many of these smaller trees exhibit a full range of forms associated with veteran condition such as rot-pockets, deadwood, fractures, broken stems and so on. In the case of Rowans, we know that some of the trees are located on very old, undisturbed structures such as prehistoric barrows, and in other cases associated with old features such as sheep-folds. (Historically in our research area, rowan was a 'witching tree' and grown to ward off the 'evil eye' that might otherwise affect the livestock.) In the case of Common Hawthorn, we can also gauge possible age of individual trees by comparison with specimens of known age from, for example, lane-sides or enclosure hedgerows taken as maybe 250 years old. By comparing form, structure and condition between the two, and with 250 years as the baseline for the younger tree, we can surmise at least a minimum age for the older specimens. For one hawthorn (Fig. 18.4) at Longshaw in the Peak District, Mel Jones and I estimated this must have been at least 500 years old prior to its death in the cold winters of 2010 and 2011. This figure would place the tree as one of the oldest living things on that estate at the time.

The multi-stemmed trees often present major problems in terms of estimating their ages, and in some cases a possible solution may be to follow the approach of Pigott in measuring or estimating linear extension of shoots from a suggested central node. This can be done for Small-leaved Lime *Tilia cordata* (Pigott 1989, 2012), for oak, willow and potentially for holly clones too, and this is work in progress to be reported in due course. However, some interim findings with oak (e.g. Vrška et al. 2016), support the suggestions by Pigott (1989, 2012) regarding lime, that these trees can be very old indeed. For oak, Vrška et al. (2016) suggest that individual coppice stools might be aged at around 800 years or more. This would relate well to the idea of areas of medieval wood-pastures being enclosed as protected coppices, discussed earlier.

A final example of an overlooked veteran tree that is being recognized as of particular significance in historical reconstruction of landscapes is Common Holly. Long-known as an important tree for production of tree fodder in medieval England, evidence is emerging of potential antiquity of holly clones in former 'holly hags' or 'hollins'. Locations such as Holmesfield in North Derbyshire for example, present tantalizing insights into the former landscape. Here there is an ancient wood within a former medieval park that was abandoned in the sixteenth century. The place-name Holmesfield may derive from 'an open areas populated by holly trees'; that is, a hollin or holly hag as a type of wooded common. Here, the holly would be either or both pollarded and coppiced, and this may leave a legacy of ancient clonal growths from abandoned coppice or veteran pollards. At Holmesfield Park Wood, some of the holly stems are growing several metres from their central point of origin. Bearing in mind the relatively slow growth-rate of holly and in view of the work on aging lime stools by Pigott (1989, 2012) and on oak stools by Vrška et al. (2016), these trees could be many centuries old. More specific work remains to be done.

Seeing the woodland through its trees

The emerging vision of these ancient landscapes and their woodlands is that we can begin to understand the lineage and descent of wooded countryside through the trees and their many

Figure 18.7 Oliver Rackham at Longshaw in 2011 at the 'Animals Man and Treescapes' conference. (P. Ardron)

forms. These reflect natural and human history sometimes over many centuries, and a better understanding of tree-form, tree age, and past management can inform our visions of past woodlands and treescapes. Combined with ecological indicators and historic archives through multidisciplinary research, the approach can join often disparate visions to forge a more robust and unified concept of landscape evolution. This informs understanding of contemporary landscapes and heritage based on historic drivers, management, and changes. Furthermore, this knowledge can help guide visions of future treescapes and rewilded landscapes.

Conclusions and a debt of gratitude

The journey to understand countryside history and specifically the nature of trees, woods and forested landscapes, began for many of us in the 1970s and 1980s with the writings of Oliver Rackham and George Peterken. In the UK, these ideas emerged in part at least from the early seminal volumes on countryside history such as Hoskins's *The Making of the English Landscape* (Hoskins 1955) and those by Darby on the English landscape in Domesday (e.g., Darby 1955), as previously mentioned. Others then developed research and ideas to follow-up and follow through, and to combine these with works across the world, especially in Europe and in North America. Oliver was not always correct and his legacy should not be seen as an endpoint in itself but as a vital part of a journey of discovery. Through a combination of meticulous research and piercing intellect, Oliver was able to address key issues to open a remarkable vision of the natural and human-formed world. However, a vital part of his legacy and influence must be through his ability to communicate his insight and his enthusiasm so eloquently to both academic and keen amateur. It was through this writing that he was able to engage a wide audience, to trigger interest and enthusiasm, and to change perceptions and opinions. This is something for which we owe Oliver a huge debt of gratitude. My concern today is how quickly some of the lessons of Rackham and his contemporaries such as Melvyn Jones are being lost on a new generation of commercially driven foresters and even on ecologically orientated conservation managers. The widespread destruction of the medieval woodlands by modernized, mechanized contemporary management, would, I believe, have horrified Oliver. I hope this volume may help re-awaken some degree of awareness of the unique and irreplaceable nature of the medieval woods and tweak the consciences of those currently wreaking such damage. We need to learn to tread softly in the footsteps of the ghosts of those woodmen and their families who shaped what are 'ancient woodlands' today.

References

Bevan, J.M.S., Robinson, D.P., Spencer, J.W. and Whitbread, A. (1992) *Derbyshire Inventory of Ancient Woodland (Provisional)* (Nature Conservancy Council, 1992), Peterborough.

Darby, H.C. (1955) *The Domesday Geography of Eastern England.* Cambridge University Press, Cambridge.

Day, S.P. (1993) Origins of medieval woodland. In: P. Beswick, I.D. Rotherham and J. Parsons (eds) *Ancient Woodlands: Their Archaeology and Ecology – a coincidence of interest.* Landscape Conservation Forum, Sheffield, pp. 12–25.

Gilbert, O.L. (1989) *The Ecology of Urban Habitats.* Chapman and Hall, London. https://doi.org/10.1007/978-94-009-0821-5

Glaves, P., Rotherham, I.D., Wright, B. Handley, C. and Birbeck, J. (2009a) *A Report to the Woodland Trust Field Surveys for Ancient Woodlands: Issues and Approaches.* Hallam Environmental Consultants Ltd., Biodiversity and Landscape History Research Institute, and Geography, Tourism and Environment Change Research Unit, Sheffield Hallam University, Sheffield.

Glaves, P., Rotherham, I.D., Wright, B. Handley, C. and Birbeck, J. (2009b) *A Report to the Woodland Trust: A survey of the coverage, use and application of Ancient woodland indicator lists in the UK.* Hallam Environmental Consultants Ltd., Biodiversity and Landscape History Research Institute, and Geography, Tourism and Environment Change Research Unit, Sheffield Hallam University, Sheffield.

Glaves, P., Rotherham, I.D., Wright, B, Handley, C. and Birbeck, J. (2009c) *A Report to the Woodland Trust Field Surveys for Ancient Woodlands: Issues and Approaches.* Hallam Environmental Consultants Ltd., Biodiversity and Landscape History Research Institute, and Geography, Tourism and Environment Change Research Unit, Sheffield Hallam University, Sheffield.

Green, E.E. (1993) Observations on the importance of old trees and their management. In: P. Beswick, I.D. Rotherham and J. Parsons (eds) *Ancient Woodlands: Their Archaeology and Ecology – a coincidence of interest.* Landscape Conservation Forum, Sheffield, pp. 91–92.

Hooke, D. (1986) *The Anglo-Saxon Landscape: The Kingdom of the Hwicce.* Manchester University Press, Manchester.

Hooke, D. (2011) *Trees in Anglo-Saxon England: Literature, Lore and Landscape.* The Boydell Press, Woodbridge.

Hoskins, W.G. (1955) *The Making of the English Landscape.* Hodder and Stoughton, London.

Jones, M. (2009) *Sheffield's Woodland Heritage*, 4th edn. Wildtrack Publishing, Sheffield

Lewis, H. (2019) Interactions between human industry and woodland ecology in the South Pennines. Unpublished PhD, University of Bradford, Bradford with University of Hull and Sheffield Hallam University.

Peterken, G. (1981) *Woodland Conservation and Management.* Chapman and Hall, London. https://doi.org/10.1007/978-1-4899-2857-3

Pigott, C.D. (1989) Factors controlling the distribution of *Tilia cordata* Mill. at the northern limits of its geographical range. IV. Estimated ages of the trees. *New Phytologist,* 112, pp. 117–121. https://doi.org/10.1111/j.1469-8137.1989.tb00316.x

Pigott, C.D. (1993) The history and ecology of ancient woodlands. In: P. Beswick, I.D. Rotherham and J. Parsons (eds) *Ancient Woodlands: Their Archaeology and Ecology – a coincidence of interest.* Landscape Conservation Forum, Sheffield, pp. 1–11.

Pigott, C.D. (2012) *Lime-Trees and Basswoods: A biological monograph of the genus* Tilia. Cambridge University Press, Cambridge. https://doi.org/10.1017/CBO9781139033275

Rackham, O. (1980) *Ancient Woodland: Its History, Vegetation and Uses in England.* Edward Arnold, London.

Rackham, O. (1986) *The History of the Countryside.* J.M. Dent and Sons, London.

Rackham, O. (2003) *Ancient Woodland: Its History, Vegetation and Uses in England,* 2nd edition. Castlepoint Press, Dalbeattie.

Rotherham, I.D. (2007) The historical ecology of medieval deer parks and the implications for conservation. In: R. Liddiard (ed.) *The Medieval Deer Park: New Perspectives.* Windgather Press, Macclesfield, pp. 79–96.

Rotherham, I.D. (2011) A landscape history approach to the assessment of ancient woodlands. In: E.B. Wallace (ed.) *Woodlands: Ecology, Management and Conservation.* Nova Science Publishers Inc., USA, pp. 161–184.

Rotherham, I.D. (2013a) *Ancient Woodland: History, Industry and Crafts.* Shire Publications, Oxford.

Rotherham, I.D. (2013b) Searching for Shadows and Ghosts. In: I.D. Rotherham, C. Handley, M. Agnoletti, and T. Samojlik (eds) *Trees Beyond the Wood: An Exploration of Concepts of Woods, Forests and Trees.* Wildtrack Publishing, Sheffield, pp. 1–16.

Rotherham I.D. (2017) *Shadow Woods: A Search for Lost Landscapes.* Wildtrack Publishing, Sheffield.

Rotherham, I.D. (2018) The magic and mysteries of Ecclesall Woods. In: M. Atherden, C. Handley and I.D. Rotherham (eds) *Back from the Edge: The Fall and Rise of Yorkshire's Wildlife,* 2nd expanded edition. Wildtrack Publishing, Sheffield, pp. 85–102.

Rotherham, I.D., Jones, M., Smith, L. and Handley, C. (eds) (2008) *The Woodland Heritage Manual: A Guide to Investigating Wooded Landscapes.* Wildtrack Publishing, Sheffield.

Rotherham, I.D., Handley, C., Agnoletti, M. and Samojlik, T. (eds) (2013) *Trees Beyond the Wood: An Exploration of Concepts of Woods, Forests and Trees.* Wildtrack Publishing, Sheffield.

Vera, F.H.W. (2000) *Grazing Ecology and Forest History.* CABI, Wallingford. https://doi.org/10.1079/9780851994420.0000

Vrška, T., Janík, D., Pálková, M., Adam, D. and Trochta, J. (2016) Below- and above-ground biomass, structure and patterns in ancient lowland coppices. *Forest,* 10, pp. 23–31. https://doi.org/10.3832/ifor1839-009

CHAPTER 19

Oliver Rackham and the Archaeology of Ancient Woods of Norfolk

Tom Williamson

Summary

The influence of Oliver Rackham's work is noted with particular reference to countryside history and especially in relation to the study of woodlands. While Rackham worked all over the world, his enthusiasm for English woodlands and especially for those of East Anglia was a feature of much of his writing. It is with this in mind that this chapter considers the ancient woodlands of Norfolk, and investigates them in much the same way that Oliver might have done.

Keywords: ancient woods, countryside history, Norfolk, primary woodland, coppices, wood-pasture, waste

Introduction

Oliver Rackham is best known for his work on ancient woodland. Indeed, while he was not solely responsible for the coinage or definition of this term, it is now more closely associated with him than, perhaps, any other researcher. Much of Rackham's work on woodland was focused on East Anglia – Norfolk and Suffolk – and on the adjacent counties of Essex and Cambridgeshire. His seminal *Trees and Woodlands in the British Landscape*, together with *Ancient Woodland* and even his wide-ranging *Woodlands*, are all replete with examples from these counties, something that reflects both Rackham's academic base in Cambridge and his own origins in the 'ancient countryside' of south Norfolk (Rackham 1976, 1980, 2006). The history and archaeology of Norfolk woods was also the subject of one of Rackham's best, but relatively little known papers, 'The ancient woods of Norfolk', published in the *Transactions of the Norfolk and Naturalists Society* in 1986, at the end of his tenure as the society's president (Rackham 1986). In this he demonstrated, with particular clarity, some of the characteristics that made him such a ground-breaking scholar: his ability to move easily in the fields of archaeology, history and ecology, to understand the complexity of ecological systems, and yet at the same time to demonstrate an impressive command of medieval Latin and of palaeography.

One of Rackham's greatest achievements was that he presented both professionals and 'amateurs' with a tool-kit for undertaking their own investigations into the subject that he

Tom Williamson, 'Oliver Rackham and the Archaeology of Ancient Woods of Norfolk' in: *Countryside History: The Life and Legacy of Oliver Rackham*. Pelagic Publishing (2024). DOI: 10.53061/ZYLE1649

loved. Chapter 6 of *Trees and Woodlands*, 'Woods as we see them today: a guide to fieldwork', although published more than 40 years ago, has yet to be bettered. In it, among other things, Rackham discussed the ways in which the evidence of earthworks could be used to tease out the history of an individual wood, making an important distinction between those archaeological features that were related to its function as a wood – boundary banks, internal subdivisions and the like – and those representing the traces of farming, settlement and other activities that had taken place within a wood's area before it had come into existence, and thus marked it out as being secondary in character, as opposed to being primary, or directly descended from the primeval vegetation (Rackham 1976) (Fig. 19.1).

The mark of a truly great researcher is not so much that he or she manages to solve all the main questions in a particular discipline (at which point these anyway cease to be matters of academic interest), but rather that they raise a host of new questions, inspire others to investigate them and provide them with an appropriate methodology for doing so. As a consequence, of course, some of their own conclusions may be modified, but in Rackham's case this has generally been along lines already hinted at or implied in his own writings. Over the last decade or so, the ancient woods in the county of Norfolk – defined for convenience as those which are identified in the Ancient Woodland Inventory – have been the subject of a programme of field survey and documentary research by the author and Dr Gerry Barnes at the University of East Anglia. The results of this work, published in 2015, have confirmed many of the conclusions reached by Rackham (Barnes and Williamson 2015). But they have also served, in certain key respects, to modify some of his suggestions, although not in ways that do violence to his wider arguments relating to woodland history.

Figure 19.1 Undated lynchet, flanked by low bank, within Middle Wood, Thorpe Abbots, Norfolk, 2014. It is one of several such features within the wood and, whether of medieval or of earlier date, indicates that parts at least of this ancient wood are secondary in character. (T. Williamson)

How ancient is ancient woodland?

Rackham's views on the extent to which 'primary' woods in England should be regarded as 'natural' shifted during the course of his career, but he always maintained a belief that the majority of examples occupied areas that had never been cleared or cultivated. Out of 50 Norfolk woods, only around 15 – 30 per cent – appear to be secondary in their entirety, having been planted, or regenerated naturally, over abandoned settlements or farmland, usually in the late medieval or early post-medieval periods. Almost all of these woods, moreover, are comparatively small. With the exception of Toombers Wood in Stow Bardolph and Stradsett (23 ha), the largest in the sample examined was Ringers Grove in Shotesham, which extends over only about 3 ha. Examples include Tivetshall Wood, the entire area of which appears to overlie a small medieval settlement and associated enclosures, and Beckets Wood in Woodton, which overlies an area of open-field arable and a number of small medieval toft sites (Barnes and Williamson 2015: 165–166, 229–230, 237–238). (A toft, from the Old Norse *topt*, was a homestead or else an entire holding, consisting of a homestead and the attached arable land.) Different patterns of development can be exhibited by secondary woods only a short distance apart. Round Grove in Hedenham, covering an area of 2.6 ha, completely overlies the area of a medieval settlement and its associated fields, while nearby Long Row appears to have developed at the expense of a wide drove road or linear common (Barnes and Williamson 2015: 213, 223–224).

Larger woods are far more likely to be primary in character, containing no evidence for the presence of former settlements, agriculture or systems of land division. Where early earthworks are present they seem to be related to the control of stock and the division of grazing before the wood was enclosed and defined, from wider tracts of wood-pasture, at some point before the twelfth or thirteenth centuries: Hockering Wood contains what may be a substantial boundary earthwork and several Norfolk woods feature enclosures, sometimes associated with moated sites, which were probably for corralling stock within wider areas of grazed wood-pasture. But for the most part these woods are empty of archaeological evidence, or at least of earthworks predating the wood's enclosure and management as coppice. Tindall Wood in Ditchingham, for example, covers no less than 44 ha and contains a single internal boundary bank, ruler-straight and evidently relating to post-medieval division or management; Sexton's Wood in Hedenham, with an area of 39 ha, contains nothing except the concrete access roads of a Second World War ammunition dump (military archaeology from the 1940s is prominent in Norfolk's woods, one of the few aspects of their heritage not really considered by Rackham).

This said, many of the county's largest woods contain, in addition to a primary core, a secondary addition. Sometimes this comprises no more than a small extension – over an abandoned adjacent lane, for example, as in the case of The Shrubbery in Tivetshall – but in some cases it is more substantial, as with Wayland Wood in Watton. Here the 'core' covers about 25 ha, but the extension, which probably regenerated (or was planted) over abandoned farmland in the late Middle Ages (and certainly, to judge from the documentary evidence, before 1595: NRO [Norfolk Record Office, Norwich] WLS IV/6), covers 7 ha. Rackham himself published, in his paper in the *Norfolk Naturalists* for 1986, the classic example of Hedenham Wood in the south-east of the county (Rackham 1986). Here, the northern part of the wood, devoid of anything but earthworks relating to its post-medieval management, is evidently primary, but the southern section, although identical in general appearance and species composition, appears to have grown up in the sixteenth century over an abandoned manorial site and associated paddocks (the site was largely, although still not entirely, tree-covered when mapped in 1617 (NRO MC 1761/2)) (Figs 19.2 and 19.3). This was only one part of a complex history, however, for in the nineteenth century, as Rackham noted, the wood was greatly reduced in area when its western section was grubbed out and converted to arable fields.

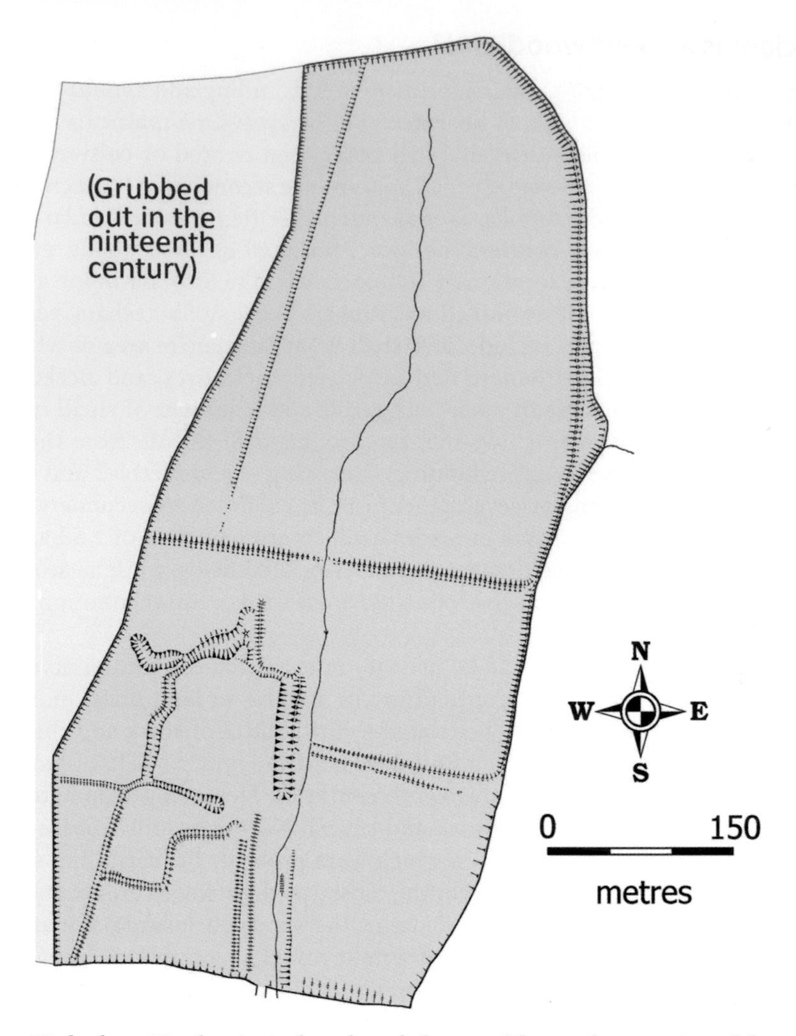

Figure 19.2 Hedenham Wood: principal earthwork features. The northern section of the wood appears to be primary in character. The wood expanded southwards in the sixteenth and seventeenth centuries, engulfing an abandoned manorial site and associated fields and roads. The distinct histories of the two parts of the wood are not reflected in their vegetation. The western part of the wood was grubbed out in the nineteenth century. (T. Williamson)

Rackham seems to have regarded Hedenham Wood as something of an anomaly: he usually emphasized the long-term stability displayed by ancient woodland. But, employing the techniques of woodland archaeology that he developed, it can be argued that this kind of dynamism and change was by no means unusual. A number of woods in Norfolk display oddities and anomalies in their archaeology that suggest a similar history of instability and shift. Hook Wood in Morley St Peter, for example, covers an area of 3.1 ha, as it did when mapped by the great surveyor Thomas Waterman in 1629 (NRO PD3/108): it was then named Park Wood, a nearby field as Park Meadow, and the outline of a small deer park (covering some 26 ha) can be picked out in the pattern of surrounding field boundaries. The wood is surrounded to south, east and west by a fairly small woodbank, between 4 and 5 m across including ditch; to the north it is bounded by a ruler-straight ditch, without a significant bank, clearly suggesting that it had been truncated in this direction, but before 1629, to judge from the evidence of Waterman's map – on which, interestingly, the field lying immediately

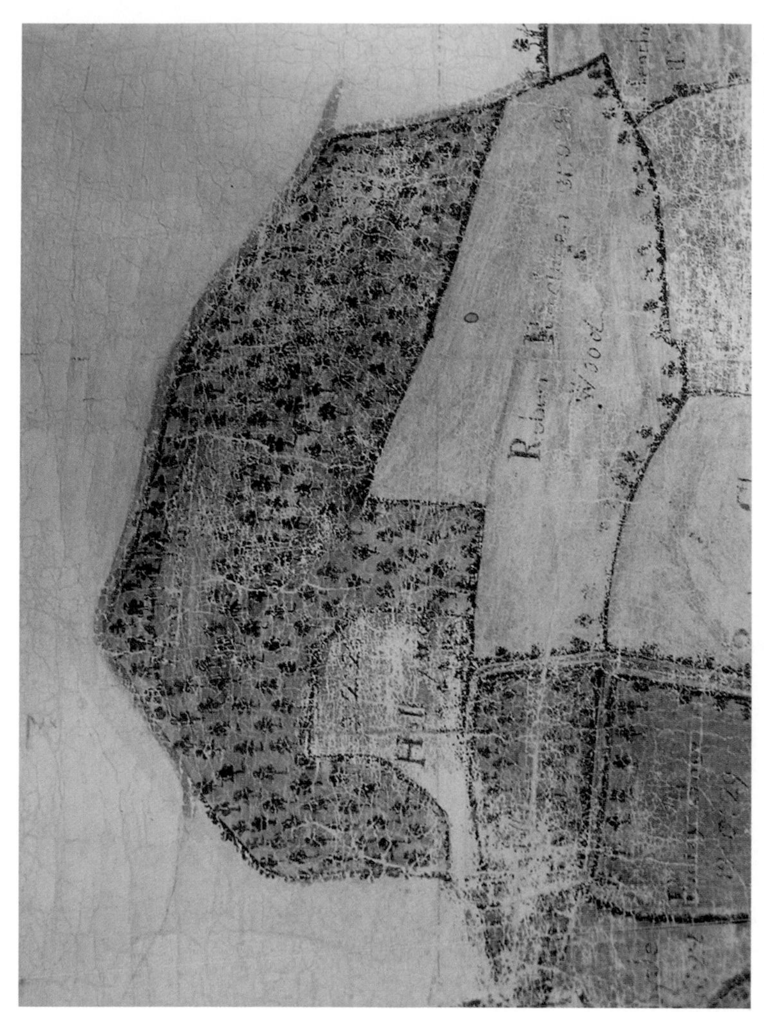

Figure 19.3 Hedenham Wood, south Norfolk, as depicted on Thomas Waterman's survey of 1617. (Courtesy of Norfolk Record Office)

to the north is named Stub'd Wood. The three older boundaries, while clearly earlier, may not be very much older – perhaps of later medieval in date – for the rather slight profiles they display is usually, in East Anglia at least, indicative of woodbanks established in the period after the thirteenth century. In contrast, a much more substantial bank and accompanying ditch, with a total width of more than 8 m, run east–west through the wood a little to the north of its present southern edge, and appear to represent an earlier northern boundary. Evidently, the boundaries of the wood had been changed – expanded to the north, truncated to the south, east and west – before a further truncation occurred to the north, probably not long before Waterman's map was surveyed in the early seventeenth century. Earsham Wood is in some ways similar. This now occupies an area of *c.* 8 ha and likewise has a massive internal bank with ditch running east–west, this time on a rather irregular course, through its centre. The fact that the current boundaries are for the most part marked by slight ditches, accompanied by at best diminutive banks, suggests that the central earthwork originally formed the northern boundary of a wood that mainly lay to the south, and was subsequently truncated in this direction and – perhaps at the same time, perhaps at a later date – extended to the north. Here the map evidence reveals further changes (Barnes and Williamson 2015: 176–82). The wood is shown with its current boundaries on an estate map of *c.* 1770 (NRO 631/2); but

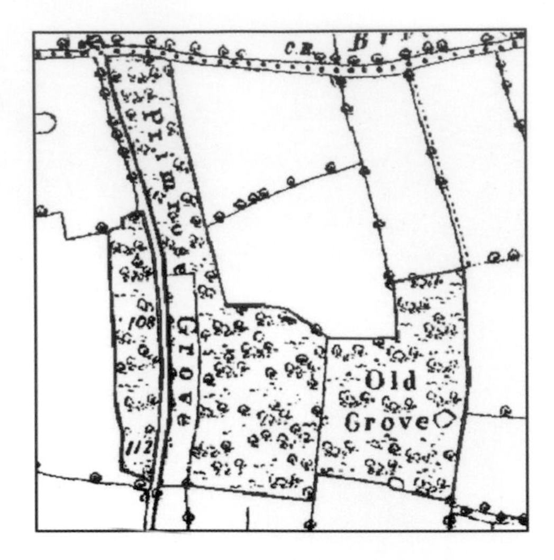

Figure 19.4 Old Grove and Primrose Grove, Gillingham, Norfolk. Right: the woods as shown on the first edition Ordnance Survey 6-inch map: 1 mile (1:10,560) map of 1884, with the Ancient Woodland Inventory area superimposed. Left: as depicted on the Tithe Award Map of 1840. (Tom Williamson, courtesy Norfolk Record Office)

when an earlier map was surveyed, in *c.* 1720 (NRO 631/1) it was significantly smaller than today, and did not extend so far to the east.

What I have written so far applies to what may be called genuine ancient woods – that is, areas of semi-natural woodland, originally managed as coppice-with-standards, which are more than 400 years old. But the official Ancient Woodland Inventory also includes a significant number of what might be called pseudo-ancient woods, a type not really recognized by Rackham or, indeed, by the wider conservation community. These are woods that display most of the characteristics of accepted 'ancient' woodland, but were planted at some point after *c.* 1600. Rackham tended, in his earlier works at least, to assume that 'traditional' management by coppicing declined in lowland England from the seventeenth century, and that when large landowners undertook afforestation schemes during the last two centuries or so they generally established plantations of forest trees, without a coppiced understorey. In fact, there are a significant number of exceptions to this, in Norfolk at least. Coppices continued to be newly planted well into the nineteenth century, in part because the market for good quality underwood held up well but partly, perhaps, because coppices made good game cover. Such woods are easily mistaken for genuine ancient woodland, especially where they are planted within fields bounded by massive, ancient hedgerows, giving the impression of a woodbank. Recent research now suggests that between a sixth and a fifth of Inventory woods in Norfolk fall into this pseudo-ancient category. Some examples are surprisingly recent in date, in spite of their appearance. The conjoined woods of Primrose Grove and Old Grove in the parish of Gillingham in the south of the county, with their coppices of European Ash *Fraxinus excelsior*, Common Hazel *Corylus avellana* and Common Hornbeam *Carpinus betulus*, were planted sometime after 1840: the tithe map for Gillingham, surveyed in that year, shows that the area they occupied then comprised three hedged fields (Stone and Williamson 2013; Barnes and Williamson 2015: 122–133, 218–219; Williamson, Pillatt and Barnes 2017: 158–160) (Fig. 19.4).

The origins of primary woodland

Ancient woodland in Norfolk is thus something of a mixed bag, including (as Rackham was well aware) much that was secondary, even if a majority – by area at least – is of primary character. Individual woods not only often contain portions of both types but also can contain areas that first became wooded at very different dates. The Inventory area of America Grove in Earsham, for example, contains what may be a primary core together with sizeable additions of probable seventeenth-century, eighteenth-century and even nineteenth-century date (Barnes and Williamson 2015: 160–161). Yet even purely primary woods may have more complex origins than is sometimes assumed, and an often tenuous relationship with the 'natural' post-glacial vegetation of the county. As Rackham was aware, most if not all represent fragments, not of untouched wildwood, but of wood-pastures that had been grazed and exploited in a range of other ways for many centuries, before they were enclosed by banks (topped with hedges or dead hedges) to exclude browsing animals, so that they could be managed more intensively as coppice-with-standards. Even at the time of Domesday, most woodland in Norfolk appears to have been grazed and uncoppiced, to judge from the fact that it is universally recorded in terms of the number of swine that could be pastured there. Most ancient woods occupy areas of poorly draining soils – a minority are found on acid sands and gravels – and their enclosure and definition went hand in hand with a process of settlement expansion and resource allocation that occurred in the county from the eleventh to the late thirteenth century, and which demonstrably created the wider matrix of commons, farmland and hamlets of which they form a part. Before parliamentary enclosure in the nineteenth century, a very high proportion of ancient woods abutted directly on common land: both developed from the same areas of wood-pasture 'waste', with the heath or green usually losing much of its tree cover over time, while the wood – invariably private property, part of the manorial demesne – retained its trees. Only occasionally do we have direct references to the act of enclosure, as at West Bradenham where, as Rackham noted, a document for 1226 records how the lord of the manor had 'about the wood ... raised one earthwork for the livestock, lest they eat up the younger wood' (Rackham 1986: 168). Enclosing woodland from the wider wastes was an act of lordly expropriation, and the woodbanks may have been intended as a barrier not just to straying livestock but also to cattle and sheep intentionally introduced by local farmers who had formerly grazed the area in question. Indeed, this may explain why early medieval examples – but not those dating from the late medieval period – are so massive. After all, in the post-medieval period, woods and plantations were protected from livestock by no more than a hedge and ditch, of the kind that served to keep grazing animals within individual fields.

One aspect of this early medieval process of enclosure and definition – not confined to East Anglia, yet not previously much discussed – is the tendency for ancient woods to occur, not anywhere within a soil type or terrain with which they are closely associated, but towards its margins. In south Norfolk, it is particularly noticeable how ancient woods are mainly located towards the edge of the boulder clay plateau that dominates the landscape, and are rarer towards its centre. Where the plateau is level and continuous, ancient woods are rare, as on the tablelands to the north of Diss; where it is dissected, they are often numerous, as in the area around Hedenham (Fig. 19.5). This pattern – the 'doughnut of woodland' – was noted by Witney in the Weald of Kent, and is clear in many districts, such as south Hertfordshire (Witney 1998; Rowe and Williamson 2013: 127–30). Witney explained it largely in terms of practical economic factors. The twelfth and thirteenth centuries saw a steady decline in the economic importance of the Wealden woods as swine pastures and grazing grounds, and a concomitant increase in the demand for, and thus in the value of, wood and timber. Areas of woodland were enclosed and more intensively managed in places where their products could be transported to markets with relative ease. In the 'central core of the Weald ... heavy

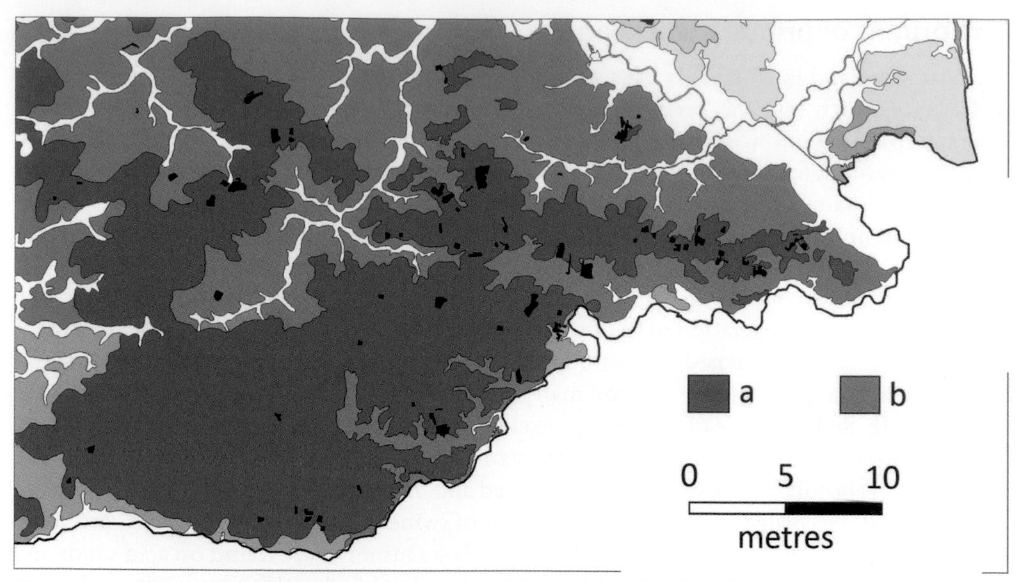

Figure 19.5 The relationship of ancient woodland (in black) to the clay plateau in south Norfolk. As elsewhere in England, woods tend to cluster on the margins of an agriculturally challenging soil type with which they are associated. (T. Williamson)

a – Beccles Association soils (heavy plateau clays).

b – Burlingham and Hanslope soils (lighter clays on valley sides).

loads were almost undisposable' because of the difficulties involved in moving laden carts in wintertime along difficult clay roads (Witney 1998: 20). Colonization was thus directed into the more remote internal districts, where the swine-woods were progressively felled and turned to farmland, or else degenerated to open commons: woods survived mainly on the periphery. Similar factors probably explain the phenomenon in Norfolk. As the population expanded through the eleventh, twelfth and thirteenth centuries, and settlement and cultivation spread onto the more challenging soils, portions of the remaining areas of woodland were enclosed and brought into more intensive management in places close to existing centres of population, and in particular manorial residence, located on the more fertile and tractable soils. A complex interplay of social, economic and environmental factors – of history and natural history – thus appears to have structured the distribution of ancient primary woods in Norfolk.

There is still a tendency among some researchers to think of ancient woods as 'islands', surviving from once more continuous stands of primeval forest. But most primary ancient woods developed in the twelfth and thirteenth centuries from grazed woodland, probably in some places quite thinly scattered with trees. Indeed, a few examples developed from wood-pastures at a much later date, in the fifteenth or sixteenth centuries, as early medieval deer parks – private wood-pastures – were converted to coppices. It is a striking fact that the four largest ancient woods in the county, each extending over an area of more than 0.75 km^2 – Foxley, Hockering, Haveringland Great Wood and Horsford – all appear to have originated in this manner. Whenever they first emerged, the fact that enclosed woods developed not directly from the primeval vegetation but from woodland pastures raises obvious questions about the character of their botanical composition, and about how far the particular combination of species found in ancient woods reflects, or is directly descended from, that found in the wildwood. Vera's arguments for a pre-Neolithic vegetation more closely resembling savanna than closed-canopy woodland remain contested (Vera 2002; Hodder et al. 2009; Yalden 2013). But even if we reject such revisionist arguments, centuries or millennia of grazing by domestic livestock, and other forms of exploitation, must have drastically modified the character of the 'wastes' from which, in early medieval times, Norfolk's primary woods were enclosed. The most obvious sign of this, in Norfolk as in many other areas, is the almost complete loss of Small-leaved Lime *Tilia cordata*, the most common tree in the early prehistoric landscape but found in only a handful of ancient woods in the county.

Enclosure and more intensive management must have led to further botanical divergence. The 'ancient woodland indicator species' – herbs such as Water Avens *Geum rivale* and Pignut *Conopodium majus*, Herb-paris *Paris quadrifolia*, Bluebell *Hycanthoides non-scripta*, and Dog's Mercury *Mercurialis perennis* – cannot survive grazing pressure well (Colebourne 1989: 74; Mitchell et al. 1990; Dolman et al. 2010). They are poorly represented in wood-pastures today, and were thus presumably also relatively rare in the grazed woodlands that predated the development of coppiced woodland in early medieval times (Rotherham 2012). It was the exclusion or restriction of grazing, together with the recurrent cycles of light and shade engendered by coppice management, which made them typical plants of ancient woodland. In much of Norfolk, as in many other areas of early enclosure, similar factors ensured that they became widely established in hedges, from which they can spread with some alacrity into secondary woods. Indeed, they can be abundant in many 'pseudo-ancient' woods, making their identification as such problematic (Fig. 19.6). Old Grove and Primrose Grove, for example, mentioned earlier, although only planted in the nineteenth century, nevertheless boast a ground flora that includes, in addition to much Dog's Mercury and Common Primrose *Primula vulgaris*, Early Dog-violet *Viola*

Figure 19.6 Lopham Grove, North Lopham, Norfolk was planted in the eighteenth century but contains mixed ash, maple and hornbeam coppice and a ground flora that includes 'ancient woodland indicator species' such as Wood Spurge *Euphorbia amygdaloides*, Wood Sedge *Carex sylvatica*, Primrose, Bluebells and Early-purple Orchid *Orchis mascula*, amid large quantities of Dog's Mercury, 2013. (T. Williamson)

reichenbachiana, Hard Shield-fern *Polystichum aculeatum* and Wood Speedwell *Veronica montana* (Barnes and Williamson 2015: 218–19).

As Rackham emphasized on a number of occasions, the dominance of oak as a timber or standard tree in most medieval and post-medieval woods – and in many still today – is less a consequence of natural factors than of practical and economic ones. Oak provides better timber than most other species, and was thus planted or encouraged by woodland managers. What he may not have emphasized sufficiently, perhaps, is the extent to which the coppice understorey may also have been modified by centuries of management. In Norfolk, hazel and ash are the most common underwood species, especially in woods on the poorly draining boulder clay soils, where they are often accompanied by Field Maple *Acer campestre*. Oak coppice occurs in a small number of woods, mainly on acid soils in the north of the county, and lime even more occasionally in woods such as Hockering Wood in the centre; Bird Cherry *Prunus padus* is occasionally prominent; but the only other species commonly found in large continuous stands is hornbeam, which is the most common underwood in woods in the south-east of the county, part of a distribution that extends into the adjacent areas of north Suffolk. What is striking is that the most abundant coppice species, in Norfolk as elsewhere, were also the most economically useful, and often occur in surprisingly pure stands, suggesting that less valuable trees and shrubs such as dogwood were for centuries weeded out to make space for those such as hazel or ash. Indeed, some documentary evidence suggests more extensive schemes of restocking. A lease for South Haw Wood in Wood Dalling in Norfolk, drawn up in 1612, bound the lessee to plant Sallows *Salix* spp. in cleared spaces following felling; the tithe files of 1836 describe how there were 35 acres of coppice wood in Buckenham in the same county, 'part of which has been newly planted with hazel' (NRO BUL 2/3, 604X7; TNA/PRO IR 29/5816). This is in line with evidence from elsewhere in England, which suggests that restocking, replanting and modification of the understorey may have been more widespread in the past than we often assume. Vancouver in 1810 noted how, in Hampshire, some of the best ash shoots were retained when the coppice was felled, and plashed 'in the vacant spaces' to form new plants; a similar practice is recorded in Surrey woods in 1809 (Stevenson 1809: 127; Vancouver 1810: 297). Boys in 1805 suggested that many of the coppices in Kent were regularly supplemented with new plants simply because 'wood, like everything else, decays and produces fewer poles every fall, unless they are replenished' (Boys 1805: 144). One Herefordshire landowner described in 1852 how 'the wood after successive fallages deteriorates as numbers of the old stools die and unless there is a considerable amount laid out in filling up the vacant places with young wood, ditching, etc. a quantity of useless stuff such as birch … and brambles grow up and consequently reduces the value of the wood' (HA]R [Herefordshire Archives and Records, Hereford A63/111/56/12).

It is in this context that we might consider the dominance of hornbeam in many south Norfolk woods. Pollen records, even those from Diss Mere in the heart of this district, suggest that hornbeam was a relatively rare species before the Middle Ages (Diss Mere lies no more than 3 km from Billingford Wood and Thorpe Wood, in both of which the underwood is today dominated by hornbeam (Peglar, Fritz and Birks 1989)). Whether entirely the consequence of natural expansion, or partly at least of deliberate encouragement, it is clear that Norfolk's hornbeam-dominated woodland has little direct connection with the natural vegetation. It is striking how the two main concentrations of hornbeam woods in England – in south Norfolk/north Suffolk, and south Hertfordshire and south Essex – are close to the country's two largest medieval cities, Norwich and London, both hungry for firewood and for charcoal. In Essex and Hertfordshire, similarly, hornbeam appears to have been a minor component of the vegetation in prehistoric times, and only gradually became important, as both a woodland coppice and a wood-pasture tree, in the course of the medieval period. But what is also striking is the manner in which, as coppicing has declined,

Figure 19.7 Neglected coppice of ash and hornbeam in Wayland Wood, Norfolk, 2014. (T. Williamson)

the proportion of hornbeam in originally more mixed woods has increased significantly, as outgrown stools have gradually shaded out other species (Fig. 19.7). Rackham's plot of the plant communities within Hedenham Wood, made in 1975, appears to show a much more diverse coppice than exists today (Rackham 1976: 174). Some of the areas which he mapped as 'mixed hazel' now seem to contain pure stands of hornbeam – another clear indication of how 'ancient woodland' is not some stable, unchanging habitat, but is instead in a state of constant change and flux.

Conclusion

It is impossible to overstate the importance of Oliver Rackham's contribution to the study of ancient, semi-natural woodland, in Norfolk as elsewhere. His ability to work with equal ease in the fields of history and ecology produced new insights and an understanding of woodland history that remains largely unchallenged. To a limited extent his ideas – especially concerning the antiquity and stability of ancient woods – have now begun to be modified and challenged by more dynamic models; but this has only happened because of the systematic application of methods and approaches that he himself pioneered.

References

Barnes, G. and Williamson, T. (2015) *Rethinking Ancient Woodland*. University of Hertfordshire Press, Hatfield.

Boys, J. (1805) *General View of the Agriculture of the County of Kent*, 2nd edition. B. McMillan, London.

Dolman, P. Fuller, R., Gill, R., Hooton, D. and Tabor, R. (2010) Escalating ecological impact of deer in lowland woodland. *British Wildlife*, 21, pp. 242–54.

Hodder, K.H., Buckland, P.C., Kirby, K.J. and Bullock, J.M. (2009) Can the pre-Neolithic provide suitable models for re-wilding the landscape in Britain? *British Wildlife*, 20(5) [special supplement], pp. 4–14.

Mitchell, F.G.J. and Kirby, K.J. (1990). The impact of large herbivores on the conservation of semi-natural woods in the British uplands. *Forestry*, 63, pp. 333–53. https://doi.org/10.1093/forestry/63.4.333

Peglar, S., Fritz, S. and Birks, H. (1989) Vegetation and land use history at Diss, Norfolk. *Journal of Ecology*, 77, pp. 203–22. https://doi.org/10.2307/2260925

Rackham, O. (1976) *Trees and Woodland in the British Landscape*. J.M. Dent and Sons, London.

Rackham, O. (1980) *Ancient Woodland: Its History, Vegetation and Uses in England*. Edward Arnold, London.

Rackham, O. (1986) The ancient woods of Norfolk. *Transactions of the Norfolk and Norwich Naturalists Society*, 27, pp. 161–67.

Rackham, O. (2006) *Woodlands* (New Naturalist 100). Collins, London.

Rotherham, I.D. (2012) Searching for shadows and ghosts. In: I.D. Rotherham, C. Handley, M. Agnoletti and T. Samojlik (eds) *Trees Beyond the Wood: An Exploration of Concepts of Woods, Forests and Trees*. Wildtrack Publishing, Sheffield, pp. 1–16.

Rowe, A. and Williamson, T. (2013) *The Origins of Hertfordshire*. University of Hertfordshire Press, Hatfield.

Stevenson, W. (1809) *General View of the Agriculture of the County of Surrey*. B. McMillan, London.

Stone, A. and Williamson, T. (2013) 'Pseudo-ancient woodland' and the *Ancient Woodland Inventory*. *Landscapes*, 14, pp. 141–54. https://doi.org/10.1179/1466203513Z.00000000016

Vancouver, C. (1810) *General View of the Agriculture of Hampshire*. Richard Phillips, London.

Vera, F. (2002) *Grazing Ecology and Forest History*. CABI, Wallingford.

Williamson, T., Pillatt, T. and Barnes, G. (2017) *Trees in England: Management and Disease Since 1600*. University of Hertfordshire Press, Hatfield.

Witney, K.P. (1998) The Woodland Economy of Kent, 1066–1348. *Agricultural History Review*, 38, pp. 20–39. https://doi.org/10.1080/07075332.1998.9640848

Yalden, D. (2013) The post-glacial history of grazing animals in Europe. In I.D. Rotherham (ed.) *Trees, Forested Landscapes and Grazing Animals: a European perspective on woodlands and grazed treescapes*. Earthscan, London, pp. 62–69.

Part V

Wider Perspectives

CHAPTER 20

Reflections from the Antipodes

Paul Adam

Summary

Through his writing, together with networks of colleagues and former students, and facilitated via collaborations worldwide, Oliver Rackham's influence was global. Furthermore, his infectious enthusiasm for his subject, and particularly for all aspects of trees, led to expeditions around the world – including, as this chapter explains, to Australia. The research here demonstrates how the application of Rackham's approaches to environmental history are transferable to the Australian situation, and help to provide a key exchange of insights from Europe to Australia and vice versa.

Keywords: Acacia, bush, mangroves, saltmarsh, soap manufacture, wattle and daub

Introduction

I went up to Cambridge in autumn 1969 to read Natural Sciences with the intention of taking Part II Botany, with a particular interest in plant ecology.

Arriving in Cambridge, new students were faced with numerous clubs and societies seeking new members. I joined what was then called the Conservation Corps – not the national body of that name, but associated with it, which carried out management of Cambridge shire and Isle of Ely Naturalists' Trust reserves, and also at other conservation sites within easy reach of Cambridge. Thus it was that I found myself one October Sunday morning outside 1 Brookside, in the basement of which the Conservation Corps kept its tools. It was there that I first became aware of the vast range of tools for the management of woodland and hedgerows, and was taught, among many things, the difference between a Yorkshire and a Suffolk billhook. (The First Fleet, arriving in Australia in 1788, included in its cargo Kent axes, not a county variant I encountered in the Conservation Corps, apparently of such low quality as to be inadequate for felling eucalypts (Brown 2019).) Tools, and the essential lunchtime refreshment of flagons of Tolly Cobbold's bitter and very rough cider – Occupational Health & Safety had not been invented at this time – were loaded into cars and we set off to Hayley Wood. There, for the first time, I encountered Oliver Rackham – in Norfolk jacket, battered hat and with spectacles that had seen better days, held together with Elastoplast, and holding what was clearly a very sharp axe. Oliver proceeded to lead us into the wood to that year's coppice plot.

Over the next nine years, there were many Sundays spent in Hayley Wood, not only coppicing but also hedge laying. Hayley was one of a number of woods in which the Conservation

Paul Adam, 'Reflections from the Antipodes' in: *Countryside History: The Life and Legacy of Oliver Rackham*. Pelagic Publishing (2024). © Paul Adam. DOI: 10.53061/XDRB9807

Corps worked. We also worked at clearing scrub from chalk grassland, and less commonly we visited fens. On one notable occasion at Chippenham Fen National Nature Reserve (NNR), the *Phragmites* caught fire, and I learnt that Wellington boot-depth water was no barrier to a good blaze!

As I advanced to postgraduate student and then research fellow, it was necessary for me to have a vehicle for fieldwork – an elderly Morris Traveller with appropriately half-timbered bodywork. Quite frequently, I transported Oliver to and from Conservation Corps worksites. During these trips I came to appreciate Oliver's vast knowledge about many subjects, and his frequently self-deprecating sense of humour accompanied by a characteristic chortle. His sense of humour also extended to his publications: in Rackham (2002) there are several examples – who but Oliver would have come up with the expression 'frog poking' (p. 28) to describe the origin of certain marks on the surface of silver (shaped similarly to, although very much smaller than, the marks referred to in Sydney as sparrow picking, or convict pick finish, the hand-etched pitting on the surface of stonework from the convict era), included a photograph of himself demonstrating how not to drink from an aurochs' horn (p. 45), or so lightly dismissed his own mathematical ability in the caption to an image of two Cambridge wooden spoons (p. 289). In Rackham (2006), he came up with the memorable phrase 'English native woods burn like wet asbestos' (p. 56).

The Rackham archive contains a considerable number of transparencies from Australia, and there are many herbarium specimens collected by Oliver when he visited Australia in 1996 as an invited speaker at the Third National Conference of the Australian Forest History Society, held at Jervis Bay in November 1996 (Dargarvel 1997). Various colleagues arranged field trips for Oliver, both in eastern Australia and the west (Rackham 1997). I was fortunate to take Oliver to Dubbo in western New South Wales. My purpose in venturing out into the wheat belt was to visit Dundullimal, believed to be the oldest surviving wooden slab construction house in Australia (Fig. 20.1).

Oliver was impressed by the mighty River Red Gums *Eucalyptus camaldulensis* along the Macquarie River. This has the widest natural distribution of any eucalypt species, being found across the mainland west of the Great Dividing Range, extending into the upper Hunter Valley on the east of the range. The species contains a number of sub-specific taxa, none sufficiently different on currently available information to warrant elevation to specific rank. It is most often associated with water, either permanent water (or as near permanent as is possible in inland Australia) as at Dubbo or on the extensive floodplain of the Murray River, or is associated with intermittent waterways, where River Red Gum flourishes on the edges of, or even in, the beds of, creeks where there is access to permanent water deep in the sand – as on the Todd River in Alice Springs in Central Australia. The occurrence of such large trees in areas of low rainfall is striking (the total distribution spans sites with average annual rainfall of 150–1,200 mm). At Geraldton in Western Australia, close to the Indian Ocean, the trees are exposed to almost constant onshore winds and have low, wind-pruned canopies.

Oliver questioned me as to the reason why the specific epithet appeared to indicate an Italian origin. But at the time, having never thought about the matter, I could no more than agree with him that it seemed strange. Today, an answer would give rise to even more questions, following investigation of the type specimen by Brooker (Brooker and Orchard 2008).

The River Red Gum was described in 1832 by Friedrich Dehnhardt, a German botanist and horticulturalist who was employed at the country estate of Francisco Ricciardi, Count of Camalduli, which was near Naples. The tree from which the type specimen was collected was over 12 m tall and more than ten years old. Zacharin (1978) reports that at least three *Eucalyptus* species were growing at Hortus Camalduli at the time, and that they had been grown from seed that might have been received from France. If this is correct, then there had been two French expeditions that could have made the original collection in the wild, those of d'Entrecasteaux (1791–4) and Baudin (1800–4) – but there is a problem in that neither expedition

Figure 20.1 Dundullimal Homestead, the oldest surviving wooden slab building in Australia. Dubbo, New South Wales, April 2021: (top) showing the slabs of Yellow Box *Eucalyptus melliodora* forming the walls; (bottom) The roof would have been originally made of wooden shingles, but is now metal – a very common roof material for Australian buildings. The design of the homestead shows some features influenced by colonial experience in India, and internally has very effective cross ventilation. (J. Green)

visited any localities at which the River Red Gum as currently understood occurs (Brooker and Orchard 2008; McDonald et al. 2009). Thus, the identity and provenance of the specimen in the herbarium in Vienna and named by Dehnhardt is unknown. Given the vast literature referring to River Red Gum and the ecological and economic importance of the species, both within Australia and as an introduced species overseas, the potential uncertainty created by the doubt as to the true identity of the type was considerable. Brooker and Orchard (2008) proposed that the situation could be resolved by conserving the name with a new type specimen. The chosen new type specimen was from Currency Creek, near the mouth of the Murray River in South Australia. A photograph of the new holotype appears as Figure 2.2 in Colloff (2014).

From Dubbo we ventured further out into the wheat belt through Narromine and toward, but not as far as, Nevertire. There were several features along this stretch of road that aroused Oliver's interest. He was very taken by the number and size of the pubs in what are now relatively small communities, a reflection of the size of the population required to provide services to farming properties and of the workforce on farms prior to the mechanical era. He was struck by the very wide main streets in the small towns – necessary to provide the turning circle for a team of oxen. Oxen provided the major form of heavy goods transport prior to the advent of motor vehicles. In many pubs and local historical society museums, there are collections of photographs of teams of oxen hauling massive logs on jinkers or drays laden with wool bales or mixed goods. The teams were controlled by bullockies, about whom there is much myth and legend, particularly about their rich vocabulary – 'to swear like a bullocky' is a description still applied today.

Alongside the roads between settlements was the Long Paddock, the name given to travelling stock routes (TSRs). These are public lands available for droving cattle and sometimes sheep, particularly in times of drought, when properties may be destocked because of lack of food and the herd may be on the road for long periods of time, sometimes many months, moving a few kilometres a day. Today, most movement of stock will be in cattle trucks, but even now there are still some landholders who rely on the availability of grazing in TSRs during drought times.

The TSRs provide important corridors of vegetation of high conservation value in largely cleared landscapes. At the time of my trip with Oliver, conditions were good and there was no stock utilizing the TSRs; instead there was good growth of the native Kangaroo Grass *Themeda australis* (sometimes known as *T. triandra*) in full flower, and considerable diversity of native herbaceous and small shrub species under an open eucalypt woodland canopy. In places there were Cypress Pines *Callitris* spp., native conifers (once an important source of termite resistant floorboards used in many houses in Sydney). Our progress was slow with numerous stops to allow study of the range of vegetation. Oliver's concept of time was flexible if there was something interesting to observe.

Saltmarsh

The focus of my research at Cambridge was saltmarsh phytosociology (Adam 1981), and I visited many sites around the British coast; I have maintained an active interest in saltmarshes not only in Britain but also around the world. Saltmarsh is not a habitat that one immediately associates with Oliver, and it received only brief mention in the final chapter of Rackham (1986), which nevertheless contains a number of insightful comments. Oliver proposed that 'The history of wetland is very largely the history of its destruction' (Rackham 1986: 375), and this is true globally. Wetlands have had a bad press for generations, and even today, while there are international treaties ostensibly offering them protection, there is still a large element in the population that regards wetlands as dangerous, unhealthy places, and holds that society would benefit from their reclamation, an insidious euphemism for destruction. 'Drain the swamp' was, after all, President Trump's mantra. 'In wetlands more than in most places, the improver has not known when to stop' (Rackham 1986: 391). In the context of the rate of loss of other wetland types, Oliver considered that: 'Least affected of all are salt-marshes. Innings from the land and rising sea-level have squeezed them to a fraction of their former extent, but they are still beautiful, have a wide range of vegetation and are internationally important for birds' (Rackham 1986: 392).

Rackham (1986) discussed examples of historic coastal erosion and the loss of saltmarshes from embanking, and recognised the possible existential threat posed by rising relative sea-level. Oliver was at pains to point out that sea-level rise per se was not the threat, and identified areas around the British coast where either subsidence or uplift of the land was occurring. In some locations, accretion may enable saltmarsh to keep pace with rising relative sea-level, although globally there are many saltmarshes where human intervention has reduced sediment supply and the ability to accrete has been reduced.

I am less optimistic than Oliver was about the future prospects for saltmarsh (Adam 2002). Before the full effects of rising relative sea-level are experienced, other factors causing loss and degradation will continue almost unabated. Despite legislation in many countries conferring high conservation status on saltmarshes, most governments provide themselves with get-out clauses in the event of issues of national interest arising (what falls into this category is rarely defined, which gives governments a great deal of scope). Globally, human population increases are greatest in the coastal zone and this brings demands for land for housing, infrastructure and industrial development. Some infrastructure, such as new or expanded ports, must necessarily occur on the coast. There is no such requirement for the siting of airports, but the strong opposition to development of inland sites has led to political pressure for the construction of new airports in estuaries.

One of my earliest publications was on the environmental history of the saltmarshes around Morecambe Bay (Gray and Adam 1974). The biggest impacts by humans on the saltmarshes occurred in the mid-nineteenth century, with the construction of the railway around the Bay. Duck (2015) has documented the extent to which Victorian railway building altered the entire British coastline. The railways along so much of the coast are a familiar feature, but their geomorphological and ecological impacts have not been widely appreciated. There is great scope for historical studies of saltmarshes and the uses humans have made of them. Some uses by humans appear to have been geographically restricted, such as the winning of clay for use in cement manufacture in the Medway, and the turf cutting industry in north-west England producing the finest turf for bowling greens (Gray 1972). When and why did these practices commence, and what was the impact on the ecology of the saltmarshes affected?

A widespread industry, over long periods, was salt production, as Rackham (1986) acknowledged. Rackham (1986; 2002) considered medieval England to be 'salt addicted'. Salt production extends back thousands of years, and before the age of refrigeration, salt was the main method for long-term preservation of meat. Although the culinary and preservative uses of salt remain important (and of concern to the medical profession), the majority of salt produced today serves as feedstock for a number of industrial chemical processes. There are two methods of salt production conducted in saltmarshes – evaporation from solar ponds and boiling seawater in large pans. The two methods are not mutually exclusive. In the British Isles, where conditions are rarely conducive for complete solar evaporation, at some locations initial concentration in ponds is followed by boiling in pans. Creation of new solar ponds is a potential threat to saltmarshes on arid coasts, where salt production occurs on a large scale. On the other hand, abandoned salt production ponds are important conservation sites.

Globally, by far the most extensive use of saltmarshes has been agricultural – for grazing and haymaking, rush cutting and, in brackish sites, reed cutting. Grazing has considerable impact on the composition of saltmarsh vegetation (Adam 1978). Cessation of grazing does not necessarily result in a return to the ungrazed state. The agricultural history of saltmarshes is not well documented, although there may be many local sources yet to be investigated.

However, their grazing draws attention to the fact that saltmarshes in northern Europe for example, would have been among the few extensive lowland grasslands after the post-glacial forest expansion and before forest clearance. Did those saltmarshes provide a refuge for large grazing animals (Levin et al. 2002)? Forest and saltmarsh would have been adjacent to each other, so large mammals would have been able to move between the two habitats. Did this provide circumstances in which the Vera (2000) hypothesis could have applied?

There are still many sites around the British coast where woodlands occur adjacent to saltmarshes, with the tree canopy overhanging the uppermost saltmarsh. Rackham (1994) included, as Walk V, a route from the Lizard plateau to the Helford River. There is a narrow fringe of saltmarsh along the shore of the river overhung by ancient woodland. Other notable examples occur in the Beaulieu River, adjacent to Chichester Harbour, along the Stour, Roudsea Wood NNR in the Leven estuary and a considerable number of sites in western Scotland.

In southern England, the flora of the woodland/saltmarsh zone is characterized by a number of species. Under the woodland canopy, characteristic species are Stinking Iris *Iris foetidissima*, Butcher's Broom *Ruscus aculeatus* and Wild Madder *Rubia peregrina*. In the upper saltmarsh under the edge of the tree canopy Marsh Mallow *Althaea officinalis* and Divided Sedge *Carex divisa* may occur. None of these species is restricted to this habitat, but nevertheless they occur together sufficiently frequently as to form a recognizable assemblage, although not one formally recognized in the National Vegetation Classification (NVC). This is not to criticize the NVC, but adds to the discussion in Rackham (2003) on the limitations of a broad classification when applied at a local scale.

Mangroves

Subtropical and temperate Australia has an extensive coastline on which both mangroves and saltmarshes occur (Fig. 20.2). There are also areas of saltmarsh in the tropics despite the prevailing myth that saltmarsh is absent from low latitudes.

Encountering mangroves was an entirely new experience for me, as it would have been for the majority of the first European colonists. Nevertheless, in the early nineteenth century, mangroves around Sydney were heavily exploited. The reason for this is a curiosity of early colonial history.

As part of light-hearted banter, Australians make disparaging comments about Poms.[1] Among these remarks are some about personal hygiene (and the lack thereof). These arose, at least in part, from the 1960s–1970s rites of passage when many young Australians travelled to London and were unimpressed by the facilities in their cheap rental accommodation. It comes, therefore, as a surprise to learn that one of the commodities much sought after by the early colonists was soap. The history of the early soap manufacturing industry in Australia is discussed by Bird (1978, 1981), from whom these details are taken.

At its simplest, soap manufacture requires fat and an alkali. The manufacture of soap in Australia may have started as early as 1796 and was certainly established by 1805. Although this was less than 20 years after the arrival of the First Fleet, cattle and sheep numbers had grown sufficiently to assure ready availability of tallow. There was a growing market for soap: convicts received a weekly ration of half a pound of soap, and presumably soldiers, bureaucrats and free settlers used at least the same amount. As Australia became a major wool-producing and exporting nation, the demand for soap would have increased as wool scours became more numerous.

---

1 *The Australian National Dictionary* (*AND*) (Moore 2016: 1184) defines Pom as: 'An immigrant from the British Isles; applied more recently to an inhabitant of the British Isles (esp. in England)'. The etymology is contested, but *AND* gives it as an abbreviation of pomegranate and dismisses suggested alternatives as fanciful.

Figure 20.2 Saltmarsh and mangroves: (top) Saltmarsh, dominated by the succulent chenopod *Sarcocornia quinqueflora*, adjacent to mangroves (*Avicennia marina*). Sydney Olympic Park, Homebush Bay, Sydney, 29 August 2020; (bottom) Mangroves (*Avicennia marina*) at Kurnell, Botany Bay New South Wales. Note the pneumatophores (upright breathing roots) showing how far from the canopy the root system extends, 1 September 2016. (P. Adam)

In England, in the late eighteenth century, soap was made by mixing tallow with alkali derived from plant ash. The source of the ash was seaweed, from Scotland and Ireland, and barilla, imported from Spain. Southern Australian coasts support a diverse algal flora, but seaweeds do not appear to have been a major source of alkali for soap-making. However, harvesting of Southern Kelp *Durvillea potatorum* – one of the world's largest brown algae – has occurred in Tasmania for many years. The algae after harvesting are dried, milled and shipped to Scotland for extraction of alginates. If algae were not the source of alkali for soap manufacturing is there an Australian substitute for barilla? In Europe, barilla production in the late eighteenth and early nineteenth centuries was a Spanish monopoly and involved the burning of coastal plants. Opposite-leaved Saltwort *Salsola soda* was the main species involved, although a number of other coastal halophytes may also have been utilized.

In Sydney in 1812, it was decided to use the Grey Mangrove *Avicennia marina* as the source of ash. (The term 'mangrove' is applied to individual trees, to species and to the community comprising mangrove trees. The suggestion by Macnae (1968) that mangal be adopted as the name of the community has not found favour.) Bird (1978) indicates that it is not known why the Grey Mangrove was burnt as the alkali content of the ash is not particularly high and the various succulent saltmarsh chenopods (now in the genera *Sarcocornia* and *Tecticornia*) would have provided a higher yield. It does not appear that cutting and burning mangroves for alkali production was practised anywhere else in the world.

However, there were 16 soap factories in Australia by 1847, including 10 in Sydney. Mangrove cutting to provide barilla peaked in the 1840s and declined soon afterwards as the new chemical industries started to produce better and cheaper products, without the need for all the hard work involved in barilla production. It is likely that most mangroves within several hundred kilometres of Sydney were cut down during the barilla harvesting period. Mangroves today are extensive and still spreading in this region. Many of the larger trees present today are multistemmed and have regrown by coppicing. Not all the multistemmed trees represent post-harvest coppicing, as similar regrowth can occur after other forms of damage. Rackham (2003) includes considerable discussion about the origin of coppicing, but does not include mangroves among his examples. Ageing of Grey Mangrove stems from ring counts is not possible because of the anomalous secondary thickening exhibited by this genus (Tomlinson 1986). Mangrove cutting in south-eastern Australia (New South Wales, Victoria and South Australia) involved only Grey Mangrove, but there was also some cutting in Queensland, including in the far north. Here mangrove forests are much taller and more species-rich, so a number of species might have been cut. Tasmania is south of the distribution limit of mangroves, but barilla was produced from a number of species, probably mostly from Shrubby Glasswort *Tecticornia arbuscula*, a succulent growing to about 2 m tall, and, on islands in Bass Strait, Grey Saltbush *Atriplex cinerea*, known locally as barilla bush (Bird 1978).

Other than barilla production, there has been little forestry in Australian mangroves. At various locations there has been collection of mangrove bark for tannin production, while mangroves were convenient locations for the cutting of stakes for use in oyster farms in adjacent waters, and small amounts of timber were taken to satisfy local small-scale needs. Today. the value of mangroves for their role in sustaining fisheries, carbon sequestration and coastline protection is increasingly recognized and any proposals in Australia for exploitation or reclamation of mangroves are likely to be rejected. There is, however, a conservation dilemma as both saltmarsh and mangroves are given legislative protection, but in south-east Australia invasion of saltmarsh by mangroves is a widespread phenomenon (Saintilan, Rogers and McKee 2018). The factors favouring invasion are not fully understood, but the question for managers is whether or not to intervene to reduce the invasion by mangroves or accept the inevitability of saltmarsh loss.

Wattles and fizgigs

> Australia is a miniature planet: its ecosystems work on different principles from the rest of the globe. Most trees (outside rainforest) are species of the vast genus *Eucalyptus*; they cast a very light shade; they are distasteful, except to koala bears;

> and fires do the recycling job that fungi do elsewhere. Most eucalyptus sprout from the base or the trunk or branches; some grow very readily from seed.
>
> Rackham 2003: 435

If asked for the botanical emblem of Australia many people would say '*Eucalyptus*', and many have written that the scent of eucalypts on the air, whether in some foreign land or on return to Australia from overseas, evokes a deep feeling of home. However, the official floral emblem of the Commonwealth of Australia is the Golden Wattle *Acacia pycnantha*, which is also the inspiration for the national sporting colours of green and gold. *Acacia* (*sensu lato*) is, like *Eucalyptus*, a large genus. However, while *Eucalyptus* is not endemic to Australia, the number of species within the genus that are endemic is very high and the number of species naturally occurring outside Australia is small and geographically restricted, occurring on islands to the north of Australia as far north as the Philippines. *Acacia* (*sensu lato*) has a much wider distribution, with many species native to Africa and the Americas.

The first species of what would now be *Acacia* (*sensu lato*) was described by Linnaeus and given the name *Mimosa scorpioides*. The genus *Acacia* was established by Philip Miller in 1754 (although acacia as a vernacular name predates this). Linnaeus's *Mimosa scorpioides* is now recognized as a synonym of Gum Arabic Tree *Acacia nilotica*. When the first specimens of the genus from Australia were named scientifically, they were assigned to *Acacia*. Why is it, then, that when *Acacia* species were first seen by convicts and settlers in Australia they were not called, informally, acacia, but were referred to as wattles? When the First Fleet arrived in Sydney Harbour in 1788, one of the first needs was to provide structures for accommodation. The ships were not laden with building materials; rather the anticipation was that there would be sufficient resources readily available, and sufficient skills among the newly arrived to rapidly construct simple but weatherproof huts. In Britain (and elsewhere in northern Europe), there was a long history of wattle and daub construction, whereby straight rods (wattles) were laid within a frame and then sealed with daub (clay) – in effect being the precursor of modern plasterboard. The term wattle applied to the rod and had no taxonomic connotations. Rackham (2003) contains photographs of a number of species used as wattle.

In 1788, within months of the arrival of the First Fleet, the name Blackwattle Bay was first recorded. (The place-name is consistently spelt as one word, while the plant is Black Wattle.) Blackwattle Bay is an embayment on the south side of Sydney Harbour, a few bays west of Sydney Cove and now within one of the most densely built up parts of Sydney. The Black Wattle that grew there is *Callicoma serratifolia* – not an *Acacia* but the only species in the genus *Callicoma*, a member of the Gondwanan family Cunoniaceae (Fig. 20.3). The species occurs in relatively moist microhabitats and is characteristic of the edge of stands of wet sclerophyll forest and rainforest. The stems are straight and pliant, with smooth dark grey bark. For anyone seeking building material it would be an obvious candidate for 'wattle', and it was soon put to the purpose—the name Black Wattle presumably reflecting the dark grey bark.

The Australian National Dictionary (*AND*) (Moore 2016) gives the origin of the use of wattle as the transference of usage from 'rods or stakes interlaced with twigs or branches of trees for fences and in the construction of buildings (Moore 2016: 1701), and the current meaning as 'any plant of the large Australian plant genus *Acacia* (family Fabaceae) of which there are in Australia nearly 800 described species, widespread elsewhere, esp. in the Southern Hemisphere' (Moore 2016: 1701) The *AND* provides many quotations of the use in Australian literature of the word wattle. The first quotation is from Watkin Tench, an officer in the Marines who arrived in Australia with the First Fleet: '32 houses ... in Sydney built of wattle plastered with clay' (Tench 1793: 78) and 'a few pannels of houses built upon the principle of ancient colonial architecture were washed down ... The crash of decayed posts and wattles was repeated' (*Sydney Gazette*, 20 November 1803).

Figure 20.3 *Callicoma serratifolia* – Black Wattle – (Cunoniaceae). Grounds of CSIRO campus, Black Mountain Canberra Australian Capital Territory, 13 November 2018. (P. Adam)

The first quotation unequivocally linking wattle to *Acacia* is from 1829 when R. Mudie stated: 'Very many species of *Acacia* are found in Australia. Locally they are known by the name of wattles, from the slender twigs being used for that purpose' (Mudie 1829: 138). Much later in 1914, E.E. Pescott wrote: 'The name wattle is purely an Australian one, and has been adapted from the practise of "wattling", or the weaving of the young pliable growth of the shrubs in the early days to make fences and even houses – "wattle and daub" or "wattle and dab" as the method was called' (Pescott 1914: 40). The first church in Australia was built in 1793 by the Rev. Richard Johnson, the chaplain of the First Fleet, at his own expense. This church was of wattle and daub construction: 'That it was built of wattle, not the tree we know by that name, but a sort of Christmas bush, *Callicoma serratifolia*' (Hassall 1902: 145).

As in England, wattle for building in the first days of colonial Australia was supplied by a number of species. They might have included species of *Acacia* – the choice of species seems to have been opportunistic, and wattle and daub was probably not intended to be a long-lasting construction material. It is also clear from the many quotes in *AND* that from the late 1820s, wattle was increasingly used to refer specifically to *Acacia*.

What would have led to the term wattle being applied to *Acacia*? The answer lies in the inflorescences. The globular inflorescences of *Callicoma* are superficially similar to those of acacia, and to those of the Sensitive Plant *Mimosa pudica* – some *Acacia* species grown in Britain in the late eighteenth century were popularly referred to as mimosa. Mimosa was sometimes applied to acacias in Australia and that association is reflected in Mimosa Rocks, on the south coast of New South Wales. *Callicoma* inflorescences are cream coloured, and although *Acacia* is the gold in green and gold, not all *Acacia* spp. have golden yellow inflorescences – the colours range from almost white through various shades of yellow and even to red (as in the Cinnamon Wattle *Acacia leprosa*).

Wattle and daub are unknown to most Australians today, so wattle is linked exclusively to *Acacia*. To those arriving on the First Fleet, wattle took its original English meaning as the timber component of wattle and daub. *Callicoma* was named Black Wattle because its stems were suitable for this use. The similarity of inflorescences between those with *Callicoma* and those of *Acacia* led to acacias being called wattle. It is not unlikely that describing acacias as wattles may have promoted the use of acacia stems in wattle and daub.

Figure 20.4 *Acacia linifolia*. The leaf like structures are not true leaves, but phyllodes. Seedlings have true bipinnate leaves, but over time, leaf development ceases and at maturity all photosynthetic structures are anatomically flattened petioles, referred to as phyllodes. Sydney Olympic Park, Homebush Bay, Sydney, August 2017. (P. Adam)

Fizgigs

An early use of wattle noted by David Collins (1804: 16) in February 1798 does not immediately seem to fit the story. The first mention of fizgigs in writing about Australia appeared in Collins (1798). David Collins arrived on the First Fleet as judge advocate of New South Wales and official secretary to Governor Phillip. In 1788 the Governor was anxious to preserve good relations with the local Aborigines and ordered that those under his command should not misappropriate the spears, fizgigs, gum and other articles belonging to the Aborigines fishing in Port Jackson (the official name for Sydney Harbour).[2]

What is a fizgig and how does it relate to wattle?

Fizgig is a word that in modern dictionaries, if it appears at all, is labelled 'archaic'. It has a number of very different meanings, probably with different etymologies. In its oldest meaning it refers to a frivolous young woman. The second meaning is a type of firework, in particular those that produce showers of coloured sparks rather than explosions. The newest meaning, first recorded in 1870 and uniquely Australian, is a police informer. None of these meanings would appear to have any relationship with wattle. Collins goes on to explain that, in the context of Phillip's orders, a fizgig was an implement with a shaft and barbed points used for catching fish. This usage most probably derives from the Spanish word for a harpoon: *fisga*. Paterson (1811) wrote that men killed fish with their fizgigs but females used hooks and line: 'The fizgig is made of the wattle; has a join in it, fastened by gum; is from fifteen to twenty feet in length and armed with four barbed prongs; barb being a piece of bone secured

2 Thomas (2012) provides an accessible account of the early days of colonization, including references to early writers, including Collins (1804).

with gum' (Paterson 1811: 104). The cited length appears excessive: representations in early artworks suggest something shorter. One of the earliest illustrations appeared in a *Journal of a Voyage to New South Wales* published in 1790 by John White, the surgeon-general who arrived in Sydney Harbour on the First Fleet. The illustration of a fizgig was in plate 63 – which is reproduced in volume 134 of *Signals* (2021: 36), the quarterly publication of the Australian National Maritime Museum.

James Atkinson (1826: 2) referred to Aborigines using the inflorescence stalks of Grass Trees *Xanthorrhoea* for the shafts of spears and 'fish gigs'. The use of fizgig in 1788 was unlikely to encompass use of *Acacia* stems (at least of those species occurring close to Port Jackson) for the shafts of fishing spears. However, the stems of *Callicoma* would have been very appropriate for the purpose.

Curiously, the use of fizgig (in any of its various spellings) for a fish spear is not recorded in *AND*. However, Collins (1804), Paterson (1811) and Atkinson (1826), all writing for an English audience, used the term without any definition or explanation, so presumably it was in relatively common usage in early nineteenth-century England, even if it has vanished from the current language.

Acacia

The genus *Acacia* in Australia contains a range of woody plants from low subshrubs and shrubs in heathland, to generally small or medium-sized trees in forest and woodland (Fig. 20.4). *Acacia* occurs across the entire continent, from sea-level to montane zones, and from the extreme arid zone to wet sclerophyll forest and rainforest.

Wattles in Australia fall into two groups in terms of their foliage: bipinnate wattles with compound leaves and phyllodinous wattles in which the photosynthetic organs are phyllodes – flattened petioles. Phyllodes vary enormously in size between species. On the edge of phyllodes are characteristically one or more extrafloral nectaries. The number, position and size of these provide useful characters to distinguish between species. All wattles after germination produce bipinnate leaves; successive leaves have fewer bipinnate leaves and more flattened petioles, until a stage is reached when only the flattened petiole is produced.

Commercial timber production from acacias is limited, although the Australian Blackwood *Acacia melanoxylon* is important as a source of high-quality timber used for cabinetmaking. This is a species of temperate rainforest and wet sclerophyll forests in south-east Australia, a very different habitat from that of the majority of wattles.

Other species of *Acacia* have been a source of firewood, fence posts and timber for carving and the manufacture of small objects such as bowls or small furniture items, but the most important commercial product has been of bark for tannin extraction. Bark of a number of species has been used, and Australian wattles were established in plantations overseas, particularly in Africa, to establish local tanning industries. A number of Australian wattles grown overseas have become major environmental weeds. Perhaps the largest single source of tannins, and a species regarded as a major environmental weed, is Black Wattle *Acacia mearnsii*, native to south-east Australia but very widely distributed around the globe as an introduced species.

While for the most part acacias are relatively minor, even if sometimes locally abundant, components of the vegetation in the high rainfall zones around the continental margin it is a different story in the arid interior. Although eucalypts are still a significant presence, the major trees are acacias. Many of the landscapes of the outback have been substantially impacted by Europeans, as a result of which today's visitors get a misleading impression of the vegetation.

Rackham (2003) referred to *Eucalyptus* as a vast genus, but while it clearly exceeds the arbitrary cut-off of five hundred species to qualify as a big genus (Frodin 2004) it is a long way from being the largest; *Acacia* is larger than *Eucalyptus* and is itself a long way from being the largest genus in the world. Frodin (2004) suggested that some genera have become large through chaining, where new taxa accrete to existing concepts without any stepping back to see whether revision is required, but also recognized that even after revision, some genera will remain large. Even if the taxonomic nettle is grasped, and good reason is found for splitting, acceptance of new genera may well be opposed by the inherent conservatism of the end-users of taxonomy.

Revisions of generic concepts in both *Acacia* and *Eucalyptus* have been proposed in recent years. In *Eucalyptus*, the recognition of a number of genera within what was *Eucalyptus* has achieved a high degree of acceptance as being biologically appropriate, even if there is some grumbling about the consequent nomenclatural changes. The largest of the genera split from *Eucalyptus* is *Corymbia* (the bloodwoods) with 113 species. This includes a number of widespread species, one of which is probably the most widely planted 'gum tree' in gardens – the Western Australian Flowering Gum *Corymbia ficifolia*. Revision of *Acacia* has been far more controversial, raising substantial issues arising from the differing views of Australian and South African botanists (Carruthers and Robin 2010; Kull and Rangan 2012).

Countryside or bush?

> By 1788 the English had long been used to trees and forests as a settled permanent part of the cultural landscape (Rackham 1990). When describing the very different landscapes of Australia they used existing words but gave them new meanings.
>
> ... most of the Australianisms seem to date from early in the colony's history, and are independent of American terminology.
>
> Rackham 1997: 12

Table 1 of Rackham (1997) lists English terms that are used differently in Australia or would be unfamiliar to Australians. However, he did not discuss one of the major differences between Britain and Australia, and that relates to the terms 'countryside' and 'country'. Oliver wrote of trees and woodland, but also extensively of the countryside. Countryside is not a term much used in Australia. The nearest equivalent in Australia is bush, a term derived from South Africa (from the Dutch word '*bosch*', which the *AND* indicates means woodland). Cape Town was a staging post for shipping from Britain to Australia, and some of the earliest European arrivals in Sydney may well have had discussion with South Africans, or even had opportunity to view the environs of the Cape. They would have seen landscapes encompassed by the term '*bosch*'. Rackham (1997 Table 1) gives the English meaning of bush as an 'individual shrub or undershrub (not used for an area of land)': in Australia a particular area might be referred to as a patch of bush. While bush is a term applied to areas of native vegetation, it is also used extensively for any area which is not urban – if a person has gone bush they are out of town.

The word 'country' has assumed great significance in recent decades. The importance to First Nation Australians of being 'On Country' is now widely recognized. Many public meetings commence with a formal welcome to Country by an elder of the local Aboriginal people, and even more meetings commence with an acknowledgement that the meeting is being held on the ancestral lands of Aborigines and paying respects to their elders past and present. Country in this sense encompasses the landscape, but also expresses a deep psychological attachment. The attachment to Country is perhaps similar to the concept expressed by Seddon (1972) as 'a sense of place', but it is even deeper.

Scrub and brush

Two terms used in Australia for rainforest are scrub and brush. Rainforest was not a word in English in 1788 – it was introduced into the language in the translation (1903) of Schimper's overview of world vegetation as rain forest – following Baur (1968) Australians spell the term as a single word, and this form is gaining currency globally. Although rainforest is now applied to oak woodlands in the extreme Atlantic fringe of Western Britain, there was no previous term in English that recognized the distinctiveness of these communities.

Rackham (1997, Table 1) gives the English meaning of scrub as 'Woodland in a young stage of development'. Scrub is used in this sense in Australia; regrowth of previously cleared woodland is often called scrub. However, there is no evidence that when areas of rainforest were called scrub they were regrowth; for example, the Big Scrub in the Northern Rivers region of New South Wales was possibly the largest contiguous area of subtropical rainforest in the world. It was mostly cleared in a very short time in the latter half of the nineteenth century for conversion to agricultural uses, but was not a vast area of regrowth prior to this.

While brushwood is a common term in English it does not carry connotations of a particular plant community. *AND* gives a number of very early references to brush (and particularly to dense brush), applying to the understorey of forest and woodland. The perception of impenetrability, conveyed by this use of brush, possibly extended to rainforest, although impenetrability is an edge effect in rainforest patches, as the understorey under a dense canopy in the centre of rainforest stands is relatively open. There are a number of stands of rainforest that were called brush in the nineteenth century and appear as such on maps – for example, Cobcrofts Brush in the upper Forbes River catchment in the mid north coast region of New South Wales, and the Yarrawa Brush, once covering 2,500 ha, but now reduced to tiny fragments on top of the Illawarra escarpment south of Wollongong. Brush as part of the compound name of rainforest species is common, as in Brush Box *Lophostemon* spp., which is characteristic of the rainforest/sclerophyll forest transition.

Atkinson (1826) provides information about the names applied to different vegetation types in the earliest years of settlement. The first chapter of the book provides a broad description of the vegetation, which is useful in identifying what names were used, but less so in providing explanations of why. What we would now call coastal heathlands he referred to as barren scrubs. He commented that these had ' a profusion of beautiful shrubs and bushes, producing elegant flowers' but that 'they scarcely seem to be susceptible to any improvement'. He also recognized areas that he called brushes, of which there where a number of different kinds. The most extensive he called coppice brushes, from a perceived resemblance to 'the coppices in England'. Coppice brushes included much of the woodland on the Cumberland Plain and Blue Mountains. Atkinson noted that the 'expense of clearing was high', and that in many places the coppice brushes were 'wretchedly poor' and 'wholly unimprovable'. Vine brush was the name Atkinson gave to what we now call rainforest. He discusses a number of the characteristic species, including *Dendrocnide* (Urticaceae) the giant stinging tree that can 'inflict a sting infinitely more painful than the nettle of Europe'. He noted that 'the soils in many of these brushes is extremely rich, but the labour of clearing is immense'. Of forest he wrote: 'It is, however, always to be understood, that forest means land more or less furnished with timber trees, and invariably covered with grass underneath, and destitute of underwood' (Atkinson 1826: 1–5). These particular criteria for distinguishing 'forest' do not seem to have been universally applied; neither then, nor subsequently.

Woodland

AND gives the meaning of wood in Australia as being obsolescent and states it to be 'Applied to a tract of naturally treed land and now superseded by bush' (*AND* 1786).

The First Fleet arrived in what is now Sydney Cove on 26 January 1788, and disembarkation took place on 27 January. David Collins recorded that: 'The confusion that ensued will not be wondered at, when it is considered, that every man stepped from the boat literally into a wood' (Collins 1804: 11). It is curious that the term 'a wood' was used. Sydney Cove is several miles from the entrance to the Harbour. As the fleet advanced from the Heads, they would have observed the vegetation on both shores, which would have been largely what would be referred to today as woodland, with some areas of heathland. There would not have been discrete patches of woodland such as to justify the title of 'a wood'. Perhaps right from the start of the British colonization the words used to describe Australian vegetation started to divert from Standard English, as Rackham (1997) had suggested.

In Australia, as Europeans proceeded to occupy more of the continent, they made a distinction between forest and woodland, based on height and extent of canopy cover. Given the diversity and novelty of the flora, it is not surprising that the early European settlers did not describe vegetation in terms of lists of species – the structure of vegetation and perceived dominants were the main criteria on which classification was based. The peak of this approach is represented by the Specht scheme (1970, 1981). Unlike most vegetation classifications, which are potentially open-ended in terms of number of entities recognized, the Specht scheme has a fixed number of top order categories, defined in relation to two axes – projective foliage cover of the highest stratum and canopy height/growth form. Utilizing projective foliage cover rather than canopy cover separates rainforest from sclerophyll forest and highlights the distinctive nature of eucalypt canopies. At maturity, the leaves of most species of *Eucalyptus* are pendent and so cast less shade than leaves held horizontally. This means that the light environment under a eucalypt canopy is very different from that under deciduous northern temperate forest.

Woodlands in the scheme have projective foliage cover below 30 per cent and are a major vegetation type across very extensive areas of Australia. For example, *White Box–Yellow Box–Blakely's Red Gum and Derived Native Grassland* is listed as a Critically Endangered Ecological Community under federal and state legislation. It originally covered millions of hectares of the Tablelands and Western Slopes from southern Queensland to Victoria, including much of the wheat/sheep belt. Today, perhaps 400,000 ha remain in good condition in the form of numerous fragments.

Oliver considered that what Australians have typically referred to as woodlands were really examples of savanna, and in Rackham (2008) stated that both Australians and Americans referred to denser savannas as woodlands. Accepting that the word 'woodland' in Australia is used differently from the original use of the terms in English, the application to a structural form of vegetation in Australia has been consistent. The term 'savanna' in Australia is frequently applied to communities across northern tropical Australia, and this is consistent with international usage. Tropical savanna falls into woodland in the Specht scheme.

Is the woodland structure of much Australian vegetation a reflection of the responses of the component plants species to underlying environmental factors such as soil nutrient status and long-term water availability or does it reflect the impact of human activities? It is likely that both sets of factors apply, and their relative importance will vary between sites. One form of woodland to which Oliver drew attention was what he called 'savanna by subtraction' (Rackham 2018: photo 3), in which survivors of clearing exist within grazing or cropping.

A gentleman's park

The early explorers in south-east Australia made frequent references to the bush they travelled through being like 'a gentleman's park'. From the numerous examples quoted by Williams (1989), it is clear the same description had been applied to landscapes in North America some 200 years earlier. Various commentators on early nineteenth-century Australia have either explicitly (McLoughlin 1999: note 21) or implicitly equated 'park' in descriptions of Australian landscapes with the landscape designs of Capability Brown and others, but this is unlikely given the much earlier references to park in the Americas, and because in the late eighteenth and early nineteenth centuries the great English landscaped gardens of today would have been in their infancy. It is more likely that the reference to 'park', by colonists in America and Australia, was to landscapes similar in appearance to deer parks in Britain. Oliver wrote extensively on deer parks, which were characterized by large scattered trees within grassland, and maintained essentially as deer farms. Feral deer, now a major threat to forest and woodland in eastern Australia, had yet to be introduced when the term 'park' was first used in Australia, so deer management was not the cause of the parklike appearance of east coast woodland. Weight is placed on the parklike appearance of the bush by Gammage (2011) to support his thesis of widespread patch burning by Aborigines. I have analysed the case made by Gammage in detail (Adam 2017a), and am not convinced that it provides an explanation in all cases – although it might in some. The critical question that is difficult to answer is what proportion of the landscape was parklike. Recent commentators such as Gammage assume it was a large proportion, but an alternative view is that it might have been limited, but was commented upon simply because its openness provided obvious routes through the landscape, and so was seen by early travellers.

There were some estates in Australia that were intended to more clearly resemble the landscaped gardens of England – for example Fernhill at Mulgoa, now on the edge of the expanding Sydney metropolis. The house at Fernhill, built in 1840, is in the Greek Revival style and set in a landscape of native species, with the distribution of trees created by elimination from the original woodland rather than by planting.

Clearing the bush

Those who arrived on the First Fleet in 1788 were faced with numerous challenges. The most pressing need was to secure food supplies. By the time it arrived in Australia, the First Fleet had little food, although it did have some livestock as well as seeds of potential crops. When Cook in the *Endeavour* landed on the east coast of Australia on 29 April 1770, it was at what is now called Kurnell in Botany Bay. Cook first named the bay Stingray Harbour, given the prevalence of those fish, but later renamed it Botanist Bay and then Botany Bay. The name change was to reflect the diversity and novelty of the flora collected by the botanists on the expedition.

The floristic richness of the vegetation at Kurnell was interpreted by Joseph Banks as reflecting the fertility of the area. Hence, the First Fleet arrived with the expectation that once land was cleared it would support crops and the colony could become self-sustaining – a vital necessity given the distance from other potential sources of food. This expectation was very soon shown to be ill founded. The first crops grown near Sydney Cove either failed or gave very poor yields. For several years, the future of the new colony hung in the balance. We now know that high species richness in plant communities is frequently associated with low soil nutrient status and there is an inverse relationship between soil nutrient status and species richness, as appears to have been apparent to Atkinson (1826). The sand- and sandstone-derived soils of the Sydney region have some of the lowest levels of phosphorus recorded anywhere but have very high species richness, particularly of sclerophyll shrubs (Adam, Stricker and Anderson 1989). It was only when more fertile soils on shales were located further inland from Sydney Cove, along the Parramatta River, that the prospects for the new colony improved.

In order to establish sites for agriculture it was necessary to clear the existing vegetation. Atkinson (1826) devoted much of his fifth chapter to the various methods employed in the earliest years of settlement. One of the early techniques applied to land clearing was the ringbarking of trees – the removal from the trunk of a circumferential band of bark to below the phloem. This results in the death of the tree, although in large trees the process may take several years. Although used from the start of colonial times, ringbarking was practised on a very large scale in the late nineteenth and early twentieth centuries (Stubbs 1998). The objective was to make the land available for grazing and/or cropping, not to provide a yield of timber. After the trees died, there was sometimes selective or total removal, by felling or fire, but dead trees subject to ringbarking decades ago can still be observed in the bush. When it came to seeking timber for immediate use, or as a resource for commercial exploitation, the first colonists were faced with a great diversity of trees, many in the early years not scientifically described and named, and whose properties were unknown. Aboriginal knowledge was of limited use to the settlers; the first Australians had stone axes, which were used for among other things cutting steps to enable the climbing of tall trees to capture arboreal mammals, but the felling of large trees was not practised. Furthermore, felling very large trees was not something the colonists had prior knowledge or experience of in Britain.

Indeed, the felling of large trees was a difficult business. In the case of eucalypts, it was often necessary to establish a platform above the flare of the butt, while in rainforests the platform was required to be above the buttresses. To create the platform springboards were used – boards about 2 m long and 18 cm wide with a steel shoe at one end. The feller cut a hole in the trunk – about 15 cm deep – and inserted the shoe of the springboard into it, then climbed onto the springboard to cut another hole higher up and to the side, in which was inserted another springboard. The process was continued until the necessary height was reached, when another board was inserted at the same level. Planks were laid across the two springboards to create a stable platform from which to work (Penfold and Willis 1961). Springboards were also known as jigger boards, as in the caption to Photo 7.12f in Colloff (2014), which shows the holes for inserting jigger boards cut in the stump of a large River Red Gum.

In a very short period of time, specific uses for a very large number of species, both of eucalypts and rainforest trees, were established – which speaks of a high degree of woodmanship and woodworking skills among the early colonists. For information on the uses of rainforest trees, see Francis (1929) and Floyd (2008), and for forest trees more generally, Boland et al. (2006). Floyd lists uses or potential uses for nearly every species of rainforest tree in south-east Australia, although for many the suggestion is for turnery or carving; he documents, for instance, the timber of the Tulip Satinwood *Rhodosphaera rhodanthema* (Anacardiaceae), as 'Soft, fine-grained and beautifully figured. Durable. Used for inlays and cabinet work. A lute made from this timber produced perfect sound and was used in concerts in Europe' (Floyd 2008: 67).

Cabinetmakers in Britain had used imported rainforest timbers for fine furniture long before 1788, so that when rainforest was encountered in Australia the potential for profitable exports was recognized. Various Australian timbers were displayed at Crystal Palace in the Great Exhibition of 1851 and a number of species became valued in Britain; by far the most important was Australian Red Cedar *Toona ciliata*, one of the mahoganies (Meliaceae). Finding cedar played a major role in opening up the eastern Australian ranges for settlement. As well as having magnificent timber, Red Cedar is unusual in the Australian context in being winter deciduous, and could be spotted from afar at that time of year. Cedar getters were often the first Europeans to locate stands of rainforest, and lived lives largely remote from contact with other humans (Vader 1987; Ritchie 2004). Red Cedar logs float (unlike those of a number of eucalypts, which settlers were disconcerted to discover had specific gravities greater than one), and after being felled, cedar logs were often transported down coastal rivers. The logs were branded by the feller and were caught in nets at the mouth of the rivers perhaps many months later. The timber merchants then paid the owner of the brand the appropriate fee – this involved trust in the merchants by the cedar getters. In Australia today, Red Cedar is commercially extinct, although juveniles are still widespread and the species is not recognized as threatened in conservation legislation. Red Cedar logs do not easily decay: there are still logs felled more than a century ago and left in the back paddock or recovered from streams, which appear at very high prices on the market.

Despite the potential utility of rainforest timbers, large areas of what was already a limited resource were cleared for agriculture in the nineteenth century. Logging for species such as Red Cedar was selective, with isolated trees being removed. In the pre-Second World War years, Australian forestry researchers were world leaders in developing timber technology, including for the cutting of very large sheets of veneer and glues to permit development of high-quality plywood. During the Second World War, Australian speciality plywoods made a major contribution to the war effort, but the consequence was that at war's end there were mills with long-term allocations of rainforest timbers but little knowledge of the size of the resource or its sustainability. This set the scene 25 years later for the rainforest conservation battles (Adam 1992; 2017b), which resulted in the virtual ending of rainforest logging and the eventual inclusion of many stands of rainforest in eastern Australia on the World Heritage List.

Environmental history in Australia

For environmental historians, there are considerable difficulties establishing an Australian historical record before 1788. There is no written record but there is a rich oral tradition, much of which has possibly been lost in the last 250 years. There are numerous language groups, with large vocabularies and complex grammars. Stories and traditions are passed down orally and in dance and song. The First Australians had a very detailed knowledge of astronomy (Norris and Norris 2009; Pascoe 2018), which both provided a calendar and a means of location and navigation. There was considerable knowledge of many aspects of flora and fauna, in marine (coastal), freshwater and terrestrial environments. Ethnobotanical studies have documented very detailed knowledge of useful and dangerous plants (Webb 1969), and of techniques to render otherwise poisonous plant products nutritious sources of food, for example cycad seeds – information largely vouchsafed to women.

In some modern accounts, the first people had an encyclopaedic knowledge of what we now call biodiversity and lived in complete harmony with nature. This is overly romantic, but certainly for tens of thousands of years Aborigines survived and thrived within the Australian environment. Population numbers and overall density were low, reflecting environmental constraints.

Today, one of the most complex and difficult issues facing land managers in Australia is the extent to which priority should be given to restoring Aboriginal practices. Australia is large and environmentally diverse. Aboriginal practices would have reflected this environmental heterogeneity; there is not a single set of practices that, in detail, would have been uniform, even though there were common themes. European Australians have wrought massive changes to the environment – clearing forests, introducing pastoralism and large-scale arable cropping (Pascoe (2018) makes a strong case for agricultural practices employed by Aborigines being widespread, but although harvesting of seeds of *Themeda* and other perennial grasses occurred there were no equivalents of annual cereals), applying fertilizers and other agricultural chemicals, establishing towns and cities and their associated infrastructure, fragmenting natural landscapes and introducing large numbers of plant and animal species. In many parts of the continent Aborigines were displaced, and their involvement in land management terminated very early in the colonial era.

My view is that there are parts of Australia where there has been continuity of forms of Aboriginal management, particularly in the tropics and arid centre, where Aboriginal management practices are appropriate both culturally and ecologically, albeit with changes to reflect both environmental change and availability of new technologies. It would be inappropriate to expect the Aboriginal owners of land to function in museums given the availability of firearms, drip torches for lighting fires, chainsaws, motor vehicles and powered boats. Practices necessarily and inevitably evolve.

In those parts of the continent most heavily transformed, and where the continuity of traditional practice has been broken, there is a need to learn as much as possible about past practices, and where relevant seek to apply them. However, the nature of the modified landscape, and society's expectations of the outcomes of management, mean that new paradigms are required, different from both those of the Aborigines and those practised by Europeans over the last two centuries.

Although there is no written evidence, Australia has a rich archaeological heritage of art and artefacts. The art includes rock engravings, stencils and paintings on rocks. This record is particularly valuable for the information it provides on the past distribution of the vertebrate fauna (Veth, McDonald and de Konning 2018), but unfortunately has little representation of flora or of events such as fires. More indirect records are provided by fossil pollen and charcoal, though compared with northern Europe or North America the pollen record is limited. Preservation of pollen requires continuously anaerobic conditions, and in a land marked by repeated drought many wetlands that might have acted as pollen traps have dried out at various periods leading to oxidation of pollen and gaps in the record. In addition, *Eucalyptus* species have very similar pollen not easily or reliably distinguished at the species level. *Acacia* release pollen in polyads; these large masses are poorly dispersed. There are some notable long-term records that are of international significance, but Australia is never likely to have a record with the fine-scale spatial and temporal resolution available in the northern hemisphere.

One of the best-known and influential environmental histories in Australia is Gammage's *The Biggest Estate on Earth* (2011) – reprinted many times and winner of major literary prizes. The book aims to demonstrate how Aboriginal management of fire transformed the vegetation and ecology of the whole continent. It is an important work, elements of which are supported by evidence. However, I have argued elsewhere (Adam 2017a) that critical aspects of the hypothesis are not well founded. A major strand of evidence relied upon is interpretation of early colonial landscape paintings, which I argue is a much more complicated field for analysis than Gammage recognizes. Gammage (2011) includes a lengthy appendix in which he criticizes non-historians for venturing into the field, but environmental history is necessarily multi- and interdisciplinary, and there are few polymaths with the breadth of knowledge and experience of Oliver Rackham. Even in single authored works it is appropriate to call upon work from many disciplines. Gammage is particularly scathing of ecologists daring to interpret the diaries and notes of the original colonial explorers, a task that he sees as pre-eminently a field for historians. The explorers did not have the benefits of GPS, and they were venturing into what for them was terra incognita. They plotted their position and course by use of sextants and chronometers; converting the observations to maps required extensive calculations carried out by hand. Whitehead (2003 – the first of a series of volumes) reanalyses the extensive field notes and survey data of early explorers, sometimes resulting in changes to the plotted locations and requiring a re-evaluation of previous historical interpretations. Whitehead was a surveyor and former Shire engineer. I doubt that many historians (or ecologists) would have the technical knowledge to do this, but a historian taking the explorers' accounts at face value could be lead astray.

All budding writers of environmental/ecological history, regardless of their geographical base and ecosystem of interest, should read Rackham (2018) and, in particular, the dot points in his conclusion section, a guide to what not to do. Absorbing this distilled wisdom would improve the quality of histories and reduce the potential for sterile debates.

In discussing broader issues of forest management and conservation more generally (as, for example, in Rackham 1998; 2003; 2008), Oliver is a voice of rationality and deep common sense. When he started his career, some conservation agencies employed ecologists and land managers. Conservation science was not a separate discipline with its own journals. One of the changes over the last half-century has been the development of conservation biology as a major subdiscipline with a substantial number of dedicated journals and university degree courses. This is to be welcomed, but Oliver on several occasions cautioned against being too wedded to current fashion in conservation thinking and urged plurality of approaches and receptiveness to empirical data and observations over the unthinking application of theory. He was concerned that narrow professionalism might threaten natural history and its rich traditions. Chapter 33 of Rackham (2003) should be essential reading for all conservation biologists: the underlying messages are relevant to all ecosystems. There are few if any polymaths whose range of knowledge approaches that of Oliver. To understand the ecological history of particular species, especially ecological communities, or particular places more fully, will require establishing networks of those with the necessary experience across the range of disciplines in the sciences and the humanities (including citizen scientists, citizen historians and what might be termed 'antiquarians'). I say this because there is one very prominent historian in Australia who argues strongly that environmental history is the province of historians, and that scientists should stay out of it.

References

Adam, P. (1978) Geographical variation in British saltmarsh vegetation. *Journal of Ecology*, 66, pp. 339–366. https://doi.org/10.2307/2259141

Adam, P. (1981) The vegetation of British saltmarshes. *New Phytologist*, 88, pp. 143–196. https://doi.org/10.1111/j.1469-8137.1981.tb04577.x

Adam, P. (1992) *Australian Rainforests.* Oxford University Press, Oxford.

Adam, P. (2002) Saltmarshes in a time of change. *Environmental Conservation*, 29, pp. 39–61. https://doi.org/10.1017/S0376892902000048

Adam, P. (2017a) Can ideas be dangerous? *Australian Zoologist*, 38, pp. 329–374. https://doi.org/10.7882/AZ.2017.008

Adam, P. (2017b) The world heritage list and New South Wales rainforest – reflections on the events of 30 years ago. *Australian Zoologist*, 39, pp. 228–56. https://doi.org/10.7882/AZ.2017.014

Adam, P., Stricker, P. and Anderson, D.J. (1989) Species richness and soil phosphorus in plant communities in coastal New South Wales. *Australian Journal of Ecology*, 14, pp. 189–198. https://doi.org/10.1111/j.1442-9993.1989.tb01426.x

Atkinson, J. (1826) *An Account of the State of Agriculture and Grazing in New South Wales.* J. Cross, London.

Baur, G.M. (1968) *The Ecological Basis of Rainforest Management.* Forestry Commission of NSW, Sydney.

Bird, J.F. (1978) The nineteenth-century soap industry and its exploitation of intertidal vegetation in Eastern Australia. *Australian Geographer*, 14, pp. 38–41. https://doi.org/10.1080/00049187808702731

Bird, J.F. (1981) Barilla production in Australia. In: D.J. Carr and S.G.M. Carr (eds) *Plants and Man in Australia.* Academic Press, Sydney, pp. 274–280.

Boland, D.J., Brooker, M.I.H., Chippendale, G.M., Hall. N., Hyland, B.P.M., Johnson, R.D., Kleinig, D.A., McDonald, M.W. and Turner, J.D. (2006) *Forest Trees of Australia*, 5th edition. CSIRO Publishing, Melbourne. https://doi.org/10.1071/9780643069701

Brooker, M.I.H. and Orchard, A.E. (2008) (1844) Proposal to conserve the name *Eucalyptus camaldulensis* (Myrtaceae) with a conserved type. *Taxon*, 57, pp. 1002–1004. https://doi.org/10.1002/tax.573039

Brown, J. (2019) Keen edge of history. *The Land*, 3 January 2019. p. 27.

Carruthers, J. and Robin, L. (2010) Taxonomic imperialism in the battles for *Acacia*: identity and science in South Africa and Australia. *Transactions of the Royal Society of South Africa*, 65, pp. 48–64. *https://doi.org/10.1080/00359191003652066*

Collins, D. (1804) *An account of the English colony in New South Wales, from its first settlement in January 1788 to August 1801.* Cadell and Davies, London. https://doi.org/10.5962/bhl.title.94762

Colloff, M.J. (2014) *Flooded Forest and Desert Creek: Ecology and history of the River Red Gum.* CSIRO Publishing, Collingwood. https://doi.org/10.1071/9780643109209

Dargavel, J. (ed.) (1997) *Australia's Ever-Changing Forests III. Proceedings of the Third National Conference on Australian Forest History.* CRES, Canberra.

Duck, R. (2015) *On the Edge: Coastlines of Britain.* Edinburgh University Press, Edinburgh. https://doi.org/10.1515/9780748697632

Floyd, A.G. (2008) *Rainforest Trees of Mainland South-Eastern Australia.* Terania Rainforest Publishing, Lismore.

Francis, W.D. (1929) *Australian Rain-Forest Trees.* Government Printer, Brisbane.

Frodin, D.G. (2004) History and concepts of big plant genera. *Taxon*, 53, pp. 753–776. https://doi.org/10.2307/4135449

Gammage, B. (2011) *The Biggest Estate on Earth: How Aborigines made Australia.* Allen and Unwin, Sydney.

Gray, A.J. (1972) The ecology of Morecambe Bay. V. The salt marshes of Morecambe Bay. *Journal of Applied Ecology*, 9, pp. 207–220. https://doi.org/10.2307/2402057

Gray, A.J. and Adam, P. (1974) The reclamation history of Morecambe Bay. *Nature in Lancs.*, 5, pp. 13–20.

Hassall, J.S. (1902) *In Old Australia: Records and reminiscences from 1794.* R. S. Hews, Brisbane.

Kull, C.A. and Rangan, H. (2012) Science, sentiment and territorial chauvinism in the Acacia name change debate. *terra australis*, 43, pp. 197–219. https://doi.org/10.22459/TA34.01.2012.09

Levin, P.S., Ellis, J., Petrik, R. and Hay, M.E. (2002) Indirect effects of feral horses on estuary communities. *Conservation Biology*, 16, pp. 1364–1371. https://doi.org/10.1046/j.1523-1739.2002.01167.x

Macnae, W. (1968) A general account of the fauna and flora of mangrove swamps and forests of the Indo-West-Pacific region. *Advances in Marine Biology*, 6, pp. 73–270. https://doi.org/10.1016/S0065-2881(08)60438-1

McDonald, M.W., Brooker, M.I.H. and Butcher, P.A. (2009) A taxonomic revision of *Eucalyptus camaldulensis* (Myrtaceae). *Australian Systematic Botany*, 22, pp. 257–283. https://doi.org/10.1071/SB09005

McLoughlin, L.C. (1999) Vegetation in the early landscape art of the Sydney Region, Australia: accurate record or artistic license? *Landscape Research*, 24, pp. 25–47. https://doi.org/10.1080/01426399908706549

Moore, B. (ed.) (2016) *The Australian National Dictionary: Australian words and their origins* (2 volumes), 2nd edition. Oxford University Press, Melbourne.

Mudie, R. (1829) *The Picture of Australia: Exhibiting New Holland, Van Dieman's land, and all settlements, from the first at Sydney to the last at Swan River.* Whittaker, Treacher, London.

Norris, R. and Norris, C. (2009) *Emu Dreaming: An introduction to Australian Aboriginal astronomy.* Emu Dreaming, Sydney.

Pascoe, B. (2018) *Dark Emu.* 2nd edition. Magabala Books, Broome.

Paterson, G. (1811) *The History of New South Wales, from its First Discovery to Present Times.* Mackenzie and Dent, Newcastle-upon-Tyne. [the first printing is by an anonymous 'Literary Gentleman' – subsequent printings identified the author as Paterson]

Penfold, A.R. and Willis, J.L. (1961) *The Eucalypts: Botany, cultivation, chemistry and utilization.* Leonard Hill: Interscience, London.

Pescott, E.E. (1914) *Native Flowers of Victoria.* George Robertson and Company, Melbourne. https://doi.org/10.5962/bhl.title.115946

Rackham, O. (1986) *The History of the Countryside.* J.M. Dent and Sons, London.

Rackham, O. (1994) *The Illustrated History of the Countryside.* Weidenfeld and Nicholson, London.

Rackham, O. (1997) European concepts of forest age. In: J. Dargavel (ed.) *Australia's Ever-Changing Forests III. Proceedings of the Third National Conference on Australian Forest History.* CRES, Canberra, pp. 11–23.

Rackham, O. (1998) Woodland Conservation: Past, present, and future. In: M.A. Atherden and R.A. Butlin (eds) *Woodlands in the Landscape: Past and Future Perspectives.* [PLACE Research Centre, the University College of Ripon and York St John] Leeds University Press, Leeds, pp. 60–78.

Rackham, O. (2002) *Treasures of Silver at Corpus Christi College, Cambridge.* Cambridge University Press, Cambridge.

Rackham, O. (2003) *Ancient Woodland: Its History, Vegetation and Uses in England*, New edition. Castlepoint Press, Dalbeattie.

Rackham, O. (2006) *Woodlands.* (New Naturalist 100), Collins, London.

Rackham, O. (2008) Ancient woodland: modern threats. *New Phytologist*, 180, pp. 571–586. https://doi.org/10.1111/j.1469-8137.2008.02579.x

Rackham, O. (2018) Archaeology of trees, woodland and wood pasture. In: A.H. Çolak, S. Kirca and I.D. Rotherham (eds) *Ancient Woodlands and Trees: A guide for landscape planners and forest managers.* Turkish Academy of Sciences/IUFRO, Ankara, pp. 39–60. https://doi.org/10.53478/TUBA.2020.039

Ritchie, R. (2004) The Red Cedar timber industry in New South Wales and Queensland. In: V. Sripathy (ed.) *Red Cedar in Australia.* Historic Houses Trust, Sydney, pp. 42–59.

Santillan, N., Rogers K. and McKee, K. (2018) The shifting saltmarsh – mangrove ecotone in Australasia and the Americas. In: G.M.E. Perillo, E. Wolanski, D.R. Cahoon and C.S. Hopkinson (eds) *Coastal Wetlands: An integrated ecosystem approach*, 2nd edition. Elsevier, Amsterdam, pp. 915–945. https://doi.org/10.1016/B978-0-444-63893-9.00026-5

Schimper, A.W.F. (1903) *Plant-Geography upon a Physiological Basis.* Translated by W.R. Fisher, edited by P. Groom and I.B. Balfour. Clarendon Press, Oxford. https://doi.org/10.5962/bhl.title.24775

Seddon, G. (1972) *Sense of Place: A Response to an Environment. The Swan Coastal Plain Western Australia.* University of Western Australia Press, Perth.

Specht, R.L. (1970) Vegetation. In G.W. Leeper (ed.) *The Australian Environment*, 4th edition. Melbourne University Press, Melbourne, pp. 44–67.

Specht, R.L. (1981) Foliage projective cover and standing biomass. In: A.N. Gillison and D.J. Anderson (eds) *Vegetation Classification in Australia.* ANU Press, Canberra, pp. 10–21.

Tench, W. (1793) *A Complete Account of the Settlement at Port Jackson in New South Wales: Including an accurate description of the situation of the colony; of the natives; and of its natural productions.* G. Nicol and J. Sewell, London.

Thomas, J.H. (2012) Close encounters of a different kind: Arthur Phillip and the early opening of Australia. *Terrae Incognitae*, 44, pp. 43–67. https://doi.org/10.1179/0082288412Z.0000000001

Tomlinson, P.B. (1986) *The Botany of Mangroves.* Cambridge University Press, Cambridge.

Vader, J. (1987) *Red Cedar: The Tree of Australia's History.* Reed Books, Frenchs Forest.

Vera, F.W.M. (2000) *Grazing Ecology and Forest History.* CABI, Wallingford. https://doi.org/10.1079/9780851994420.0000

Veth, P., McDonald, J. and de Koning, S. (2018) Archaeology and rock art of the North-West Arid Zone with a focus on animals. In: H. Lambers (ed.) *On the Ecology of Australia's Arid Zone.* Springer, Cham, pp. 283–305. https://doi.org/10.1007/978-3-319-93943-8_11

Webb, L.J. (1969) Australian plants and chemical research. In: L.J. Webb, D. Whitelock and J. Le Gay Brereton (eds) *The Last of Lands: Conservation in Australia.* The Jacaranda Press, Milton, pp. 82–90.

White, J. (1790) *Journal of a Voyage to New South Wales.* J. Debrett, Piccadilly.

Whitehead, J. (2003) *Tracking and Mapping the Explorers. Volume 1. The Lachlan River exploration, 1817, Oxley–Evans–Cunningham.* John Whitehead, Coonabarabran.

Williams, M. (1989) *Americans and their Forests: A historical geography.* Cambridge University Press, Cambridge.

Zacharin, R.R.F. (1978) *Emigrant Eucalypts: Gum trees as exotics.* Melbourne University Press, Melbourne.

CHAPTER 21

Managing Pollards and the Last Forest

Vikki Bengtsson

Summary

This chapter explains how approaches to veteran trees and their management came about through the emergence of the Ancient Tree Forum and the influence of the work and writing of Oliver Rackham – in particular regarding the challenges in managing lapsed pollards. Innovations in the veteranization of individual trees are also discussed; veteranization may prove to be a powerful tool in nature conservation. Some of the work was undertaken at Hatfield Forest, a site very close to Oliver's heart.

Keywords: Pollards, Sweden, veteranization, Epping Forest, Burnham Beeches, firewood, Hatfield Forest

Hedgerow pollards and wood-pasture trees

Veteran trees, and in particular pollards, have interested me throughout my working life. I first encountered pollards in the hedgerow trees of Essex, one of Oliver Rackham's old stomping grounds. The interest grew, and I have been involved with the management of ancient and other veteran trees since 1992, when I also first met Ted Green and became involved in the Ancient Tree Forum.

I now live and work in Sweden as a nature conservation consultant, primarily advising and training people on the management of wood-pastures, wooded meadows and ancient trees. I first visited Sweden in 2002 on an Ancient Tree Forum trip and fell in love with the wonderful wooded meadows with pollards and flower-rich meadows. I moved in 2003, and have been lucky enough to work both in England and Sweden, visiting sites that are some of the finest in Europe for their veteran tree populations.

I first came into contact with Oliver Rackham's work *The History of the Countryside* (Rackham 1986) and *Trees and Woodland in the British Landscape* (Rackham 1979) when I undertook my degree dissertation focusing on changes in the countryside and the loss of hedgerows over the last 100 years. His books were bibles, and truly helped in our appreciation of how trees could help us understand the history and use of our landscape. I became fascinated by the biological and cultural significance of hedgerows and the pollards they contain. Pollarding, as a management technique, involves cutting trees at about head height on a regular basis. How regularly the trees are cut, and what species, depends on the product that the cutting aims to produce. Fodder was a common product from pollarding in more northerly climates

Vikki Bengtsson, 'Managing Pollards and the Last Forest' in: *Countryside History: The Life and Legacy of Oliver Rackham*. Pelagic Publishing (2024). © Vikki Bengtsson. DOI: 10.53061/ZPLM6676

and trees were then cut on a four to five year rotation. Popular tree species included ash, lime, elm and maple. Firewood was another important product and involved cutting the trees on a slightly longer rotation; somewhere between 10 and 15 years was more common. With the onset of the Industrial Revolution and coal mining, pollarding declined and in many places was completely abandoned. Pollarding was often undertaken at a local level by farmers for their own local needs. It was also undertaken on an industrial scale, at places such as Epping Forest and Hainault Forest, where hornbeam was cut and transported into London to fire the bakers' ovens. This is also true in the north of Spain, where in the Basque region pollarding was undertaken for the iron industry (Read et al. 2010; Read and Bengtsson 2019).

As I was working in the east of Essex, it was frightening to see how quickly the hedgerows and pollards in the landscape were disappearing. Oliver Rackham's work helped me see the importance of going back to historical records, maps and photographs to help measure and follow these changes (Fig. 21.1). My studies reflected those of others: that there were huge losses since the Second World War, but primarily in the two decades before 1990, when the loss was at a peak and accounted for a third of the entire post-war loss (Barr et al. 1991; Mabey 1996).

Figure 21.1 Oliver Rackham speaking at an Ancient Tree Forum event at Staverton, Suffolk, 2010. (V. Bengtsson)

Figure 21.2 The Cage Pollard at Burnham Beeches, 2011. (V. Bengtsson)

Burnham Beeches and Oliver Rackham

Following my graduation from the University of Essex, I was lucky enough to get a job working at Burnham Beeches in Buckinghamshire. On my first day driving to work, in October 1992, it was 'raining' orange beech leaves – and I felt that all my Christmases had come at once! At that time, the City of London had organized a number of pollard conferences to share knowledge and try to develop knowledge on how to manage these trees, despite the fact that they had not been pollarded for many decades (Fig. 21.2). A conference was arranged at Burnham Beeches by Helen Read and Mark Frater, to which Oliver Rackham, among others, was invited. The day involved lots of lively discussions and debates regarding how to deal with lapsed pollards and overgrown wood-pasture. At that time, the debate was focused on how to stop these old trees from falling apart and how they could be retained as a cultural and historical feature of the landscape. Discussing how to create new pollards was also important, particularly as the historical knowledge regarding how to do this had been all but lost by this time in the UK.

I found myself sitting next to Oliver Rackham at lunch. I felt so nervous and overwhelmed. That sensation of wanting to make a good impression, wanting to say something intelligent and wanting to take the opportunity to ask those questions you have stored up and have suddenly disappeared from your brain. Nothing came to me, so we both sat there just eating our lunch. I tried several times to start a conversation, all of which ended in Oliver saying either 'yes' or 'no'. Finally, I said something about the Berkshire pigs that Burnham Beeches had introduced into the wood-pasture area and he lit up! There was no stopping him after that.

Ashtead Common

The following year, I moved jobs within the City of London and spent four wonderful years working at Ashtead Common in Surrey, a 200-ha site on heavy clay soils, with over

Figure 21.3 A few of the old, lapsed oak pollards at Ashtead Common, 2009. (V. Bengtsson)

2,300 old oak pollards (Fig. 21.3). Part of the Manor of Ashtead, it was probably used for providing wood and grazing land for the local people. This was a wonderful place to work, but there were also many challenges. Many of the old pollards had been killed or badly damaged in a series of fires that had taken place in the 1980s, before the City of London acquired the site. It was then an ungrazed, overgrown wood-pasture with a huge Bracken *Pteridium aquilinum* problem. The dry bracken in the early spring was a massive fire hazard and was the reason why the fires, when they occurred, spread so widely. My time there involved working with volunteers, undertaking practical work, ecological survey work and environmental education. It was a great mixture, and I got to know many extremely skilled and knowledgeable people.

Ted Green and John Smith were contracted into survey and map all the old pollards. This was a way to get a better understanding of the population of trees that we had at Ashtead Common and to understand the management needs. It would also allow us to be able to monitor the loss rates of the trees in the future, and hopefully to be able to measure the impact of the work to be done. Over 1,300 old oaks were still alive, and the value of these was huge in terms of their historical importance and their value for biodiversity – but also in their own right. The real challenge, however, was how to prolong their lives and bridge the age gap. Ashtead Common, like many wooded commons in England has a big age gap between the old oaks and the next generation.

Pollarding probably ceased around 150 years ago, and recently a work programme to reduce the crowns of the trees has started to try to stop them falling apart from the heavy weight of the branches. The management of lapsed pollards, such as those at Ashtead Common, is something that our ancestors did not have to deal with. Historically, these trees were probably left uncut, and any that died would not have been a major concern as new ones could be planted and subsequently cut. Today however, our lapsed pollards are precious because they are remnants of a landscape and culture that is no longer active, and the numbers of trees are substantially lower than in the past and declining. This has required developing new knowledge and skills in dealing with these trees, which has built upon experiences from several sites including Burnham Beeches and Hatfield Forest in the 1980s and 1990s, which were not always positive (Read et al. 2010; Read and Bengtsson 2019).

For many wooded commons, such as Ashtead Common, reintroduction of grazing animals is the single most important management option for conserving these valuable habitats. This is potentially more important than any work that may be required on the individual trees. It has a twofold effect. First, it helps keep down the competition from new trees, and secondly, it can also help keep the bracken in check. Restoring wood-pastures should, however, be carried out in stages, to make sure that the old trees are not suddenly shocked by fast changes to their environment (Read 2000; Lonsdale 2013; Read and Bengtsson 2019).

The Last Forest

In 1997, I was persuaded by Ted Green to apply for the job of property manager for the National Trust at Hatfield Forest. I was very fortunate to get the job and spent the next six years working at 'The Last Forest'; a site made famous by Oliver Rackham's book (Rackham 1989). Hatfield Forest is considered to be the most intact example of a medieval hunting forest remaining in the UK today. It was a compartmentalized forest, meaning that there were different areas that were clearly marked out and where the tree management was rigidly defined. About 220 ha of the total 400 ha was coppice cut on an 18-year cycle. The cut areas were fenced to keep all browsing animals out for the first six years, after which deer were admitted, and then after a further three years all animals were admitted. The remaining area of the forest was open 'plains' of grassland with pollarded trees and scrub (Rackham 1989). Today, some 800 pollards remain (Fig. 21.4), including more tree species as pollards than any other site in England and with Common Hornbeam *Carpinus betulus* the most frequent. Oliver Rackham was also the first person to highlight the value of the hawthorn pollards at Hatfield Forest that previously had been overlooked.

Working at Hatfield Forest meant that I met Oliver Rackham on a more regular basis. He brought students from his Ancient Woodland course that he ran from Flatford Mill each year,

Figure 21.4 One of the many old hornbeam pollards at Hatfield Forest, 2012. (V. Bengtsson)

and we also arranged a day each year to take a walk around the forest to discuss management. These experiences were in complete contrast to my first meeting at Burnham Beeches; there was rarely a moment's silence. We shared a deep fascination for the amazing survival strategies exhibited by the old trees at Hatfield Forest; the layering, the phoenix trees, the internal roots. We also shared the fascination of the many layers of history that could be read from the landscape.

Much was learned from the early work on the old, lapsed hornbeam and oak pollards at Hatfield Forest; about how not to restore them. In the late 1970s and early 1980s, there was a perception that the old pollards were falling apart. Many pollards were then re-pollarded, involving removal of several large limbs back to the bolling. Many of these limbs were as large as trees themselves. Initially, many of the pollards responded by producing lots of new growth from around the cut surfaces. However, after three to five years all of the oak pollards cut in this way died, and around 20 per cent of the old hornbeams. When I started working at Hatfield Forest, my observation was that the trees that had been cut hard 25 years or so before were the ones that were the most fragile. This was likely because of the huge amounts of decay that had developed as a consequence of the large wounds that had been created by the removal of large limbs and the subsequent dysfunction. The cutting had the objective of reducing the likelihood of collapse of the old pollards, and this may have been achieved. It did, however, also provide lessons in what happens when old trees are cut hard, and that it can take more than 20 years before we can genuinely evaluate the success of any tree work.

It was a challenging time to be in charge of Hatfield Forest. Two major incidents occurred during my years as property manager. The first was an air-crash in December 1999, when a Boeing 747 carrying cargo crashed on the edge of the forest. It was a tragic accident in which five people lost their lives, and it also left a huge amount of debris in the area. The suggestion was made that a large area of soil and trees containing the majority of the debris should be removed, but this felt completely inappropriate. As Oliver had identified in his book, Hatfield Forest was truly a 'Last Forest', and that includes the soil in which it grows. An area of the forest was, for the first time, fully enclosed to allow removal of the debris and to keep visitors safe. This is an area that we both visited at the time: there were many fantastic old Field Maple *Acer campestre* coppice stools that had remarkably survived any significant damage, despite containing large pieces of aircraft debris. The second incident was foot-and-mouth disease, which had a terrible impact on the grazier at Hatfield Forest, but also meant the forest had to be shut to visitors.

When dealing with the consequences of the air-crash and foot and mouth disease, it became clear that the full value of wood-pasture for biodiversity has not always been fully appreciated. The key elements in wood-pasture that are so important for biodiversity are the old trees and the decaying wood they contain and produce, as well as the soil in which the trees grow. The fact that Hatfield Forest had never been ploughed means that as well as having many old trees, it also has valuable undisturbed soil.

Decaying wood in old trees provides conditions that are suitable for a wide range of species (especially fungi and insects); many of these are very rare and some have very precise requirements. Some species may have difficulty colonizing new sites if they are too far apart. As described earlier, many wood-pasture sites are fragmented, with few old trees remaining, and there may be a large generation gap between the old trees and their successors.

Sometimes it is desirable to attempt to create some decaying wood habitats to help close up this generation gap. Pollarding is one reliable technique in which, through regular cutting, the trees develop hollows at a younger age than trees that have not been pollarded (Sebek et al. 2013). Other techniques to create decaying wood habitats in younger trees, known as veteranization (Bengtsson, Niklasson and Hedin 2015), involve damaging younger trees in

Figure 21.5 An oak that was intentionally damaged with a sledgehammer in 1998 at Hatfield Forest, to encourage basal decay as an alternative to felling. There is now a cavity the size of a fist, 2018. (V. Bengtsson)

a variety of ways by mimicking nature. The principle is never to do this to the old trees but to those that are being felled or cut for other management reasons. This is something that we started at Hatfield Forest, in particular where we were restoring an area of overgrown wood-pasture. It seemed sensible to try and make use of the younger trees and 'damage' them rather than remove them all. A variety of different techniques have been attempted, such as the creation of holes using a drill or a chainsaw, making larger hollows within the trunks of live trees and mimicking bark damage by horses or sheep at the base of a trees using chainsaws or sledgehammer. Interestingly, 20 years on, several of the trees that were damaged intentionally, or veteranized, have now developed decay (Fig. 21.5). Once again, this is an indication that when dealing with trees, we need to be aware of tree-time rather than human time.

Sweden and Ash Dieback

In 2003, I moved to Sweden at a time when the management of trees with high biodiversity values was high on the nature conservation agenda. An Action Plan for Trees with High Nature Conservation Values was produced in 2004, and this initiated a nationwide survey of these valuable trees. Some of the same issues were highlighted in terms of management. Shade was identified as one of the greatest threats to the majority of the veteran trees in

Figure 21.6 A lime pollard still cut on rotation from a wooded meadow in Sweden, 2013. (V. Bengtsson)

Sweden. Another issue was how to deal with old pollards out of a regular pollarding cycle (Fig. 21.6). In 2006, 'The Oak Conference' was organized by the County Administrative Board of Östergötland, and Oliver Rackham was one of the key speakers, along with Ted Green and Frans Vera. This conference highlighted some of our joint history and our common history, resulting from the high numbers of old trees still present in both Sweden and the UK (Green 2013).

More recently, it has become clear that many of our old trees are suffering from new pests and diseases: the numbers of pests and diseases on trees has increased tenfold over the last two decades (Forestry Commission 2018). A significant example that is having an impact on the old trees in Sweden and Britain is Ash Dieback. Tree diseases and how this might shape our future treescapes was an area of real concern to Oliver in his final years. He was, however, pleased to know that our old trees, and in particular old ash pollards, appeared to be more robust than younger trees (Bengtsson and Stenström 2017) (Fig. 21.7, overleaf).

Figure 21.7 An old ash pollard, being cut in cycle and without symptoms of Ash Dieback in Sweden, 2020. (V. Bengtsson)

References

Barr, C., Howard, B., Bunce, B., Gillespie M. and Hallam, C. (1991) *Changes in Hedgerows in Britain between 1984 and 1990*. Institute of Terrestrial Ecology, Grange-over-Sands.

Bengtsson, V., Niklasson, M. and Hedin, J. (2015) Tree veteranisation. Using tools instead of time. *Conservation Land Management*, Summer 2015, pp. 14–17.

Bengtsson, V. and Stenström, A. (2017) Ash dieback – a continuing threat to veteran ash trees? In: R. Vasaitis and R. Enderle (eds) *Dieback of European Ash (*Fraxinus *spp.): Consequences and guidelines for sustainable management*. Swedish University of Agricultural Sciences, Uppsala, pp. 262–272.

Forestry Commission. (2018) *Pest and Pathogens, Problems Threatening Britain's Trees*. www.forestry.gov.uk/forestresearch.

Green, T. (2013) Ancient trees and wood pastures. In: Rotherham, I.D. (ed.) *Trees, Forested Landscapes and Grazing Animals: A European Perspective on Woodlands and Grazed Treescapes*. Routledge, London and New York, pp. 127–142.

Lonsdale, D. (2013) *Ancient and Other Veteran Trees: Further Guidance on Management*. The Tree Council, London.

Mabey, R. (1996) *Flora Britannica*. Sinclair Stevenson, London.

Rackham, O. (1976) *Trees and Woodland in the British Landscape*. J.M. Dent and Sons, London.

Rackham, O. (1986) *The History of the Countryside*. J.M. Dent and Sons, London.

Rackham, O. (1989) *The Last Forest: The story of Hatfield Forest*. J.M. Dent and Sons, London.

Read, H.J. (2000) *Veteran Trees: A Guide to Good Management*. English Nature, Peterborough.

Read, H.J., Wheater, C.P., Forbes, V. and Young, J. (2010) The current status of ancient pollard beech trees at Burnham Beeches and evaluation of recent restoration techniques. *Quarterly Journal of Forestry*, 104(2), pp. 109–120.

Read, H.J. and Bengtsson, V. (2019) The management of trees in the wood pasture systems of South East England. In: F.A. Álvarez (ed.) *Silvicultures: Management and Conservation*. Intechopen. https://doi.org/10.5772/intechopen.86100

Sebek, P., Altman, J., Platek, M. and Cizek, L. (2013) Is active management the key to the conservation of Saproxylic biodiversity? Pollarding promotes the formation of tree hollows. *PLoS ONE* [online] 8(3): e60456. https://doi.org/10.1371/journal.pone.0060456

CHAPTER 22

The Value and Meaning of Traditional Natural Resource Use Systems in Satoyama Landscapes in Japan

Katsue Fukamachi

Summary

Japan's rural landscape or *satoyama* is a combination of diverse elements including settlements, forests, rivers and streams. From a case study on the traditional *satoyama* landscape on the shore of Lake Biwa, we gained knowledge and hints as to how future societies in rural farming and fishing villages could look and function. Recently, the traditional systems of use and management of natural resources in *satoyama* areas have either dramatically changed, have become a mere shell or have been completely lost. It is crucial to create a system of use and management of natural resources that looks at entire regions as comprehensive landscapes. To find the most appropriate system, it is necessary to understand the uniqueness of each *satoyama* landscape in addition to being aware of its general features.

Keywords: *satoyama* landscape, natural resource, sustainable system, biocultural diversity

Introduction

Japan's rural landscape, known as *satoyama*, is a combination of diverse elements including settlements, forests, rivers and streams (Fukamachi, Oku and Nakashizuka 2001). Traditionally, the use and management of *satoyama* natural resources in rural areas has been based on the link between forested areas (woodlands and grasslands), settlements (villages and cultivated land) and waterside areas (seaside, lakeside and riverside).

Satoyama rural areas were recently defined as Japan's socio-ecological production landscapes and seascapes. It has been shown that in these landscapes, lifestyles have been shaped in close connection with the local natural features, and that in each *satoyama* over time a unique system for the use and application of local natural resources developed (Fukamachi et al. 2011).

A careful analysis of use and management systems traditionally implemented in specific *satoyama* landscapes and the clarification of the impact of ecological services and local culture on the sustainable use of natural resources has been undertaken. This helps us to identify the essential factors that have over time contributed to the development of 'smart' sustainable

Katsue Fukamachi, 'The Value and Meaning of Traditional Natural Resource Use Systems in Satoyama Landscapes in Japan' in: *Countryside History: The Life and Legacy of Oliver Rackham*. Pelagic Publishing (2024).
DOI: 10.53061/KONR7906

systems in a large number of rural farming and fishing villages (Fukamachi, Oku and Rackham 2003). If we aim to establish a future vision for a *satoyama* landscape, we must shed light on the history and traditions of the area, and identify traditional wisdom and techniques that are still alive and useful today. In addition, we must examine how the *satoyama* landscape is structured and how it connects to the outside. Finally, it is necessary to look for additional new techniques and systems that may be useful in the landscape, and to identify the special characteristics that make it unique. Based on these data, a specific system that is grounded in local ecology and culture and meaningful in today's society can be established.

Insights gained from a case study of a lakeside satoyama landscape near Lake Biwa

Using the example of a rural *satoyama* landscape that extends between the Hira Mountain Range and the shores of Lake Biwa (Fig. 22.1), we identified the local traditional system of use and management of land and natural resources. This analysis was based on the history of local lifestyles and activities since the late Meiji period (1868–1912) up to the early Showa period (1926–89). The target of the study, Moriyama (120 m above sea level), is a village that lies within the larger area of Otsu City. It covers a total area of about 360 ha between the lake shore (85 m) and the summit of Mount Horai (1,174 m). The whole region has traditionally been used quite intensively for daily life and work. At the end of the Meiji period, the village had 70 households. Major local income sources were forestry, the use and production of stone, and the cultivation of rice. Moriyama lies on the traditional trading route that links the ancient Japanese capital Kyoto to the northern Hokuriku region at the Japan Sea.

Figure 22.1 *Satoyama* landscape of Moriyama on the shores of Lake Biwa, Shiga, Japan, 2018. (K. Fukamachi)

The traditional use of space by local residents

Figure 22.2 shows a historical map of Moriyama in the Edo period (1603–1868). Local residents in Moriyama perceived space as a combination of nature and the social environment. They used special names for certain key areas, such as lake shore, rice paddies, plots where vegetables, tea or bamboo were grown, and the village. Forests that were close to the settlement and located in the alluvial fan that reaches up to an elevation of about 300 m were mainly privately owned and called *jiyama*, while forests that were located at higher elevations up to about 500 m, where the local Konpira Shrine is located, were called *herayama*. *Herayama* forests were found on

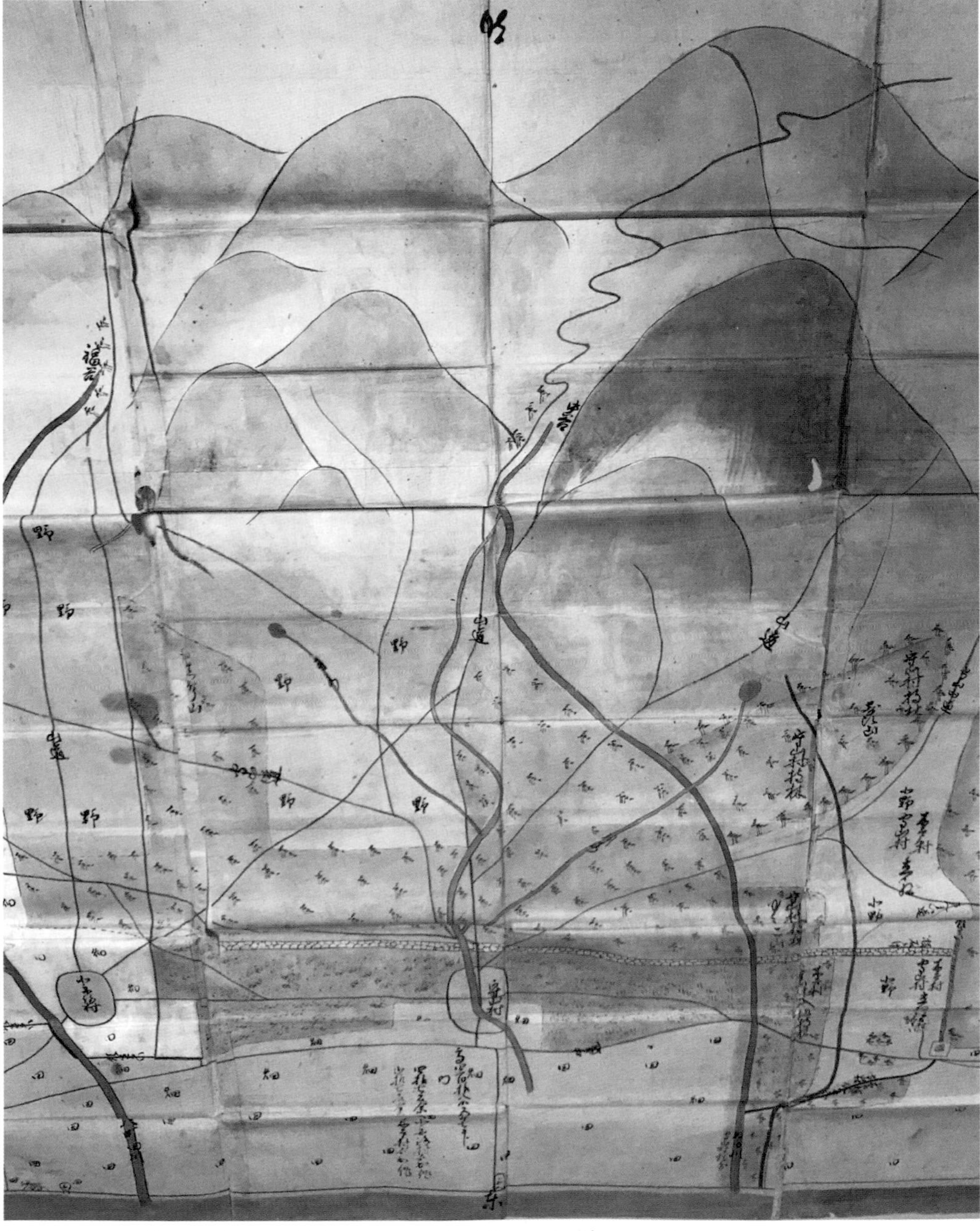

Figure 22. 2 Historical map of Moriyama in the Edo period (1603–1868). (Moriyama Property Ward, Otsu City)

steep slopes and were mainly owned in common. Forests lying between Konpira Shrine and the mountain summit were commons and were called *sannai*, while those close to the summit were called *dake*. In *sannai* areas, local residents had the right to freely gather firewood and to collect red clay for their clay-walled houses. Water could be drawn for use in rice paddies and other cultivated areas from the Norikogawa stream near Konpira Shrine.

On the boundary between the *herayama* and the *jiyama*, there was a resting place called *koba*. At the border between the *jiyama* and the *zaisho* settlement, there was a goods collection point called *maki* and a sacred site for worship of the mountain deity called *yama no kami*. Areas at the border between private forests and common forests were called *iekujiyama*; those bordering on neighbouring villages or near riversides were called *wariyama*. In *iekujiyama* common areas, the right to use trees was divided equally among all households in the village. If a tree was taken from the village commons, it had to be returned or paid for. *Wariyama* were areas owned by selected village households and served the purpose of determining the exact border with neighbouring villages. The transfer of ownership of *wariyama* to non-residents was not allowed.

Furthermore, there were forests of limited size such as *yashiroyama*, which the village maintained for the use of temples, and *yoritoyama*, which were managed by *yorito* members. *Yorito* were families that had the exclusive authority to conduct the local *gokamatsuri* spring festival, an important religious ritual. *Yorito* members had exclusive rights to the use of *yoritoyama*.

Water drawn from hillside streams and springs was used to irrigate nearby terraced rice paddies. Stone walls between paddies were built using local stone. Plots that were located inconveniently for irrigation were used by farmers to plant vegetables and tea for home use. Rice paddies located between the lake shore and the village were called *satoda*, and those located on the hillside *yamada*. On the border between cultivated land and forests, fences were built to fend off wild animals such as wild boar. The lake shore or *hama* was used in common by the locals. An inlet at the lake shore served as a landing place to dock boats and was called *shima*. It played an important role during the *gokamatsuri* spring festival.

The traditional use of natural resources

The schematic diagram in Figure 22.3 shows the main flow of natural resource use in Moriyama in around 1900. We can see that land-use took into account environmental and topographical features, soil, ownership, the management system and other factors. Cultivated land was neatly separated from forests by fences to fend off boar and other wild animals. Forests supplied many essential natural resources for daily life. From the lake shore, forest resources and rocks could be transported by ship to customers along the shore of Lake Biwa or to Kyoto. Freshwater algae collected along the shore of the lake were used as fertilizer on paddies and fields. Reeds harvested at the lake shore were used to cover roofs (Fig. 22.4). *Shijimi* clams, *ayu* fish, *unagi* eel and other edible shellfish and fish were caught in local springs, streams, creeks, in rice paddies, at the lake shore and in the lake.

Red Pine *Pinus resinosa* groves (Fig. 22.5) were planted within the *jiyama* privately owned forest. Red Pine large logs were used as timber or as material to build transportation carts. Branches and other parts of the tree were used to produce firewood and brushwood that could be sold to a roof tile-burning factory on the east side of the lake. Fallen pine needles were gathered to light fires, and pine roots were used to make pine root oil. Pine mushrooms or *matsutake* and other mushrooms such as *aburabon* were harvested in abundance and enriched the people's daily diet.

All parts of the Red Pine – the stem, branches, needles and roots – were used as natural resources and helped support the daily life of residents. According to a diary written by a local resident in the late Meiji period, there were 45 techniques and ways of using Red Pine, including cutting pine logs into round slices, cutting branches, ripping off pine bark,

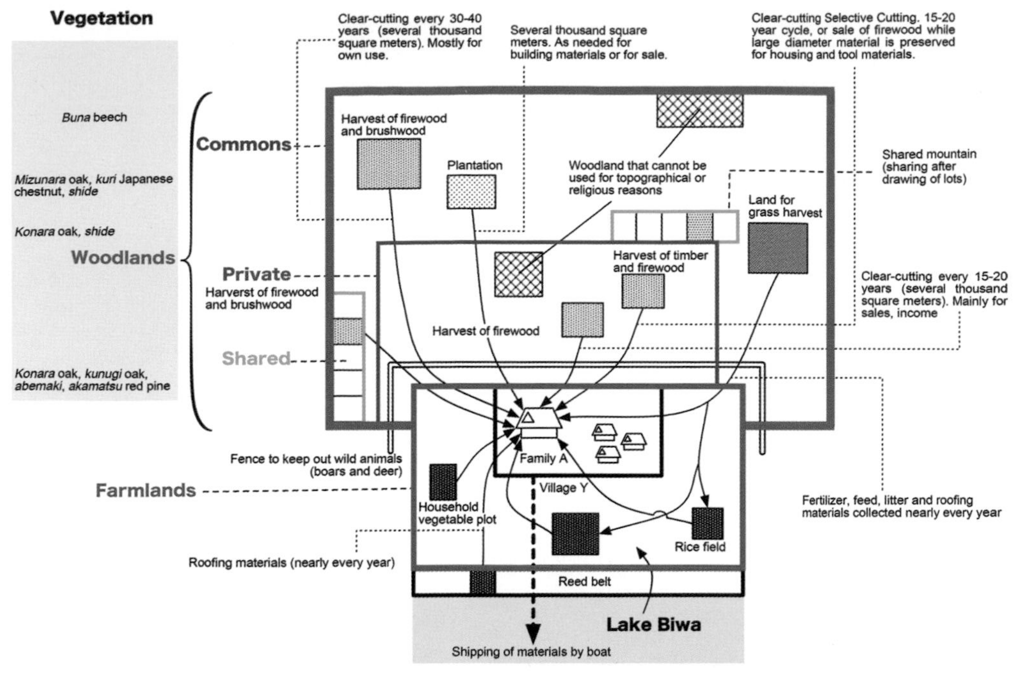

Figure 22.3 Main flow of natural resource use in Moriyama in around 1900. (Fukamachi and Oku, 2011)

Figure 22.4 Traditional farmhouse with reed thatched roof with Oliver Rackham, 2003. (K. Fukamachi)

Figure 22.5 Red Pine grove in *satoyama* landscape with Oliver Rackham, 2003. (K. Fukamachi)

collecting pine cones, digging out pine roots and collecting pine needles. At the time, local people used their wisdom and abundant technical knowledge about how to exploit Red Pine natural resources to facilitate and enrich their lives.

Kunugi and *konara* oak from the *jiyama* privately owned forests were employed both for home use and for sale in the form of firewood or brushwood. Oak was harvested in cycles of 15–20 years. Some of the *kunugi* oak trees were planted; others were wild. Dominating species in the *sannai* forest were *mizunara* oak and *shide* Japanese hornbeam species. Firewood made from these slow-growing hardwood species was called *kenkera* and was appreciated as high-quality firewood for home use. Small-scale plantations of conifers, especially *sugi* Japanese Cedar *Cryptomeria japonica* and *hinoki* Japanese Cypress *Chamaecyparis obtusa*, were used to produce timber. Areas in which trees had been felled for lumber or where firewood and brushwood had been harvested turned into grasslands or *yamakusa* with budding broadleaf trees and with grasses and herbs such as *itadori* Japanese Knotweed *Reynoutria japonica*, or *susuki* Japanese Pampas Grass *Miscanthus sinensis*, that could be harvested for various uses. For example, *susuki* Pampas Grass from the mountains was used as roof cover. Grasslands also existed on the mountaintops, close to streams and on the linear levees that surround rice paddies. Work related to the harvest of *yamakusa* grass included 15 different activities. Among these, the *hotorakari* grass harvest took place during the rice-planting season, the *yokusakari* harvest during July and August, and the *kariboshi* harvest for the production of dried grasses during August and September. *Yokusa* referred to the harvest of grass to feed cattle and to use for fertilizer in rice paddies; *hotora* referred to grass that was used directly as fertilizer in rice paddies; *kariboshi* referred to grass that was dried in the mountains during the hot season and carried to the valley to feed cattle in the autumn.

Figure 22.6 Traditional stone walls between paddies, 2018. (K. Fukamachi)

Finally, local rocks were also an important resource that could be harvested. The local *Moriyama-ishi*, a chert with a characteristic stripe pattern found along streams and in other places in Moriyama, has been widely used for gardening, for building stone walls between paddies (Fig. 22.6) and for lining waterways. Since 1890, the year in which the Lake Biwa Canal was opened, rocks have been carried to Kyoto by boat, where they are sold for garden landscaping.

The traditional common management of natural resources

The village of Moriyama is linked to the summit of Mount Horai by an ancient 1-m-wide mountain road or *daidomichi* that leads up to the summit through a steep and narrow mountain valley. In the past, forest resources and rocks were carried out of the forest on this road by human power or by using wooden carts built for that purpose. In several locations along the road, ditches were dug and firewood and earth were piled up to help prevent soil to be washed away by rainwater. Above Konpira Shrine, there was a steep path called *yuri* winding its way up the mountain on a steep slope. At a key point of the path, there was a group of trees called *yurigi*, which served the dual purpose of preventing the road from collapsing and stopping people from slipping. The *yurigi* was composed of *mizunara* Japanese Oak *Quercus mongolica* var. *crispula*, *inubuna* Japanese Blue Beech *Fagus japonica*, Mountain Cherry *Cerasus jamasakura* and other large broadleaf trees that grow wild. It was forbidden to fell those trees.

Diverse sets of residents were involved in the management of natural resources, among them groups of villagers and groups of shrine members. Depending on the group in question, tasks were distributed according to age or according to the location of the family's house. Tasks were distributed among groups and their members in proportion to location, size and importance of the natural resource. Together, the groups operated a well-functioning system of natural-resource management.

One aspect of the common management concerned restrictions on the use of natural resources in common areas. For instance, it was forbidden to harvest grass for *yokusakari* or cattle in the spring before a certain day of the year that was called *kama no kuchi*, a term referring to the use of the sickle in the grassland. In addition, there was a day called *yamamodori* on which it was forbidden to enter the *sannai* area, and there was a rule that any firewood had to be brought down to the *maki* goods collection place on the very day it was harvested. It was the task of the villagers' association or *murakata*, led by the village head or *kucho*, to set out rules and determine penalties when these were violated. However, supervision of the mountain and catching violators on the spot was the task of members of the local youth organization, the *sanchokata*. Since all common forests except the high-altitude *sannai* were located on the steep mountain slopes behind the village, individuals were not permitted to fell trees there in order to avoid risks and disasters. As a rule, some plots were planted with *sugi* Japanese Cedar and *hinoki* Japanese Cypress for sale to provide income for the village, while most other forests were allowed to grow old and were maintained. Commonly managed areas were mainly places and traffic ways that were used by all for daily life and work. They included common forested land, forest and farm roads, and waterways. Some additional places were also managed in common, for example the entrance area to the village from the lake shore, border areas between different land-uses or different owners, temple and shrine precincts, and spaces reserved for religious rites such as the ritual for the safe journey of the spirits of the dead.

The timing of each management activity was mostly set in advance each year. If necessary, though, the *murakata* village association members would get together and consult whether special measures for a new timing needed to be taken. Depending on the type and scale of the common management, there were cases when all households were obliged to take part and others when groups of households would take turns doing the work. Among management work that needed to be done was the maintenance of roads and waterways (repairs, cleaning and mowing of weeds), and the maintenance and upkeep of areas bordering on neighbouring villages. The common management of waterways was strictly regulated. Taking water for agricultural use from a mountainside water point near Konpira Shrine was called *yugakari*. All households that made use of the water came together once a year for *mizosarae* or the cleaning of waterways and irrigation channels. In the period of rice growth between mid-June and late August, villagers took turns to do *mizuban* or irrigation channel patrols: they had to make the round of the paddies to make sure that water was evenly distributed to all. It was strictly forbidden to pollute or redirect the waterways that carried water from the mountain through the village and paddies down to Lake Biwa. Villagers cooperated in doing waterway patrols and any necessary repairs. When there was a risk of damage owing to a natural disaster, the drums of the village's Wakamiya Shrine were beaten. This was a sign for all villagers to go and check the local streams and waterways, and, if necessary, to pile up soil or to shift water from irrigation channels back into streams in order to mitigate the risk of floods. In the vicinity of the main mountainside water point, there was a big rock called *chogoroiwa*. There was a legend in the village that the rock had the power to shift the direction of the stream and to spare the village from floods. To express their gratefulness, at the beginning of each year the *murakata* villagers made an offering of rice wine and dedicated it to the deity of the rock.

Moriyama has an abundance of large and small sacred spaces dedicated to rituals in its forests, in the village and on the lake shore (Fukamachi and Rackham 2012). These spaces have always been managed in common. In preparation of seasonal religious events, villagers cleaned and prepared the precincts of Wakamiya Shrine, which serves as the location for the spring festival *gokamatsuri*, and cleaned and repaired the lake shore road, which is used to carry *mikoshi* portable shrines during religious rituals. On ordinary days, the sacred spaces were excellent spots for gathering mushrooms, fruits and nuts, and they served as an ideal playground for local children. Besides the spring festival, there were other rituals and events that used sacred space. For instance, there was the so-called *dondoyaki* event, when a group of eight shrine members came together twice a year to make offerings to the mountain deity to pray for the safety of the local forestry workers. During this ritual, a big bonfire of bamboo, firewood and Japanese Wisteria *Wisteria floribunda* vines was lit, and offerings of vegetables and *donko* fish from Lake Biwa were made. The regular spiritual events were an opportunity to make good and regular use of natural resources from the forests, from the lake and from the cultivated land. Although the amount of resources needed for rituals was limited, the events motivated the villagers to take good care of natural resources, and to manage diverse spaces in their village sustainably.

The link between natural resources and children's daily life and play

The daily life and play of children was closely related to the local natural resources. Seventy-two types of play were found to have been practised by local children in the early Showa period. Among these, 69 took place outside or made use of natural resources. Most frequent were games that involved plants. They included plucking and eating wild *akebi* fruit, making pine chewing gum and crafting skis from bamboo. Play involving animals included catching eels, playing with *kuwagata* beetles and crafting small birds. In the Red Pine grove, where villagers often worked, a large variety of plants and mushrooms such as *matsutake* were found. Gathering them was part of a child's play experience. There were 36 types of play recorded in which children gathered natural resources to make tools they could use in play. Materials used were bamboo, stone and wood, mostly the same materials that were used by adults. For some games, children used tools that were quite difficult to make. Children played in a great variety of places, among them the mountains, rivers, riversides, channels, the lake, the lake shore, the lake inlet where boats docked, roads, shrines, temple and shrine groves, paddies and fields, paddy levees, springs and free-standing trees. While play differed according to gender, season and the location of a child's home, most play took place in forests and at waterside locations close to the village. This was in an environment where natural features and wildlife were the direct result of the regular use and management of natural resources by the villagers, such as harvesting firewood and grass from the mountains, or using waterways for the irrigation of rice paddies. The villagers' regular involvement with natural resources in their daily life and work naturally laid the groundwork for a great variety of places where children could play, and for many types of play they could engage in. Through their involvement with natural resources from a young age, children naturally became skilled at recognizing and using all kinds of materials.

Insights: a look to the future

From this case study on the traditional *satoyama* landscape on the shore of Lake Biwa, we can gain knowledge and insights as to how future societies in rural farming and fishing villages could look and function.

First, we can obtain important knowledge from looking at the special terminology that local people used to name certain key locations in their village, and at the way they made distinctive use of various spaces in the natural and social environment, which shows their keen awareness of space. Based on traditional wisdom, decades of experience and careful assessment of the land situation in their village, *murakata* or villagers made sure that no more trees than necessary were felled in the forests so that resources would never dry up, and so that natural disasters would cause less damage. They had strict rules about the location, quantity and frequency of natural resource use, and they had a highly functioning system of land ownership and organizational management.

Based on this system, all essential strategies for the use and management of the village's land were covered. Appropriate and sustainable use of natural resources in the entire *satoyama* landscape of the village naturally led to the area being efficiently managed. The whole *satoyama* area had a close-knit network of roads and waterways, which connected the spaces where daily life and work, religious rituals, children's play and other activities took place. It can be said that this integrated network of spaces, roads and waterways used in daily life effectively doubled as a safety traffic network that allowed access to a variety of places in case of emergencies. Furthermore, there were diverse regulations that supported the sustainable use of the village's land and natural resources, and helped to equally distribute the village's limited natural resources among the villagers. Such rules included putting border areas to neighbouring villages into common ownership and setting up *iekujiyama* and *wariyama* areas.

The village's highly differentiated use of space, the efficient organization of use and management, and village-specific rules were also highly significant in terms of disaster prevention and disaster response in a mountainous area with steep slopes where there was always a risk of landslides or floods.

Local residents became familiar with natural resources as children through play and would later have many opportunities to get in direct touch with natural resources. Through this they developed a deep understanding of them. During exchanges with other villagers, traditional wisdom and techniques were also acquired.

So how does the situation look today? During the last few decades, there have been drastic changes in the social environment of *satoyama* landscapes. Nationwide, the traditional systems of use and management of natural resources in *satoyama* areas have either dramatically changed, have become a mere shell of what they once were or have been completely lost. The formerly close relationship of residents with their natural environment and with local natural resources has been tremendously weakened. In addition, in many villages there are spaces where the land ownership is unknown and spaces that have not been managed for a long time. Furthermore, there is a general lack of concern regarding natural disaster prevention and the preservation of biodiversity.

With the purpose to promote sustainable resource use and biocultural diversity, our efforts in *satoyama* landscapes on the shores of Lake Biwa have focused on: (1) making good use of diverse local natural resources during daily life and everyday work; and (2) making good use of the arrangement in space and of existing links between the *satoyama* landscape elements of lake, farmland, village and forest.

In order to counter the current situation, it is necessary to create a system of use and management of natural resources that looks at entire regions as comprehensive landscapes. To find the most appropriate system, we must understand the uniqueness of each *satoyama* landscape in addition to being aware of its general features. To this end, we need to identify and carefully review the traditional system and the local culture of each *satoyama* landscape. Based on these data, we can then examine which insights and technologies of our time can be added in order to bring to fruition the vision of sustainable rural farming and fishing villages in Japan's *satoyama* landscapes.

During our research on the connection between Japan's culture and Japan's ecology, I received invaluable advice from the late Professor Oliver Rackham. More than 20 years have gone by since I first met him when he came to Japan. After that, we frequented each other's research sites, conducted fieldwork and research together, and collaborated in publishing our results in books and other publications. A special common field of interest was the comparison between the British countryside and the Japanese *satoyama* landscape. By means of a combination of methodologies and based on a variety of data showing the particularities or common features of the traditional countryside of each country, I deepened my understanding of the characteristics of the respective landscapes. Oliver carefully took notes while walking in the *satoyama* landscape at the shores of Lake Biwa. In the preface to the Japanese edition of Professor Rackham's book, *The History of the Countryside* (originally published in 1986), which we published in Japan in a process that extended over several years, he remarked:

> When I first wrote this book I had no notion that it would be of interest to anyone outside Britain or possibly Ireland. Japan is a very different country, from which British culture must seem as remote and independent as that of the Aztecs. However, both countries are lands of ancient settled civilization; their plants and animals have something in common; and to some extent their human inhabitants have encountered similar problems and have dealt with them in similar ways.
>
> Rackham 2012: v

While pointing out the unique characteristics of the culture and ecology of each country, Professor Rackham shed light on features found in both landscapes and drew attention to various problems we face today, thus teaching us how important it is to think about ways to find solutions to them.

References

Fukamachi, K., Miki, Y., Oku, H. and Miyoshi, I. (2011) Distribution of isolated trees and hedges in a Satoyama landscape on the west side of Lake Biwa in Shiga Prefecture, Japan: a case study from the viewpoint of biocultural diversity. *Landscape Ecology and Ecological Engineering*, 7(2), pp. 195–206. https://doi.org/10.1007/s11355-011-0164-1

Fukamachi, K. and Oku, H. (2011) Attempt of comparative satoyama theory – from the fieldwork of the Tango Peninsula mountainous area, the west side of Lake Biwa, and the Keihanna hills. In: Osumi and T. Yumoto (eds) *Environmental History of Sato and Woodlands*, Bunichisogoshuppan, pp. 209–38.

Fukamachi, K., Oku, H. and Nakashizuka, T. (2001) The change of satoyama landscape and its causality in Kamiseya, Kyoto Prefecture, Japan between 1970 and 1995. *Landscape Ecology*, 16, pp. 703–17. https://doi.org/10.1023/A:1014464909698

Fukamachi, K., Oku, H. and Rackham, O. (2003) A comparative study on trees and hedgerows in Japan and England. In: H. Palang and G. Fry (eds) *LANDSCAPE INTERFACES Cultural Heritage in Changing Landscapes*. Academic Publishers, Kluwer, pp. 53–69. https://doi.org/10.1007/978-94-017-0189-1_4

Fukamachi, K. and Rackham, O. (2012) Sacred groves in Japanese satoyama landscapes: a case study and prospects for conservation. In: G. Pungetti, G. Oviedo and D. Hooke (eds) *Sacred Species and Sites: Advances in biocultural conservation*. Cambridge University Press, Cambridge, pp. 419–23.

Rackham, O. (1986) *The History of the Countryside*. J.M. Dent and Sons, London.

Rackham, O. (2012) *I-gi-ri-su-no Kan to-rii-sai-do* [translation of O. Rackham (1986) *The History of the Countryside* into Japanese, supervised by H. Oku, H. Ito, D. Sakuma, K. Shinozawa, K. Fukamachi and K. Showado]. Showado Press, Kyoto, Japan.

CHAPTER 23

Pollard Beech Trees in Snowy Areas of Japan

Tohru Nakashizuka, Hideo Miguchi and Tomohiko Kamitani

Summary

Various tree species are managed as pollards in Japan and a special type of pollard beech tree, Japanese Beech *Fagus crenata*, is found in some secondary forests located in areas with heavy snowfall. The wood of this tree is used for firewood or charcoal production; pollards are cut 1 to 3 m above the ground, while there is snow on the ground, and the wood is carried by traditional sled to reduce yarding labour. Since beech does not continually produce shoots if all the shoots are cut at one time, a portion of the pollard stems were repeatedly cut at 20- to 40-year intervals to keep the individual trees alive and productive after cutting. Such management processes create bollings, which are long-lived and can survive as ancient trees.

Keywords: Japanese Beech *Fagus crenata*, pollarding, Japanese forest, bolling, forest management, coppicing

Introduction

Pollarding is a system that utilizes trees for fodder or timber by cutting off the upper parts of the tree, up to several metres above the ground. It has a different purpose and management scheme from coppicing. Rackham (1989, 1994, 2003) described pollard systems in England and other countries for various tree species, such as maple, ash, holly, oak, beech, elm, willow, yew, black poplars, hawthorn, juniper, hornbeam, hazel and aspen. Some of these species are different from those used for coppice, and they vary in terms of their use and management. Pollard systems require more labour than coppice systems as ladders are used to cut stems at a higher position above the ground. Thus, like-for-like, it naturally has higher labour costs than coppicing.

In England, pollarding is used for firewood rather than timber, and sometimes thinner branches and leaves are used for fodder for livestock, which are cut high above the ground to avoid herbivory by animals (Rackham 1989, 1994). Pollards are frequently found at the borders between grasslands and/or forests as the large bollings can make the borders clearer, and even after their shoots have been cut they can physically act as fences, regulating the behaviour of animals. Such management processes increase the longevity of trees, resulting in pollards surviving as 'ancient trees' (Rackham 2003). Thus, the uniqueness of pollard

Tohru Nakashizuka, Hideo Miguchi and Tomohiko Kamitani, 'Pollard Beech Trees in Snowy Areas of Japan' in: *Countryside History: The Life and Legacy of Oliver Rackham*. Pelagic Publishing (2024). © Tohru Nakashizuka, Hideo Miguchi and Tomohiko Kamitani. DOI: 10.53061/MASW5073

management systems is related to the species used, the purposes for use and expected effects of the management system.

Pollarding is a traditional system that was used in Japan until several decades ago (Nakashizuka et al. 2000; Rackham 2003; Suzuki 2009), although it is seldom seen in operation today. The species used, and the aims and management of pollard systems in Japan, were different from those in England or other European countries (Nakashizuka et al. 2000; Suzuki 2009). In particular, Japanese Beech *Fagus crenata* forests, known as Agariko, are one of the best-known examples of pollarding in Japan (Fig. 23.1). These pollarding systems are unique, and adapted to snowy climates. In this chapter, we review existing knowledge about the pollarding system (distribution, utilization and management) of beech and other species in Japan, and discuss how it differs from pollarding practices in England and other countries.

Pollard tree species in Japan and their utilization

Many species were used for pollards in Japan. Mulberry *Morus* spp. was used as fodder for silkworm, and there used to be large areas of mulberry farmland across Japan until the 1960s. Some oak species, such as the Sawtooth Oak *Quercus acutissima*, were also used as fodder for wild silkworms (Suzuki 2009), although this particular species was also employed for firewood (Nozaki 1994). Fresh leaves and young shoots were taken and the pruning height was relatively low, up to about 1 m above the ground, which made harvesting easier and reduced labour costs (Suzuki 2009).

Willows *Salix* spp. were utilized as pollards for domestic mammal fodder, and coniferous tree species, such as Japanese Cedar *Cryptomeria japonica* and Sawara Cypress *Chamaecyparis pisifera*, were managed as pollards for the continuous production of thin logs (Suzuki 2009).

Pollard systems prevent animals from eating the young shoots and leaves, and many pollard trees are found in grasslands and borders in England and European countries (Rackham 1989). In Japan, some pollards, for example oaks and willows, may also avoid herbivory by domestic mammals, although it is not clear if this was an aim of the foresters (or farmers) (Nozaki 1994). There are some traditional pastures or grasslands in Japan, which are mostly

Figure 23.1 Pollard beech trees in Japan. a) A typical pollard tree at Mount Chokai, Yamagata Prefecture, Japan. The girth of the bolling is 762 cm, and it is famous as an ancient tree named Agariko-daio (king of pollard), July 2018. (T. Nakashizuka)

Figure 23.1 (*continued*) b) Beech pollard with many stems at Ohshirakawa, Niigata Prefecture, Japan, April 2017. (T. Kamitani)

Figure 23.1 (*continued*) c) Pollard beech with stems sprouted at more than 3 m above the ground at Tainai, Niigata Prefecture, August 1993. (T. Nakashizuka)

Figure 23.1 (*continued*) d) A forest dominated by pollards at Mount Chokai, July 2018. (T. Nakashizuka)

located in areas with low snowfall, possibly because of the ease of using fire to maintain the grasslands (Nakashizuka and Iida 1995). Fires are used in winter or early spring when the climate is dry on the Pacific side of Japan; on the western side of the country, where beech trees dominate, the land is still covered by snow during this period.

Pollards of some tall broadleaf tree species, such as Japanese Zelkova *Zelkova serrata*, Japanese Horse Chestnut *Aesculus turbinata*, Painted Maple *Acer mono*, Japanese Lime *Tilia japonica*, oaks (Mizunara *Quercus crispula* and Jolcham Oak *Q. serrata*), Japanese Big-leaf Magnolia *Magnolia obovata* and beech were utilized mostly for firewood or charcoal production (Nozaki 1994; Nakashizuka et al. 2000; Suzuki 2014). The pollards of these tall broadleaf tree species are mostly solitary and distributed in secondary forests, sometimes with coppices of trees other than beech, although there are some forests dominated by pollards, for instance Mount Chokai, Yamagata Prefecture, Japan (Fig. 23.1d).

Some pollard systems may be able to prevent soil erosion on steep slopes, for example those for Japanese Zelkova and Japanese Horse Chestnut (Suzuki 2009). These species tend to be distributed on such topographies for other ecological reasons (probably being dependent on geomorphic disturbances), and people may have noted that the existence of bollings contributed to protection of the soil and land, even after the pollards were cut.

Distribution of pollard trees in Japan

Beech pollards are distributed in areas of heavy snowfall, most with a maximum snow depth of more than 1 m (Fig. 23.2). The natural distribution of beech is biased toward snowy areas, and beech dominance increases with maximum snow depth (Honma 1997). However, the distribution of pollards is restricted to areas of heavier snow. Even in the warmer western parts of Japan, pollarded beech trees are distributed among the areas of heaviest snowfall, such as mountain summits (Fukamachi et al. 1999; Ogura 2018). This association between the distribution of pollarded beech and snow is unique to Japan, and is related to utilization and management.

Pollards of species other than beech are sometimes found in snowy areas, even though they have their distribution biased to snowy climate. Pollards of mulberry, willows and oaks

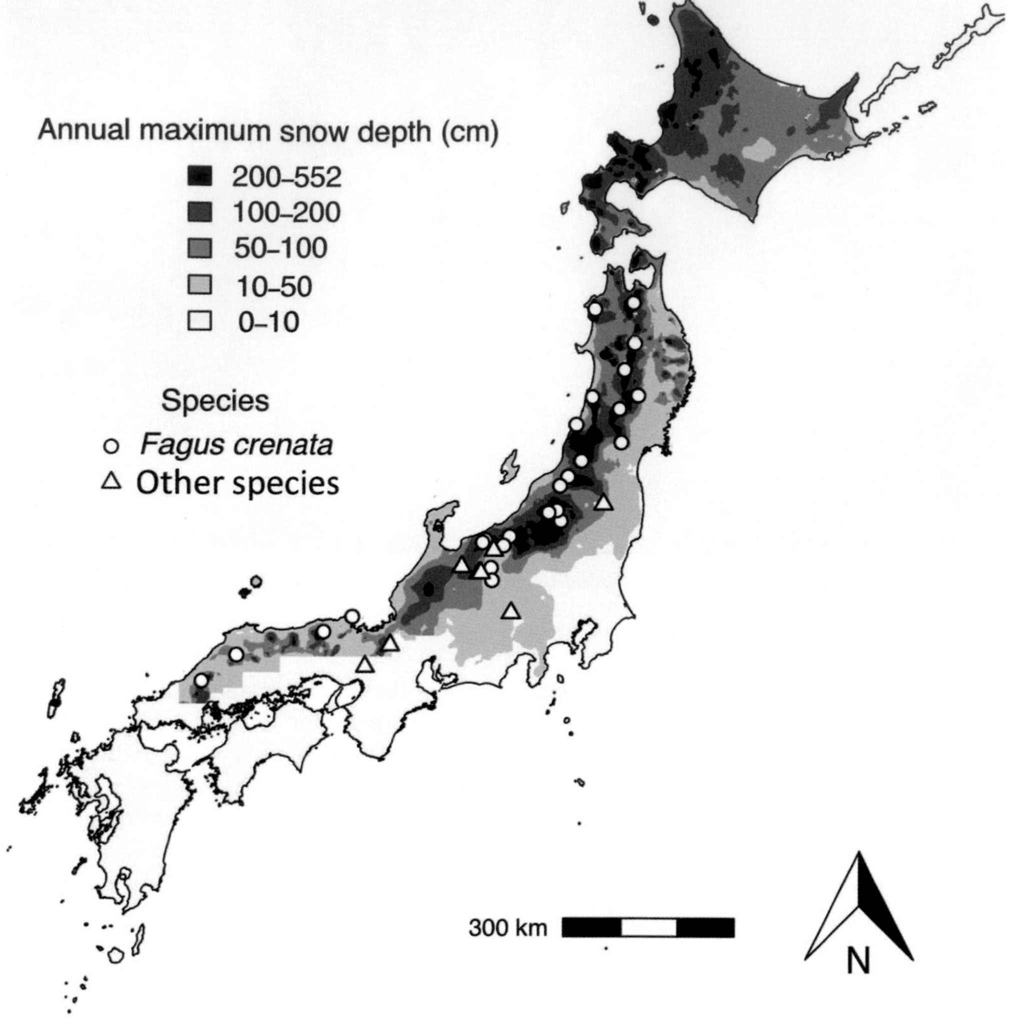

Figure 23.2 Distribution of pollards and maximum snow depth in Japan. The distribution of pollards of tall trees species used for firewood and/or charcoal production are shown; the distribution of mulberry pollard, which was used for silkworm fodder, is not shown since it is distributed across Japan. The pollard trees of beech are distributed only in areas of heavy snowfall. (the authors)

are sometimes distributed in both snowy and non-snowy regions (on the Pacific Ocean side of Japan). In snowy regions, pollards tend to be isolated in secondary forests, suggesting that they were utilized for firewood and/or charcoal production. On occasion, they were also used for construction if the area lacked tree species that would be more suitable for construction. In general, conifers such as *Cryptomeria* and *Chamaecyparis* are the most suitable species for structural timber in Japan, although they are not found in snowy regions, or tend to have non-straight stems in snowy regions because the pressure of the snow acts to bend them. In such cases, beech pollard could be used to construct farmers' residences (Ida et al. 2010). Thus, pollards in snowy regions were not used for fodder, but rather for firewood, charcoal and timber, which require thicker diameter shoots than fodder.

The pollards of Painted Maple, Japanese Lime, oaks (Mizunara and Jolchum), Japanese Big-leaf Magnolia and beech, which were frequently used for firewood or charcoal, are often encountered in snowy regions (western side of Japan). In some cases, old charcoal kilns are found in such forests (Fig. 23.3), or pollard trees are present among coppices. Pollards are

Figure 23.3 Charcoal kiln in a pollard forest at Mount Chokai, July 2018. (T. Nakashizuka)

generally isolated and distributed among secondary forests, suggesting that some particular characteristics of an individual tree render it suitable for use as a pollard. In some exceptional areas, pollard beech trees dominate a forest, such as in Mount Chokai (Fig. 23.1d) (Nakashizuka et al. 2000). Thus, it is suggested that the pollards in snowy regions in Japan had different purposes and management systems from those in non-snowy regions or European countries.

Management of pollard systems in Japan

Information about traditional coppice management has been recorded to some extent (see Kamitani 1986a,b). However, we do not know much about pollard management. Coppice forests in Japan used to be logged every 20 to 30 years, and the logs were utilized for firewood or charcoal production. Oaks, chestnut and hornbeams *Carpinus* spp. were the species used most frequently as coppice. Since the 1970s, demand for firewood and charcoal has reduced, and the area of coppice management has greatly decreased. Forests once used as coppice can be 60 years old or more. However, information on pollard management for firewood or charcoal has not been recorded scientifically, and includes many estimates (Nakashizuka et al. 2000).

It is thought that beech tree pollards were logged in late winter, based on foresters' records of working in freezing conditions (Saito 1979), which describe logging beech for firewood from February to April at Mount Chokai. During this season, there is still deep snow cover (sometimes more than 3 m) in forests, and the foresters had to cut shoots high above the ground. On Mount Chokai, where beech pollards are frequently observed, many pollarded stems branch from bollings up to 4 m above the ground (Fig. 23.4). The maximum snow depth of this area is over 3 m, and logs were carried using sleds or else by sliding them directly on the snow to reduce labour. A traditional tool called the *Ippon-zori* (one-bladed sled) is a special yarding tool that allows timber to be carried over the snow by one person (Fig. 23.5).

The utilization and management of beech pollards are also related to the ecology of beech and its habitat. In the case of coppice management, most tree species have the ability to produce shoots after they are cut back, and this reshooting ability is lost when the tree reaches a species-specific size (Kamitani 1986a; Shibata et al. 2014). Beech is very different from oak, and loses its ability to produce shoots at a relatively smaller size, before it grows to

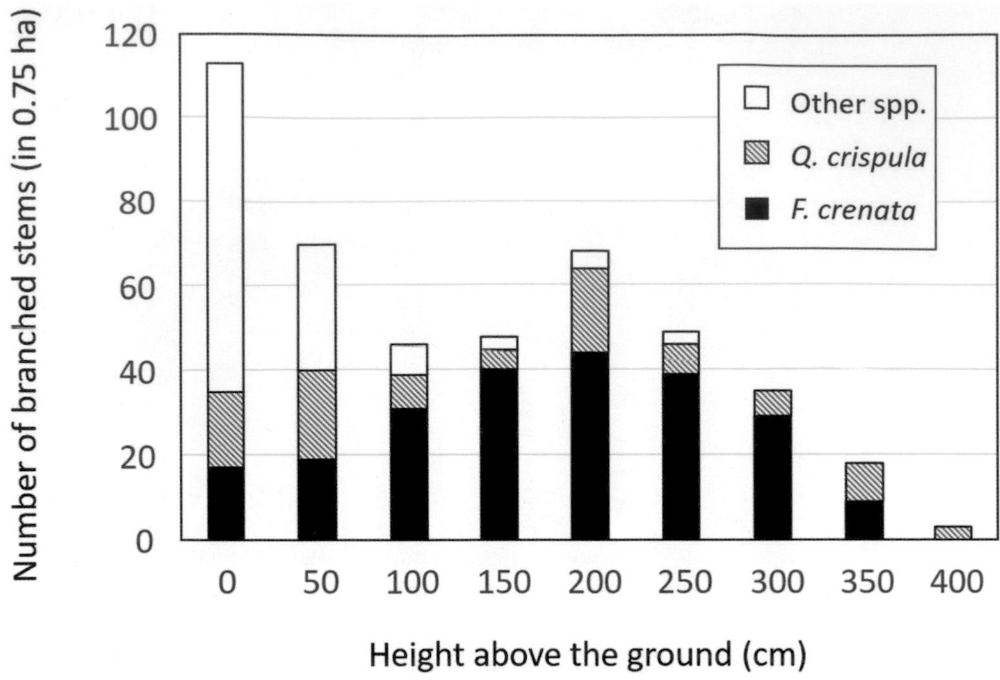

Figure 23.4 Height of pollard branched stems. Height includes trees with single stems and coppices. (Data from Nakashizuka et al. (2000) were reanalysed)

Figure 23.5 Revived traditional tools: an *Ippon-zori* (one-blade-sled) used for yarding in the snow in Ohshirakawa, Niigata, Japan, May 2018. (T. Kamitani)

a size suitable for firewood or charcoal production (Kamitani 1986a). Therefore, beech tends to be excluded from coppice forests, and if beech is to be utilized for firewood or charcoal it must be regenerated with seeds or managed as pollard. Beech regeneration via seed is relatively difficult because of their mast seeding habit and the dense forest-floor vegetation in Japan (Nakashizuka 1987). Pollard systems are more effective to harvest beech wood in a continuous way, because shoots are produced even after a tree grows to a sufficient size to produce firewood or charcoal (Kamitani 1986b). This may be the reason that beech pollards exist among coppices of other species that can be utilized as coppice for firewood and charcoal.

In the case of large beech pollards, if all the stems are cut at one time, the whole bolling will die after it has produced shoots once after being cut (Mitchell 1989; Fig. 23.6; Nakashizuka et al. 2000). Therefore, to sustainably maintain pollard beech trees, branches should not be cut all at once. Tree-ring analysis on Mount Chokai showed that only some parts of pollard stems were cut, and one pollard tree consisted of stems sprouted after two or three prunings (Fig. 23.7). Stems from a bolling were repeatedly cut at 20- to 40-year intervals when they reached about 5 to 10 cm in radius, leaving several trees unlogged. Stems of 5 to 10 cm in radius are suitable for firewood and charcoal production. Cutting during the winter has advantages for beech trees as new shoots are produced the following spring (Nakashizuka et al. 2000; Mitchell 1989).

Pollard management may have originated because the natural trees were damaged by snow (Tanimoto 1993). Trees with trunks damaged by heavy snow were still able to produce multiple shoots after being damaged. Traditional foresters might have learned to utilize such trees to harvest these new stems, and they developed a system to manage the pollard in a sustainable way. This may be why pollard trees are mostly distributed in a solitary fashion within secondary forests. Repeated cuttings are an effective way to produce wood suitable for firewood and charcoal, and such schemes made the pollard trees thicker and resulted in bollings with many knuckles and fists, which are typical of veteran and ancient trees.

Figure 23.6 Dead bolling after all the branched stems were cut, July 1997. The bolling produced shoots once. (T. Nakashizuka)

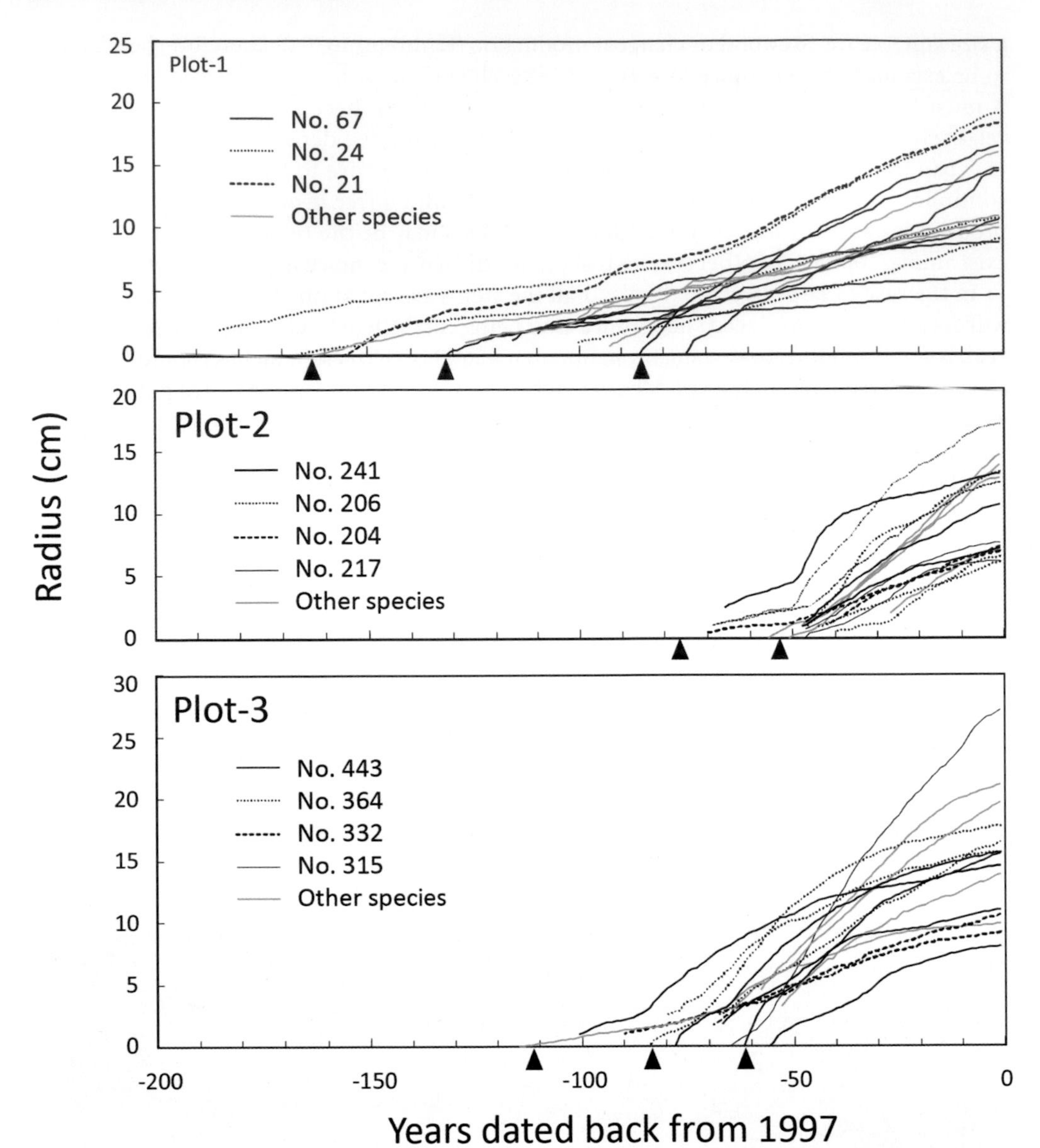

Figure 23.7 Radial growth of pollard stems by annual ring analyses. The growth curves of pollards from the same individuals (indicated in numbers) are shown by the same symbols. Triangles under the x-axis indicate the estimated time of cutting, showing several cuttings even for stems from the same individual. Trees other than beech were also cut at the same time as beech pollards. (After Nakashizuka et al. 2000)

Conclusions

The pollard beech in Japan is rather different from pollards found in England and other European countries (Rackham 1989, 1994, 2003), both in its purpose and in terms of the management system used. Some tree species other than beech have been used for pollards to provide fodder and/or to avoid herbivory from animals. However, in snowy regions, pollard (especially beech) was developed so that logs could be cut in winter and the snow helped people to handle the logs with less labour. Partial and repeated cuttings in winter were used to maintain the pollards for sustainable wood production, based on the sprouting ecology of beech. These pollards are now long-lived, resulting in ancient trees.

Acknowledgements

We thank Dr Rackham for his many suggestions regarding traditional management systems in Japanese forests, including information on pollards, during his stay in Japan. We appreciate the visits to many forests that showed us rich examples of human-forest interactions. This chapter introduces an example of the discussions we had with Dr Rackham, and we hope to further develop his ideas on human-forest interactions, which we believe include many important implications for sustainable forest management.

References

Fukamachi, K., Oku, H., Shimomura, A., Kumagai, Y. and Yokohari, M. (1999) Relations between forest management methods and ecological features of the SATOYAMA beech forests in the Kamiseya and lkaga Districts, Kyoto. *Journal of the Japanese Institute of Landscape Architecture*, 65, pp. 687–92 (in Japanese with English summary). https://doi.org/10.5632/jila.62.687

Homma, K. (1997) Effects of snow pressure on growth form and life history of tree species in Japanese Beech forest. *Journal of Vegetation Science*, 8, pp. 781–88. https://doi.org/10.2307/3237022

Ida, H., Shoji, T., Goto, A., Ikeda, C. and Tsuchimoto, T. (2010) Comparison of species composition of traditional farmhouse structural timbers and surrounding forest in central Japan. *Journal of Japanese Forestry Society*, 92, pp. 139–44 (in Japanese with English summary). https://doi.org/10.4005/jjfs.92.139

Kamitani, T. (1986a) Studies on the process of formation of secondary beech forest in a heavy snowfall region (II). The relationship between stump age and the reproductive capacity for coppice sprouts of main woody species. *Journal of Japanese Forestry Society*, 68, pp. 127–34 (in Japanese with English summary).

Kamitani, T. (1986b) Studies on the process of formation of secondary beech forest in a heavy snowfall region (III). Seed production in secondary beech forests with six different diameter classes. *Journal of Japanese Forestry Society*, 68, pp. 447–53 (in Japanese with English summary).

Mitchell, P.L. (1989) Repollarding large neglected pollards: a review of current practice and results. *Arboricultural Journal*, 113, pp. 125–42 (in Japanese with English summary). https://doi.org/10.1080/03071375.1989.9756411

Nakashizuka, T. (1987) Regeneration dynamics of beech forests in Japan. *Vegetatio*, 69, pp. 169–75. https://doi.org/10.1007/BF00038698

Nakashizuka, T. and Iida, S. (1995) Composition, dynamics and disturbance regime of temperate deciduous forests in Monsoon Asia. *Vegetatio*, 121, pp. 23–30. https://doi.org/10.1007/BF00044669

Nakashizuka, T., Izaki, J., Matsui, K. and Nagaike, T. (2000) Establishment of a pollard beech forest, 'Agariko,' in Mt. Chokai, northern Japan. *Journal of Japanese Forestry Society*, 82, pp. 171–78 (in Japanese with English summary).

Nozaki, E. (1994) Ecological studies of the forest vegetation in the Hokusetsu Mountains. Kinki District. I. Vegetation of the secondary oak forests dominated by *Quercus acutissima* and *Q. serrata*. *Kobe College Studies*, 41, pp. 135–46 (in Japanese with English summary).

Ogura, J. (2018) Vegetation landscape changes and their backgrounds in the West Chugoku Mountains region since the rapid economic growth period as the turning point: cases of Yawata Highland, northwest Hiroshima prefecture and Akiyoshidai, Yamaguchi Prefecture. *Research Report of National Museum of Japanese History*, 207, pp. 43–77 (in Japanese).

Rackham, O. (1989) *The Last Forest: The story of Hatfield Forest*. Phoenix, London.

Rackham, O. (1994) *The Illustrated History of the Countryside*. Weidenfeld and Nicolson Ltd, London.

Rackham, O. (2003) *Ancient Woodland: Its history, vegetation and uses in England*, New edition. Castlepoint Press, Kirkcudbrightshire.

Saito, H. (1979) Freezing death of a charcoal maker. *Culture of Kisakata*, S–53, pp. 53–54 (in Japanese).

Shibata, R., Shibata, M., Tanaka, H., Iida, S., Masaki, T., Hatta, F., Kurokawa, H. and Nakashizuka, T. (2014) Interspecific variation in the size-dependent resprouting ability of temperate woody species and its adaptive significance. *Journal of Ecology*, 102, pp. 209–20. https://doi.org/10.1111/1365-2745.12174

Suzuki, W. (2009) Genealogy of pollard in Japan. *Rihgyo Gijutu (Forestry technology)* 803, pp. 2–6 (in Japanese).

Tanimoto, T. (1993) Sprouting of beech (*Fagus crenata* Blume) in the regeneration of the beech forests and its environmental condition. *The Journal of Japanese Society of Forest Environment*, 35, pp. 42–49 (in Japanese with English summary).

CHAPTER 24

So Human a Landscape: Oliver Rackham's Influence on a New England Ecologist

Henry W. Art

Summary

An intellectual and collaborative association with Oliver Rackham commenced in the 1980s as the author began to unravel the patterns and processes in a post-agricultural landscape in Western New England, USA. Oliver Rackham's attention to the details of field evidence and the reconstruction of woodlands and forests from diverse sorts of historical data reinforced the need to widen perspectives in both 'interviewing' and interpreting the landscape. Ultimately, the most important perspective came from his considering humans and their activities as integral parts of nature, whether in small, defined tracts or at the scales of landscapes and regions.

Keywords: nature, forest, the Hopkins Memorial Forest, storms

Introduction

> It is debatable whether 'virgin forest' or 'primaeval forest' unaffected by mankind exists anywhere in the world, or whether it is one of those phantoms, like 'primitive man', that haunt the scholarly imagination.
>
> Rackham 2006: 14

By the late twentieth century, the magnitude of human activities altering the biosphere had become so dominant that some declared the death or end of nature as we had known it for millennia. Carolyn Merchant (1980) argues philosophically that the death of the human perception of a beneficent, maternal nature ended during the Scientific Revolution of the sixteenth and seventeenth centuries, which allowed humans to control and alter the natural world in ways and on a scale that had not previously been possible. Our transition from a natural magic to a mechanical world-view, coupled with mercantile capitalism and the rise of nation-states, according to Merchant, changed our fundamental relationship with what we considered to be nature. This sentiment is extended by Bill McKibben (1989), who posits that human-influenced global environmental modifications, primarily greenhouse gas emissions, have become so pervasive and rapid that there is no place on planet Earth that has not been affected by unnatural environmental changes.

Henry W. Art, 'So Human a Landscape: Oliver Rackham's Influence on a New England Ecologist' in: *Countryside History: The Life and Legacy of Oliver Rackham*. Pelagic Publishing (2024). © Henry W. Art. DOI: 10.53061/IGHM6239

Has nature really ended? I think not, and I am sure that Oliver Rackham also thought not. Rackham's investigations of the nature of woodlands and forests in Great Britain and Greece clearly show humans as part of the natures of places, and human alteration and interactions with the environment as integral elements of natural systems long before the Scientific Revolution or the Nuclear Age. The distribution of individual plants and animals that combine into biotic communities was, and remains, greatly influenced by humans as components of nature.

I had the privilege of coming to know Oliver by hosting him in Williamstown, MA for three days in late May 1981 and conducting him through the Hopkins Memorial Forest (HMF), a post-agricultural landscape of diverse, largely deciduous woodlands. This contact ultimately led to Oliver sponsoring me as a Visiting Fellow at Corpus Christi College, Cambridge during 1987–8, when we were able to further share our mutual interest in how land-use history has influenced the patterns and processes of forest and woodland ecosystems.

As an incipient forest ecologist educated in North American colleges and universities in the middle of the twentieth century, I had been strongly influenced by more experienced ecologists embracing the notion that there was a premium on discovering the patterns and processes of communities and ecosystems in what was then referred to as their 'natural state'. I was told that the best ecological research, characterizations of community dynamics and studies of ecosystem function occurred within that holiest of grails, the Primeval Forest, untouched by human hands.

The Hopkins Memorial Forest

In the early 1970s, upon my arrival at Williams College, I commenced the investigation of the components of a landscape situated in the north-west corner of Williamstown, Berkshire County, Massachusetts, USA. As a newly appointed faculty member, I happened to be living in Williams College housing located in the HMF. This was a 656-ha (1,625-acre) post-agricultural landscape stretching from the Hoosic River, 182 m (600 ft) msl, to the Massachusetts–New York state border near the crest of the Taconic Range, 732 m (2,400 ft) msl. The land is named for Amos Lawrence Hopkins, who assembled the tract from small subsistence farms in the late nineteenth century. Hopkins merged these tracts into his Buxton Farms, an agricultural showplace that was a fully active farm until 1924 and thereafter only partially active into the mid-1930s (Fig. 24.1).

In 1934, A.L. Hopkins's widow gave the land to Williams College as a memorial to her late husband. The following year, Williams deeded the land for $1 to the US Forest Service (USFS) to be used as a research facility, with the provision that the land would revert back to the College upon the completion of the USFS's tenure (Art 1994).

In the mid-1930s, the USFS's research intention was to provide New England woodland owners the means for more scientific management of their forest resources. In establishing the Lawrence Hopkins Experimental Forest, they gridded the landscape into 2.02-ha (5-acre) cells, and as they traversed the tract, they recorded observations of probable previous land-uses (Fig. 24.2). At the intersection of the North–South and East–West grid lines they established permanent 0.1-ha (0.25-acre) plots, and wherever there were trees present, they recorded the species and measured the diameter at breast height (dbh) >1.25 cm (>0.5 in) to the nearest inch. For plots situated in old fields, hay meadows and pastures, the land conditions were noted as they conducted their surveys. The USFS intended to revisit these plots at regular intervals in the future, and to use some of them for forest management and timber harvest experiments.

With the advent of the Second World War, the Hopkins Memorial Experimental Forest was put into administrative dormancy, to be reactivated in 1947 after the war. When the forest re-emerged at mid-century, the USFS had shifted its research interest to forest genetics,

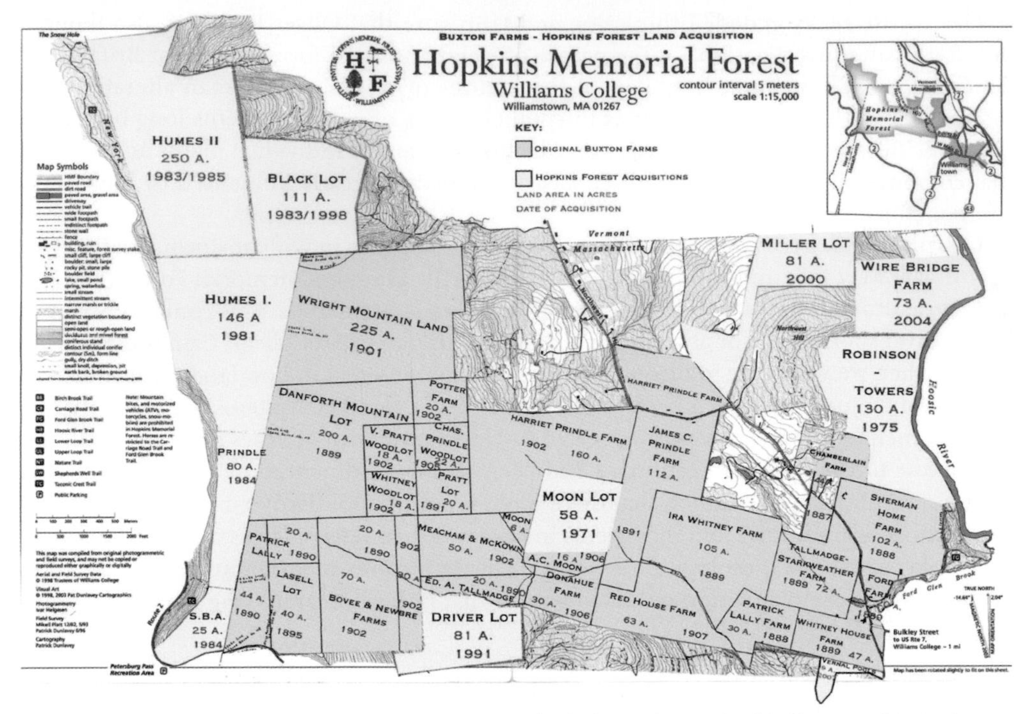

Figure 24.1 Property acquisition in the HMF: grey-shaded parcels acquired by A.L. Hopkins prior to 1906; yellow-shaded parcels acquired by Williams College after 1970. Tract name is in black type, tract land area (acres) in blue type and year of acquisition in red type. (H.W. Art)

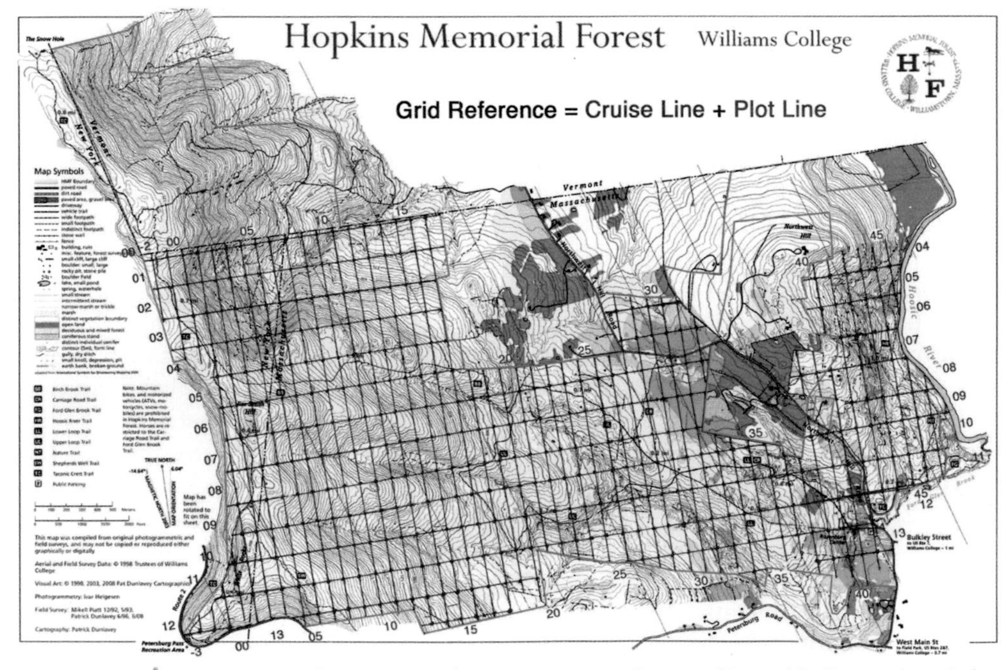

Figure 24.2 HMF topography (5 m. contours) and permanent plot sampling grid of 5-acre (2.02-ha) cells. (H.W. Art)

and plans to commence timber harvests and active management experiments had been shelved. Numerous forest genetic plantations were established in abandoned pastures and hay meadows that had persisted for a decade or so.

In the late 1960s, USFS administrators in Washington, DC made the decision to consolidate much of the research activities taking place in small facilities spread across New England into a single location, the University of New Hampshire in Durham, New Hampshire. The Lawrence Hopkins Experimental Forest ceased operation in 1968, and the land reverted to Williams College ownership. Arriving at Williams College two years later, I had the opportunity to immerse myself into the ecology of an entire landscape, and a search for pristine ecosystem patterns and processes commenced.

The field notes taken and the permanent plots established by the USFS in 1935 became the foundation for interpreting the changing forested landscape, looking both backward toward the Pre-Colonial era and forward to a future of changing climate and environment. In the early 1970s, the original permanent plot data were recovered from the USFS, their plot markers were found and the tree census was expanded to areas that had become recently reforested.

In 1971, the HMF underwent a renaissance as a landscape for education, research and passive recreation. Williams College incorporated the property into its newly formed Center for Environmental Studies, made it available for interdisciplinary research and restarted the long-term monitoring of the permanent plots established 35 years earlier. Between the mid-1930s and the early 1970s, the landscape had changed significantly: nearly all of the previously open fields had either been transformed to genetic plantations or had been invaded by shrub and tree species re-establishing woodland cover of one sort or another (Fig. 24.3).

The field observations we were making were corroborated by the quantitative data from the initial and subsequent permanent plot surveys. We simultaneously commenced a long-term project of assembling a complete deed history of the parcels that comprised the HMF, in the hope that descriptions of the parcels and the lands adjacent would better enable us to reconstruct the history of the forest. Tax records would supplement this data, although these were limited to after the mid-nineteenth century because a fire destroyed earlier records.

The circumstances of the HMF included much of a serendipitous nature. In 1970, as the USFS made available to Williams College much of the data that they had collected over the previous 35 years, the Forest Service personnel suggested that we should contact Mr Arthur E. Rosenburg, a then 80-year-old farmer in Williamstown, whom they claimed knew more about the Forest than anyone else. They were correct. In 1910, Rosenburg, aged 19, started to work on A.L. Hopkins's Buxton Farms as a farmhand. Although of limited formal education, he was a keen observer of nature and collector of not only mental notes, but photographs and artefacts as well. Rosenburg became a primary resource, a human library of information about the early twentieth-century agricultural and woodland landscape that was to become the HMF. The role of human activities of the past in shaping the nature of the contemporary landscape was coming into sharper focus.

In the late 1970s, I encountered Oliver Rackham's (1976) *Trees and Woodlands in the British Landscape*. It was a mind-expanding experience to realize that Oliver had been conducting detailed interviews of the British landscape incorporating intensive field investigations, historical records including church and civil economic data, climate data and dendrochronology. His historical perspective of British woodlands extended back to the era of Domesday Book and for some tracts even to Roman times. It was inspirational to see how he brought together apparently disparate elements as he gave deeper insights to the ecology of woodlands. It was both breathtaking in scope and intimidating to realize that my immediate historical window of 250–400 years in the HMF was a relatively brief moment in comparison.

Figure 24.3 Aerial photographs of the HMF. A. US Forest Service, 13 October 1935; B. Agricultural Stabilization and Conservation Service, US Department of Agriculture, 12 September 1970; C. Col-East Aerial Photography, North Adams, MA, 1 October 2013.

Interpretation of trends seen in the HMF

From the original 220 plots that were forested and censused by the USFS in the 1930s, the system grew to 424 plots by the 1990s as the woodlands expanded into meadows and pastures. The size of the HMF also has grown since the mid-twentieth century, now covering more than 1,012 ha (2,500 acres) in Massachusetts, New York and Vermont. The permanent plot system had been censused in the mid-1930s, the early 1970s, mid-1990s and early 2010s, using techniques identical to those originally used by the USFS.

During the first decade of Williams College research into the HMF, we came to realize that dramatic changes in the nature of the forest landscape had taken place since the original census of permanent plot system in the mid-1930s. The open agricultural lands, some of which had persisted since the nineteenth century, had undergone transformation from pastures and hay-lands to woodlands. Many of the woodlands of the first third of the twentieth century were dominated by the same species 35 years later, but many others showed significant changes in species dominance as expressed by relative basal area (Fig. 24.4). Sugar Maple *Acer saccharum*, American Beech *Fagus grandifolia*, Red Oak *Quercus rubra*, White Ash *Fraxinus americana* and Quaking Aspen *Populus tremuloides* all showed increases in the number of plots in which they dominated. However, Paper Birch *Betula papyrifera* had undergone a significant decline in the number of plots where it dominated, a trend that continues to this day.

In looking to field observations to explain why these patterns of changing species dominance occurred, we found that areas that had been hay meadows or pastures grazed by sheep or cattle in the first third of the twentieth century had usually undergone a transition to young woodlands dominated by Red Maple *Acer rubrum*. On the other hand, lands that had been in cultivation earlier in the century, had often been invaded by Quaking Aspen (and/or Paper Birch) (Fig. 24.5). Tracts of land in the HMF that had been used as woodlots to produce firewood, timber, charcoal or other forest products were observed to have different patterns that generally reflected the establishment or increase of Striped Maple *Acer pensylvanicum*, Yellow Birch *Betula alleghaniensis* or Black Birch *B. lenta*, along with pre-existing species.

It quickly became abundantly clear that the HMF could not provide the Holy Grail of a forest free from human manipulation. We could not even with certainty conclude that the

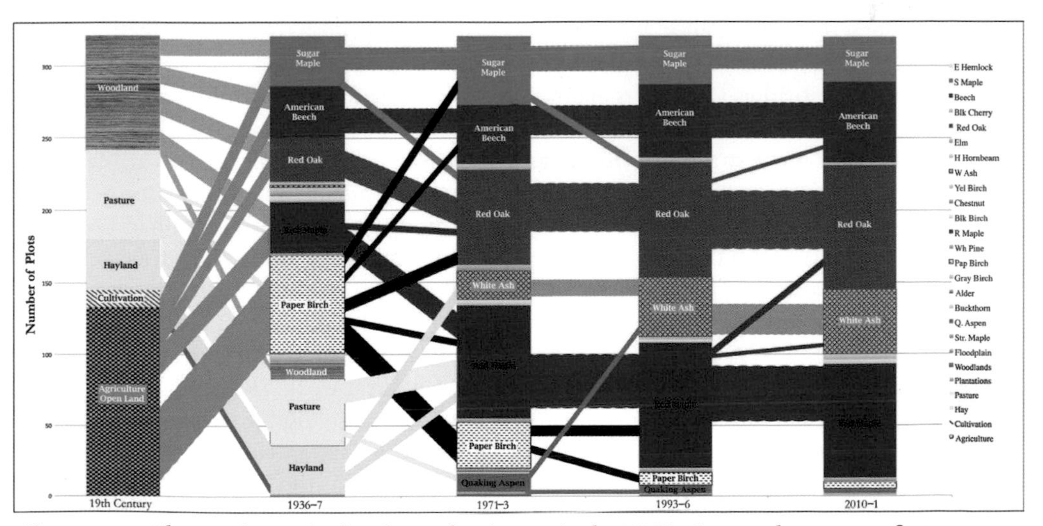

Figure 24.4 Changes in species basal area dominance in the HMF, nineteenth to twenty-first centuries. The vertical histogram is the number of plots in the Massachusetts section of the HMF that are dominated by given species or land-use types. The width of the connecting lines between stack elements is proportional to the number of plots undergoing transitions during adjacent time periods. (H.W. Art)

Figure 24.5 Field last ploughed in 1924 invaded by Paper Birch *Betula papyrifera*, Moon Lot, HMF, 21 March 1973. Snow is seen filling the old furrows but has melted off the ridges. (H.W. Art)

12-acre tract known as The Beinecke Stand that had extremely large (1– 1.3 metres d.b.h.) trees such as Sugar Maple, American Beech, White Ash and Eastern Hemlock *Tsuga canadensis*, all common to the pre-colonial forests of the region, had not been subjected to some tree cutting or stripping of hemlock bark to be used in tanning of hides.

Oliver Rackham's Visit to Williamstown, MA

My first interaction with Oliver Rackham was also serendipitous. Sheafe Satterthwaite, a colleague at the Williams College Center for Environmental Studies, had previously met with Oliver in Cambridge. Oliver was planning a trip to New England in late spring 1981, and Satterthwaite put him in contact with me, thinking that the research I was conducting in the HMF would be of interest to him. Oliver came to Williamstown, and spent 21–23 May 1981 there before continuing to southern Berkshire County and north-western Connecticut. Together we walked across the HMF as he queried me about what I thought had been the history of the different vegetated patches that were sewn together by stone walls in a patchwork of forested landscape.

Oliver's observations were recorded in a small, red field book, which I must admit I did not realize at the time was Volume 277 in a veritable library of field notes forming a natural history catalogue of his lifetime encounters with the natural world (Rackham 1981). These volumes are being scanned and made available to the public as a digital archive (https://cudl.lib.cam.ac.uk/collections/rackham) (see also the Introduction in *this volume*).

He was fascinated by features that I was well aware of, such as the large cavities chiselled-out of a Sugar Maple tree by a Pileated Woodpecker *Dryocopus pileatus* (Fig. 24.6).

Oliver was also a set of fresh eyes looking at the subtle curves in the trunks of trees growing on the slopes of the Taconic Range, referring to them as 'Snee-form', or perhaps it was *Schnee*-formed from the German for 'snow'. 'How were they so shaped?' Oliver wanted to know – was

Figure 24.6 Oliver Rackham observing a hollowed-out sugar maple. He notes, 'Sugar maple very extensively hollowed-out by pileated woodpecker. Great pile of chips. Living and healthy tree. Big black bird with scarlet cap. Peculiar cry' (Rackham 1981). HMF, 22 May 1981. (H.W. Art)

it snow slowly moving downhill deforming them as saplings? Was it some other biological or physical force such as ice deposition early in the life of the tree? I had not really noticed this before, but haven't ceased to notice it since.

Throughout his visit, he made notes on the patterns of Dutch Elm Disease on American Elm *Ulmus americana*. This was at a time when the disease was nearing the end of the first wave of reducing the elm tree from a floodplain dominant to sapling status in New England between the 1950s and 1970s. During that period, populations of American elms in regional forests had declined by over 80 per cent. However, as a shade tree on Williamstown streetscapes, American Elm was still relatively common in 1981. Twenty-five years later, Oliver published a photograph taken during his 1981 visit of an American Elm succumbing to Dutch Elm Disease on the Williams College campus in his *Woodlands* (Rackham 2006: 433).

His species-level observations were balanced by his examination of the processes that led to the structure of the old-growth Beinecke Stand, ranging from those of the Great Hurricane of September 1938 to the falling of individual red oaks during typical winter storm events and oak replacement by Sugar Maple and beech. Soil processes on the marble bedrock that Oliver referred to as 'limestone pavement' underlying the Beinecke Stand and soil organic matter distribution were contributing to the processes that he was interested in.

Spongy Moths *Lymantria dispar* were just starting the beginning of their third successive year of eruption, and had recently begun to emerge when Oliver arrived in Williamstown. While they preferentially defoliated oaks, other tree species ranging from Quaking Aspen to old apple trees had also been heavily defoliated in previous years. Of particular interest to Oliver was the elevational distribution of egg masses on trees of different species, and by the end of his short visit he had provided a catalogue of Spongy Moth distribution at the close of their eruptive phase.

What I called 'hedgerows' Oliver called 'tree lines'. He was curious about the species composition and rates of erosion of abandoned roads, which he referred to as 'holloways', that the trees lined. It would be several years later that I would come to truly appreciate British hedgerows and how they differed from the remnants of the pre-colonial forests in New England that were left behind to demark property boundaries of roadsides during the process of converting a forested landscape into an agricultural one.

If only we had consulted with Oliver, or had more extensively read his work on coppice forest management history, we would have been more successful with an experiment that we had started in February 1978, in which a 40 × 25 m Red Maple-dominated woodlot was clear-cut to produce firewood. In the subsequent three years, deer had browsed the saplings to the ground. We had neglected to protect the developing stools with fencing or wattles, which I think amused Oliver as much as it embarrassed me. He gave me a copy of his book *Hayley Wood, Its History and Ecology* (1975) upon his departure, 'with the author's gratitude and good wishes'. Had I read it earlier, our coppice experiment in the HMF would undoubtedly have had a different outcome.

The impact of Oliver's writings and visit on the HMF endured beyond the spring of 1981. His attention to the details of the interactions among the landscape features, soils and species (which certainly include *Homo sapiens*) over time directly contributed to the formulation of the model to interpret successional trends in the HMF (Fig. 24.7). The initial land-use conditions, be they cultivation, woodlot or hay meadows and pastures, set the selective environments for species that are either present on the site or have seeds that arrive there. The human use of the land appears to be the most important feature in determining the initial success of various species, with cultivated fields providing bare mineral soil conducive to the colonization by light-seeded species requiring abundant sunlight for their establishment. The established grassland communities of pastures and hay meadows tend to be invaded by heavier-seeded species such as meadowsweets *Spiraea*, Red Maple and White Pine *Pinus strobus*. Woodlots were harvested mostly in a selective manner, cutting individual trees to open small gaps in

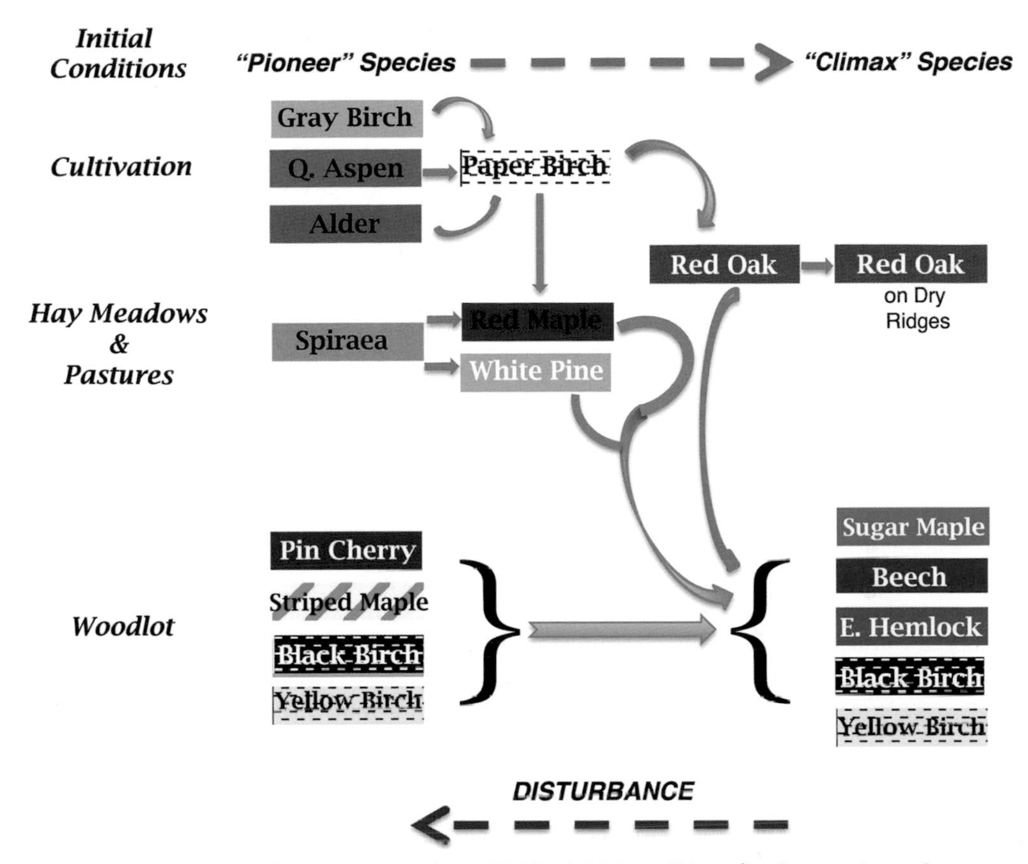

Figure 24.7 HMF Secondary Successional Model. The initial conditions (land-use at time of abandonment) create environments that favour colonization by various species. The transition from one phase to another is dependent on the frequency and intensity of disturbance. (Modified from Art and Dethier 1986)

the stand. Under these conditions, species that require organic seedbeds and a patchy light environment flourish. Over time, provided there is minimal disturbance, there are transitional phases as shorter-lived and less shade-tolerant species are replaced by longer-lived and more shade-tolerant species.

Corpus Christi College Cambridge 1987–8

Oliver and I remained in contact after his visit, and when I was on holiday in Britain in the summer of 1986 I contacted him to see if we could visit Hayley Wood together; but alas he was leaving for Crete before I would arrive in Cambridge. However, he made arrangements for me to visit Hayley Wood on my own. Oliver also planted the seed that I should consider applying for a visiting fellowship to Corpus Christi College for my upcoming sabbatical in 1987–8.

With Oliver as my sponsor, I was successful in being granted the Visiting Fellowship at Corpus Christi and affiliation with the Botany School at Cambridge University. I attended Oliver's Lent-Term lectures, and having purchased an automobile for the year acted as Oliver's driver-cum-field assistant for research excursions to Hatfield Forest, Hayley Wood and other sites of ecological interest. Oliver was meticulous in the recording of field data in notebooks and on data cards of his own design. The research would typically continue over lunch in a pub, where Oliver would carefully examine the beams, especially their transverse sections to determine the growth-ring patterns. This activity always attracted the attention of the publican, who would then be engaged in discussions of his experiences, the history of the building and ultimately the history of the woods that were used in its construction, their species, age, provenance and so on.

More serendipity interjected on Friday, 16 October 1987. Oliver and I had planned to revisit Hatfield Forest on that day to conduct soil sampling that would contribute to *The Last Forest*, which he was writing (Rackham 1989). However, winds of 120–160 kph swept across southern England in the early morning, an event that came to be known as The Great Storm, the most extreme winds on record here. I had suggested to Oliver that we cancel our excursion to Hatfield, but he suggested that we attempt a late morning departure. By then, any wind-downed trees that had blocked major roads were largely cleared, and we travelled to Hatfield to survey what the storm had wrought. Over the course of the remainder of 1987 and spring of 1988, we worked our way down through south-eastern England to Kent, investigating the impacts of a relatively large regional disturbance – Hatfield Forest, Hayley Wood, Sevenoaks, Knowle, Chartwell, Box Hill, Staverton and Newmarket as well as other sites in East Anglia, West Essex, Surrey and Kent (Rackham 1987a,b; 1988a,b).

We made observations on the sizes of trees and their root plates, their species, exposures, growth rates and prior disease states. In the weeks following the storm event, Oliver expressed dismay at the rate of 'tidying-up' of uprooted and wind-impacted trees, supposing that the trees that were partially, or even mostly, uprooted were not necessarily dead (Fig. 24.8). Native species that had established themselves showed less damage than exotic species that had been intentionally planted, especially the exotic conifer species in Forestry Commission plantations. The plantations showed greater damage of trees growing in the centres of the blocks than on the edges, presumably because the trees on the periphery had previously formed stronger wood under the more frequent swaying of trunks exposed to winds in the past. Most of the root plates we observed were remarkably shallow, less than 1 m in depth, or

Figure 24.8 Oliver Rackham standing on a beech root-plate tipped up during the Great Storm of 16 October 1987, Box Hill, Surrey, 9 March 1988. (H.W. Art)

Figure 24.9 Henry Art standing by a beech trunk blown over by the Great Storm of 16 October 1987. Cowden, Sevenoaks, Kent, 24 March 2015. (P.B. Art)

for some now less than 1 m in thickness, since the root masses had been shifted to vertical by the Great Storm.

I revisited some of the sites in Kent in March 2015, the month following Oliver's death. Indeed, many of the trees we had seen blown over 28 years earlier were still alive and had established their own coppices of branches that had become trunks (Fig. 24.9). Oliver's caution that tidying-up was premature was certainly vindicated: the areas that were left to recover on their own showed greater structural, habitat and species diversity than those managed by humans with chainsaws.

The Great Storm of 1987 was instructive in our examination of an intense wind-generated disturbance that occurred in part of the HMF on 29 May 2012. In the late afternoon, a small, but intense derecho (widespread, long-lived, straight-line windstorm) event with gusts in excess of 160 kph (100 mph) cut a swath approximately 300 m (1,000 ft) wide and 4 km (2.5 miles) long from west to east through the HMF. We had completed the fourth census of the permanent plot system just the year before, so we could assess the impacts of the disturbance event with great precision (Dickhaus 2014).

As with the Great Storm of 1987, large individuals of fast-growing tree species such as Quaking Aspen suffered the greatest breakage, with entire tree crowns snapped off and blown a distance of up to 40 m (130 ft) downwind from the trunk. Tall overstorey trees, mostly Red Oaks in excess of 30 m (100 ft) in height and over 125 years in age, were especially susceptible to uprooting, excavating large pits upwind and forming large mounds of soil attached to their root plates downwind (Fig. 24.10). The intense disturbances were similar in creating the microtopographic diversity that rejuvenates soil nutrients and creates greater microhabitats and niche diversity. Whether the North American species of oaks, poplars and others will be as successful in resprouting after being partially uprooted as their European counterparts remains to be seen, as we continue to monitor changes in the permanent plot system of the HMF.

Figure 24.10 Red Oaks blown over by the Derecho of 29 May 2012 in the HMF, Williamstown, MA. (H.W. Art)

What I learned through working with Oliver

There are many continuing influences of Oliver Rackham on my ways of thinking about and interacting with nature. As a colleague, he infused patience in paying attention to details of what one observes, openness to information of all sorts – yet filtered by critical eyes, and an organic sense of humour. Other lessons are:

- Don't merely repeat what is in the literature without your own field verification.
- The ecology of the woodland itself is the best evidence upon which to base management of that natural resource.
- Each woodland tract is an individual case having its own history, although generalizations can be made by examining the numerous woodlands of a region.
- The best descriptions of how woodlands work are the result of holistic data collection that does not ignore the insights of non-professionals: interview the landscape, as many of the elements as possible.
- Most importantly, humans are an integral part of nature, not separate from it.

Acknowledgements

Initial permanent plot censuses were conducted by Norman E. Borlaug, M.B. Dickerman, Walter K. Starr, E.B. Chamberlain, G.W. Hawkins, W.L. Robbinette, Karl F. Wenger, Messrs Berlough, Blake, Blakeney, Boureau, Cameron, Curtin, Drysgola, Exford, Gould, Rancourt, Sweet, Walden, Walker, Williams, Woodman and Civilian Conservation Corpsmen. Further data collections by the USFS in the 1940s through 1960s were done by Frank E. Cunningham, Raymond Lavigne, P.R. Ledger, Samuel Wescott and Messrs Borneman and Hutnik. The

1970s to 1980s permanent plot data was collected by James M. Affolter, Taber C. Allison, David M. Blanchard, Donald A. Campbell, William G. Constable, Cecelia M. Danks, Claude de Pamphilis, Lee C. Drickamer, Peggy Duesenberry, Jeffery T. Erickson, Anne M. Forrestel, Patricia A. Friedman, Richard E. Geier, Jeffery Glitzenstein, Gordon M. Greene, Kathleen A. Haas, Christine Henry, Scott Highleyman, Jan L. Hitchcock, David Houle, Martin E. Immerman, Thomas Kalt, John A. Kruse, Barbara E. LeBarron, Melinda Lindquist, Kenneth S. Liu, David B. Lupke, Elisabeth C. Marr, David R. Marrs, Peter B. Mc Chesney, Donald Keith McInnis, Hervey E. McIver, James W. Norton, Jennifer J. Quinn, Jeffery M, Reardon, Catherine W. Richardson, Kathryn A. Saterson, David H. Sprague, William H. Stahl, Rebecca B. Stevens, Gail P. Stuart, Stephanie A. Symmes, Geraldine L. Tierney, Elizabeth O. Titus, Peter Walford, Donald C. Weber, Alan S. White, Julie A. Woodward. The 1990s permanent plot data was collected by Dawn Biehler, Timothy Billo, Nadine E. Block, Christopher M. Brookfield, Lawrence D. Callanan, Jr., Jonathan C. Cluett, Daniel F. Currie, Gail T. French, Caleb E. Gordon, Emilie B. Grossmann, Susan M. Halbach, Michele N. Koppes, Maya E. Kumar, Anna Malkowski, Benjamin R. Montgomery, Kristian S. Omland, Marcangelo B. Puccio, Amy K. Smith, Charles H. Stevenson, Charles C. Wall, Christopher S. Warren, Jessica L. Wege, Chara A. Williams and Jonah C. Wittkamper. The 2000s to 2010s permanent plot data was collected by Flynn Boonstra, Jasmine S. Smith, T. Michael Gallagher, Ryan Buchanan, Claudia Carona, Matthew Cranshaw, Wade Davis, K.B. DiAngelo, Eric Hagen, Laurel Hamers, David Hansen, Nick Lee, Mari Lliguicota, Julio Luquin, Eric Outterson, Mark Lyons, Abbie Martin, Dan Nachun, Alex Peruta, Jackie Pineda, Sarah Rowe, Amelia Simmons, Jennifer Turner, Jamie Dickhaus, Laura Stamp and Alice Sterns. Faculty collaborators include Sharron Macklin, Chris Warren, Andrew Jones, and Professors Edward Flaccus, David Dethier, William T. Fox and David Backus.

Funding for the HMF research was provided by the USFS, Williams College, the National Science Foundation, The Richard King Mellon Foundation, the Andrew W. Mellon Foundation, the Edward John Noble Foundation, Howard Hughes Medical Institute and the Holloman-Price Foundation.

References

Art, H.W. (1994) *The Amos Lawrence Hopkins Memorial Forest: An Eclectic History of its First Century 1987–1987*. Williams College Center for Environmental Studies, Williamstown, MA. https://hmf.williams.edu/history/eclectic-history/

Art, H.W. and Dethier, D.P. (1986) *The Influence of Vegetative Succession on Soil Chemistry of the Berkshires*. Water Resources Research Center. UMASS Publication 153, Amherst.

Dickhaus, J.L. (2014) Short-term ecological responses to an intense storm event in the Hopkins Memorial Forest. BA thesis, Williams College, Biology Department.

McKibben, W. (1989) Reflections (the end of nature). *The New Yorker*. 11 September: 48–105.

Merchant, C. (1980) *The Death of Nature: Women, Ecology, and the Scientific Revolution*. Harper and Row, San Francisco.

Rackham, O. (1975) *Hayley Wood: Its History and Ecology*. Cambridgeshire and Isle of Ely Naturalists' Trust Ltd, Cambridge.

Rackham, O. (1976) *Trees and Woodlands in the British Landscape*. J.M. Dent and Sons, London.

Rackham, O. (1981) Notebook 277, covering 21 to 24 May 1981. (CCCC14/6/2/1/277). https://cudl.lib.cam.ac.uk/view/MS-CCCC-00014-00006-00002-00001-00277/1.

Rackham, O. (1987a) Notebook 356, covering 16 September to 20 October 1987. (CCCC14/6/2/1/356). https://cudl.lib.cam.ac.uk/view/MS-CCCC-ss00014-00006-00002-00001-00356/1

Rackham, O. (1987b) Notebook 357, covering 21 October to 11 December 1987. (CCCC14/6/2/1/357). https://cudl.lib.cam.ac.uk/view/MS-CCCC-00014-00006-00002-00001-00357/1

Rackham, O. (1988a) Notebook 360, covering 27 January to 9 March 1988. (CCCC14/6/2/1/360). https://cudl.lib.cam.ac.uk/view/MS-CCCC-00014-00006-00002-00001-00360/1

Rackham, O. (1988b) Notebook 361, covering 9 March to 3 April 1988. (CCCC14/6/2/1/361). https://cudl.lib.cam.ac.uk/view/MS-CCCC-00014-00006-00002-00001-00361/1

Rackham, O. (1989) *The Last Forest: The Story of Hatfield Forest*. J.M. Dent and Sons, London.

Rackham, O. (2006) *Woodlands*. (New Naturalist 100), Collins, London.

Part VI

Legacy, Archive and Publications

CHAPTER 25

Conclusions: The Legacy of Oliver Rackham

Jennifer A. Moody and Ian D. Rotherham

Summary

Oliver Rackham's archives are central to maintaining his legacy. Equally important are the practitioners, teachers, woodsmen, scholars, friends, and fans that continue to promote and expand on his work and ideas.

One of Oliver's greatest achievements was in changing the way that people around the world, including other scholars, thought about landscape and history. This was especially the case in Britain and Europe in relation to the countryside and particularly woodlands. Work with collaborators such as George Peterken revolutionized perceptions of treescapes and of the history and heritage contained physically within them but also in the archives, where they exist, of their lives and management. Triggered by Oliver's penetrating scholarship and his boundless enthusiasm, others began to look at their landscapes and countryside anew. The process continues today as a living, vibrant testament to his achievement. However, Oliver's work not only stimulated and informed that of his peers, but his remarkably accessible popular writing projected this insight to a much wider public audience. A challenge for future generations will be to maintain the awareness and passion for these wonderful and irreplaceable landscapes that Oliver described as being akin to illuminated medieval manuscripts there to be deciphered and read. Beyond this is the need to protect the countryside and its ancient woods from what Oliver previously described as 'The Locust Years', where modern machine-led management erases all that has been before. Such damage is swift and irreparable, and it was Oliver that brought this threat to our attention in the late twentieth century. A key part of his legacy must surely be in the effective recognition and conservation of such places for their own sake, and also for that of future generations.

Introduction

The chapters gathered in this volume are a testament to Oliver Rackham's ongoing global impact on woodland and historical ecology scholarship. They range geographically from Japan to Texas to Britain to Greece and topically from the Australian bush to British shadow woods to a Bavarian Abbey. Oliver's energetic lecturing and teaching circuit when he was alive, together with his prolific publications (see Rackham's bibliography *this volume*), inspired

Jennifer A. Moody and Ian D. Rotherham, 'Conclusions: The Legacy of Oliver Rackham' in: *Countryside History: The Life and Legacy of Oliver Rackham*. Pelagic Publishing (2024). © Jennifer A. Moody and Ian D. Rotherham.
DOI: 10.53061/ETMV8679

several generations. Now, through his extensive archive, his insights, intellect and humour can enthuse new generations today and tomorrow.

From ecology to archives – opening our eyes to countryside history

The preservation of ecological archives had long concerned Oliver. In the second edition of *Ancient Woodland*, he makes a 'plea for a proper archive of ecological records'. He laments:

> What would I give for a sight of the notebooks and photographs of Sir Arthur Tansley? Ecological records are part of the archives of local history, and should be treated with the respect that is given to tithe apportionments, churchwarden's accounts, and court rolls.
>
> Rackham 2003: 530

Three years later, when he wrote *Woodlands*, he was cautiously optimistic:

> Notebooks and photographs belonging to deceased ecologists are part of the nation's archives and need to be properly housed and catalogued. As yet this is not systematically provided for, although some Biological Records Centres have made a valiant beginning.
>
> Rackham 2006: 523

What would the reader give to explore Oliver's own fieldbooks, letters and photographs?

Happily, Oliver's many colleagues, fans and friends have been working to preserve, curate and digitize his archives, the bulk of which are at the University of Cambridge. The richness of Oliver Rackham's archives is witness to his creativity and genius. Preserving them and making them as freely available to the public as possible is one of the best ways his ideas will continue to spread locally and around the world. Future generations who consult these resources are indebted to the foresight in 2015 of Tim Harvey-Samuel, Stuart and Sibella Laing, James Buxton, Lucy Hughes, Beverly Glover, Christine Bartrum, Valerie Cooper, Peter Grubb, Christopher Preston, Simon Leatherdale, Paula Keen, David Morfitt, Philip Oswald, Wick Dossett, Nick and Cara Grant, Philippa and Tony Sims, Guy Corbett-Marshall, Susan Ranson, Jo Burgon, Camilla Lambrick, Keith Kirby, Peter Carolin, George Peterken, Beccy Speight, Louise Bacon, Adrian Cooper and a number of others.

Oliver's physical archive consists of four main parts: (1) fieldbooks, (2) images (slides, photographs and digital images), (3) herbarium, and (4) miscellaneous papers, files, maps, drawings and notes. The bulk of this material is held in the archive at Corpus Christi College, Cambridge, and the Cambridge University Herbarium, with smaller collections at Small Woods Association's Green Wood Centre and in Crete and Texas.

Corpus Christi College, Cambridge

The Oliver Rackham Archive at Corpus Christi College, Cambridge (CCCC), can be divided into four major categories:

- *Personal / biographical* – Oliver took his role as a citizen very seriously and was not shy about expressing his opinions strongly, whether it be potholes on a local road or the globalization of plant diseases. These and many other largely unpublished letters and notes are held in the archive at Corpus Christi College, Cambridge. Also included are letters to and from his father; letters to and from friends; emails, photographs of himself and/or friends; sketches and jottings from his early life; interviews and profiles that appeared in the press.
- *Academic* – The largest part of the Rackham Archive at CCCC relates to his research and a significant part consists of his fieldbooks. Oliver kept two types of field notebooks: Red notebooks that contain his original notes taken on the spot in the field, and 'Blue' notebooks which are usually compilations and further thoughts he

had on specific locations (often individual woods) based on his original field notes (Fig. 25.1). There are 698 red fieldbooks and 440 blue ones. Lucy Hughes, the Corpus archivist at the time Oliver died and who helped retrieve his archive from his house, comments:

> Part of the pleasure of reading the notebooks derived from their format and scale (a roomy pocket-sized) and the way that sketches and words combined together to build up a picture of a place. Sketch maps, diagrams (particularly of churches or timber constructions), drawings of fungi or leaves, all added visual interest.

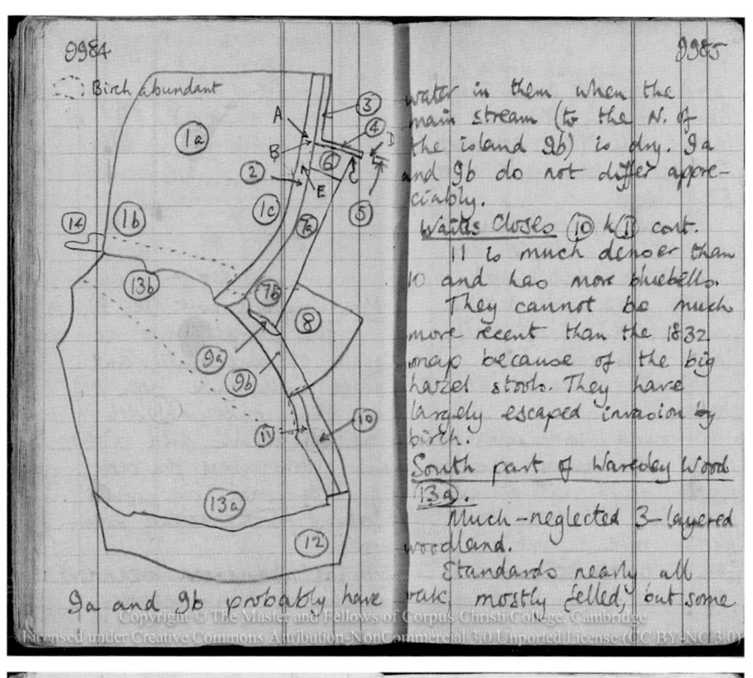

Figure 25.1 Oliver Rackham's Fieldbooks: (top) *Red Notebook* 130 (24 January to 19 March 1968). Sketch of woodland near Great Gransden, Huntingdonshire; (bottom) *Blue Notebook, Madingley Wood 3* (13 December 1995 to 29 September 2002), including a photograph of a tree from 1950. (© Corpus Christi College, Cambridge)

> These kinds of features are what give the notebooks (both the red and the blue series) their particular appeal, and a kind of unselfconsciously 'aesthetic' value: they are expressive, as well as informative – a response to a particular place at a particular time.
>
> Hughes 2022: 17

Another large part of Oliver's academic archive at Corpus is his slides. The collection contains over 26,000 slides dating between 1965 and 2005. The majority are from Britain and Greece (especially Crete), but there are significant collections from Japan, Australia, Tasmania, USA (mainly Texas), Italy, France and other parts of Europe.

Oliver went digital in 2005. His digital images have not yet been catalogued but copies are on hard drives stored at Corpus and Boutsounaria, Crete. The originals are in Texas with Jennifer Moody, who holds the copyright.

Other records include his PhD thesis on plant transpiration; woodland record cards; notes on topography, history and other subjects such as the college silver; maps; correspondence with readers and colleagues; some audio material (talks, interviews); and hemispherical photographs of Madingley Wood.

- *Administrative* – This collection mainly consists of college-related business, including meeting minutes of the various committees on which Oliver sat; PCC minutes of St Benet's Church; studies and reports on college buildings (and on the Eagle Inn), and some memos advising on the upkeep of the same.
- *Books and Off-prints* – A selection of books from Oliver's home was kept for the Parker Library and are identified with an 'ex libris' book plate bearing the initials 'OR' in his distinctive calligraphic style; these are mostly on local history and antiquarian topics; they are searchable on the Newton catalogue. Off-prints of Rackham's periodical publications and pamphlets are also kept at the Parker Library.
- *Computer files* – There are copies of the contents of Oliver's computers, hard drives, and flash drives in the Corpus Archive. The originals are in Texas with Jennifer Moody, who holds the copyright to all the files that were on these devices at the time of Oliver's death.

For more information about the Rackham Archive at Corpus Christi College, Cambridge, please email archivist@corpus.cam.ac.uk.

Cambridge University Herbarium

Oliver collected plants from the places he visited, most of which he mounted on acid-free paper in acid-free boxes that he made himself. The earliest mounted sample is of Nettle-leaved Bellflower *Campanula trachelium* collected from a 'bombed site in Fisher's Lane, Norwich' on July 1950, when Oliver was 10 years old (Fig. 25.2).

Oliver's extensive herbarium collection of over 17,000 samples was a surprise to many of his colleagues and friends. In 2015 when Christine Bartrum, then acting curator of the Cambridge University Herbarium (CUH), which is part of the Department of Plant Sciences, first saw Oliver's home collection housed in his bedroom, she was stunned and remarked that his herbarium was not stored in his bedroom, it was Oliver that slept in his herbarium! Her vision allowed Oliver's specimens to remain together at CUH rather than be slotted into the general collection because, as she notes:

> [He] produced a collection that is unique in its 'voice' and which is a joy to digitize. Volunteers databasing Oliver's material never leave without saying they have learned something new from working with his herbarium. On a cold, grey, Cambridge afternoon it is delightful to be momentarily transported to the 'laurisylvan sacred forests' of Japan or the 'great Eucalyptus rainforests' of Tasmania

Figure 25.2 Rackham herbarium sheet of Nettle-leaved Bellflower *Campanula trachelium* with notes added later. (© University of Cambridge Herbarium)

> even if certain shrubs, as he noted on the *Coprosma* he collected, 'smell of drains'. Detailed annotations describing plant physiology, soil, underlying rock, patterns of grazing, and even the palatability of the plants he tasted himself, make this collection not just a taxonomic tool but an invaluable resource for teaching and research.
>
> Bartrum 2017: 6

The cataloguing, curation and digitizing of Oliver's herbarium continues under Lauren Gardener, the Curator of the Cambridge University Herbarium since 2017. Most specimens come from Britain and Greece (especially Crete) but there are also numerous samples from Texas, Japan and Australia. In addition to the thousands of herbarium sheets, there are a number of boxes containing unmounted specimens that were found folded in newspapers at his home. Under Lauren's guidance Euan McKenzie completed the cataloguing and digitisation of Oliver's 2,800 British samples in 2022 (McKenzie 2022).

For more information about the Rackham Herbarium, please email herbarium@plantsci.cam.ac.uk.

University of Cambridge Digital Library

The digitization of Oliver's fieldbooks and slides has been a collaboration between Corpus Christi College Cambridge (CCCC), Cambridge Digital Library (CDL) and the Friends of Oliver Rackham (FOR); it was spearheaded in 2016 by Lucy Hughes (CCCC), Huw Jones (CDL) and Jennifer Moody (FOR). To date, 459 notebooks are available online. In 2020, during covid, CDL launched a crowd-sourcing project, 'Alone in a crowd ... transcribing together', to transcribe Oliver's scanned and uploaded fieldbooks, allowing word searches and other data-processing on the content. So far, about 140 notebooks have been transcribed by volunteers (Fig. 25.3). CDL also hosts some of the Rackham slide collections that have been digitised (Jones 2022).

These resources can be accessed here: https://cudl.lib.cam.ac.uk/collections/rackham.

Digitizing Oliver's notebooks, slides and herbarium allows people from around the world to access these resources at the click of a button. In 2019, according to CDL, Oliver's notebooks were accessed over 31,000 times. The curation of electronic archives, however, is still in its infancy and should not replace a physical archive of papers, photographs and objects – as Oliver cautioned:

> Neglect is death to electronic archives in a matter of decades, but does not greatly matter to paper copies, which need only to be kept dry and can be scanned back into electronic forms.
>
> Rackham 2006: 524

Small Woods Association's Green Wood Centre

Since 2018 the eclectic materials Oliver kept in his home carpentry workshop have been stored at the Small Woods Association's home at The Greenwood Centre in Coalbrookdale,

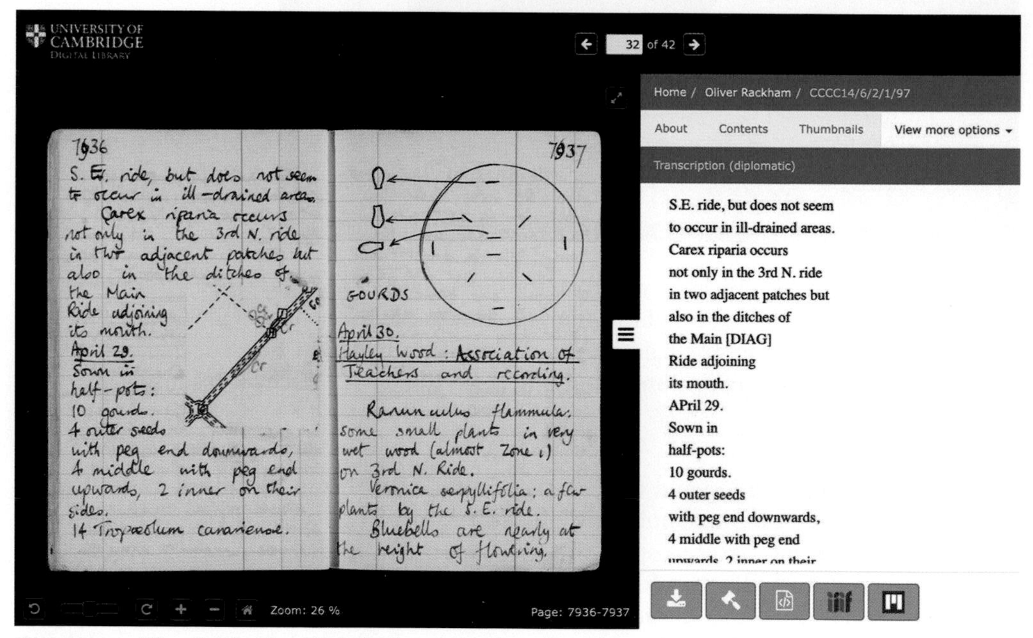

Figure 25.3 The Cambridge Digital Library window showing pages in Oliver's notebook (left) and the transcribed text (right). (Image courtesy CDL)

Figure 25.4 Elements of the Rackham Collection at the Green Wood Centre. (S. Niemann)

Shropshire (Niemann 2019). Items range from his original research fellowship thesis for Churchill College, Cambridge, written in February 1964, to home-made contraptions designed specifically for carrying out research experiments – which baffle us today – to paint tins, carpentry tools, a huge sideboard, and a framed photograph of a dog (Fig. 25.4).

Curating of the Rackham collection continues. All the objects have been described, but there is still much to be done in terms of conservation and presentation. There are also several mysterious items of unknown function that may yet be identified.

The Centre's goal is to create engaging exhibits that use examples from Oliver's work that will inspire the next generation, such as the recent *Oliver Rackham – Roots and Branches*, which ran from 8 October 2021 to 31 January 2022 (Reese 2022a and b).

For more information about the Rackham Collection at Small Woods Association, please email office@swa.org.uk.

Boutsounaria, Chania, Crete (Greece)

Oliver's books, photographs and papers stored in Crete deal with his research there and in other parts of Greece. They include files and binders on medieval architecture, Venetian archives, Greek plants, landscapes and weather. His collection of Cretan landscape photographs, begun in 1968 and repeated at 10-year intervals, is also there and was last repeated in 2018.

Oliver's books on the Eastern Mediterranean (some annotated) are also housed at Boutsounaria in a wooden bookshelf he designed and built.

A charcoal reference collection of specimens collected mainly in Crete by Oliver Rackham and Jennifer Moody in 2012–2014 also resides there. The collection includes *Acer sempervirens, Arbutus unedo, Berberis cretica, Castanea sativa, Ceratonia siliqua, Crataegus* sp., *Cupressus sempervirens, Juniperus phoenicea, Juniperus macrocarpa, Nerium oleander, Olea europea, Olea europea* var. *sylvestris, Phillyrea latifolia, Pistacia lentiscus, Platanus orientalis, Prunus webbii, Pyrus spinosa (amygdaliformis), Quercus pubescens* subsp. *brachyphylla, Quercus coccifera, Quercus ilex, Quercus ithaburensis* subsp. *macrolepis, Styrax officinalis, Vitis vinifera,* and *Zelkova abeliceae.*

Copies of all the files on Oliver's computers, hard drives and flash drives are also stored in Crete.

For more information about the Rackham Collection in Crete, please email Jennifer Moody at hogwildjam@mac.com.

Celebrating and promoting Oliver's legacy

Within days of Oliver Rackham's death on 12 February 2015, tributes were published not only in major and minor British newspapers, journals and webpages, but also in Hungarian, Italian and Greek ones. Soon numerous events were being organized in his memory. To keep Oliver's colleagues, friends and fans informed, a newsletter, *The Friends of Oliver Rackham (FOR)*, began to be electronically circulated in August 2015 by Jennifer Moody. To date, twenty newsletters have been sent out.

Conferences

A number of conferences have been held in Oliver's memory. To list but a few:

- 'ECSLAND', a conference on European Sacred Landscapes held in Sassari, Italy in April 2015 was dedicated to Oliver's memory.
- 'Yorkshire Woodlands', a conference hosted by PLACE of Yorkshire in April 2016 and dedicated to Oliver. It is now available as a book (Atherden and Wallace 2017).
- 'The Oliver Rackham Memorial Symposium', a two-day event hosted by Corpus Christi College, Cambridge, in August 2016 and attended by over 200 people (Moody and Dossett 2016). Richard Mabey, renowned nature author, gave the keynote address. There were lectures on 'Oliver at Corpus', his archive and herbarium; 'Oliver and British woods'; 'Oliver and the Mediterranean'; and an international panel discussed 'Woodland and Landscape Conservation and Management'. Peter Carolin guided guests around Corpus's Old Court (Rackham and Carolin 2020, Carolin 2021) and Peter J. Grubb, with significant input from Louise Bacon, Ed Tanner, Richard Dowsett, Vince Lea, Rob Jarman, Stephen Tomkins, David Coomes, Jenny Mackay and Mark Ricketts, orchestrated excursions to Hayley Wood (Fig. 25.5). In the Parker Library, there was an exhibit of college silver – Oliver had for many years been the Keeper of the College Plate and wrote a book on it (Rackham 2002). The *Ely Coucher Book* that played such an important role in Oliver's approach to ancient woodland was also on view, kindly lent by Gonville and Caius College. In recognition of Oliver's deep faith, there was a moving recital and Communion Service in Corpus Christi chapel, where Oliver is buried. The present book was partly inspired by this symposium.
- 'The Forests of Essex', hosted by Place Services in February 2018, celebrated woodland in Essex and Oliver's impact on it. The keynote address was by Tom Williamson.
- 'Trouble in the Woods', 2024, dedicated to Oliver Rackham, Melvyn Jones, David Hey, Donald Pigott and Frank Spode.

Guided walks, lectures, work parties and exhibits

In 2015 and 2016, guided walks took place at Wheatfen, Tiger Hill and Bradfield Woods – all places pivotal to Oliver's thinking and career.

The Cambridge Conservation Forum sponsored Oliver Rackham Memorial Guided Walks at Hayley Wood in the spring of 2015–2018 and may do so again.

The Cambridge Conservation Volunteers sponsor an annual work party in the fall where Oliver is remembered and his birthday celebrated.

The Woodland Trust (WT) also sponsors Rackham Memorial Events. In May 2015, Clive Anderson, president of WT since 2004, gave the Hay Festival Rackham Memorial Lecture. The next WT Rackham Memorial Event was at Markshall Estate, Essex in May 2017 and consisted of

Figure 25.5 The Oliver Rackham Memorial Symposium at Corpus Christi College Cambridge. Excursion to Hayley Wood led by Edmund Tanner amidst coppice-poles, 14 August 2016. (W. Dossett)

lectures, lunch and a guided walk. Since then, WT focused on the completion and publication of two unfinished manuscripts found on Oliver's computer: *The Ancient Woods of the Helford River* (Rackham 2019) and *The Ancient Woods of South-East Wales* (Rackham 2022). To that end, they hosted several events to launch and promote the two books. The last event was an online book launch for *The Ancient Woods of South-East Wales* in July 2022 hosted by Natalie Buttriss, director of Woodland Trust Wales, and attended by 100+ people. It included talks by George Peterken, Jennifer Moody and Paula Keen (Keen 2022b).

The botanical artist Sarah Morrish dedicated her exhibit at the Royal Horticultural Society Botanical Art Show in February 2016 to Oliver. The art exhibit 'Trees Observed' held at Fen Ditton Gallery in September and October 2018 also celebrated Oliver's life and work. Part of the latter's proceeds went to the publication of his two unfinished books.

In October 2021, the Small Woods Association displayed some of the Rackham Collection they hold at the Green Wood Centre in the exhibit *Oliver Rackham – Roots and Branches* (Fig. 25.6) (Reese 2022a and b).

Tree dedications

Some of the most moving tributes have been tree dedications in Greece and Britain.

- In 2015, Guy Sanders planted two wild almonds in his garden at Corinth, Greece. Later that summer, a great beech growing in a sacred forest in Epirus, Greece was dedicated to him by his friends and colleagues, Kalliopi Stara and Rigas Tsiakiris (Tsiakiris et al. this volume p. 215, fig. 13.10).
- In 2018, fifty years after he set foot on the island of Crete, a veteran, endemic elm pollard (*Zelkova abeliceae*) growing in the White Mountains near Chania was dedicated to Oliver,

Figure 25.6 *Oliver Rackham — Roots and Branches* exhibit at the Green Wood Centre: (top) overall view of the exhibit; (bottom) close-up of notes and Rackham's negatives of leaves, demonstrating a variety of heliotropic actions and notes. (D. Reese)

organized by Kalliope Pediaditi Prud'homme and Jenny Moody, the Municipality of Platanias, the Samaria National Park Management Body, the Forestry Service of Chania, and the Mediterranean Agronomic Institute Chania (Moody 2018a and b, *this volume* Figs 12.3–4).

Later that year, three great Selakano oaks were also dedicated to his memory, organized by the Cultural Association of Selakano with support from the Municipality of Ierapetra, the Natural History Museum of Crete and Eptastiktos (Moody 2018a and b, *this volume* Fig. 12.2).

- In September 2021 the Woodland Trust dedicated to Oliver the 'Curley Oak' (*Quercus robur*) in Wentwood, Wales – a snaggly pollard and the only named tree on the 1881 British Ordnance Survey map (Fig. 25.7). Despite the soggy conditions and pouring rain, a good crowd attended (Keen 2022a).
- In February 2022, Ashley Arbon MBE organized the planting of three *Populus nigra* subsp. *betulifolia* cuttings in Oliver's memory in the grounds of the Hamilton Kerr Institute in Whittlesford, Cambridgeshire.

Books

Although it is not surprising that numerous books have been dedicated to Oliver since his death, some are rather unexpected and indicate just how wide his influence was. For example, the *Atlas of the Predaceous Water Beetles (Hydradephaga) of Britain and Ireland* (Foster et al. 2016) was co-dedicated to him because he 'understood historical landscape'.

References to Oliver's work also appear in unexpected places, such as in an article 'Ars Ponendi Lucum: Groves in Poetry and Art' (Prosperetti 2022). Here, the author, an art historian at the University of Houston, says:

> Rackham's fable [about the origins of Hayley Wood] sets the stage for the groves of Homer and Virgil's luci and inspires us to wonder about the appearance of the groves in landscape, both real and as imagined by landscape painters.
>
> Prosperetti 2022: 314

A tremendous contribution to Oliver Rackham's legacy has been the completion and publication of projects and manuscripts he left unfinished at the time of his death. The late Philip Oswald organized the publication of Oliver's manuscript on the fungi of Brandon Park (Rackham 2015). Gloria Pungetti wrapped up the Esland project on island landscapes begun with Oliver in 2007 (Pungetti 2017, *this volume*). Paula Keen and David Morfitt with their teams of volunteers and the sponsorship of WT, CCCC, and FOR, and the collaboration of Little Toller Books, finished the last two of Oliver's books *The Ancient Woods of the Helford River* (Rackham 2019) and *The Ancient Woods of South-East Wales* (Rackham 2022). There are several more Rackham publications and research projects in the works (see Rackham's bibliography *this volume*).

A unique approach to scholarship

Beyond Oliver's brilliant mind and engaging writing style, an often-overlooked part of his legacy is his humanity. An unexpected example came to light 1 December 2020, nearly six years after his death, during Morning Prayers by the Very Reverend Robert Willis, then Dean of Canterbury Cathedral. It was International Tree Week and each day the Dean talked about a tree species that he was sitting beside in the deanery garden. On this day it was the ash tree, and he mentions Oliver's book on that tree and the man himself, who was his friend. The Dean uses the threat of Ash Dieback detailed by Oliver as a parable for making sound judgements. He cautions that ash trees may seem strong but if they are not properly cared for and tended, they will die – as English elms did in the 1970s decimated by Dutch Elm Disease – because of poor policies and rash judgements regarding the importation of exotic plants.

Figure 25.7 The 'Curley Oak' dedication: (top) Big smiles from Paula Keen and Jenny Moody by the 'Curley Oak' after the dedication; (bottom) The 'Curley Oak' dedication marker to Oliver Rackham, erected by the Woodland Trust, 28 September 2021. (W. Dossett)

Oliver opened our eyes to the secret lives of trees and woods, celebrating their individuality. He cautions against creating tidy categories where diversity exists:

> Is a variant of ragged robin evolution's answer to a thousand years of coppicing in Hayley Wood, but forget-me-not in Hempstead Wood and wood-spurge in Bradfield Woods? Is this why dandelion and cowslip are coppicing plants on Øland, but are not woodland plants in Britain?
>
> Is the thought of woods as islands more than just a metaphor? Is island biogeography at work, producing different coppicing ecotypes in different woods? Conservationists do not disapprove of real islands: they never recommend 'conserving' endemic plants by translocating those of Crete to Cyprus and vice versa. Why propose the equivalent with the special floras of ancient woods?
>
> Rackham 2006: 552

He was long concerned with truth, writing passionately about the pitfalls of pseudo-history and factoids.

> ***pseudo-history*** which has no connection with the real world, and is made up of ***factoids.*** A factoid looks like a fact, is respected as a fact, and has all the properties of a fact except that it is not true.
>
> ...
>
> Pseudo-history is a fascinating subject, part of the anthropology of human motives and attitudes to woodland. But it does matter. It is all the history that most of the public — and some Cabinet Ministers — ever read; much of what passes for conservation is based upon it. We must try to get the story right.
>
> Rackham 1990: 23, 25

He raised this alarm in nearly every book he wrote. In *Woodlands* he even included sections on 'How to write pseudo-history' and 'How to write pseudo-ecology'. The first begins with 'Do no fieldwork' and the latter ends with 'Never admit you don't know something' (Rackham 2006: 185–86). Oliver's call for truth and transparency in scholarship has galvanised researchers across many fields, including most of the contributors to this volume, e.g., Szabó *this volume*; Newton *this volume*, etc.

A passion for his subject beyond measure

Oliver Rackham loved trees. But in truth he loved all parts of the natural world, marvelling at their beauty and bemoaning their loss. In *History of the Countryside*, first published in 1986 and possibly Oliver's most-read book, his lament is as true today as it was then:

> There are four kinds of loss. There is the loss of beauty, especially that exquisite beauty of the small and complex and unexpected, of frog-orchids, or sundews or dragonflies. There is the loss of freedom, of highways and open spaces, which results from the English attitude to land-ownership .... I am especially concerned with the loss of meaning. The landscape is a record of our roots and the growth of civilization. Each individual historic wood, heath, etc., is uniquely different from every other, and each has something to tell us.
>
> Rackham 1986: 25–26

From a young age, Oliver was intensely curious about how the natural world worked. He grasped tiny details, asked new questions and dismissed pseudo-history. He wove his questions and myriad observations into books that surprised, informed and delighted. But he was much more than a great intellect; his humanity – his passion, care for others and humour – further endeared him to thousands around the world (Fig. 25.8). He was a much-loved genius. He was the perfect storm.

Figure 25.8 Oliver Rackham 2006, Chania, Crete, Greece. (J. Moody)

References

Atherden, M. and Wallace, V. (2017) *Yorkshire Woodlands*. PLACE, York.

Bartrum, C. (2017) "'Smells of drains' and 'tastes like cucumber': a unique herbarium voice". *The Friends of Oliver Rackham Newsletter #9, February 2022*, pp. 5–7.

Carolin, P. (2021) Oliver Rackham's paper on Corpus Christi College Old Court republished. *The Friends of Oliver Rackham Newsletter #16, September 2021*, pp. 8–10.

Foster, G.N., Bilton, D.T. and Nelson, B.H. (2016) *Atlas of the Predaceous Water Beetles (Hydradephaga) of Britain and Ireland*. Field Studies Council for the Centre for Ecology and Hydrology's Biological Records Centre.

Hughes, L. (2022) Working on the Rackham Archive. *The Friends of Oliver Rackham Newsletter #17, February 2022*, pp. 13–17.

Jones, H. (2022) Transcribing together: Digitizing Oliver Rackham's Archive. *The Friends of Oliver Rackham Newsletter #17, February 2022*, pp. 19–22.

Keen, P. (2022a) 28 September 2021 – Woodland Trust Memorial Event 2021: dedication of the Curley Oak, Wentwood, to Oliver Rackham, *The Friends of Oliver Rackham Newsletter #17, February 2022*, pp. 25–29.

Keen, P. (2022b) 5 July 2022 – Official online launch of Oliver Rackham's book *The Ancient Woods of South-East Wales*, sponsored by the Woodland Trust. *The Friends of Oliver Rackham Newsletter #18, October 2022*, pp. 25–28.

Niemann, D. (2019) The Stuff of Genius. *Smallwoods* (January 2019), pp. 24–25 [Small Woods Association members quarterly newsletter]

McKenzie, E. (2022) Digitisation of Oliver Rackham's British Herbarium Collection. *The Friends of Oliver Rackham Newsletter #17, February 2022*, pp. 17–19.

Moody, J. (2018a) Oliver Rackham: his legacy and the ancient trees of Crete. *Kentro, Newsletter of the INSTAP Study Center for East Crete*, 21 (Fall 2018), pp. 16–19.

Moody, J. (ed.) (2018b) '16 September 2018' and 'August 2018'. *The Friends of Oliver Rackham Newsletter #12, October 2018*. pp. 11–19

Moody, J. and Dossett, W. (2016) The Oliverium: the Oliver Rackham Commemorative Symposium. *The Letter*, 95, pp. 36–39.

Prosperetti, L. (2022) Ars Ponendi Lucum: Groves in Poetry and Art. In: G. Lamsechi and B. Trinca (eds) *Spiritual Vegetation. Vegetal Nature in Religious Contexts across Medieval and Early Modern Europe.* V&R Unipress, Göttingen, Germany pp. 309–321. https://doi.org/10.14220/9783737014267.309

Pungetti, G. (2017) *Island Landscapes: An expression of European Culture.* Routledge, Abingdon and New York. https://doi.org/10.4324/97813155590110

Rackham, O. (1990) *Trees and Woodland in the British Landscape* (revised edition). J. M. Dent and Sons, London. Reprinted (paperback) 1993; 1995; 1996; 2001.

Rackham, O. (2002) *Treasures of Silver at Corpus Christi College, Cambridge.* Cambridge University Press, Cambridge. [With photographs by John Cleaver.]

Rackham, O. (2003) *Ancient Woodland: Its history, vegetation and uses in England* (new edition). Castlepoint Press, Kirkcudbrightshire.

Rackham, O. (2006) *Woodlands.* (New Naturalist 100), Collins, London. Reprinted 2010.

†Rackham, O. (2015) Brandon Park fungi 1959–2014: change and stability in occurrence. *Suffolk Natural History: Transactions of the Suffolk Naturalists' Society,* 51, pp. 26–32 and Plates 13–15.

†Rackham, O. (2019) *The Ancient Woods of the Helford River.* Little Toller Books, Beaminster Dorset.

†Rackham, O. (2022) *The Ancient Woods of South-East Wales.* Little Toller Books, Beaminster Dorset.

†Rackham, O. and Carolin, P. (2020) *The Courts of Corpus Christi.* Corpus Christi College, Cambridge.

Reese, D. (2022a) Oliver Rackham – Roots and Branches. *The Friends of Oliver Rackham Newsletter #17, February 2022*, pp. 11–13.

Reese, D. (2022b) October 8, 2021 – January 8, 2022 – Oliver Rackham – Roots and Branches. The early work of the late Dr Oliver Rackham: Polymath, Green Wood Centre (Station Road, Coalbrookdale, Telford TF8 7DR, UK). *The Friends of Oliver Rackham Newsletter #17, February* 2022, pp. 23–25.

Oliver Rackham: Publications, newspaper letters and unpublished lectures

Compiled by Jennifer A. Moody, with contributions from †Philip Oswald, Peter Grubb, Christopher Preston and Susan Ranson

Introduction

Among Oliver Rackham's computer files were the beginnings of a list of his publications. Oliver's list has been greatly expanded with contributions from many of his friends and colleagues, especially the late Philip Oswald, Peter J. Grubb, Christopher Preston and Susan Ranson.

Given Oliver's proclivity to sometimes publish in obscure, hard-to-find series, booklets and journals, it is likely that some publications are missing from this list. The same is true for letters and reviews published in newspapers and magazines. Over time and with more research these documents will come to light.

We further note that there have been three books and eleven articles published posthumously where Oliver is either sole author or co-author. Several more are 'forthcoming' or 'in preparation'. We are all grateful to the teams of colleagues who have volunteered their time to bring these important documents to press.

Books

1975 Evans G.C., Bainbridge R. and Rackham O. (eds) (1975) *Light as an Ecological Factor II*. Blackwell Scientific Publications, Oxford.

Rackham, Oliver (1975) *Hayley Wood: Its history and ecology*. Cambridgeshire and Isle of Ely Naturalists' Trust, Cambridge. Reprinted 1990.

1976 Rackham, Oliver (1976) *Trees and Woodland in the British Landscape*. Archaeology in the Field Series. J. M. Dent and Sons, London. Reprinted 1978; 1981 (paperback); 1983.

1980 Rackham, Oliver (1980) *Ancient Woodland: Its history, vegetation and uses in England*. Edward Arnold, London.

1982 D. Moreno, P. Piussi and O. Rackham (eds) (1982) *Boschi: storia e archeologia, Quaderni Storici* 4a (1), Il Mulino, Bologna.

1986 Rackham, Oliver (1986) *The Ancient Woodland of England: The woods of South-East Essex*. Rochford District Council, Rochford.

	Rackham, Oliver (1986) *The History of the Countryside: The full fascinating story of Britain's landscape.* J. M. Dent and Sons, London. Reprinted 1987 (paperback); 1993; 1995; 1997.
1989	Rackham, Oliver (1989) *The Last Forest: The story of Hatfield Forest.* J. M. Dent and Sons, London. Reprinted (paperback) 1993; 1998.
1990	Rackham, Oliver (1990) *Trees and Woodland in the British Landscape* (revised edition). J. M. Dent and Sons, London. Reprinted (paperback) 1993; 1995; 1996; 2001. Reprinted by Weidenfeld and Nicolson 2020.
1991	Grove, A.T., Moody, J. and Rackham, O. (1991) *Crete and the South Aegean Islands: Effects of changing climate on the environment* [report on the European Economic Community project, 1988–90].
1992	Grove, A.T., Moody, J. and Rackham, O. (eds) (1992) *Stability and Change in the Cretan Landscape*, Cambridge University Geography Department, Cambridge.
1994	Rackham, Oliver (1994) *The Illustrated History of the Countryside.* Weidenfeld and Nicolson, London. Reprinted 1997; 2000 (paperback). Revised edition 2003. [With photographs by Tom Mackie.]
1996	Rackham, O. and Moody, J. (1996) *The Making of the Cretan Landscape.* Manchester University Press, Manchester.
2000	Moore, J., Rackham, O., Greef, G. and Bailey, E. (2000) *Winterbourne A.D. 2000.* Winterbourne Parochial Church Council, Winterbourne. [Booklet]
	Rackham, Oliver (2000) *The History of the Countryside: The classic history of Britain's landscape, flora and fauna.* Phoenix Press, London. Reprinted by Weidenfeld and Nicolson 2020.
2001	Grove, A.T. and Rackham, O. (2001) *The Nature of Mediterranean Europe: An ecological history.* Yale University Press, New Haven, CT and London. Reprinted with corrections 2003.
	Rackham, Oliver (2001) *Trees, Wood and Timber in Greek History.* [The Twentieth J. L. Myres Memorial Lecture, 1999.] Leopard's Head Press, Oxford.
2002	Bury, P. (2002) *A Short History of the College of Corpus Christi and the Blessed Virgin Mary in Cambridge.* (3rd edition, heavily revised by Oliver Rackham.) Corpus Christi College, Cambridge.
	Rackham, Oliver (2002) *Treasures of Silver at Corpus Christi College, Cambridge.* Cambridge University Press, Cambridge. [With photographs by John Cleaver.]
2003	Rackham, Oliver (2003) *Ancient Woodland: Its history, vegetation and uses in England* (new edition). Castlepoint Press, Kirkcudbrightshire.
2004	Rackham O. and Moody, J. (2004) Η δημιουργία του Κρητικού τοπίου [Greek translation of *The Making of the Cretan Landscape*]. Crete University Press, Herakleion.
2005	Rackham, Oliver (2005) *The Building of Winterbourne Medieval Barn.* Winterbourne Parochial Church Council [Gloucestershire]. Online at https://stokegiffordhistory.wordpress.com/0225-winterbourne-barn/
2006	Rackham, Oliver (2006) *Woodlands.* (New Naturalist 100) Collins, London. Reprinted 2010.
2007	Rackham, Oliver (2007) *Transitus Beati Fursei: A translation of the 8th century manuscript Life of Saint Fursey.* Fursey Pilgrims, Norwich.
2013	Rackham, Oliver (2013) *I-gi-ri-su-no Kan-to-rii-sai-do* [Japanese translation of *The History of the Countryside*]. Showado Press, Kyoto, Japan.
2014	Rackham, Oliver (2014) *The Ash Tree.* A Little Toller Monograph. Little Toller Books, Dorset.

2019 †Rackham, Oliver (2019) *The Ancient Woods of the Helford River.* [edited by David Morfitt, with contributions from George Peterken, Simon Leatherdale and Paula Keen] Little Toller Books, Beaminster Dorset.

2020 †Rackham, O. and Carolin, P. (2020) *The Courts of Corpus.* Langham Press for Corpus Christi College Cambridge, Cambridge. [Booklet]

2022 †Rackham, Oliver (2022) *The Ancient Woods of South-East Wales.* [edited by Paula Keen, with contributions from Simon Leatherdale, David Morfitt and others] Little Toller Books, Beaminster Dorset.

In preparation Nixon, L., Moody, J., †Price, S., †Rackham, O. and Francis, J. *The Sphakia Survey,* Vols 1 and 2. Oxford University Press, Oxford.

Articles and Chapters

1961 Rackham, O. (1961) Ecological significance of hybridisation between *Rumex sanguineus* and *R. conglomeratus. Proceedings of the Botanical Society of the British Isles* 4, p. 332.

1966 Rackham O. (1966) Radiation, transpiration, and growth in a woodland annual. In: in G.C. Evans, R. Bainbridge and O. Rackham (eds) *Light as an Ecological Factor.* Blackwell Scientific Publications, Oxford, pp. 167–185.

1967 Rackham, Oliver (1967) The history and effects of coppicing as a woodland practice. In: E. Duffey (ed.) *The Biotic Effects of Public Pressures on the Environment,* Monks Wood Experimental Station Symposium, 3. Natural Environment Research Council, Abbots Ripton, pp. 82–93.

1968 Rackham, Oliver (1968) Medieval woodland areas. *Nature in Cambridgeshire,* 11, pp. 22–25.

Rackham, Oliver (1968) The armed ponds of Cambridgeshire. *Nature in Cambridgeshire,* 11, pp. 25–27.

1969 Rackham, Oliver (1969) Knapwell Wood. *Nature in Cambridgeshire,* 12, pp. 25–31.

1970 Rackham, Oliver (1970) Historical studies and woodland conservation. In: E. Duffey (ed.) *The Scientific Management of Animal and Plant Communities for Conservation,* British Ecological Society Symposium 11. Blackwell, Oxford, pp. 563–580.

1971 Rackham, Oliver (1971) Charcoal. In: C. Renfrew, M. Gimbutas and E.S. Elster (eds) *Excavations at Sitagroi, a Prehistoric Village in Northeast Greece,* Monumenta Archaeologica 13. Cotsen Institute of Archaeology Press, Los Angeles, pp. 55–62.

1972 Rackham, Oliver (1972) Charcoal and plaster impressions. In: P. Warren (ed.) *Myrtos: An Early Bronze Age settlement in Crete,* British School at Athens Supplementary Volume 7. British School at Athens, London, pp. 299–304.

Rackham, Oliver (1972) Grundle House: on the quantities of timber in certain East Anglian buildings in relation to local supplies. *Vernacular Architecture,* 3, pp. 3–8.

Rackham, Oliver (1972) The vegetation of the Myrtos region. In: P. Warren (ed.) *Myrtos: an Early Bronze Age settlement in Crete,* British School at Athens Supplementary Volume 7. British School at Athens, London, pp. 283–298.

Rackham, Oliver (1972) Responses of the barley crop to soil water stress. In: A.R. Rees, I.E. Cockshull, D.W. Hand and R.G. Hurd (eds) *Crop Processes in Controlled Environments.* Academic Press, London, pp. 127–138.

Rackham, Oliver (1972) Water relations: drought effects in barley (Nuffield Project). *Annual Report of the Plant Breeding Institute.* Plant Breeding Institute, Cambridge, pp. 158–159.

1974

Rackham, Oliver (1974) The oak tree in historic times. In: M.G. Morris and F.H. Perring (eds) *The British Oak: Its history and natural history*. Classey, Faringdon, pp. 62–79.

1975

Evans, G.C., Freeman, P. and Rackham, O. (1975) Developments in hemispherical photography. In: G.C. Evans, R. Bainbridge and O. Rackham (eds) *Light as an Ecological Factor II*. Blackwell Scientific Publications, Oxford, pp. 549–557.

Rackham, Oliver (1975) Temperatures of plant communities as measured by pyrometric and other methods. In: G.C. Evans, R. Bainbridge and O. Rackham (eds) *Light as an Ecological Factor II*. Blackwell Scientific Publications, Oxford, pp. 423–450.

1976

Rackham, Oliver (1976) The Hayley Wood deer count. *Nature in Cambridgeshire*, 19, pp. 32–34.

Rackham, O., Coombe, D.E., Dymond, D.P. and Hooper, M.D. (1976/1970) 'The hedge's worth and age'. In: K. Gregory (ed.) *The First Cuckoo: A selection of the most witty, amusing and memorable letters to The Times 1900–1975*. George Allen and Unwin and Times Books, London, pp. 303–305.

1977

Rackham, Oliver (1977) Hedgerow trees: their history, conservation, and renewal. *Arboricultural Journal*, 3, pp. 169–177.

Rackham, Oliver (1977) Neolithic woodland management in the Somerset Levels: Garvin's, Walton Heath, and Rowland's tracks. *Somerset Levels Papers* 3, pp. 65–71. Produced by J.M. Coles, B.J. Orme and others associated with the Somerset Levels Project from 1975 to 1988.

1978

Rackham, Oliver (1978) Archaeology and land-use history. In: D. Corke (ed.) *Epping Forest – the Natural Aspect?*, Essex Naturalist NS 2. Essex Field Club, Basildon, Essex, pp. 16–57.

Rackham, Oliver (1978) Sunday, 8 October, fungus foray in the Bradfield Woods. *Nature in Cambridgeshire*, 22, p. 14.

Rackham, Oliver (1978) The flora and vegetation of Thera and Crete before and after the great eruption. In: C. Doumas (ed.) *Thera and the Aegean World I*. Distributed by Aris and Phillips Ltd, London, pp. 755–764.

Rackham, Oliver (1978) Woodlands and their management. In: J. MacConnell (ed.) *Landscape History and Habitat Management*. South Essex Natural History Society, Essex, pp. 8–16.

Rackham, O., Blair, W.J. and Munby, J.T. (1978) The thirteenth-century roofs and floor of the Blackfriars Priory at Gloucester. *Medieval Archaeology*, 22, pp. 105–122.

1979

Rackham, Oliver (1979) Documentary evidence for the historical ecologist. *Landscape History*, 1, pp. 29–33.

Rackham, Oliver (1979) Neolithic woodland management in the Somerset Levels: Sweet Track I. *Somerset Levels Papers* 5, pp. 59–61, 65–71. Produced by J.M. Coles, B.J. Orme and others associated with the Somerset Levels Project from 1975 to 1988.

1980

Rackham, Oliver (1980) Sunday, 12 October, fungus foray at The Lodge, Sandy, Bedfordshire. *Nature in Cambridgeshire*, 24, p. 16.

Rackham, Oliver (1980) The medieval landscape of Essex. In: D.G. Buckley (ed.) *Archaeology in Essex to AD 1500*, CBA Research Report 34. Council for British Archaeology, London, pp. 103–107.

1981

Rackham, Oliver (1981) Land-use and the native vegetation of Greece. In: M. Bell and S. Limbrey (eds) *Archaeological Aspects of Woodland Ecology*, BAR International Series 146. British Archaeological Reports, Oxford, pp. 177–198.

Rackham, Oliver (1981) Saturday, 3 October, fungus foray at Buff Wood, East Hatley. *Nature in Cambridgeshire*, 25, p. 15.

Rackham, Oliver (1981) The Avon Gorge and Leigh Woods. In: M. Bell and S. Limbrey (eds) *Archaeological Aspects of Woodland Ecology*, BAR International Series 146. British Archaeological Reports, Oxford, pp. 171–176.

1982 Rackham, Oliver (1982) Boschi e storia dei sistemi silvo-pastorali in Inghilterra. In: D. Moreno, P. Piussi and O. Rackham (eds) *Boschi: storia e archeologia, Quaderni Storici* 4a (1), Il Mulino, Bologna, pp. 16–48.

Rackham, Oliver (1982) Sunday, 10 October, fungus foray in the Bradfield Woods, Suffolk. *Nature in Cambridgeshire*, 26, p. 13.

Rackham, Oliver (1982) The growing and transport of timber and underwood. In: S. McGrail (ed.) *Woodworking Techniques before A.D. 1500*, BAR International Series 129. British Archaeological Reports, Oxford, pp. 199–218.

1983 Rackham, Oliver (1983) Observations on the historical ecology of Boeotia. *The Annual of the British School at Athens*, 78, pp. 291–351.

1984 Rackham, Oliver (1984) The forest: woodland and wood-pasture in medieval England. In: K. Biddick (ed.) *Archaeological Approaches to Medieval Europe*. Medieval Institute Publications, Western Michigan University, Kalamazoo, pp. 70–104.

1985 Rackham, Oliver (1985) Ancient woodland and hedges in England. In: S.R.J. Woodell (ed.) *The English Landscape – Past, Present, and Future*, Wolfson College Lectures for 1983. Oxford University Press, Oxford, pp. 48–67.

1986 Bridge, M.C., Hibbert, A. and Rackham, O. (1986) Effects of coppicing on the growth of oak timber trees in the Bradfield Woods, Suffolk. *Journal of Ecology*, 74, pp. 1095–1102.

Rackham, Oliver (1986) The ancient woods of Norfolk [the Presidential Address to Norfolk & Norwich Naturalists' Society for 1985]. *Transactions of Norfolk & Norwich Naturalists' Society*, 27, pp. 161–177.

1987 Rackham, Oliver (1987) The countryside: history and pseudo-history. *The Historian*, 14, pp. 13–17. [Historical Association (Great Britain)]

1987–8 Rackham, Oliver (1987–8) The making of the Old Court, Parts 1 and 2. *Letter of the Corpus Association*. Corpus Christi College, Cambridge.

1988 Nixon, L., Moody, J. and Rackham, O. (1988) Archaeological survey in Sphakia, Crete. *Echos du monde classique/Classical Views (EMC/CV)*, 32 n.s. 7, pp. 157–173.

Rackham, Oliver (1988, 3rd edition 1999) Medieval woods. In: D. Dymond and E. Martin (eds) *An Historical Atlas of Suffolk*. Suffolk County Council, Ipswich, Suffolk, pp. 50–51.

Rackham, Oliver (1988, reprinted 2004) Trees and woodland in a crowded landscape – the cultural landscape of the British Isles. In: H.H. Birks, H.J.B Birks, P.E. Kaland and D. Moe (eds) *The Cultural Landscape: Past, Present and Future*. Cambridge University Press, Cambridge, pp. 53–78.

Rackham, Oliver (1988) Wildwood. In: M. Jones (ed.) *Archaeology and the Flora of the British Isles: Human influence on the evolution of plant communities* [papers read at a joint conference of the Botanical Society of the British Isles and the Association of Environmental Archaeology, held at St Anne's]. Oxford Committee for Archaeology Monograph 14. Oxford University School of Archaeology, Oxford, pp. 3–6.

1989 Nixon, L., Moody, J., Price, S. and Rackham, O. (1989) Archaeological survey in Sphakia, Crete. *Echos du monde classique/Classical Views (EMC/CV)*, 33 n.s. 8, pp. 201–215.

Rackham, Oliver (1989) *Hedges and Hedgerow Trees in Britain: A thousand years of agroforestry*. Social Forestry Network Paper 8c. Overseas Development Institute (ODI), London. https://odi.org/en/publications/hedges-and-hedgerow-trees-in-britain-a-thousand-years-of-agroforestry/ [last accessed March 2023]

Rackham, Oliver (1989) The life of trees. In: R. Malbert and J. Drew (eds) *The Tree of Life: New images of an ancient symbol.* South Bank Centre, London, pp. 48–53.

1990

Nixon, L., Moody, J., Price, S., Rackham, O. and Niniou-Kindeli, V. (1990) Archaeological survey in Sphakia, Crete. *Echos du monde classique/Classical Views (EMC/CV)*, 34 n.s. 9, pp. 213–220.

Rackham, Oliver (1990) Ancient landscapes. In: O. Murray and S. Price (eds) *The Greek City from Homer to Alexander.* Clarendon Press, Oxford, pp. 85–111.

Rackham, Oliver (1990) Observations on the historical ecology of Santorini. In: D.A. Hardy, J. Keller, V.P. Galanopoulos, N.C. Fleming, T.H. Druitt (eds) *Thera and the Aegean World III, vol 2: Earth Sciences.* The Thera Foundation, London, pp. 384–391.

Rackham, Oliver (1990) The greening of Myrtos. In: S. Bottema, G. Entjes-Nieborg and W. Van Zeist (eds) *Man's Role in the Shaping of the Eastern Mediterranean Landscape.* Balkema, Rotterdam, pp. 341–348.

1991

Rackham, Oliver (1991) Introduction to pollards. In: H.J. Read (ed.) *Pollard and Veteran Tree Management.* Corporation of London, London, pp. 6–10.

Rackham, Oliver (1991) Landscape and the conservation of meaning [Reflection Riding Memorial Lecture for 1990]. *Royal Society of Arts Journal*, 139, pp. 903–915.

Rackham, Oliver (1991) Listing the countryside. *Country Life* 19 September, pp. 98–101.

1992

Hall, J.A., Atherden, M., Rackham, O. and Grove, A.T. (1992) A pollen diagram from the White Mountains – a preliminary interpretation. In: A.T. Grove, J. Moody and O. Rackham (eds) *Stability and Change in the Cretan Landscape.* Cambridge University Geography Department, Cambridge, pp. 40–41.

Hayden, B.J., Moody, J.A. and Rackham, O. (1992) The Vrokastro Survey Project, 1986–1989: research design and preliminary results. *Hesperia*, 61: 293–353.

Moe, D. and Rackham, O. (1992) Pollarding and a possible explanation of the neolithic elm-fall. *Vegetation History and Archaeobotany*, 1, pp. 63–68.

Rackham, Oliver (1992) Conservation in the cultural landscape: the historical context and the story of Crete. In: *So that God's Creation Might Live: The Orthodox Church responds to the ecological crisis*, Proceedings of the Inter-Orthodox Conference on Environmental Protection, The Orthodox Academy of Crete, November 1991). The Ecumenical Patriarchate of Constantinople (with ICOREC, WWF and Syndesmos), Istanbul, pp. 89–94. [Also in Greek, pp. 107–114]

Rackham, Oliver (1992) Gamlingay Wood. *Nature in Cambridgeshire* 34, pp. 3–14.

Rackham, Oliver (1992) Lodges and standings. In: M.W. Hanson (ed.) *Epping Forest through the Eye of the Naturalist*, Essex Naturalist 11. Essex Field Club, Basildon, Essex, pp. 8–17.

Rackham, Oliver (1992) Mixtures, mosaics and clones: the distribution of trees within European woods and forests. In: G.R. Cannell, D.C. Malcolm and P.A. Robertson (eds) *The Ecology of Mixed-species Stands of Trees*, British Ecological Society Special Publication 11. Blackwell Scientific, Oxford, pp. 1–20.

Rackham, Oliver (1992) The Essex landscape in Morant's time and today. In: K. Neale (ed.) *Essex Heritage: Essays presented to Sir William Addison.* Leopard's Head Press, Oxford, pp. 249–270.

Rackham, Oliver (1992) The excursion. In: *So that God's Creation Might Live: The Orthodox Church responds to the ecological crisis,* Proceedings of the Inter-Orthodox Conference on Environmental Protection, The Orthodox Academy of Crete, November 1991). The Ecumenical Patriarchate of Constantinople (with ICOREC, WWF and Syndesmos), Istanbul, pp. 104–110. [Also in Greek, pp. 124–132]

Rackham, Oliver (1992) Vegetation history of Crete. In: A.T. Grove, J. Moody and O. Rackham (eds) *Stability and Change in the Cretan Landscape.* Cambridge University Geography Department, Cambridge, pp. 29–39.

Rackham, Oliver (2012) Woodland ecology in recent and historic aerial photographs. *Photogrammetric Record,* 14, pp. 227–239.

Rackham, O. and Moody, J. (1992) Terraces. In: B. Wells (ed.) *Agriculture in Ancient Greece,* 7th International Symposium of the Swedish Institute at Athens. Skrifter utgivna av Svenska Institutet I Athen, 4°, 42, Stockholm, pp. 123–130.

Rackham, O., Moody, J.A. and Grove, A.T. (1992) Notes on the excursions [from Khaniá to Thérisso and Souyia]. In: A.T. Grove, J. Moody and O. Rackham (eds) *Stability and Change in the Cretan Landscape.* Cambridge University Geography Department, Cambridge, pp. 77–88.

1993

Grove, A.T. and Rackham, O. (1993) Threatened landscapes in the Mediterranean: examples from Crete, *Landscape and Urban Planning,* 24, pp. 278–92.

Rackham, Oliver (1993) The Rivenhall woods. In: W.J. and K.A. Rodwell (eds) *Rivenhall* [Essex]*: investigations of a villa, church and village, 1950–1977 vol. 2,* CBA Research Report 80. Council for British Archaeology, London, pp. 120–125.

1994

Nixon L., Moody J., Price S. and Rackham O. (1994) Rural settlement in Sphakia, Crete. In: P. Doukellis and L. Mendoni (eds) *Structures rurales et sociétés antiques (Actes du colloque de Corfou 14–16 mai 1992),* Annales littéraires de l'Université de Besançon. Les Belles Lettres, Paris, pp. 255–264.

Rackham, Oliver (1994) Woods and the Soul of England. *Countryside Campaigner* Spring, pp. 12–13.

1995

Rackham, Oliver (1995) Looking for ancient woodland in Ireland. In: J.R. Pilcher and S. Mac an tSaoir (eds) *Woods, Trees and Forests in Ireland.* Royal Irish Academy, Dublin, pp. 1–12.

Rackham, O. and Jarman, R. (1995) Welshbury hillfort. *Dean Archaeology,* 8, pp. 36–42.

1996

Grove, A.T. and Rackham, O. (1996) History of Mediterranean land use. In: P. Mairota, J.B. Thornes and N. Geeson (eds) *Atlas of Mediterranean Environments in Europe.* John Wiley and Sons, Chichester, pp. 76–78.

Moody, J.A., Rackham, O. and Rapp Jr, G. (1996) Environmental archaeology of prehistoric NW Crete. *Journal of Field Archaeology,* 23, pp. 273–297.

Rackham, Oliver (1996) Ecology and pseudo-ecology: the example of ancient Greece. In: G. Shipley and J. Salmon (eds) *Human Landscapes in Classical Antiquity.* Routledge, London, pp. 16–43.

Rackham, Oliver (1996) Greece (geography), In: S. Hornblower and A. Spawforth (eds) *The Oxford Classical Dictionary 3rd edition* (revised 2003). Oxford University Press, Oxford, pp. 648.

Rackham, Oliver (1996) Landscapes (ancient Greek). In: S. Hornblower and A. Spawforth (eds) *The Oxford Classical Dictionary 3rd edition* (revised 2003). Oxford University Press, Oxford, p. 813–814.

Rackham, O. and Coombe, D.E. (1996) Madingley Wood. *Nature in Cambridgeshire,* 38, pp. 27–54.

1997

Rackham, O. (1997) European concepts of forest age. In: J. Dargavel (ed.) *Australia's Ever-Changing Forests III: Proceedings of the Third National Conference on Australian Forest History*, CRES, Canberra, pp. 11–23.

1998

Moody, J., Nixon, L., Price, S. and Rackham, O. (1998) Surveying poleis and larger sites in Sphakia. In: W. G. Cavanagh, M. Curtis, J. N. Coldstream, A. W. Johnston, R. Curtis and J. Johnston (eds) *Post-Minoan Crete. Proceedings of the First Colloquium on Post-Minoan Crete*, British School at Athens Studies 2. British School at Athens, London, pp. 87–95.

Rackham, Oliver (1998) Savanna in Europe. In: K.J. Kirby and C. Watkins (eds) *The Ecological History of European Forests.* CABI, Wallingford, pp. 1–24.

Rackham, Oliver (1998) The abbey woods. In: A. Gransden (ed.) *Bury St Edmunds: Medieval art, architecture, archaeology and economy.* British Archaeological Association, London, pp. 139–160, Plates XXXIII and XXXIV.

Rackham, Oliver (1998) Woodland Conservation: past, present and future. In: M.A. Atherden and R.A. Butlin (eds) *Woodlands in the Landscape: Past and Future Perspectives*, PLACE Research Centre, the University College of Ripon and York St John. Leeds University Press, Leeds, pp. 60–78.

1999

Cevasco, R., Moreno, D., Poggi, G. and Rackham, O. (1999) Archeologia e storia della copertura vegetale: esempi dall'Alta Val di Vara. *Memoria della Accademia Lunigianese di Scienze "Giovanni Capellini"*, 69, pp. 241–261.

Rackham, Oliver (1999) Boundaries and country planning, ancient and modern. In: P. Slack (ed.) *Environments and Historical Change: The Linacre lectures.* Oxford University Press, Oxford, pp. 96–117.

Rackham, Oliver (1999) The woods 30 years on: where have the Primroses gone? *Nature in Cambridgeshire*, 41, pp. 73–87.

2000

Rackham, Oliver (2000) Aegological note on Eupolis Fr. 13. In: D. Harvey and J. Wilkins (eds) *The Rivals of Aristophanes: Studies in Athenian old comedy.* Classical Press of Wales, Swansea, pp. 349–350.

Rackham, Oliver (2000) Climate, Ecology, Forestry, Landscape, Plants. In: G. Speake (ed.) *Encyclopedia of Greece and the Hellenic Tradition.* Fitzroy Dearborn, London, pp. 358f, 525f, 628, 919f, 1330.

Rackham, Oliver (2000) David Coombe (1927–1999) *Nature in Cambridgeshire*, 42, pp. 71–73.

Rackham, Oliver (2000) Prospects for landscape history and historical ecology. *Landscapes*, 2, pp. 3–15, Plates 1–3.

Rackham, Oliver (2000) Purposes and methods of ecological history. *Transactions of Suffolk Naturalists' Society*, 36, pp. 1–28.

Rackham, Oliver (2000) The parish's setting: its appearance. In: J. Moore, O. Rackham, G. Greef and E. Bailey (eds) *Winterbourne A.D. 2000.* Winterbourne Parochial Church Council, Winterbourne, pp. 27–40.

Rackham, Oliver (2000) Woodland. In: T. Kirby and S. Oosthuizen (eds) *An Atlas of Cambridgeshire and Huntingdonshire History.* Anglia Polytechnic University, Cambridge, pp. 69–70.

Rackham, Oliver (2000) Woodland in the Ely Coucher Book. *Nature in Cambridgeshire*, 42, pp. 37–44, 61–67.

2001

Rackham, Oliver (2001) Forests. In: P.J. Crabtree (ed.) *Medieval Archaeology: An encyclopedia.* Garland, New York, p. 111.

Rackham, Oliver (2001) Hedges. In: P.J. Crabtree (ed.) *Medieval Archaeology: An encyclopedia.* Garland, New York, p. 159.

Rackham, Oliver (2001) Land-use patterns, historic. In: S.A. Levin (ed.) *Encyclopedia of Biodiversity, vol. 3*. Elsevier Academic Press, pp. 675–687.

Rackham, Oliver (2001) Parks. In: P.J. Crabtree (ed.) *Medieval Archaeology: An encyclopedia*. Garland, New York, p. 252.

Rackham, Oliver (2001) Woodland. In: P.J. Crabtree (ed.) *Medieval Archaeology: An encyclopedia*. Garland, New York, pp. 381–382.

2002

Rackham, Oliver (2002) Observations on the historical ecology of Laconia. In: W. Cavanagh, J. Crouwell, R.W.V. Catling and G. Shipley (eds) *Continuity and Change in a Greek Rural Landscape: The Laconia survey*, British School at Athens Supplementary Volume 27. British School at Athens, London, pp. 73–119.

Rackham, Oliver (2002) The Holy Mountain [Athos]. *Plant Talk*, 27, pp. 19–23.

Rackham, Oliver (2002) The Blean woods and timber-framed buildings. In: W. Holmes and A. Wheaten (eds) *The Blean: The woodlands of a cathedral city*. White Horse Press, Whitstable, pp. 77–81.

Rackham, Oliver (2002) The trench along Watling Street. In: W. Holmes and A. Wheaten (eds) *The Blean: The woodlands of a cathedral city*. White Horse Press, Whitstable, p. 48; 77–81.

Rackham, Oliver (2002) Trees with personality. *The Spectator*, 11 October.

2003

Fukamachi, K., Oku, H. and Rackham, O. (2003) A comparative study on trees and hedgerows in Japan and England. In: H. Palang and G. Fry (eds) *Landscape Interfaces*. Kluwer Academic Publishing, Dordrecht, pp. 53–69.

Rackham, Oliver (2003) Oliver Rackham (Fellow) 1958. In: M.E. Bury and E.J. Winter (eds) *Corpus within Living Memory: Life in a Cambridge College*. Third Millennium Publishing, London, pp. 186–191.

Rackham, Oliver (2003) The physical setting. In: D. Abulafia (ed.) *The Mediterranean in History*. Thames and Hudson, London, pp. 33–66.

Rackham, Oliver (2003) Why Corpus Christi? In: M.E. Bury and E.J. Winter (eds) *Corpus within Living Memory: Life in a Cambridge College*. Third Millennium Publishing, London, pp. 9–17.

2004

Rackham, Oliver (2004) Our Lady's Garden: the historical ecology of the Holy Mountain. *Annual Report of the Friends of Mount Athos*, 2004, pp. 48–57.

Rackham, Oliver (2004) Pre-existing trees and woods in country-house parks. *Landscapes*, 5, pp. 1–16.

Rackham, O. and Clark, J. (2004) On the historical ecology of Pseira. In: P.P. Betancourt, J.A. Clark, P.M. Day, W.R. Farrand, R.H. Simpson, T. Howard, O. Rackham and C.H. Stearns, *Pseira VIII: The archaeological survey of Pseira Island, part I*. INSTAP Academic Press, Philadelphia. 55–62.

2005

Rackham, Oliver (2005) Medieval woodland. In: T. Ashwin and A. Davison (eds) *An Historical Atlas of Norfolk*. Phillimore, Chichester, pp. 54–55.

2006

Rackham, Oliver (2006) All roots and branches, *The Spectator*, 27 January.

Rackham, Oliver (2006) Mountains, woods, and waters in the European Mediterranean: a summary for the last 200 years. In: M. Armiero (ed.) *Views from the South: Environmental stories from the Mediterranean world (19th–20th centuries)*. Istituto di Studi sulle Societa del Mediterraneo, Naples, pp. 225–237.

2007

Rackham, Oliver (2007) Child of the New Forest, *The Spectator*, 10 August.

Rackham, Oliver (2007) Long live the weeds and the wilderness, *The Spectator*, 31 August.

2008

Price, S., Rackham, O., Kiel, M. and Nixon, L. (2008) Sphakia in Ottoman census records: a *Vakif* and its agricultural production. In: Antonis Anastasopoulos (ed.) *The Eastern Mediterranean under Ottoman Rule: Crete, 1645–1840. Halcyon Days in Crete VI (a Symposium held in Rethymno, 13–15 January 2006).* University of Crete Press, Rethymno, pp. 69–99.

Rackham, Oliver (2008) Ancient woodlands: modern threats (Tansley Review). *New Phytologist* 180, pp. 571–586.

Rackham, Oliver (2008) Holocene history of Mediterranean island landscapes. In: I.N. Vogiatzakis, G. Pungetti and A.M. Mannion (eds) *Mediterranean Island Landscapes: Natural and cultural approaches.* Springer Dordrecht, Berlin, pp. 36–60.

Vogiatzakis, I.N. and Rackham, O. (2008) Crete. In: I.N. Vogiatzakis, G. Pungetti and A.M. Mannion (eds) *Mediterranean Island Landscapes: Natural and cultural approaches.* Springer Dordrecht, Berlin, pp. 245–270.

2009

Nixon L., Price S., Rackham O. and Moody J. (2009) Settlement patterns in mediaeval and post-mediaeval Sphakia: issues from the environmental, archaeological, and historical evidence. In: J. Bintliff and H. Stöger (eds) *Medieval and Post-medieval Greece: The Corfu papers,* BAR International Series 2023. British Archaeological Reports, Oxford, pp. 43–54.

Rackham, O. Foreword. In: Pollard, E. and Brawn, E. (eds), *The Great Trees of Dorset.* The Dovecote Press, Wimborne.

2010

Rackham, Oliver (2010) Animals without backbones, *The Spectator,* 2 July.

Rackham, Oliver (2010) Foreword. In R. Oaks and E. Mills, *Coppicing and Coppice Crafts: A comprehensive guide.* The Crowood Press Ltd, p. 7.

Rackham, Oliver (2010) Leaves on the line: British trees face a growing range of threats, *The Spectator,* 17 December.

Rackham, Oliver (2010) Seeing the wood from the trees, *The Spectator,* 3 September.

Rackham, O., Moody, J., Nixon, L. and Price, S. (2010) Some field systems in Crete. In: O. Krzyszkowska (ed.) *Cretan Studies in Honour of Peter Warren.* British School at Athens Studies 18. British School at Athens, London, pp. 269–284.

2011

Rackham, Oliver (2011) Forest and upland. In: J. Crick and E. van Houts (eds) *A Social History of England, 900–1200.* Cambridge University Press, Cambridge, pp. 46–55.

Rackham, Oliver (2011) Foreword. In: P.H. Oswald and C.D. Preston (eds and translators) *John Ray's Catalogue (1660).* Ray Society, London. vii.

Rackham, Oliver (2011) The history of Cretan landscapes and their special plants. *Mediterranean Garden,* 63, pp. 5–14.

Rackham, Oliver (2011) The woods [of South Weald]: history and composition. *Essex Naturalist,* NS 28, pp. 129–131.

2012

Fukamachi, K. and Rackham, O. (2012) Sacred groves in Japanese satoyama landscapes: a case study and prospects for conservation. In: G. Pungetti, G. Oviedo and D. Hooke (eds) *Sacred Species and Sites: Advances in biocultural conservation.* Cambridge University Press, Cambridge, pp. 419–423.

Pungetti, G., Hughes, P. and Rackham, O. (2012) Ecological and spiritual values of landscape: a reciprocal heritage and custody. In: G. Pungetti, G. Oviedo and D. Hooke (eds) *Sacred Species and Sites: Advances in biocultural conservation.* Cambridge University Press, Cambridge, pp. 65–82.

Rackham, Oliver (2012) History of the countryside as a model for the future. *Suffolk Natural History,* 48, pp. 61–78.

Rackham, Oliver (2012) Island landscapes: some preliminary questions. *Journal of Marine & Island Cultures*, 1, pp. 87–90.

Rackham, Oliver (2012) Of knowledge, life, good and evil, *The Spectator*, 11 December.

Rackham, Oliver (2012) The ghosts at the ends of the earth: tree-land in four hemispheres. In I. Rotherham, M. Jones and C. Handley (eds) *Working & Walking in the Footsteps of Ghosts – Volume 1: The wooded landscape*. Wildtrack Publishing for the South Yorkshire Biodiversity Research Group and the Landscape Conservation Forum, Sheffield, pp. 118–132.

Rackham, O. and Moody, J. (2012) Drivers of change and the landscape history of 'Cavo Sidero'. In: T. Papayannis and P. Howard (eds) *Reclaiming the Greek Landscape*. Med-INA, Athens, pp. 219–232.

2013

Rackham, Oliver (2013) Richard Rigby, 1722–88, Fellow Commoner. *The Letter*, 92, pp. 20–26.

Rackham, Oliver (2013) The Man Who Planted Trees, by Jim Robbins – review, *The Spectator*, 5 June.

Rackham, Oliver (2013) The Tradescants' Orchard, by Barry Juniper – review, *The Spectator*, 8 May.

Rackham, Oliver (2013) Woodland and wood-pasture. In: I.D. Rotherham (ed.) *Trees, Forested Landscapes and Grazing Animals*. Routledge, London, pp. 11–22.

2014

Rackham, Oliver (2014) The Pseudo-Marlowe portrait. A wish fulfilled? *The Letter*, 93, pp. 30–3.

Rackham, Oliver (2014) Warning: the beautiful trees in this book may very soon be extinct, *The Spectator*, 28 May.

2015

†Rackham, Oliver (2015) Brandon Park fungi 1959–2014: change and stability in occurrence. *Suffolk Natural History: Transactions of the Suffolk Naturalists' Society*, 51, pp. 26–32 and Plates 13–15.

†Rackham, Oliver (2015) Greek landscapes: profane and sacred. In: L. Käppel and V. Pothou (eds) *Human Development in Sacred Landscapes*. V&R Unipress, Göttingen, Germany, pp. 35–50.

2017

†Rackham, Oliver (2017) Island landscape history: the Isles of Scilly, UK. In: G. Pungetti (ed.) *Island Landscapes: An expression of European culture*. Routledge, Abingdon and New York, pp. 21–37.

†Rackham, Oliver (2017) Landscape history of Cyprus: a preliminary report. In: G. Pungetti (ed.) *Island Landscapes: An expression of European culture*. Routledge, Abingdon and New York, pp. 47–58.

†Rackham, Oliver (2017) Saaremaa: tackling landscape history in Estonia. In: G. Pungetti (ed.) *Island Landscapes: An expression of European culture*. Routledge, Abingdon and New York, pp. 38–46.

2018

†Rackham, Oliver (2018) Archaeology of trees, woodland, and wood-pasture. In: A. Çolak, S. Kirca and I. Rotherham (eds) *Ancient Woodlands and Trees: A guide for forest managers and landscape planners*, IUFRO [International Union of Forest Research Organizations] World Series Vol. 37, Vienna, pp. 39–60.

2019

Ntinou, M., †Rackham, O., Moody, J. and Ważny, T. (2019) The wood charcoal remains from the Kamilari tholos tomb. In: L. Girella and I. Caloi (eds) *Kamilari: una necropoli di tombe a tholos nella Messará (Creta)* [Monografie Scuola Archeologica Italiana di Atene e delle Missioni Italiane in Oriente 29] All'Insegna del Giglio, Athens, pp. 639–646.

2020 †Rackham, Oliver (2020) Our Lady's Garden: the historical ecology of the Holy Mountain. In: P. Howorth and C. Thomas (eds) *Encounters on the Holy Mountain, Stories from Mount Athos*, Brepols, Turnhout Belgium, pp. 264–269 [originally published in 2004 in *Annual Report of the Friends of Mount Athos*, pp. 48–57]

†Rackham Oliver (2020) The making of the Old Court [revised by P. Carolin]. In: O. Rackham and P. Carolin, *The Courts of Corpus*. Langham Press for Corpus Christi College Cambridge, Cambridge, pp. 8–25.

Ważny, T., Tzigounaki, A., †Rackham, O., Moody, J., Helman-Ważny, A., Pearson, C., Giapitsoglou, K., Troulinos, M., Fraidhaki, A. and Apostolaki, N. Trees, Timber and tree-rings in historic Crete, Byzantine to Ottoman. *Archaiologika Erga Kritis (AEK)* 4, pp. 339–349.

2023 †Rackham, O. (2023) Archaeology of trees, woodland, and wood-pasture. In: A. Çolak, I. Rotherham and S. Kirca (eds) *Ancient Woods, Trees and Forests: Ecology, History and Management*. Pelagic Publishing, Exeter, pp. 31–71. [Originally published in 2018 in A. Çolak, S. Kirca and I. Rotherham (eds) *Ancient Woodlands and Trees: A guide for forest managers and landscape planners*, as above]

Forthcoming Moody, J., †Rackham, O. and Desmond, S. (forthcoming) Preliminary study of the archaeological charcoal from Alonaki. In: A. Katretsou (ed.) *The Alonaki Excavations on Mt Juktas*. INSTAP Academic Press, Philadephia, PA.

Moody, J., †Rackham, O. and Desmond, S. (forthcoming) The making of the Juktas landscape. In: A. Karetsou and J. Moody (eds) *Mount Juktas. An Introduction to its Natural Environment and the Minoan Peak Sanctuary*. Cultural Association of Archanes and the Department of Crete, Heraklieon.

Ważny, T., †Rackham, O. and Moody, J. Dendrochronology of the Cretan Cypress.

Newspaper articles and letters

1970 Coombe, D.E., Dymond, D.P., Hooper, M.D. and Rackham, O. (1970) The hedge's worth and age, *The Times* 6 April.

1989 Rackham, Oliver (1989) New Year lessons of Great Storm, *The Times* 3 January.

1995 Rackham, Oliver (1995) The trouble with trees, *The Independent* 6 October, http://www.independent.co.uk/news/uk/the-trouble-with-trees-1576326.html. [last accessed March 2023]

2008 Rackham, O. and Moody, J. (2008) Harsh course of nature, *The Guardian* 5 March, http://www.theguardian.com/environment/2008/mar/05/endangeredhabitats.endangeredspecies. [last accessed March 2023]

2012 Rackham, Oliver (2012) What can I tell you about ash disease? Easy – I told you so!, *Daily Mail* 3 November and *The Mail on Sunday* 4 November.

Radio; TV; Speeches

2000 Rackham, Oliver (2000) [Response by Dr Oliver Rackham to honorary degree ceremony Oration at Essex University. Overview of the achievements of conservation work in Essex 1970–2000.]

2006 *Trees and Me* (TV documentary). Directors J. Hughes and L. Swingler. Oliver appears as himself. http://www.imdb.com/title/tt0930628/?ref_=nm_flmg_slf_1. [last accessed March 2023]

2012 — *The Making of the European Mediterranean Landscape.* Talk given by Oliver Rackham to the Mediterranean Garden Society, Portugal, March 2012. https://vimeo.com/48385302 [last accessed March 2023]

Cambridge Conservation Volunteers' 50th Anniversary Bash. 'CCV celebrates 50 years of green space conservation. Featuring: Hayley Wood, 50 or so CCV volunteers (and that's not only the age of some...), some grey - if not white – beards, a furious bonfire, a mad axe man, a stunning cake, and Oliver Rackham. https://www.youtube.com/watch?v=oA9LYjVAmXA. [last accessed March 2023]

Woodland Communication Day 2012: Woods Past. Talk given by Oliver Rackham posted 10 November 2013 https://www.youtube.com/watch?v=7_-WbvsUoTk. [last accessed March 2023]

Saving Species: Ash Dieback/ Managing woodland. (20 November 2012) BBC Radio 4 programme with Brett Westwood, Peter Marren, Julian Roughton, Oliver Rackham [to hear Oliver go to 11.48–13.46 minutes] https://www.bbc.co.uk/sounds/play/b01nxh4q. [last accessed March 2023]

Unpublished lectures and manuscripts available at the Rackham Archive, Corpus Christi College, Cambridge

1979 — Grant, D.R. and Rackham, O. (1979) Drought in Barley 1. Methods for monitoring and measuring soil moisture in field experiments [manuscript submitted to the *Journal of Agricultural Science* but returned with edits Oliver refused to accept].

Rackham, Oliver (1979) Drought in Barley 2. The effects of artificial drought and irrigation on yields and principal yield components in a field crop [manuscript submitted to the *Journal of Agricultural Science* but returned with edits Oliver refused to accept].

Rackham, Oliver (1979) Drought in Barley 3. Growth analysis [manuscript submitted to the *Journal of Agricultural Science* but returned with edits Oliver refused to accept].

1998 — 'Japanese notes' [unpublished typescript].

1999, 2013 'Park Wood, Chrishall & Wenden Lofts' [short manuscript].

2006/7 — 'Ancient Woodland in Herefordshire' [partly written, many photos (ancient documents folder with photos)].

2007 — 'Woodland ownership in history'. Lecture given at Small Woods Association National Conference in Sussex, 20 September.

2007, 2013 — 'Crete to California: exchanges of plants and plant diseases' [Text and lecture; possible book discussed in email, 15 June 2007].

2009, 2014 — 'Grimesthorpe Park' [consultation project? Report written 2014; photos].

2012 — 'Ancient Greek farming in its ecological context. Development of Greek cultural landscapes'. For the conference 'Rethinking Ancient Greek Agriculture', Aarhus, Denmark.

'Not all ancient woods (nor all tropical forests) have always been woodland'. Lecture given at the University of Kent, 24 February 2012 [lecture notes, Powerpoint].

'Racecourse Plantation, Thorpe-next-Norwich'. Report on its history [manuscript written November].

'Some landscapes of Welsh Herefordshire: change or stability?' Hellens, Much Marcle, 15 May [lecture notes].

	'The landscape of the English. Part of the *Via Francigena*'. Lecture in Dover, August [Powerpoint].
	'Veteran trees around the world'. Lecture given August [lecture and Powerpoint].
2013	'Ethiopia lecture' [Powerpoint].
	'European culture as expressed in island landscapes outside Europe: the case of Tasmania' [manuscript July 2013. prepared for the Island Landscape project but then not used as papers were restricted to locations in Europe].
	'Historical Ecology and Sacred Groves'. Lecture given in Ioannia, Greece, February 2013.
	'Landscape history and landscape along the Via Francigena'. Lecture given in Rimini, January 2013.
	'Methods in the History of Island Landscapes', Lecture Aug 2013, Sassari, 2013 [lecture and Powerpoint].
	'Old Broom, Risby' [short report April 2013].
	'The significance of sacred trees and groves'. Lecture given in Manchester, September 2013 [Powerpoint].
	'The story of Little Gidding' [manuscript written March 2013].
	'Trees, forests and people in Ancient and Modern Greece'. Lecture given in Ioannia, Greece, 13 February [lecture and Powerpoint].
	'What is wrong with woodland?' [article written 16 November for *Country Life*. They returned it with edits that Oliver refused to accept].
2013, 2014	'Conservation vs re-creation' [Powerpoint].
2014	'Cerne Abbas Park' [short manuscript, 26 August 2014].
	'Ecology and pseudo-ecology'. Lecture given at Uppsala University, 17 March and Konitsa, August 2014 [lecture and Powerpoint].
	'English sacred landscapes'. Lecture given in Venice [lecture and Powerpoint].
	'History of sacred landscapes'. Lecture given in Venice [lecture and powerpoint].
	'Mountains and ecological history'. International conference, 'Biodiversity, conservation and sustainable development in mountain areas of Europe: the challenge of interdisciplinary research', 20–4 September 2005. Ioannina, Epirus, Greece [manuscript rewritten July 2014].
	'Sacred landscape history – methods and question' [notes and text, December 2014].
	'Woods and woodland bats'. Lecture given at the *Woodland Bat Symposium*, Rugby, Warwickshire, November 2014 [lecture and PowerPoint].
2015	'Sacred landscapes: examples and case-studies' [manuscript written 14 January 2015, includes Stonehenge, Mount Sinai, Mount Athos, Little Gidding].

Acknowledgements from the Editors

We thank David Hawkins and the team at Pelagic Publishing for their incredibly hard work over the several years of this project. We also thank the authors for their contributions and Professor Peter Grubb for his foreword and his encouragement. We are grateful to Professor Christopher Kelly, Master Corpus Christi College Cambridge, for his generous support and to Dr Edmund Tanner for his help in meticulously proofreading the text.

Finally, we acknowledge the great debt we owe Oliver Rackham for his life's work, the inspiration for this book.

General Index

References to figures appear in *italic* type; those in **bold** type refer to tables; and the letter 'n' indicates a footnote.

Aboriginal Australians 309, 310, 311; knowledge 314–16; land management 316, 317 *see also* First Nation Australians
Acacia species 307–8, 310–11 *see also* Australian Blackwood; Wattle; Gum Arabic Tree
Acacia linifolia (White Wattle) 309
account books 245–6, 262–3; 'Book of sales of the woods of Champagne', France (thirteenth century) 262; Choustník woodland account, Bohemia (1447) 263; 'Gotahelm's inventory', Tegernsee Abbey (1023) 251–7; *see also* documentary evidence
Acer opalus (Italian Maple) 201, 211
Act of Commons (Statute of Merton AD 1235) 276–7
Action Plan for Trees with High Nature Conservation Values (Sweden) 326–7
Action Programme for the Environment of DG Research (European Commission) 140
agariko (Japanese) *see* Beech, Japanese
Agariko-daio (king of pollard) (giant Japanese Beech pollard) *341*
ageing trees *see* annual rings; dendrochronology
Aghía Marína, Choudalinaná Kíssamos, Crete *187*
Aghios Antónios, Mathés, Crete *183*, 184
Aghios Nikólaos, Samariá Gorge, Crete 172, *172*
Agia Paraskevi sacred forest, Vovoussa (Stavinere) 201, 203, **204**, 208–10, *210*
Agra, Western Lesvos **163**
Agricultural Stabilization and Conservation Service (US Department of Agriculture) *354*
agriculture: Crete 186; crop wastage 45; exploitation 101; grain cultivation 127–8; and grasslands 99, 101; intensive 23, 28, 37, 128, 279; landscape evolution 144; and saltmarshes 303–4
agro-silvo-pastoral systems 160, 200, 226, 230
agroforestry 159–63, *162*; ancient tree cultivation 160; oak 160–3;
Aidonochori/Aidhonokhóri, Epirus, Greece *193*, 213
alder (*Alnus*) 85, 110, 127, 226, 264
Alder, Common *Alnus glutinosa* 223, 264, 280
Alder, Green *Alnus viridis* 128, 132
Alder, Grey *Alnus incana* 128
algae: for fertilizer 332; for soap-making 306
Ali Paşa Tepelenë era (1744–1822) 198
Alíakes spring, Thérisso, Crete *166*, *178*
alkali for soap-making 305–6
Alpine Arc 124, 125 *see also* Eastern Alps
Alps 125 *see also* Central Alps; Eastern Alps
Amári valley, Crete *166*, 186
ambelitsiá (Greek) *see* Elm, Cretan
America Grove, Earsham 291
Anatolí, Crete *166*
Ancient Greeks 156
ancient semi-natural woods (ASNW) 27, 69–70
Ancient Tree Forum 10, 11, 320, *321*
ancient woodland 17, 19–21, 65, 287–90; *Alte Wälder* 125; and ancient woods 20–1, 273–6; and clayey soils 291; ecology of woodlands 28; historical significance 67; legalistic and the biological understanding 27; Norfolk 291; plantations on ancient woodland sites (PAWS) 27; primary and secondary woods 23–4, 34; protection from development 274–5; reassessing 26–7; and wildwood 34 *see also* old-growth forest; primary woodland
Ancient Woodland course (Flatford Mill) 324–5
ancient woodland indicators 21–3; Foresta Demaniale del Lerone 222; grazing pressure 293; ground flora trait analysis 68; Lopham Grove, North Lopham,

Norfolk 293; saw-wort 228–30; 'slow colonizing species' 26–7; studies of ground flora 65; understanding of ancient woods 273–5 *see also* coppice-associated species; ground flora; seed-banks
Ancient Woodland Inventory 19, 25–6, 286, 290, *290*
Ancient Woodland (Rackham book) 4–5, 8, 31, 65, 78, 109–12, *275*; appearance of landscapes 111; Australia 306–7; Domesday Book 110–11, 236; ecological archives 368; *Eucalyptus* 310; forest management and conservation 317–18; lime in woods 82; meaning of 'wood,' 'woodland' and 'forest' 110; Norfolk 285; oak regeneration 112–13; overlooked by Forestry Commission 49; 'The prehistory of the woodland' 111–12; regeneration of trees 110; rosaceous trees 58; second edition 54, 59, 65, 273, *275*, 368; shade in coppices 111; source of guidance 52–4; 'Uncompartmented forests' 110, 113; vegetation theory (Vera) 54; wattle 307; versus wildwood 19, 25, 52–54; Whitty Pear *Sorbus domestica* 59; 'wood-pasture' 110; woodland history and management practices 34
ancient woods: age 290; and ancient woodland 20–1, 273–6; as 'islands' 292; 'meaning' 52; as medieval survivals 19; Norfolk 285–95; Oxlip 65; primary and secondary woods 23–4; pseudo-ancient 290; and woodland pasturage 24
'The ancient woods of Norfolk' (Rackham article) 285
The Ancient Woods of South-East Wales (Rackham book) 375, 377
The Ancient Woods of the Helford River (Rackham book) 375, 377
Andredes leag (Kentish Weald) 37, 39
Angathopí Sphakiá, Crete *173*, *174*
Anglo-Saxon: charters 37, *40*; landscapes 37–43; 277–8
annual rings 169–70, 265–6; anomalous growth patterns 170, *170*; and coppicing 2656; earlywood (spring growth) 265; false rings 265; Grey Mangrove 306; latewood (summer growth) 265; partial rings 170, *170* *see also* dendrochronology; cambial age; tree-ring analysis
Ano Flória, Crete *166*, *180*
Ano Pedina, Epirus, Greece *193*
Ano Voúves Monumental Olive 170, *171*, 186
Anógiea, Crete *166*
Anópolis, Crete *166*, 175
Anston Stones, South Yorkshire 240
Aoos river, Epirus, Greece *193*, *195*
Aosta Valley, Italy 130
apple (*Malus*) 60, 358; wild 59
Apple, Wild *Malus sieversii* 60
aquifers: depletion 157; replenishment 205
archaeology 27, 273; wood 265; and woodland 27, 265, 273 *see also* charcoal
archives 5, 34 *see also* documentary evidence
archives (Rackham) *see* Rackham, Oliver archives
Archivio di Stato di Genova 219, 226
Archivio Fotografico Comando Coorte di Genova della Milizia Nazionale Forestale 222
Arden (*arddu* 'high land') 37, 39
Arenzano, Upper Lerone Valley, Italy 223, 226–8
Argan *Argania Spinosa* 156
Arslan-Bop, Uzbek village 60
ash (*Fraxinus*) 23, 60, 294; coppice 265–6, 295; disease *see* Ash Dieback; woodland 18
Ash Dieback 71, *74*, 326–7, *326–7*, 377
Ash, European *Fraxinus excelsior* 34, 74, 128, 290
Ash, Manna *Fraxinus ornus* 199, 200, 206, 223
Ash, White *Fraxinus americana* 355, 356
The Ash Tree (Rackham book) 11, 12
Ashtead Common, Surrey 322–4, *323*
Asteroúsia Mountains, Crete *166*, 179
Attica 154
Aurochs *Bos primigenius* 54, 101, 108, 117
Australia 9, Chapter 20 (environmental history); aboriginal management and heritage 316–17; botanical emblem 307; British Australian (Pom) 304n1; cedars 315; environmental history 316–18; European Australians 316; felling large trees 315; first crops 314; First Nation Australians 311, 316, meaning of 'On Country' 311; forest and woodland distinction 313; fossil pollen and charcoal records 317; logging 315–16; mangroves 304, *305*; 'park' 313–14; Rackham visits 300–2; rainforests 9, 311–12, 315–16; scrub and brush (rainforest) 311–12; soap manufacturing 304–6; *Sydney Gazette* 307; Third National Conference of the Australian Forest History Society (Jervis Bay 1996) 300; wattle and daub 307, *308*; wattles and fizgigs 306–11; woodlands as savanna 313 *see also* bush

Australian Blackwood *Acacia melanoxylon* 310
Australian National Dictionary (*AND*): brush 312; bush ('*bosch*') 311; wattle 307; wood 312
Australian National Maritime Museum 310
Australian Red Cedar *Toona ciliata* 315
Austria 8, 130–1; Chapter 8 (alpine forests); deadwood *133*; forest area *128*; forest naturalness *132*; forest ownership *130*; tree species *129*
avalanches vii, 127, 131, 203, 207, 213 *see also* protection forests
Avon Valley, England 39
axes 254–5, 299, 315; double 254–5
Ayiá, Crete *166*
Ayu (Japanese) fish 332

'Bacino del Rio Lerrone vicino a Ponte Negrone' 222–3; 1927 photograph *224*; 1984 photograph *225*; 2019 photograph *227*
badlands 157, 197, 199, 207–8
bailivaux (French) standard trees 262, 266
Balkan Aegean Dendrochronology programme 169, 174
banks/ditches (*haga*) 39, *40*, 42; boundary 20 *see also* earthworks
Bank Vole *Clethrionomys glareolus* 93
Bannwald 213
barilla 305–6 *see also* soap manufacturing
bark beetles (*Phloeosinus* species) 174
barrows (artificial mounds) 116, 281
Basque region, Spain 321
Baudin expedition 300–2
Bavaria 250–7; Chapter 16 (Bavarian Abbey); cellarer's inventory 253–7; protection forests 130; reconstructing subalpine landscapes 250–1; Tegernsee Abbey 251–3, *252*
Bayerische Staatsbibliothek (BSB) 253
Beauchief Abbey, Sheffield 239
Beccles Association soils 292
Beckets Wood, Woodton 287
beech (*Fagus*) 109; ancient (Greveniti forest) 215, *215*; coppicing 345–7; disease *see* Beech Dieback; fallen in Great Storm (1987) *360*, *361*; Rackham tree dedication (Epirus) 215, *215*, 375; regeneration via seed 347
Beech Dieback 57, *58*
Beech, American *Fagus grandifolia* 355, 356
Beech, European *Fagus sylvatica* 24, 32, 56–7, 108, 126, 128, 211, 280
Beech, Japanese *Fagus crenata* (*agariko*) 340, 341; Agariko-daio (king of pollard) *341*
Beech, Japanese Blue *Fagus japonica* (*inubuna*) 335
beech pollard: Agariko-daio ('king of pollard', giant Japanese Beech pollard) *341*; Cage Pollard (Burnham Beeches) *322*; harvesting 347; height of branched stems *346*; logging 345; management 345–8; Mount Chokai, Yamagata Prefecture, Japan *341*; Ohshirakawa, Niigata Prefecture, Japan *342*; snowy climate 343–4; Tainai, Niigata Prefecture *342*
Beigua Regional Park, Italy 223
Beijing 148
The Beinecke Stand (HMF) 356–8
Belgium 266
Belovezhskaya Pushcha National Park (Belarus) 115
Benediktbeuern monastery 255
Berkswell Court Rolls 83
Berkswell Enclosure Plan 83
Bersano, Tortona, Italy 219
Bertignana terraces, Varese, Italy 219
Białowieża National Park 115
Białowieża Primeval Forest (BPF), Poland *100*, 115–22; Chapter 7 (Białowieża Forest); anthropogenic pressure 119; 'artificial forest' factoid 117–20, 121; Biala Wieża (White Tower) 116; continuous afforestation 118; 'factoids' and 'pseudo-history' 116, 116–21; hunting ground 116–17, 121; old growths *120*; primeval European lowland forest 120; '*pushcha*' (empty land) 116; 'The Royal Oaks Route' 121; unconfirmed species 117; Zamczysko (Old Castle Place) 116
Big Scrub, Northern Rivers region, NSW 312
Binley Common Wood, Warwickshire 85–92, *86*
Binley Little Wood, Warwickshire 85, *86*
biocultural diversity 143, 147–8; and farming 127, 129, 132
biocultural landscapes 138–9, 147–8
biodiversity: and ancient woodland 28; landscape intensification and abandonment 161, 196; and nature reserves 134; and old trees 323, 325
birch (*Betula*) 20, 57, 60, 128, 255; 278, 280, 294; tar 119
Birch, Black *Betula lenta* 355
Birch, Paper *Betula papyrifera* 355, *356*
Birch, Yellow *Betula alleghaniensis* 355
Birchley Woods, Brinklow, Warwickshire *80*, 83, 85–7, 89–92, *90–1*
birds 32, 103, 105; and coppicing 259; seed dispersal 54; and veteran trees 169, 202,

214 *see* Great (White) Egret; Greylag Goose; Jay; Osprey; Pileated Woodpecker; Sea Eagle
bison *see* European Bison; large herbivores
black barks (mature timber trees 40–60 years old) 238, 241–2, *241*, 248 *see also* lordings
Blackthorn *Prunus spinosa* 56, *110*, 111
Blackwattle Bay, Sydney Harbour 307
bloodwoods (*Corymbia* species) 311 *see* Western Australian Flowering Gum
Bluebell *Hyacinthoides non-scripta* 18, 21, *21*, 23, 65, *66*, 72, *272*, 293, *293*
Board of Agriculture Reports 1790–1813 (England) 19 *see also* documentary evidence
Bodleian Library, Oxford 86
Bohuslavice forest district, Czech Republic *261*
bolling 325, 340, *341*, 343, 345, 347, *347* *see also* pollarding
Borage *Borago officinalis* 67
Borkener Paradise, Germany 107, *110*
Borrowdale, Cumbria 24
boscus forinsecus (Latin) 'foreign' wood 87–9
botanical indicator species *see* ancient woodland indicators
Botany Bay, NSW (Stingray Harbour/ Botanist Bay) 314
botes (Middle English) manorial rights 237
boulder clay plateaux: Cambridge 18; south Norfolk 'doughnut of woodland' 291, *292*
Boutsounária, (Kydonias) Chania, Crete *166*,179, 373–4
Box Hill, Surrey *360*
Bracken *Pteridium aquilinum* 22, 36, 323
Bradfield Woods, Cambridge 51, 374
Bramble *Rubus fruticosus* 69, *70*, 73, 94
Brandon Park, Binley, Warwickshire 87–9
'Brandon Park fungi 1959–2014' (Rackham article) 377
British coastline and railway building 303
British Ecological Society 273, 274
British forestry: forestry policy and management 69–71; medieval wooded landscapes 12 *see also* Forestry Commission
Brno, Czech Republic *266*
Broadleaf Cattail *see* Great Reedmace
Brocton Coppice, Cannock Chase, Staffordshire 44
Bronze Age (Greece) 153, 156–7, 185; bark beetles 174; charcoal *170* *see also* Minoan art
brugo (Italian) *see* Tree Heather
brush (Australian rainforest) 311–12, 315
Brush Box *Lophostemon* species 312
brushwood 312; and firewood 332, 334 *see also* 'Ramell'
Buckenham, Norfolk 294
bullockies 302
Burlingham and Hanslope soils, Norfolk 292
Burnham Beeches, Buckinghamshire 322, *322*
Buschwald (German) coppice 262
bush (Australian) 311–12; clearing 314–16
Butcher's Broom *Ruscus aculeatus* 304
Buxton Farms, New England 351

cabinetmakers 315
cambial age 265, 267 *see also* annual rings; tree-ring analysis
Cambridge Centre for Landscape and People (CCLP) 140, *141*, 143 *see also* European Culture Expressed as Landscape
Cambridge Digital Library (CDL) 199, 372
Cambridge University Herbarium (CUH) 368, 370–2 *see also* Rackham herbarium
Cambridgeshire 285
Cambridgeshire and Isle of Ely Naturalists' Trust 79, 299
'Camelot Zone' 182
Canker Stain Disease (CSD) 179
Canklow Wood, Sheffield 241, *241*
Cannock Forest, Staffordshire 43
canopy woods *see* high forest
Cape Town 311
Carob *Ceratonia siliqua* *183*, 184 *see also* Holy Carob
Carrifran Wildwood Project, Scottish Borders 53–4
Cartulary, Binley, Warwickshire 87, 89
Casanova mountains, Italy 229
cascina (Italian) courtyard farmhouses 219
Case Bottini, Liguria, Italy 222
Cassego, Upper Vara Valley, Liguria, Italy 219
cattle *see* livestock
cave-chapels, Panayía, Kourí forest, Greece 206
Cavo Sidero (Crete) campaign 8
Cedar of Lebanon *Cedrus libani* 223
cedro (Italian) coppice 261
Center for Biocultural Landscape and Seascape (CBLS) (University of Sassari, Sardinia) 147–8
Center for Environmental Studies (Williams College, USA) 353
Central Alps 252
Central Asia *see* Kyrgyzstan
Central Europe 134
Cerbaie Hills, Liguria, Italy 229–30
Chalkney Wood, Essex 65, *66*

Chaniá, Crete *166*, 174, 187, *380*
Chapeltown furnace, Ecclesall 245–6
charcoal: archaeological 6, 170, 174–5, 179, 187, 230, 264, 267; hearths 279; kiln (Japan) 345; production 119, 170, 186, 199, 239–40, 242–6, 248, (Japan) 344–7, *344*, 355; reference collection (Crete) 373
charter(s) Anglo Saxon 31, 37, 40, 110; Kent 39; Pre-Conquest 37; research 2, 37–41
Chatsworth House and Park, Derbyshire 276
Cherry, Bird *Prunus padus* *18*, 294
Cherry, Cornelian *Cornus mas* 199
Cherry, Mountain *Cerasus jamasakura* 335
chestnut (*Castanea*): charcoal (*Castanea*-type) 179, (wood) 345; coppice cycles 260; coppice stools *180*; coppicing 345; disease 179; hollowing 170; landmarks in Mediterranean 160; long-lived trees 171; pollards 179; 'Ntoulia' great pollard (Crete) *180*; Valle Lagorara 219 *see also* Horse Chestnut
Chestnut, Sweet *Castanea sativa* 34, 179
Chippenham Fen National Nature Reserve (NNR) 300
chogoroiwa, a named sacred rock (Moriyama, Japan) 336
Choudalianá, Crete *166*
Christmas bush *see* Wattle, Black
churches 202–6
Cinque Finestre, Italy 219
climate change 57, 67, 71, 157, 350–1, and 'risk dialogue' 135 *see also* human activities
climax vegetation forests 99–101, *100*, 103
close-to-nature silviculture (Pro Silva) 125, 134, 135
closed-canopy forests *see* high forest
Coal Measures, South Yorkshire 237
coal: decline 321; merchants 246; mining 242, 245; from wood 242; *see also* charcoal
coasts 144, 157; developments 303 *see* Australia
Cobcrofts Brush, Forbes River, NSW 312
Col-East Aerial Photography, North Adams, MA 354
Collyweston Great Wood, Northamptonshire 23
Combe Cartulary (1255) 87–9, 92
Common Elder *Sambucus nigra* 102
Common Primrose *Primula vulgaris* 23, 72, 293, *293*
Common Reed *Phragmites australis* 101
Common Sainfoin *Onobrychis vicifolia* 229
Common Yew *Taxus baccata* 39
commons 291; 'ancient' (Crevari) 228; Italy 227–8; grazed 90; Japan 332; management 336; residues of unenclosed land 111; wooded 1, 237, 278–80; 323–4 *see also* Act of Commons; Ashtead Common
compartmentalized forests 324 *see also* coppice management
The Computer Mapped Flora of Warwickshire 82
comunaglie (Italian) common land 227
coniferization 84; impact 65, 360; policies 69; reversal vii; *66*, *70*, 94; and scouts 83
'Consegne dei Boschi,' commune di Arenzano 226
conservation 34; Alps 127; and Ancient Woodland 25–8; and biodiversity 57, 61; Crete 186–7; changes since 1980 34; and coppicing 65–6, 259–60; and grazing 117, Landscape Conservation Forum 10, 273; Malvern Hills 46–7; Oostvaardersplassen experiment 105; and pseudo-history 379; Rackham OBE 1; Rackham's voice 4, 11–12; 36; 112; 122, 317–18; and rewilding 53–4, 67, 72; sacred forests 214; secular versus sacred forest conservation 213; and veteranization 320, 325–6; woodland and wood-pasture distinction 36 *see also* conventions and protocols
Conservation Corps (University of Cambridge) 299–300
conventions and protocols: Alpine Convention 135; 'Alpine Farming' protocol (1994) 125; Broadleaves Policy (UK) 25–6; Convention on Biological Diversity (CBD) 138; Countryside Stewardship Agreements (Defra) 37; European Landscape Convention (ELC) 138–9, 145, 230; forest code of Charles V of France (AD 1376) 262; Framework-Agreement of the Alpine Convention (1991) 124, 125; Mountain Forest and Soil Protection protocols of the Alpine Convention ('Mountain Forest' protocol, 1996) 124, 125, 135; 'Nature Conservation and Landscape Maintenance' protocol (1994) 125; World Heritage Convention (UNESCO) 145
Copper Age Hungary 263
coppice: ancient 24, 259, 276; *Buschwald/Hochwald* (Germany) 262; commons 111; deer parks/private wood-pastures 292; Dutch 266; enclosed 24; as rich habitat 259; hazel 111–12; Japan 345; Kent 294; names for 261–2; and shrubland 199 *silva caedua* 262; *silva minuta* (England) 24, 237, 262; Watson-Wentworth estate, County Wicklow 246–8, *247*; Wye Gorge 24

coppice-associated species 71, 260n2, 379
 see also ancient woodland indicators; ground flora; seed-banks
coppice brush (Australia) 312 *see also* vine brush
coppice management: ancient semi-natural woodland 69–70; ancient woodland 65–7; ancient woods 24; bans on pasturing 262; compartments 239, *241*, 241–2, 260, 324; cutting cycles 238, 248, 260–3, *261*, 262; Hatfield Forest 324; HMF experiment 358; Japan 345–7; pollen record 263–4; Roman 24; South Yorkshire 238, 239–42
coppice stools 264–5, 281; lignotuber *186*; stool size and age 92
coppice-with-standards *18*, 277, 290; coppiced underwood and timber 244; 'cut-and-come-again' resource 277; Ecclesall Woods, Sheffield 244–6; medieval records 238–9; *Mittelwald* 262; products of 244; South Yorkshire 238–44, *238*; as 'spring' 238 *see also* spring woods
coppicing 67, 258–68; Chapter 17 (coppice history); annual ring pattern 265–6; and Blackthorn 111; and cambial age and wood diameter 267; decline and revival 258–9, 290; history 259–61; products 258–9, 242, 245–6, 248; prolonging life of trees 171–2; and 'scientific' forestry 258–9
Coronata white wine 219
Corpo Forestale delle Stato 222
Corpus Christi College, Cambridge (CCCC) *145*, 368–70, 372 *see also* Rackham, Oliver archive; University of Cambridge
Costa di Begato, Liguria, Italy 219
cotica secca con graminacee (Italian) turf 222
Cotinus coggygria (Smoke Tree) 199
Cotswold scarp 45, *46*
Council of Europe 138, 145
'countryside' and 'country' (Australia) 311
County Kildare 246, *247*
County Wicklow 246, *247*
Coventry Abbey wood 87
Cowden, Sevenoaks, Kent, *361*
cows 32, 60, 106, 107–8, *110* *see also* large herbivores
Crataegus pontica (Turkish Haw) 60
Craven Estates, Binley, Warwickshire 86, 89
Cretan Tree-Ring Group 169, *173*, 174, 175
Crete 6, 8, 153, 165–88; Chapter 12 (veteran trees); agriculture 186; ancient, historic and veteran trees 169; 'ancient' wild trees 165; archives 368; Bronze Age 185, *185*, *186*; conservation 186–7; coppicing 171–2; dendrochronology 169–70; Forest Directorates 176; long-lived trees 171–82; meaning in landscape 182–5; 'miraculous and healing' trees 182, 184; pollarding 171–2; Prefecture 186; Rackham tree dedications 165, 165n3, *167–9*, 375, 377; veteran trees 169 *see also* Cretan Tree-Ring Group; Monumental Olives
Crevari mountains, Italy 229
Critical Forest Management 106, 107
Crown plantations, New Forest 19
cultural heritage 140–5
cultural landscapes 120, 127, 148; relict 196, 202, 280; value 140, 148 *see also* biocultural landscapes
'Curley Oak' (*Quercus robur*) 377, *378*
Currency Creek, Murray River, South Australia 302
'cut-and-come-again' resources *see* coppice-with-standards
cypress 169–70; ancient 172–4, *172*, *173*; charcoal 170; disease *see* Cypress Canker; opportunistic growers 170
Cypress, Japanese *Chamaecyparis obtusa* (*hinoki*) 334, 336, 344
Cypress, Mediterranean *Cupressus sempervirens* 172–4
Cypress, Sawara *Chamaecyparis pisifera* 341
Cypress Canker disease 174
Cypress Pines *Callitris* species 302
Czech forest law (1995) 259
Czech Republic 9; Chapter 17 (reconstructing coppice histories), 259; medieval and historic documents 260–3, *261*, 262; coppicing history 265, *266*

daidomichi (Japanese) mountain roads 335
Dark Peak, South Yorkshire 236
δάσος (Greek) 'forest' 198
deadwood *133*, 200, 206, 211
deer 36, 65; culling vii; browsing 50, 358; enclosure (*haie*) 39; exclusion fencing 70; parks (private wood-pastures) 288, 292, 313-14; over population 68–9, 94 *see* Fallow Deer, Muntjac, Red Deer, Roe Deer
deforestation 118, 154, 157
dendrochronology: Cretan Elm 179; Black Pines 206; Crete 169–70; multi-stemmed trees 170, 281; olive trees 182; pines 175 *see also* annual rings; tree-ring analysis
Dendrocnide (*Urticaceae*) 312
Denmark 262, 264

Denny Wood, New Forest 57
Department for Environment, Food and Rural Affairs (Defra) 12, 37
derecho windstorms 361, *362*
deserts and desertification 154, 157
development planning protection 274–5
Díkti Mountains, Crete 165, *166*, 179 *see also* Lasíthi Mountains
diseases (animals) 106; foot and mouth 45, 325
diseases (trees): 3, 11, 34, 71, 327, 358, 377; and globalized trade 11, 12; and pests 12, 50, 67, 327 *see also* Ash Dieback; Beech Dieback; Canker Stain; Cypress Canker; Dutch Elm Disease; Olive Quick Decline; scale-insect
Diss Mere, Norfolk 294
ditch(es) 20–1 *see also* banks/ditches
Divided Sedge *Carex divisa* 304
documentary evidence 5, 260–3, 276–7, 291, 294; Chapter 15 (Yorkshire documents and woodland); Chapter 16 (Bavarian Abbey); early maps 81, 83, 89, 219, 241, 288–9, *331*; handbill 243, *244*; historical records 235, 321; importance of 235–6, 249; medieval estate maps and records 19–20 *see also* account books; archives; charters; Tithe Maps; travellers' accounts
'Documentary Evidence for the Historical Ecologist' (Rackham article) 5
Dog's Mercury *Mercurialis perennis* 65, *66*, 293, *293*
Doles Wood, Hampshire 23
Domesday Book: Binley, Worcester 87–8; composition of woods 110–11; *haie* (deer enclosure) 39; identifiable woods with Bluebells 18; Norfolk 21, 291; Piles Coppice, Warwickshire 89; pre-Domesday landscape 277; South Yorkshire 236–7, *237*; wood-pastures/shadow woods 279
dondoyaki a sacred ritual (Moriyama, Japan) 337
droveways 39, *41*; traveling stock routes (TSRs) 302
Dubbo, New South Wales 300
Dundullimal Homestead, Dubbo, NSW 300, *301*
Dutch Elm Disease 358, 377

Early Dog-violet *Viola reichenbachiana* 65, 293–4
Early-purple Orchid *Orchis mascula* 293
Earsham Wood, Norfolk 289
earthworks 86, *88*, 287, *288* *see also* banks/ditches; woodbanks
East Anglia 19–20, 23, 264, 285, 291
East Midlands 19–20
Eastern Alps 8; Chapter 8 (alpine forests) 124–35; anthropogenic influences 127–8; 'Centers of Plant Variety' (World Wildlife Fund (WWF) and International Union for Conservation of Nature (IUCN)) 126; clearings (slash-and-burn fields) 129–30; cultures and landscapes 132; deadwood 133, *133*; diversity 125, 126; exploitation of forests 128; farming 127–30; forest area 128–9, *128*; forest ownership 130; habitats and species 132; high-forest climax communities 132; management-model 133; mountain economies 132; Natura 2000 sites 134; naturalness and natural forests 132, 133–4; nature conservation 127, 134; nature reserves 134; old-growth forests 130–1; pastoral farming and grain cultivation 127–8; population 125; protection forests 130–1, 133–5; secular versus sacred forest conservation 213; silviculture 127–9; study area 126; timberline 126–8; tree seeds 130; tree species distribution 128, *129*; as untouched nature 127; virgin forests 129, 134; wood-pastures 129–30
Eastern Hemlock *Tsuga canadensis* 356
Eastern Zagori *194*
Ecclesall Woods, Sheffield 236, 241, 244–6, *245*, *246*
ecology *see* historical ecology; forest ecology; physiological ecology
Ecumenical Patriarchate of Constantinople archives 203
Edlington Wood, South Yorkshire 245
Eleftherokhóri, Epirus, Greece 203
Elighiás Gorge, Sphakiá, Crete *166*, *175*
elm (*Ulmus*) 109, 179; disease *see* Dutch Elm Disease; mid-Holocene decline 263 *see also Zelkova*
Elm, American *Ulmus americana* 358
Elm, Cretan *Zelkova abelicea* 165, *168*, 171, 176–9; *ambelitsiá* *169*, 179; conservation in Crete 176; pollards 179; Rackham tree dedication (White Mountains, Crete) 165, *169*, 375, 377
Elm, English *Ulmus procera* 32
Elm, Sicilian *Zelkova sicula* 176
Elos, Crete *166*, 179
Ely Coucher Book (1251) 18, 374
enclosure: Berkswall Enclosure Plan (1802) 83; of coppice 24, 90; early medieval 275, 277, 279–80, 291–3; of fields 32, 128; hedged

and walled landscapes 32; medieval encoppicements 19, 20; of woods 91–2, 287, 291 *see also* banks/ditches; earthworks
England: Chapter 1 (ancient woodland); Chapter 2 (Anglo Saxon woodland); Chapter 3 (wildwood and forest ecology); Chapter 4 (woodland ground flora); Chapter 5 (Warwickshire limewoods); Chapter 15 (Yorkshire documents and woodland); Chapter 18 (shadow woods); Chapter 19 (Norfolk ancient woods); Chapter 21 (Ashstead Common, Burnham Beeches, Hatfield Forest); feudal landscape 277; The Great Storm (1987) *360*, 360–1; Old English terms 37; pollard systems 340; woodland/saltmarsh flora 304
English Nature *see* Natural England
d'Entrecasteaux expedition 300–2
environmental change 57–8 *see also* climate change; human activity
environmental history 250, 316–18 socio-natural sites 250
environmental stress 280–1
Epirus, Greece 7; Chapter 13 (relict landscapes) *193*; deadwood 200; environment 196–7; fires 200–1; grazing 199–200; landscapes 194–6; place names 197–8; Rackham tree dedication 215, *215*, 377; sacred forests 202–14; savanna 198; shrubland and coppice-woods 199; tree species 199, 203; woody ground vegetation 201–2
Epping Forest, Essex 113, 321
eresztvény (Hungarian) coppices 261
erosion 154, 156–7; coastal 303; gullies 132, 157, 197; and pollards 343; and trees 131 *see also* badlands
Essex 285, 294, 320; 'The Forests of Essex' conference (2018) 374
estovers (*estoveir*, 'be necessary') 237
Ethiopia 5, 9
ethnographic research 198, 203–4
eucalypts (*Eucalyptus*) 300, 307, 310–11, 315; canopy 313; fire adaptation 9; pollen 317; shade 306–7, 313; White Box–Yellow Box–Blakely's Red Gum and Derived Native Grassland ecological community 313 *see* Red River Gum; Yellow Box
Europa myth 184, *184*
European Bison *Bison bonasus* (Wisent) 54, 101, 107, 117 *see also* large herbivores
European Commission 7, 140, 141
European Culture Expressed in Landscape (EUCEL) Initiative 139, 140–1, *141*, 143; partnerships *142*; Sassari conference (2015) 143, *144*, 374; programmes: European Culture Expressed in Agricultural Landscapes (EUCALAND) 140–1, *141*, *142*, 144, *145*; European Culture Expressed in Island Landscapes (ESLAND) 141, *141*, 145–6, *146*; European Culture expressed in Sacred Landscapes (ECSLAND) 141–3, *143*, 147
European Landscape Convention (ELC) 138–9, 145, 230
European landscapes 6–9, 143–5; island landscapes 145–6, 146
European Larch *Larix decidua* 128
European Wildcat *Felis silvestris* 117
Evangelístria church, Mazi, Epirus, Greece 205
excommunication of forests 203, 211
exhibits *see* Fen Ditton Gallery; Green Wood Centre; Sarah Morrish at Royal Horticultural Society Botanical Art Show
Exposed Coalfield zones, South Yorkshire 236

factoids 116–22, 235, 379 *see also* pseudo-history
Fallow Deer *Dama dama* 107
False Wood-brome *Brachypodium sylvaticum* 73
Fen Ditton Gallery, 'Trees Observed' exhibit 375
Fernhill, Mulgoa, NSW 314
field books *see* Rackham field books
field surveys 20, 279; Kyrgystan 60; New Forest 56; Norfolk 286; pollards 323; value of 61, 235
fires 9; aboriginal management (Australia) 317; effect on Saw-wort 229; litter and ground-litter fires 201, 206; pros and cons 219; sacred forests 200–1, 208
firewood: cooking systems 222; coppicing 259; *kenkera* 334; pollarding 321, 340; production 294, 343, 344
First Fleet (to Australia) 307, 312, 314
First Nation Australians 311, 316; meaning of 'On Country' 311 *see also* Aboriginal Australians
First World War 43
fizgigs 309–10
Flatford Mill, Suffolk 324–5
Flória, Crete 179
fodder *see* leaf fodder
Fontanabuona Valley, Genoa 222
foot-and-mouth disease 45, 325
forage 32, 200; acorns and beech-mast 33, 34; Common Sainfoin 229; and transhumance 32, 39; and wood-pasture 34

forest: anthropogenic degradation 156; compartmentalized 324; dense forest 161, *162*; high forest 24; and retrogressive succession 99, 100; scientific concept 108–9; Tansley mantra 108; and woodland in Australia 313; uncompartmentalized 110, 113 *see also* high forest
forest ecology 49–50
Forest of Dean 24
Forest of Dorstone 43
forestry: policies 26, 69, 203 'scientific' forestry 117, 258–9 *see also* agroforestry; Forestry Commission
Forestry Commission 49, 52, 65, 69, 273, 274, 360
Forestry Service, Greece 205, 206
Fournés, Crete *166*
Foxley ancient wood, Norfolk 292
France: 'Book of sales of the woods of Champagne', France (thirteenth century) 262; Chapter 8 (alpine forest); deadwood *133*; forest area *128*; forest code of Charles V of France (1376) 262; forest naturalness *131*; forest ownership *130*; protection forests 130–1; oak coppice 267; Olive Quick Decline 182; shredding 205; tree species *129*; Via Francigena 230
Friends of Oliver Rackham (FOR) 165n3, 372; newsletter 374
fuelwood 262–3

'*galaverna*' weather (Italy) 219, 220
Gardom's Coppice, Peak District 278
Genoa 219, 222
'gentleman's park' (bush, south-east Australia) 313–14
Geraldton, Western Australia 300
German-Austrian Alpine Association 134
Germany: Chapter 8 (alpine forests); Chapter 16 (Bavarian Abbey); deadwood *133*; forest area *128*; forest naturalness *131*; forest ownership *130*; protection forests 130; tree species *129*; 261, 261–2, 267
ghost woods 279; *see also* lost woods; shadow woods
Gillfield Wood, Sheffield 239, *239*
goats 156, 160, 200; climbing trees 8, 156; trampling 199, 200
gokamatsuri (Japanese) spring festival 332
Goldilocks Buttercup *Ranunculus auricomus* 65
Gorse *Ulex europaeus* 56
Gortyn, Crete *166*, 184–5, *184*
Granowe Spring coppice wood 240
Grass Trees *Xanthorrhoea* 310
grassland 9, 276; ancient 222; Chapter 6 (Oostvaardersplassen marsh and the origin of grassland); as a human-made habitat 99, 101; hay meadows 195; Japan 334, 336, 341, 342; and large wild herbivores 32, 54; steppe tundra succession 101; tree succession *55*, 54–7, 358 *see also* Vera theory and debate
grazing: abandonment and intensification 161, *162*; agroforestry systems in the Mediterranean 159–63; ancient woodland indicator species 293; and close-canopy woodland 32; conservators 47; foot-and-mouth disease 325; and large herbivores 72; and oak regeneration 112; and retrogressive succession 32, 99–101; sacred forest of Mazi 200; saltmarshes 303–4; and succession 103–5; and tree regeneration 110, 113, 161; wood-pastures 33; and woodland composition 90 *see also* succession theory; Vera theory and debate
Great Dividing Range, Australia 300
Great Exhibition (1851) 315
Great Gransden, Huntingdonshire *369*
Great Hurricane (1938) 358
Great Reedmace (Broadleaf Cattail) *Typha latifolia* 101, 102, *102*
The Great Storm (1987) 67, *68*, 280, 360–1, *360*, *361* *see also* wind-downed trees
Great (White) Egrets *Ardea alba* 105, 106
Greece 6–7, 33, *33*, Chapter 10 (critique of Grove and Rackham 2001); Chapter 11 (agroforestry) **103**; Chapter 12 (Crete veteran trees); Chapter 13 (Epirus relict landscapes); Rackham archives 370, 371–2; Rackham tree dedications 165, *167–9*, 215, *215*, 375, 377 *see also* Crete; Epirus
Greek-Swedish Excavations, Chaniá 187
green manure: from leaves 129, 227; from Tree Heather 219
Green Olive *Phillyrea latifolia* 199
Green Wood Centre, Coalbrookdale 368; exhibit 'Oliver Rackham – Roots and Branches' 373, 375, *376*; Rackham Collection 372–3, *373*, 375, *376* *see also* Small Woods Association
Greveniti, Epirus, Greece *193*, **204**, 211, *212*, *215*
Grey Saltbush *Atriplex cinerea* (barilla bush) 306
Grey Wolf *Canis lupus* 107
Greylag Goose *Anser anser* 102–8, *102*, *103*
Groton Wood, Suffolk 25
ground flora 64–74; Chapter 4 (woodland ground flora); climate change 71;

forestry policy and management 69–71; observational and descriptive ecology 73; Rackham's views 65–8; rewilding 71–2, *72*; Wytham Woods, Oxfordshire 73–4 *see also* ancient woodland indicators; coppice-associated species; herbaceous plants
groundwater 157
The Grove, Binley, Warwickshire 85–7, *90*, 91, *91*
Gruppo di Voltri 222
Gum Arabic Tree *Acacia nilotica* (*Mimosa scorpioides*) 307

Hadulf wood (Munechet), Binley, Warwickshire 87
haga (Domesday) banks/ditches 39, *40*, 42 *see also* earthworks; woodbanks
haie (Domesday) 'ha-ha' enclosures 39, 42
Hainault Forest, Essex 321
Hamilton Kerr Institute, Whittlesford, Cambridgeshire 377
Hampshire *40*, 294
Hanging Wood, Claverdon 82–3
Hard Shield-fern *Polystichum aculeatum* 294
harpoon (*fisga*, fizgigs) 309–10
Hartshill Hayes, Hartshill, North Warwickshire *80*, 84–5
Hatfield Forest, Essex ('The Last Forest') 4, 34–6, 52, 324–6, *324*, *326*, 359–60
Haveringland Great Wood, Norfolk 292
hawthorn (*Crataegus*) *110*, 110–11, *278*, *279*, 280–1, 340; pollards 324 *see also Crataegus*
Hawthorn, Common *Crataegus monogyna* *56*, 276, *276*, 280, 281
Hayley Pear 58–9
Hayley Wood (Rackham book) 2–4, 5, 358
Hayley Wood, Cambridgeshire 2–4, *66*, 79; Conservation Corps 299–300; coppice cycle 260; coppice stools (ash) 265–6; deer populations 68–9; Ely Coucher Book 18; oaks 266; Oliver Rackham Memorial Symposium (CCCC 2016) *375*; Pear *Pyrus communis* 58–9; research and recording 359–60
haymaking 222, 303
hazel (*Corylus*) 54, 111–12, 199, 294; coppice 226, 264, 290; pollen 263; prehistoric trackways 264
hazel-Pendunculate oak wood 79, 83, 87
hazel-Sessile oak wood 86
Hazel, Common *Corylus avellana* 32, 51, 128, 223, *226*, 264, 290
Hedenham Wood, Norfolk 287–8, *288*, *289*, 291, 295
hedgerow 237, *239*; ancient 290; and ancient woodland 82–5; destruction 158; pollards 320–1; as seed-banks 293
Hellenistic silver stater coins *184*, 185
Hempstead, Essex 65
Heráklieon, Crete *166*
herayama forests (Moriyama, Japan) 331–2, 332
Herb-paris *Paris quadrifolia* 22, *22*, 67, 293
herbaceous plants 201n2, 228, 302 *see also* ground flora
herbivores and herbivory 54–6, *55*, *56* *see also* large herbivores
'herdyng wood' (Rollestone Wood) 239
Herefordshire 294
Heritage Lottery Fund 273, 277
Hertfordshire 291, 294
high forest (canopy woods) 24, *100*, 109, 112, 245, 258, 276–7; Alps 131; 'single-tree management' 131, 135; suppression of diversity 71
high-forest climax communities 132
High Tauern National Park, Austria 131
Highland pinewoods 19, 24
hinoki (Japanese) *see* Cypress, Japanese
historical ecology 33, 50–1, 78–9, 230; Historical Ecology Discussion Group 19
The History of the Countryside (Rackham book) 4, 5, 139–40, 320; Japanese edition 9, 338–9; loss 182, 379; meaning 52; saltmarshes and wetlands 302–3; *Tilia cordata* and ancient woods 82; willow 249
Hockering Wood, Norfolk 287, 292, 294
Hohe Tauern National Park, Austria 134
holly *Ilex* 39, *45*
Holly, Common *Ilex aquifolium* 34, 51, 280, 281
Holmesfield, North Derbyshire 281
Holy Carob, Aghios Antonios *183*, 184
Hook Wood, Morley St Peter, Norfolk 288
hop hornbeam (*Ostrya*) 208, 209, 211
Hop Hornbeam, European *Ostrya carpinifolia* 223
Hopkins Memorial Forest (HMF) 351–6, *352*, *354*, *355*, 358, 361; Chapter 24 (post-agricultural landscape); Secondary Successional Model 358, *359*
hornbeam (*Carpinus*) 32, 109, 205, 206, 294–5, *295*, 321, 345; coppice 293, *295*; pollards (Hatfield Forest) *324*
Hornbeam, Common *Carpinus betulus* 32, 108, 290, 324, *324*
Hornbeam, Japanese (*shide*) 334
Hornbeam, Oriental *Carpinus orientalis* 199, 206
Horse Chestnut *Aesculus hippocastanum* 197n1

Horse Chestnut, Japanese *Aesculus turbinata* 343
horse *see* konik pony; Wild Horse
Horsford ancient wood, Norfolk 292
Hortus Camalduli, Naples 300
hotora (Japanese) grass used as fertilizer 334
hotorakari grass harvest (Moriyama, Japan) 334
Hudnalls, Gloucestershire 24
human activities: altering the biosphere 350–1; ancient world 154; anthropogenic forest degradation 156; saltmarshes 303 *see also* climate change
Humberhead Levels, South Yorkshire 236
Hundred Rolls (Warwickshire) 87, 89
Hutcliff Wood, Sheaf valley, Sheffield 239
Hymenoscyphus fraxineus (Ash Dieback) 71

iekujiyama (Japanese) private–common forest border 332
Ierápetra, Crete 165
Imbros, Crete *166*
Industrial Revolution 273, 321
International Association of Landscape Ecology (IALE) 148; 8th World Congress (Beijing) 148; Working Group on Biocultural Landscape 148
inubuna (Japanese) *see* Beech, Japanese Blue
Ioannina Castle 198
Ioannina, Epirus, Greece *193*, *204*; BioScene project conference (2005) 191
ippon-zori (Japanese) one-bladed sled 345, *346*
Ireland 246–8 *see also* Watson-Wentworth Irish estate
iron industry 4, 242, 321
islands: habitats 22; initiatives *see* European Culture Expressed in Landscape; landscape history 145–6 *see also* Crete; Isle of Arran; Sardinia
'Island landscape history, the Isles of Scilly, UK' (Rackham chapter) 146
Isle of Arran, Scotland 59
itadori (Japanese) *see* Japanese Knotweed
Italy 6–7; Chapter 8 (alpine forests) 126; Chapter 14 (Liguria historical landscape) 219; deadwood *133*; forest area *128*; forest naturalness *132*; forest ownership *130*; protection forests 130; tree species *129*; Rackham archive 370 *see also* Liguria, Italy; Sardinia; University of Genoa; University of Sassari
Itanos Peninsula, Crete 8

Japan 9; Chapter 22 (satoyama landscapes); Chapter 23 (beech pollards in snow); charcoal production 343, 345; children's play 337; coppice forests 345; firewood 332–4, 336, 337, 343, 345; management of natural resources 335–7; pollard beech trees 340–9, *341*, *342*; pollard broadleaf trees 343, *343*; pollard management 345–8; pollard tree species 341–3; pollards avoiding herbivory 341; pollards in snowy climates 343–4, *344*; reed-thatched roofs *333*; snowy areas 340–9; uses of natural resources 332–5
Japanese Big-leaf Magnolia *Magnolia obovata* 343
Japanese Cedar *Cryptomeria japonica* (*sugi*) 334, 336, 341, 344
Japanese Knotweed *Reynoutria japonica* (*itadori*) 334
Japanese Pampas Grass *Miscanthus sinensis* (*susuki*) 334
Japanese Wisteria *Wisteria floribunda* 337
Jay *Garrulus glandarius* 32, 54
jigger boards (springboards) 315
jiyama forests (Moriyama, Japan) 331–2, 334
juniper (*Juniperus*) 60
Juniperus communis 199
Juniperus oxycedrus 199

kaldirími (Greek) mule-roads 195
Kalkalpen National Park, Austria 134
kama no kuchi (Japanese) day after which grass can be harvested 336
Kamilári, Crete *166*, *170*
Kándanos, Crete *166*, 179
Kangaroo Grass *Themeda australis* 302
kariboshi harvest (Moriyama, Japan) 334
Kasrti, Aoos gorge, Greece 206
Kastélli Kíssamos, Crete *166*, 187, *188*
Kato Pedina, Epirus, Greece *193*, 198
Kavalari, Epirus, Greece *193*
Kavoúsi Ierápetra, Crete 166, *181*, 186
kenkera (Japanese) firewood 334
Kent 39, 294
Kentish Weald (*Andredes leag*) 37, 39
Kiveton Park, South Yorkshire 242
kladera (Greek) groups of or isolated shredded trees 196
Knepp Estate, Sussex *72*
koba resting place (Moriyama, Japan) 332
Kommós, Crete *166*
konara (Japanese) oak 334
konik pony (Polish) 107–8, *108*
Konitsa, Epirus, Greece *193*, 196, 198, 199–200
Konitsa Kourí sacred forest, Epirus, Greece 191, **204**, 206, *207*, 215
Konitsa Summer School 191, *195*

Konpira Shrine, Moriyama, Japan 331–2, 335, 336
kourí sacred forests, Epirus, Greece 197–8, **204** *see also* Mazi, Kourí sacred forest, Epirus, Greece; sacred forests
Kouroútes Amári, Crete *166, 181*
Koutsopétra, Omalós Plain, Crete *166, 169*
kunugi (Japanese) *see* Oak, Sawtooth
Kurnell, Botany Bay, NSW *305*, 314
Kyrgyzstan 9, 59–61, *61*; as a 'paradise' 60; Walnut-fruit forests 60–1, *61*

La Stevinieri, Vovoúsa/Voiása village 208–10
Laboratory of Tree-Ring Research (University of Arizona). 182
Lago di Bargone, Liguria, Italy 219–22
Lake Biwa, Moriyama, Japan 329–39, *330*
Lake District 45
Lake Ijsselmeer, Holland 101
Lake Rezzo, upper Val d'Aveto, Liguria, Italy 229
Lákkoi, Imbros Sphakiá *166, 176*
Lákkoi Kydonías, Crete *166*, 176, *177*
LANDIS II computer model 55–6, *56*, 57, 60
landscapes 250; agricultural 144; change and continuity 161–2; conference 'Permanent European Conference for the Study of the Rural Landscape', Limnos (2004) 33, *33*; contemporary 156, 196. 228, 282, 353; ecology 31, 277; heritage landscapes 8; history 49; individual landscapes 218, 230; planned 32–3; pre-feudal 277; pre-Neolithic 32; rural 328 *see also* cultural landscape; Mediterranean landscape; satoyama landscapes; treescapes
Landscape Conservation Forum (LCF) 10, 273
'Landscape history of Cyprus' (Rackham chapter) 146
'Landscape history of southern Europe' (Rackham chapter) 144–55
landslides vii, 204, 211 *see also* avalanches; protection forests
large herbivores 27–8, 72, 106–9, *108, 110*; Białowieża Primeval Forest (BPF) 117; domestic livestock 32, *55*, 60, *72*, 101, 106, 107–8, *110*, 113; role of herbivory 54–6, *55*, *56*; wild ungulates 32, 54, 101, 113 *see also* Vera theory and debate
Lasithi mountains, Crete 6
The Last Forest (Rackham book) 4, 34–6, 324, 360
Lavenham, Suffolk 51
Lawrence Hopkins Experimental Forest 351–3
leaf fodder 34, 240, 263 195, 320–1; ash, lime, elm and maple 321; and elm decline 262; as green manure 129, 227; holly 34, 40, 43, 281; mulberry, oak and willow (Japan) 341; for silkworms 344 *see also* coppicing; pollarding; shredding
lēah (Old English) 'wood-pasture' 37–9
Lerone Forest State Domain 228
Lerone Valley, Liguria, Italy 222–3, *223*, 228
lichens 3, 11, 51, 57, *58*
Liechtenstein: Chapter 8 (alpine forests); deadwood *133*; forest area *128*; forest naturalness *132*; forest ownership *130*; tree species *129*
Light as an Ecological Factor (Bainbridge, Evans and Rackham book) 50
Light Detection and Ranging (LiDAR) 23–4
lignotubers *186 see also* coppice stool
Liguria, Italy Chapter 14 (historical landscape) 218–30, *223*; fieldwork 219–22; repeat photography 222–8; saw-wort 228–30
Lily of the Valley *Convallaria majalis* 229
lime (*Tilia*) 108, 109, 263–4; and ancient woods 25, 51–2; and grazing 89–92; limewoods 22, 24, 25, 84; misidentifying 82; pollards *327*; reproduction 92–5; seedlings 93; stools 85, 91, 93, 265, 281
Lime, Common *Tilia* x *europaea* 82
Lime, Japanese *Tilia japonica* 343, 344
Lime, Large-leaved *Tilia platyphyllos* 18, 25, 32, 82
Lime, Small-leaved *Tilia cordata*: 25, 82; estimating ages 281; Hanging Wood, Claverdon 82–3; Hudnalls, Gloucestershire 24; Marks Hall Woods, Essex 94, *94*; meaning 52; Norfolk 293; northern limit 92; Peak District 18; Piles Coppice, Warwickshire *93*; relict species 82; saplings 93–4; shade-tolerant species 32; Silk Wood, Gloucestershire 92; stools *93*; Swanton Novers Great Wood, Norfolk *18*; Warwickshire 82–3
Limnos, Greece 33, *33*
Lincolnshire 22, 23, 24, 65
Little Toller Books 377
livádi (Greek) meadows 198
livestock, domestic 32, *55*, 60, *72*, 101, 106, 107–8, *110*, 113 Hecke cattle *108*; Scottish Highland Cattle 107 *see also* large herbivores; swine
Loch Awe, Argyll, Scotland 24
'Locust Years' (Rackham term) 12, 367 *see also* 'New Locust Years'
lōh (Old High German) copse, grove, woodland, undergrowth, scrub 37
London 262–3, 294

Long Paddock, Australia 302
Long Row, Norfolk 287
longevity of trees 57–8, 200; coppicing and pollarding 90, 171–2, 280, 340
Longshaw, Peak District *276*, 281
Lopham Grove, North Lopham, Norfolk *293*
Lost Eden *see* Ruined Landscape theory
lost woods 278, 279–80 *see also* ghost woods; shadow woods
lordings (mature timber trees 60+ years old) 238 *see also* black barks
Lower Wye Valley 25
lupinella (Italian) *see* Common Sainfoin
lynchets (earth terraces) *286*

Macquarie River, Australia 300
Madingley Wood 3 (*Blue Notebook*, Rackham) *369*
maerimium (Latin) timber 238
Magnesian Limestone, South Yorkshire 236–7, 240
mahoganies (Meliaceae) 315
maki goods collection point (Moriyama, Japan) 332
The Making of the Cretan Landscape (Rackham and Moody book) 6, 153, 160, 199
Makrokhórafo, Epirus, Greece 203
Málles, Crete *166*
Malvern Hills 45, *46*
Manassi, Epirus, Greece *193*
Mangrove, Grey *Avicennia marina* *305*, 306
mangroves 304–6, *305*
maple (*Acer*) 60 *see also Acer*
maple-deciduous oak pastoral wood 211
Maple, Field *Acer campestre* 294, 325
Maple, Montpelier *Acer monspessulanum* 199, 201
Maple, Painted *Acer mono* 343, 344
Maple, Red *Acer rubrum* 355, 358
Maple, Striped *Acer pensylvanicum* 355
Maple, Sugar *Acer saccharum* 355, 356, *357*
Maple, Sycamore *Acer pseudoplatanus* 128, *279*
Marks Hall Woods, Essex 94, *94*
Markshall Estate, Essex 374–5
Marsh Mallow *Althaea officinalis* 304
marshland *see* Oostvaardersplassen; saltmarsh; wetland
Mathés, Crete *166*, *183*, 184
Mazi, Kourí sacred forest, Epirus, Greece *193*, 199–200, 204–6, **204**, 205
Meadowsweet *Filipendula ulmaria* 66
meadowsweets (brideworts) *Spiraea* species 358
MEDALUS Project 140
Mediterranean Agronomic Institute of Chania (MAICH) 176
Mediterranean landscapes 6–8, *7*; agroforestry 159–60; deforestation 154; dense forest 161, *162*; grazing 160–3; maquis forest 161; resilient landscapes 157; Ruined Landscape theory 153, 154–8, 161; savanna 159
Mediterranean woodland degradation hypothesis 222
Meiji period (1868–1912) 330, 332–4
Mereao sacred forest, Palioseli, Epirus, Greece **204**, 208, *209*
Mesovouni, Epirus, Greece *193*
Micheldever Wood, Hampshire 23
Middle Ages: Austria (protection forests) 131–2, 213; Bavaria Chapter 16 (abbey inventory); coppicing 260; Hatfield Forest 36
Middle Wood, Thorpe Abbots, Norfolk *286*
Midland England 21
Millstone Grit 237
mimosa 308 *see* Sensitive Plant
Mimosa Rocks, New South Wales 308
Mimosa scorpioides see Gum Arabic Tree
Minoan art 185; 'Ring of Minos' *186*; trees *185*, 185, *186*; Vapheio cup *185*
miracle and healing trees *166*, 182–5, *183*, *184*
Miriáouwa, Palioseli, Greece 200, 206–8
Mittelwald (German) coppice-with-standards 262
mizosarae/mizuban (Japanese) irrigation channel cleaning/patrols 336
mizunara (Japanese) *see* Oak, Japanese
Moccas Park, Shropshire 43, *43*
mochros (Welsh) moor/place of pigs 43
Monks Wood Experimental Station, Cambridgeshire 17
Montagna di Fascia, Liguria, Italy 222
Montbenoît, France 260
monte bajo (Spanish) coppices 262
Monte Porcile, Liguria, Italy 219
Monte Reixa, Liguria, Italy 222, *223*
Monte Tardia, Liguria, Italy 222, *223*, 228, *228*
Monumental Olives of Crete *166*, *181*, 182, 186
Moose *Alces alces* 101
Morecambe Bay 303
moriyama-ishi (Japanese) chert rock 335, *335*
Moriyama, Otsu City, Japan: Chapter 22 (satoyama landscapes) 330–9; common management of natural resources 335–7; in the Edo period 331–2, *331*; natural resource use 332–5, *333*, 337–8; sacred spaces 336–7; satoyama rural landscape 330, *330*
Morocco 156
Moulianá, Crete *166*

Mount Æminádhia, Greece 206
Mount Chokai, Yamagata Prefecture, Japan *341*, *343*, 345, *345*, 347
Mount Kédros, Crete 179
Mt Timfi, Epirus, Greece *193*
Mountain Ash *Sorbus aucuparia* 59
mulberry (*Morus* species) 341; pollards 343–4
Mulschachen, Czech Republic 260
Muntjac *Muntiacus reevesi* 94
murakata (Japanese) village association 336
Murray River, Australia 300
Murrough shingle beach, Wicklow 248
Museo Contadino, Cassego, Liguria 219
Myrtos, Crete 6, *166*; Early Minoan 6; Fournoí Korifoí (excavation) 165.

Narromine, NSW, Australia 302
National Trust 34, 324
National Vegetation Classification (NVC) 68, 79n2, 304 *see also* vegetation classification
Natura 2000 nature protection areas 115, 134; Sites of Community Importance (Natura 2000) 134; Sites of Special Scientific Interest 25
Natural England (English Nature) xi, 34, 45, 52, 65, 273
Natural History Museum of Crete 165
natural resource use (Moriyama, Japan) 332, *333*, 337–8; and children's play 337, 338
Nature Conservancy Council (NCC) 4, 18, 19, 34, 64; county-by-county inventories of ancient woodland (NCC) 34; debate 18–9
The Nature of Mediterranean Europe (Grove and Rackham book) 7, 140, 219; Chapter 10 (critique of book) 154, *155*, 156–8, 161
Needwood, Staffordshire 39
Neolithic: Hungary 263; Mediterranean agroforestry systems 160; Somerset 264
Netherlands: Chapter 6 (Oostvaardersplassen marsh and the origin of grassland)107; prehistoric trackways 264
Nettle-leaved Bellflower *Campanula trachelium* 370, *371*
New Biggin Wood, Kent *73*
New Close Woods, Brinklow, Warwickshire *80*, 85–7, 89–92, *90–1*
New England 8, 351, 358 *see also* Hopkins Memorial Forest (HMF)
New Forest, Hampshire 17, *35*, 45; 'Ancient and Ornamental' woods 19; ancient woodlands 19, 56, 57; Beech Dieback 57, *58*; boundary banks 20; European Beech 56–7; impact of herbivory on woodland dynamics 56, *56*; limes 25; medieval encoppicements 19, 20; oak 56; regeneration of tree species 56; Vera's theory 55, 107
'New Locust Years' 12 *see also* 'Locust Years'
New South Wales (NSW): Big Scrub, Northern Rivers region 312; first governor 116
Nídha Plain, Crete *166*
Niederwald (German) coppices 261
nitrogen deposition 71, *71*
Norfolk 23; Chapter 19 (ancient woodland) 285–95; ancient coppices 24; ancient woodland indicators 68; ancient woods 286, 291–5, *292*; boulder clay plateau 291, *292*; Domesday Book 291; hazel and ash underwood species 294; hornbeam-dominated woodland 294–5; *Norfolk Naturalists* (journal) 287; oak coppices 294; oddities and anomalies in archaeology 288; pollen records 25, 287, 294; primary woods 293; secondary woods 287, 293; *Transactions of the Norfolk and Naturalists Society* 285; wood-pasture 'waste' 291; woodland management 292
Norman Conquest 277 *see also* Domesday Book
Norton, South Yorkshire 238, 239
Norway Spruce *Picea abies* 126, 128
Norwich 294, 370
nurse species *see* thorny plants

oak (Pedunculate)-Hazel wood 83
oak (*Quercus*) ancient (Brocton Coppice) 44; basal decay 326, *326*; churches and sacred forests 202; coppice *see* oak coppice; deciduous 176, *177*, 199, 202, 205; grazing 112–13, 160–1, **163**; Hayley 266; *jiyama* forests, Japan 334; 'medusiod' (multi-stemmed) 280; New Forest 56–7; pollards *see* oak pollards; pollen diagrams 108, 109; protected by hawthorn 110, *110*; Rackham tree dedication 'Curley Oak' (Wales) 377, *378*; Rackham tree dedication Prickly Oak (Crete) 165, *167*, 377; regeneration 108–10, 112–13; restoring oak stands 69, *70*; savanna 160–1; shadow woods 280–1; Snowdonia 24; 'spyres' (spears/spires) 239; timber 198, 294; wood-pasture trees *45 see also Quercus*
oak (Sessile)-limewood wood 84
oak coppice: Czech Republic 265; France 267; Japan 345; Norfolk 294
oak pollards: Ashtead Common, Surrey 323, *323*; Japan 344; 'The Oak Conference'

(Östergötland 2006) 327; Psiloríti Mountains, Crete 176, *177*
Oak, Downy *Quercus pubescens* 205
Oak, Holm *Quercus ilex* 223
Oak, Hungarian *Quercus frainetto* 205
Oak, Japanese *Quercus mongolica* var. *crispula* or *Quercus crispula* (*mizunara*) 334, 335, 343, 344
Oak, Jolcham *Quercus serrata* 343, 344
oak, *konara* (Japanese) (*Quercus glandulifera*) 334
Oak, Macedonian *Quercus trojana* 205
Oak, Pedunculate *Quercus robur* *26*, 32, 51, 109; 'Curley Oak' Rackham dedication 377, *378*
Oak, Prickly (Kermes) *Quercus coccifera* 6, 165, *167*, 171, 175–6, *176*, 199; Great Prickly Oak stool (Lákkoi, Imbros Sphakiá) *176*; Rackham tree dedication Selákano 165, *167*, 377
Oak, Red *Quercus rubra* 355, 361, *362*
Oak, Sawtooth *Quercus acutissima* (*kunugi*) 341
Oak, Sessile *Quercus petraea* *18*, 32, 84, 109, 205–6
Oak, Turkey *Quercus cerris* 205, 208, *221*, *222*, 223
oakwoods 24; hanging 280
Oberholz (Czech) standard trees *261*
Ohshirakawa, Niigata Prefecture, Japan *342*
Old English/Old Norse woodland clearance names 236, *239*
Old English terms 37, 39, 42
Old Grove and Primrose Grove, Gillingham, Norfolk 290, *290*, 293–4
old-growth forests (ancient forest) 124–35; Chapter 8 (alpine forests); alpine landscape 126–7; deadwood 133, *133*; defining 125; farm forest utilization 129–30; origin and development 127–9; protection areas 130–1; protection forests 134; Southern Limestone Alps *126*; tree species and naturalness 131–3, *132* *see also* ancient woodland; primary woods
Old World Flying Squirrel *Pteromys* 117
olive (*Olea*) 179–82; 'ancient' trees *171*, *181*, 187–8; Byzantine olive (Kouroútes Amari) *181*; conservation 186–7; disease *see* Olive Quick Decline Syndrome; growth rates 170; Monumental Olives of Crete *166*, *181*, 182, 186; opportunistic growers 170; 'Saving the Ancient Olive Trees of Crete' 186; Society of Cretan Olive Municipalities 186; varieties (Chondroliá, Koronéiki, Mastoeídis-Mouratoliá, Mastoeídis-Tsounáti, Throumboliá)182; Venetian olive (Tsourounianá, Crete) *188*; wild Cretan olives 182
'The Olive in Crete' (Rackham and Moody lecture) 182
Olive Quick Decline Syndrome (*Xylella fastidiosa*) 182
'Oliver Rackham – Roots and Branches' exhibit *see* Green Wood Centre
Oost-Flevoland polder, Netherlands 103
Oostvaardersplassen, Netherlands 101–8, *103*, *108*; ecological experiment 105; farmers 106–7; geese 102–3, 105–6; Nature reserve 101, 105
Opposite-leaved Saltwort *Salsola soda* 306
Ordnance Survey (OS) 19, *86*, 290
Orthodox Christian Church 203
Osprey *Pandion haliaetus* 105
Östergötland County, Sweden 327
Ottomans 197, 203
overgrazing 177, 200; goats 156; and upland sheep farming 45; *see also* grazing
Overhall Grove, Cambridgeshire *74*
overstorey trees 361
Owler Bar, Peak District *272*
oxen 302
Oxlip *Primula elatior* 23, 65–6, *69*

Palaioséli/Palioseli, Epirus, Greece *193*, 199–200, **204**, 206–8
Pale Sedge *Carex pallescens* 22
Paliurus spina-christi (Jerusalem thorn) 199
Panayía, Kourí forest, Greece 206
Páno Amhélia, Crete 166, 176, 177
Parker Library 370, 374
Passo di Cento Croci, Liguria, Italy 222
pastoral farming 127–8 *see also* wood-pasture
Peak District 18
pear (*Pyrus*) 58–60 *see also* Hayley Pear; *Pyrus*
Pear *Pyrus communis* 58–9
Pear, Wild *Pyrus pyraster* 58–9
Pendulous Sedge *Carex pendula* 74
Pennine Fringe, South Yorkshire 236
photographs: aerial *73*, 198, 206, *354*; historical 198, 218–19, 222–3, *224*, *225*, *227*, 321; repeat 222–8, *224*, *225*, *227*
physiological ecology 50
Pian Brogione, Casanova, western Apennines 229
Pignut *Conopodium majus* 83, 293
pigs *see* swine
Pileated Woodpecker *Dryocopus pileatus* 356, *357*
Piles Coppice, Warwickshire *80*, 85, 86–7, *89*; Domesday Book 89; Large Small-leaved Lime stools *93*; Lime age and regeneration 92–5; as Munechet wood

89, 92; tree communities and earthworks 86, *88*
Pindos mountains, Greece 192, *193*
pine (*Pinus*) autochthonous origin 226; dendrochronology 175; Highland pinewoods 19, 24; Kónitsa, Epirus, Greece 206
Pine, Aleppo *Pinus halepensis* 174
Pine, Alpine *Pinus cembra* 128
Pine, Austrian *Pinus nigra* 223
Pine, Black *Pinus nigra* 128, 201, 206
Pine, Cretan *Pinus brutia* 174–5, *175*
Pine, Dwarf Mountain *Pinus mugo* 128
Pine, Maritime *Pinus pinaster* 223, 226
Pine, Red *Pinus resinosa* 332–4, *334*
Pine, Scots *Pinus sylvestris* 57, 128
Pine, Swiss *Pinus cembra* 127
Pine, White *Pinus strobus* 358
place-names 5, 36, 41–2, *42*, 197, 211, 236
planatoria/boumscapun (Latin) bill hook 255
plane (*Platanus*) *178*, 179; disease *see* Canker Stain Disease; evergreen (Gortyn, Crete) 184–5; on Hellenistic silver stater coins *184*, 185
Plane, Oriental (evergreen) *Platanus orientalis* var. *cretica* 179; healing tree (Gortyn, Crete) 184–5
plantations 159–60, 290; and agroforestry 160; forest genetic plantations 353; and The Great Storm 360; Japan 334; plantations on ancient woodland sites (PAWS) 27; restoration 69; Spain 201; Yorkshire 236 *see also* coniferization
plumed thistles (*Cirsium* species) 67
Poland 9; Chapter 7 (Białowieża Forest) *100*; logging policies 117 *see also* Białowieża Primeval Forest
Polcevera Valley, Liguria, Italy 219
polder 101–5
Polish-Lithuanian Commonwealth 117
pollard: ancient (Moccas Park) 43; conference City of London 322; Crete 171–2; Hatfield Forest, Essex 324–6, *324*; lapsed 323–4, *323*; Japan 340–8, *344*; management 320–1, 340–1, 345–7; Mount Chokai, Japan *343*; mulberry, willows and oaks 343–4; pollen profiles 263–4; and soil erosion 343; radial growth *348*; uses 320–1, 340, 344; veteran trees 320
pollarding: bolling 325, 340, *341*, 343, 345; dead bolling 347, *347*; hollowing 325–6; Japan 340–8, *344*; reasons for 321; re-pollarding 325; and tree longevity 171–2
pollen 24–25, 32, 60, 228, 263–4, 294, 317; diagrams 108, 109; palynology 25, 179
pollution 67
Poms (Australian) immigrants from/inhabitant of British Isles 304
Ponte Negrone, Liguria, Italy 228
Populus nigra subsp. *betulifolia* (Black Poplar) 377
primary woods: 129, 287; and ancient woodland 23–4, 292; defining 34; as 'natural' 287; Norfolk 291–5; pre-Neolithic woodland 19, 25, 28; Rackham's view 287 *see also* ancient woodland; old-growth forests
primeval Europe: forests 34, 99, *100*, *120*, 292; grasslands 107–8; landscapes 120, 275–6; saltmarshes 303–4 *see also* Białowieża
Pro Silva-movement 125, 135
Prophétes Elías church, Epirus, Greece 204–5
protection forests: alpine old-growth forests 130–1, 133–5; avalanches 127, 131, *207*; Kourí sacred forest 206; landslides 204; sacred forests 204, 213; *yurigi* Konpira Shrine, Japan 335
protocols *see* conventions
pseudo-history 116, 235, 379 *see also* factoids
Psiloríti Mountains, Crete *166*, 176
Pyrus cajon 59
Pyrus korshinskyi 59
Pyrus tadzhikistana 59

Q-pits 273
Quaking Aspen *Populus tremuloides* 355, 358; fast-grown 361
Quercus dalechampii (Dalechamps oak) 209
Quercus pubescens var. *brachyphylla* 176 *see also* Downy Oak

Rackham Collection *see* Green Wood Centre
Rackham, Oliver aphorisms: 62, 362
Rackham, Oliver archives 367–74; computer files archive 370; Corpus Christi College, Cambridge 368–70; Crete 373–4; slides 370, 372; Texas 370; Woodland Record Cards 370 *see also* Green Wood Centre; Rackham field books; Rackham herbarium; Small Woods Association
Rackham, Oliver field books: Blue 368–70, *369*; digitization 372, *373*; recording data 359–60; Red *7*, 9, *9*, 145, 165n2, 356, 368–70; *130* (1968) *369*; transcribing 372, *372* *see also Madingley Wood 3, Blue Notebook*
Rackham, Oliver herbarium 370–2, *371*
Rackham, Oliver Memorial Events 374–7; 'Oliver Rackham Memorial Symposium'

(Corpus Christi College, Cambridge 2016) 374, *375*; *see also* Woodland Trust
Rackham, Oliver method 49–51, 198, 219–22, 235, 250–1, 284–6, 362; coppice restoration 65; field work 65, 194–211, 356, *357*, 358, 360, *360*; fungus hunting 51; historical approach vii, 18, 218, 230, 321; minimal intervention 67, 135; 'reading' a landscape 272, 325; recording field data *177*, *194*, *210*, 359–60; repeat photography 222–8, *224*, *225*, *227*; sediment cores 51–2; Woodland Record Cards 85 *see also* documentary evidence
Rackham, Oliver publications and presentations 4, 382–95; 'The ancient woods of Norfolk' 285; *The Ancient Woods of South-East Wales* 375, 377; *The Ancient Woods of the Helford River* 375, 377; *The Ash Tree* 11, 12; Brandon Park fungi 1959–2014 377; 'Island landscape history' 146; 'Landscape history of Cyprus' 146; 'Landscape history of southern Europe' 144–55; *The Last Forest* 4, 34–6, 324, 360; *Light as an Ecological Factor* (with Bainbridge and Evans) 50; *The Making of the Cretan Landscape* (with Moody) 6, 153, 160, 199; *The Nature of Mediterranean Europe* (with Grove) 7, 140, 154, *155*, 156–9, 161, 219; 'The Olive in Crete' (with Moody) 182; 'Saaremaa: tackling landscape history in Estonia' 146; 'Sacred groves and cultural landscape in Europe' (with Pungetti) 148; 'Sacred groves in a Japanese satoyama cultural landsape: a scenario for conservation' (with Fukamachi and Oku) 148; Sacred groves in Japanese satoyama landscapes (with Fukamachi) 339; 'Significance of sacred trees and groves' 148; *Trees and Woodland in the British Landscape* 4, 31, 32–4, 64, 78, 285–6, 320, 353, 379; *Trees, Wood, and Timber in Greek History* 154 *see also Ancient Woodland* (Rackham); *The History of the Countryside* (Rackham); 'Sacred Natural Sites: Sacred forests in Epirus' (Rackham); *Woodlands* (Rackham)
Rackham, Oliver quotes: 6, 7, 8, 9, 10, 11–12, 34, 36, 50, 67, 82, 110, 111, 112, 113, 116, 122, 144, 159, 169, 182, 192, 194–209, 211, 213–14, 249, 291, 300, 302, 303, 306–7, 311, 339, 350, 368, 372, 379
Rackham, Oliver terminology: 10–11; 'applied historical ecology' 34; 'Ancient Countryside' 39; 'barren scrubs' 312; 'coppicing plants' 260n2, 379; 'frog-poking' 300; 'holloways' 358; 'Locust Years' 12, 367; 'long-term ecologists' 50; 'Mediterranean savanna' 159; 'plantation' 159; 'savanna' 159; 'Trees grow again' 154n1; 'tree lines' 358; 'wildwood' 52–3; 'woodland' 159; 'wood-pasture' 159
Rackham, Oliver tree dedications 375, 377; Crete 165, 165n3, *167–9*, 375, 377; Epirus 215, *215*, 375; Wales 377, *378*
Rackhamian worldview 51–2
radiocarbon dating 170
railways 245, 303
rainforests 311–12, 315
'Ramell' (small brushwood) 242
Red Deer *Cervus elaphus* 101, 107, *108*
reeds 101–3, *102*, 332, *333* *see* Greater Reedmace
're-naturalization' (Italy) 223 *see also* rewilding
reclamation 104–5; mangrove 306; wetlands 302 *see also* polders
regeneration of trees *see* tree regeneration
Regional Forestry Inspectorate, Liguria, Corpo Forestale dello Stato 222, 224
Réthymno, Crete *166*
rewilding 53; Carrifran Wildwood Project (Scotland) 53–4; ground flora 71–2, *72*; Oostvaardersplassen experiment 105; Rackham's view 37, 67, 112
rice paddies 331–4, *335*
Ríchtis Gorge, Crete *166*
ride(s) 18, 21–22, 83, 111 *see also* tracks
ridge-and-furrow cultivation 20–1
Rijksdienst voor de IJsselmeerpolders (RIJP) 101, 105–6
Ringers Grove, Shotesham 287
River Red Gums *Eucalyptus camaldulensis* 300, 302
River Severn *41*
River Teme 39, *41*
roccia madre nuda (Italian) exposed rock under layer 222
Roche Abbey, South Yorkshire 240
Rockingham Forest, Northamptonshire 23
Roe Deer *Capreolus capreolus* 107
Rollestone Wood ('herdyng wood') 239
Roman period: Mediterranean 154, 156, 157; Rockingham Forest, Northamptonshire 23–4; woodland management 24
root plates 280–1, 360–1, *360*
roots 280–1
rosaceous trees 58–61 *see also Sorbus* species
Rosebay Willowherb *Chamaenerion angustifolium* 74
roses *Rosaceae* 60

Roudsea Wood, Cumbria *67, 70*
Rough Close, Berkswell, Warwickshire *80*, 83–4, *84*
Round Grove, Hedenham, Norfolk 287
Round-Leafed Birthwort *Aristolochia rotunda* 229
Rowan, European *Sorbus aucuparia 279*, 280, 281 *see also Sorbus*
rubetum (medieval Czech) coppices 262
Ruined Landscape theory (Lost Eden) 153, 154–8, 161
Rule of St Benedict 255, 256 *see also* Tegernsee Abbey
Ryton Wood, Warwickshire 79–83, *80, 81*; Fetherston-Dilke family records (thirteenth to twentieth century) 79

'Saaremaa: tackling landscape history in Estonia' (Rackham chapter) 146
Sable *Martes zibellina* 117
sacred forests: Chapter 13 (relict landscapes Epirus); churches 202–3, deaths from trespassing 203, ethnographic research 203–4, *kourí* sacred forests 197–8, **204**, protection forests 204, 213, as sites of minimal intervention 213–14, supernatural punishment for cutting trees 209
'Sacred groves and cultural landscape in Europe' (Rackham and Pungetti lecture) 148
'Sacred groves in a Japanese satoyama cultural landscape: a scenario for conservation' (Fukamachi, Rackham and Oku lecture) 148
'Sacred groves in Japanese satoyama landscapes' (Fukamachi and Rackham chapter) 339
'Sacred Natural Sites: Sacred forests in Epirus' (Rackham unpublished manuscript) 192–214; Agia Paraskevi sacred forest, Vovoussa **204**, 208–9; conservation 214; dead trees 200; ecological and historical significance 213–14; environment 196–7; fire 200–1; grazing history 199–200; inland Epirus 194–6; Kourí forest, Kónitsa town **204**, 206; Kourí forest, Mazi village 204–6, **204**; Miriáouwa forest, Palioseli village **204**, 206–8; place names 197–8; sacred forests 202–4, 213–14; savanna and forest definition 198; shrubland and coppice-woods 199; Toúfa forest, Greveniti village **204**, 211, *212*; tree species 199; uses 213; woody ground vegetation 201–2
Sallows *Salix* species 294
salt production 303
saltmarsh (Australia) 302–6, *305*; and airports 303; destruction 303
Samariá Gorge, Crete *166*
sanchokata (Japanese) youth organization 336
Sanicle *Sanicula europaea* 65
sannai (Japanese) forests 332, 334, 336
Sarcocornia quinqueflora (Beaded glasswort) *305*
Sardinia 141, 147
sarjerdő (Hungarian) coppice 261
Sássalo, Crete *166*, 179
satoda (Japanese) rice paddies 332
satoyama landscape, Moriyama, Japan 329–37, *330*
savanna 8–9, 11, 64, 67, 159, 198–9, 276–7; Australia 313; Mediterranean 159; Rackham's view 8–9, 11; tropical 313 *see also* wood-pasture
Savernake Forest, Wiltshire 23
Saw-wort *Serratula tinctoria* 218, 228–30, *228*
scale-insect *Marchalina hellenica* 175
Scotland: early maps 19, 69; Roy maps (*c.* 1750) 19; soap-making 305–6; *see also* Carrifran Wildwood Project
scrub (Australian rainforest) 311–12, 315
Sea Eagle *Haliaeetus albicilla* 105
Second World War 196, 204, 206, 351
secondary woodland 19, 23–4, 27, 287
seed-bank species 66–7 *see also* ancient woodland indicators; coppice-associated species
seed dispersal agents 54
seedlings: ash 205; lime 93–5; oak 32, 109, 200, 202, 205, 208; pine 207; survival 93–5, 109, 200 *see also* thorny plants
Seiridium cardinale fungus (Cypress Canker Disease) 174
Selákano, Crete 165, *166, 167*
Selva di Pessino ancient wood, Liguria, Italy 222
Sembronás, Crete *166*, 179
semi-natural habitats 27
Sensitive Plant *Mimosa pudica* 308
service tree *see Sorbus*
Service Tree, Arran *Sorbus pseudofennica* 59
Service Tree, Wild *Sorbus torminalis* 22, 85
settlements: altitudinal limit (Epirus) 195; expansion 39, 291; medieval Norfolk 287; nucleation 33; and protection forests 131, 134
Sexton's Wood, Hedenham, Norfolk 287
Shabbington Woods, Buckinghamshire *70*
shade-tolerance 101, 194, 359; and coppicing 65–7, 68, 111, 113, 293; hazel 111; lime 93, 95;

Oxslip 65; trees 32, 57, 108; Walnut 60; veteran trees (Sweden) 326
shadow woods 24, 271–83; ancient woods and ancient woodland 273–6; historical sources 276–7; lost woods 277–9; smaller veteran trees *279*, 280–1; upland 280 *see also* ghost woods
Sheaf valley, Sheffield 239
sheep 160; upland farming 45 *see also* grazing
Sheffield Archives 236
Sheffield City Council 272–3
Sheffield conferences (1992) 'Ancient Woodlands' 273, *274*, 280; (2003) 'Working and Walking with Ghosts' 4–5, 273, 276, 280; (2011) 'Animals, Man and Treescapes' 5; (2013) 'Trees Beyond the Wood' 5, 277
Shelley, Suffolk 83
shide (Japanese) *see* hornbeam, Japanese
shijimi (Japanese) clams 332
ship timber 248
Shotover and Stowood forest, Oxfordshire 24
shredding trees 195, 197, 200, *221*, 222, 265; fuel for limekilns 226–7; hornbeams 205, 206; *kladera* (Epirus) 196; for leaf fodder 156, 240, 265; oaks 219, *221*
Shrewsbury, Earl of 236, 240, *240*
The Shrubbery, Tivetshall, Norfolk 287
Shrubby Glasswort *Tecticornia arbuscula* 306
shrubland 198–9, *205*
Sidling's Copse, Oxfordshire 24
Signals (Australian National Maritime Museum) 310
'Significance of sacred trees and groves' (Rackham lecture) 148
Silk Wood, Gloucestershire 92
silkworm fodder 341, *344*
silva caedua (Domesday) coppice woods 262
silva minuta (Domesday) 'coppice' 237, 262
silva modica (Domesday) meaning unclear 237
silva pastilis (Domesday) 'wood-pasture' 237
silva (Domesday) woodland 237
Silver Fir *Abies alba* 126, 128
silviculture 127–9 *see also* agroforestry
silvo-pastoralism 61, 160, *162*, **163**, 199 *see also* agro-silvo-pastoralism; wood-pasture
Skonízo, Crete *166*, 179
Slovenia: Chapter 8 (alpine forests); deadwood *133*; forest area *128*; forest naturalness *132*; forest ownership *130*; pro-*Silva* movement 125; selective harvesting 130; tree species *129*
Small Woods Association 372–3, 375 *see also* Green Wood Centre
'Snee-form' (*Schnee*-formed, snow) 356–8
snow: damage 127, 131, 347; depth 345; and pollards 143–9, *145*
Snowdonia (Eryri), Wales 24
snowy climates *see* Japan
soap manufacturing 304–6
Somerset Levels 264, 267
Somerset trackways 24, 264
Sorbus persica 60
Sorbus species 59 *see also* Mountain Ash; Rowan; service tree; whitebeam; Whitty Pear
South Africa 311
South Flevoland, Netherlands 101, 104
South Haw Wood, Wood Dalling, Norfolk 294
South Yorkshire: archaeological survey 273; Chapter 15 (documents and woodland); coppice-with-standards 238–44, *238*; Domesday woodland 236–7, *237*; Ecclesall Woods 244–6; iron industry 242; post-medieval period 242–3; spring wood management 242; timber 246–8; timber merchants (Newton, Chambers & Co, Windle and Baker) 246; wood-pastures 237; woodland cover 236, *237*
Southern Kelp *Durvillea potatorum* 306
Southern Limestone Alps *126*
Southern Uplands, Scotland 53–4
Spain 262, 321
spears and 'fish gigs' 310
Specht classification scheme 313
Spongy Moths *Lymantria dispar* 358
Spoonbills *Platalea leucorodia* 106
spring woods 238–43 *see also* coppice-with-standards
springboards (jigger boards) 315
le Spryng bosci woods, Norton, South Yorkshire 238
'spyre' (spear/spire, young oak standard) 239
squirrels 184
stálos (Greek) livestock shelters 205
stand dieback 57
Stanmore Common, Middlesex *20*
Stara Białowieża 121, *121*
Statute of Merton 1235 (Act of Commons) 277
Staverton Park, Suffolk *26*
Staverton Thicks 51
steppe *see* grassland
Steppe Polecat *Mustela eversmanii* 117
Stinging Nettles *Urtica dioica* 102
Stingray Harbour (Botanist Bay/Botany Bay, NSW) 314
Stinking Iris *Iris foetidissima* 304
Stour Wood, Essex *68*

Strafford, second Earl of 242
Strovlés, Crete *166*, 179
succession theory 99, 101, 106, 109; HMF Secondary Successional Model 358, *359* *see also* Vera theory and debate
Suffolk 285, 294
sugi (Japanese) *see* Japanese Cedar
SUMESLAND Summer School (Sardinia, 2013) 141, 146, *147*
suoli scheletrici (Italian) stone-rich soils 222
supernatural punishments 209
Surrey woods 294
surskog (Swedish) coppices 262
susuki (Japanese) *see* Japanese Pampas Grass
Sutton Park, Birmingham 43, *45*
Swanton Novers Great Wood, Norfolk *18*
Sweden 320; Ash Die Back 326–7; ash pollard *328*; lime pollard *327*; *surskog* (coppices) 262
swine 32, 38–40, 43, *72*, 111, 196, 291–2
Swithland Wood, Leicestershire 82, 92
Switzerland 130, 134, 262; Chapter 8 (alpine forests); forest area *128*; forest naturalness *131*; forest ownership *130*; Neolithic pollarding 205; protection forests 130–1; tree species *129*
Sydlings Copse, Oxfordshire 25
Sydney Cove, NSW 307, 312, 314
Sydney Olympic Park, Homebush Bay, Sydney *305*, *309*

Taconic Range, NY State/New England 351, 356–8
taillis (French) coppice 261
Tainai, Niigata Prefecture, Japan 342
Tarpan *Equus ferus ferus* 101
Tasmania 9
Tegernsee Abbey, Germany 251–7, *252*; cellarer's inventory (AD 1023) 253–7
Texas 8–9, 368
THALIS forest inventory 202
THALIS-SAGE project (2012–15) 191–4, 198–9, 200, 214; 'Conservation through religion: the sacred groves of Epirus' 192
Thorncliff Spring coppice wood 240
Thorny Burnet *Sarcopoterium spinosum* 161
thorny plants: in early charters 39; enclosure toppings 39; as nurse plants 9, 32, 54, *55*, *56*, 57, 60, 106, 110–11 *see also* Blackthorn; hawthorn
Tiger Hill 374
timber 238–48; Alps 128; Australia 315–16; felling 43, 111, 119, 196; handbill advertising (1821) 243, *244*; immature ('spyres' and wavers) 238–9, 241, *241*, 242; as *maerimium* 238; mature (black barks and lordings) 238, 241–2, 248; merchants 246; oak 39, 198, 239; pricing 156, 242–43; ship 248; from standard trees 244–5; 'top and lop' 242–3; trade spreading disease 11, 12, 34 *see also* coppice-with-standards; high forest
Tindall Wood, Ditchingham, Norfolk 287
Tinsley Park, South Yorkshire 245
Tithe Maps (1840) 290, *290*; (1841) 83, *84*
Tivetshall Wood, Norfolk 287
Todd River, Alice Springs, Central Australia 300
Toombers Wood, Stow Bardolph and Stradsett 287
Totley, South Yorkshire 239
Toúfa sacred forest, Greveniti, Epirus, Greece **204**, 211–12, *212*
tracks/trackways 21, 27; prehistoric 264; Somerset 24, 264 *see also* rides
transhumance 39, 132; and linked estates *38*
travellers' accounts 197–8; 314 *see also* documentary evidence
travelling stock routes (TSRs) 302
Trebbia-Aveto watershed, Liguria, Italy 229
tree cultivation *see* agroforestry; silviculture
Tree Heather *Erica arborea* (*brugo*) 219
tree regeneration: artificial (planting) 109, 129–30; grazing 110, 112–13, 161; livestock 60; natural (no planting) 109, 135; New Forest 56–7; and pedoclimatic factors 161; seedling survival 201–2, 205; woodland periphery 57 *see also* rewilding
tree-ring analysis 265–7, 347; coring 281; Mount Chokai 347; radial growth of pollard stems *348* *see also* annual rings; cambial age; dendrochronology
Trees and Timber in the Mediterranean World (Meiggs book) 156
Trees and Woodland in the British Landscape (Rackham book) 4, 31, 32–4, 64, 78, 285–6, 320, 353, 379
Trees, Wood, and Timber in Greek History (Rackham lecture & booklet, Twentieth J.L. Myres Memorial Lecture at Oxford) 154
treescapes 122, 159, 276–7, 327, 367; relict 280 *see also* Sheffield conferences
Tsourounianá, Crete *166*, 186; Venctian olive tree *188*
Tulip Satinwood *Rhodosphaera rhodanthema* 315
Tuscany 229
Tyrol, Austria 130

unagi (Japanese) eel 332
'Uncompartmented forests' (Rackham section in *Ancient Woodland*) 110, 113
understorey 159–60, 161, 294
underwood 242, 244, 260, *261*; charcoal production *338*, 339–40, 248; management 238; species 112, 294; rights to 239 *see also* coppice
UNESCO: Biosphere Reserve 115; Global Geoparks 197; National (Hellas) Intangible Cultural Heritage Index 215; 'Linking Biological and Cultural Diversity in Europe' (Florence conference 2014) 139; and Secretariat of the Convention on Biological Diversity (SCBD) Joint Programme 138–9; World Heritage Convention 145
United States 8–9; Chapter 24 (Hopkins Memorial Forest); Rackham visits 8–9, 356–9, *357* *see also* Texas
University of Cambridge: Botany School field-trip (1967) 6; Conservation Corps 299–300; digitizing archives 368; herbarium 370–2 *see also* Corpus Christi College, Cambridge (CCCC)
University of Cambridge Digital Library 199, 372
University of Genoa 230; Historical Geography course 230; Laboratorio di archeologia e storia ambientale (LASA, University of Genoa) 8, 230 *see also* Diego Moreno
University of Sassari 141, 143; Center for Biocultural Landscape and Seascape 148; ECSLAND conference (2015) *see also* European Culture Expressed in Landscapes; Gloria Pungetti
unterholz (Czech) underwood 261
Upper Lerone Valley, Liguria, Italy 222–3, *223*, 228
Upper Vara Valley, Ligurian Apennines 218
US Forest Service (USFS) 351–3, *354*

Val d'Aveto, Rezzoaglio, Italy 229
Val Petronio, Liguria, Italy 222
Valia Calda, Epirus, Greece 201
Valle Lagorara, Italy 219
Valletti, Varese Ligure, Italy 219, *220*
vegetation: climax 99–100, *100*; linear succession 99, 103; natural 108–9; prehistoric 112, 292; surveys 264; theory (Vera) 54–6, *55*
vegetation classification systems 313; National Vegetation Classification (NVC) 68, 79n2, 304; problems with 79n2, 304; Specht scheme 313; stand-type system 79n1; Rackham's caution 379
Venice forestry law (1476) 260
Vera theory and debate 25, 64, 101; cyclical turnover of vegetation theory (break-up, scrub, park and grove phases) 54–6, *55*, 57; grazing and grassland 8–9, 32; nature of primeval Europe 276, 277; New Forest 55; open wildwoods 67; pre-Neolithic vegetation as savanna 292; prehistoric landscape 71–2; Rackham's view 8–9, 54, 67; retrogressive succession 99, 100; tree regeneration 57 *see also* grazing; large herbivores; succession theory
veteran trees 10, 24, 133, 278–81; ageing 169–70, 280–1; basal decay 326, *326*; Crete 6, 167–70, *178*; decaying wood habitats 325–6; hollow trees 170, 280–1; Minoan art 185, *185*; pollards 320; sacred forests 208, 214; shadow woods *278*, 279–80, 279; small trees 276, 278–81
veteranization 325–6
Victoria County Histories 19
Vikos–Aoos UNESCO Global Geoparks 197
Vikos canyon, Epirus, Greece *193*
vine brush (Australia rainforest) 312
virgin forest *see* primary woodland
Vitsa, Epirus, Greece *193*
Vlátos Kíssamos, Crete *166*, *180*
Voltri, Genoa *223*
Vovoussa, Epirus, Greece *193*, **204**, 208–9, *210*

Wales: *The Ancient Woods of South-east Wales* (Rackham book) 375, 377; Rackham tree dedication 377, *378*; Woodland Trust Wales 375
Walnut *Juglans regia* 59–60, *61*
wariyama (Japanese) border with neighbouring villages 332
Warwick County Record Office (WCRO) 79
Warwickshire Arden 39
Warwickshire Feldon 39
Warwickshire Lime Woodlands Chapter 5 (limewoods) 78–95, *80*; Binley 85–94, *86*; Hartshill Hayes, Hartshill 84–5; Rough Close, Berkswell 83–4, *84*; Ryton Wood 79–83; *terra incognita* 79
Warwickshire Vice-County *80*
Warwickshire Wildlife Trust (WWT) 79, 83, 85
Wasp Ford Close (Tithe Map 1841) 83, *84*
Waspern Fields, Beechend 83–4
Water Avens *Geum rivale* 293

Watson-Wentworth Irish estate (County Wicklow) 246–8, *247*
wattle and daub 307, 308
wattle (tree) 306–11; bipinnate 310; inflorescences 308; phyllodinous 310 *see also Acacia*
Wattle, Black *Callicoma serratifolia* 307, 308, *308*
Wattle, Cinnamon *Acacia leprosa* 308
Wattle, Golden *Acacia pycnantha* 307
wattles and fizgigs 306–11
wavers (sapling timber trees *c.* 30 years old) 238–9, 241, *241*, 242 *see also* coppice-with-standards
Wayland Wood, Watton, Norfolk 287, *295*
Weald of Kent 291–2
Wealden woods, Kent 291
Wentworth Woodhouse estate (South Yorkshire) 242–6
Weogorenaleage (Anglo Saxon Worcestershire) 39
West Bradenham, Norfolk 291
West Riding of Yorkshire 236
Western Australian Flowering Gum *Corymbia ficifolia* 311
wetlands 302–3 *see also* Oostvaardersplassen; saltmarsh
Wheatfen woods 374
White Edge, Peak District *278*, *279*
White Mountains, Crete 165, *166*, 174, 175, 179, 182
whitebeam *see Sorbus*
Whitebeam, Arran *Sorbus arranensis* 59
Whitebeam, Rock *Sorbus rupicola* 59
Whitty Pear *Sorbus domestica* 59
Wild Apricot *Armeniaca vulgaris* 60
Wild Boar *Sus scrofa* 107
Wild Garlic *Allium ursinum* 65, *71*
Wild Horse *Equus ferus* 54, 101, 107, *108*
Wild Madder *Rubia peregrina* 304
wild ungulates *see* large herbivores
wildwood 25, 52–4, 61, 67, 71–2, 275; blackthorn 111; hazel 111–12; lime 25; oak 111
Williams College, MA 351, *352*, 353
Williamstown, MA 351, 356–9
willow (*Salix* species) 128, 226, 249, 280-1, 294, 341; ancient 280; coppice 264; pollards 249, 343–44 *see also* Sallows
wind-downed trees 67, *68*, 360–1, *360*, *361*, *362*; and fast-grown trees 361; and woodland periphery 360 *see also* derecho; The Great Storm
Windsor Great Park 280
Wisent *Bison bonasus* (European Bison) 54, 101, 107, 117
Wistman's Wood, Devon 24
Wither's Wood, Suffolk 83
Wolverine *Gulo gulo* 117
wood 264–7; cambial age and diameter 265, 267; from coppiced underwood 244; meaning in Australia 312–13
wood anatomy 264, 265; reactions to cutting 265
Wood Anemone *Anemone nemorosa* 22, *68*, 72
Wood Millet *Milium effusum* 65
wood-pasture (*silva pastilis*) 24–25, 31–2; Alps 129–30; *Ancient Woodland* (Rackham) 109–12; as ancient woods 24; Borkener Paradise in Germany *110*; Burnham Beeches, Buckinghamshire 322; Cotswold scarp 45, *46*; decline 42–3; defining 278; distinction with woods 36; Domesday woodland, South Yorkshire 237; grazing 33; large herbivores 72; large-scale extensive grazing 72; legacy and future direction 43–7; Malvern Hills 45–6, *46*; New Forest 45; Norfolk 291; south-west England 42; South Yorkshire 237; uncompartmented 110, 113; understorey for cultivation or grazing 159; wood and timber supply 277; and woodland 36; Wye Gorge 24 *see also* pollard; savanna
Wood Sedge *Carex sylvatica 293*
Wood Sorrel *Oxalis acetosella* 65
Wood Speedwell *Veronica montana* 294
Wood Spurge *Euphorbia amygdaloides* 66, *293*
woodbanks 79–89, 279, 289 *see also* banks/ditches; earthworks
wooden slab buildings 301
woodland: Australia 312–13; enclosed 277–8, 291–3; grazed 292; and hedgerow management tools 299; and plantations 19; and saltmarshes 304; as savanna 313; 'single-tree management' 131, 135; thorny plants as nurse species 110; and wood-pasture 36
woodland archaeology *see* archaeology
woodland clearance 10–1, 82; ringbarking 314; slash-and-burn fields 129–30; South Yorkshire 236, 239
woodland communities 54, 71, 79n2, *88*, 90, 295; long-term 132
woodland heritage 273
woodland management: cambial age and wood diameter 267; constructional timber 246–8; forensic observations 51; reconstructing from accounts 244–6
woodland margin 24, 228, 229
Woodland Record Cards 85, 370
woodland/saltmarsh zone flora 304

Woodland Trust (WT) 45; aims 34; ancient woodlands 26, 69, 273–5; Piles Coppice, Warwickshire 85; Rackham Memorial Events 374–75; Rackham tree dedication ('Curley Oak') 377, *378*; South Wales studies 64; trees outside enclosed woods 277
Woodlands (Rackham book) 11–12, 249, 285, 300; and the BTF 122; ecologists' notebooks and photographs 368; electronic archives 372; pseudo-history and pseudo-ecology 379; 'virgin forest' unaffected by mankind 350
woodlots (USA coppice plots) 358–9
'Woods as we see them today' (Rackham, chapter 6 in *Trees and Woodlands in the British Landscape*) 286
Worcestershire: droveways 39, *41*; as *Weogorenaleage* 39
World Agroforestry Center (ICRAF) 159–60
World War II 204, 206, 351
Wychwood, Oxfordshire 45
Wye Gorge 24
Wytham Woods, Oxfordshire 65, *71*, 73–4

xoklissi (Greek) outlying church 196
Xylella fastidiosa (Olive Quick Decline Syndrome) 182

yama no kami (Japanese) mountain deity 332
yamada (Japanese) hillside rice paddies 332
yamakusa (Japanese) grass 334
yamamodori day (Moriyama, Japan) 336
Yarrawa Brush 312
yashiroyama forests (Moriyama, Japan) 332
Yellow Archangel *Lamiastrum galeobdolon 73*
Yellow Box *Eucalyptus melliodora 301*
yokusa harvest (Moriyama, Japan) 334
yokusakari harvest (Moriyama, Japan) 334, 336
yorito members (Moriyama, Japan) 332
York University, 'Humans and Ecosystems before Global Development' colloquium (1996) 251
Yorkshire: South Yorkshire *see* Chapter 15 (documents and woodland); Yorkshire Archaeological Society Archive 240; 'Yorkshire Woodlands' conference (2016) 374
yugakari (Japanese) water for agriculture 336
yuri (Japanese) steep paths 335
yurigi (Japanese) protection trees 335

Zagori municipality, Epirus, Greece 191, *193*, 196, 215
zaisho (Japanese) settlement 332
Zamczysko, Białowieża Primeval Forest (BPF) 116; medieval Slavonic cemetery 116
Zelkova see Elm, Cretan; Elm, Sicilian
Zelkova, Japanese *Zelkova serrata* 343

Index of Persons

Abramov, Shlomo *187*
Adam, Paul ix, Chapter 20
Agnoletti, Mauro 6–7
Alexander, Keith 11
Anderson, Clive 374
Angelini, Paolo 133
Arbon, Ashley 377
Art, Henry ix, Chapter 24, *361*
Atkinson, J. 310, 312, 314

Bacon, Louise 368
Bainbridge, R. 50
Baines, Chris 10
Balzaretti, Ross 218
Banks, Joseph 314
Bannister, Nicola 23
Barnes, Gerry 27, 286
Bartrum, Christine 370
Bätzing, Werner 127
Baur, G.M. 311
Beeckman, H. 266
Beevor, Sir Hugh 18, 21
Bengtsson, Vikki ix, Chapter 21
Bernard, V. 265
Billamboz, André 267; 'dendrotypology' 267
Bird, J.F. 304, 306
Birks, John and Hilary 6
Bleicher, N. 265
Boys, J. 294
Brooker, M.I.H. 302
Brown, Lancelot ('Capability' Brown) 43, 313
Bunting, E. 263–4
Burgon, Jo 368
Buttriss, Natalie 375
Buxton, James 386

Carandini, Andrea 219
Carlisle, Alan 19
Carolin, Peter 374
Cevasco, Roberta ix, 6, Chapter 14, 218, 219
Chaworth, William 239
Cicero *De Natura Deorum* 154
Clements, Frederic Edwards 99, 103, 105
Colbert, Jean-Baptiste *Ordonnance des eaux et forêts* 260
Coleridge, Samuel Taylor 53
Collins, David 309–10, 312
Cook, James 314
Cooper, Adrian 368
Cooper, Valerie 368
Copley, Lionel 242
Cotes, John 239
Cotta, Heinrich 262
Crone, Anne 267

Daszkiewicz, Piotr ix, Chapter 7
Darby, H.C. 283; *Domesday Geography* 24, 283; *The Domesday Geography of Eastern England* 271
Dargavel, John 153
Darwin, Charles 50
Davis, Bailey *172*
Davy, Lynn *178*
Day, S.P. 275
Deakin, Roger 60
Deforce, K. 266
Dehnhardt, Friedrich 300, 302
Déjeant-Pons, Maguellonne *144*
Dickenson, William 239
Dossett, Wick 368, *375*, *378*
Duck, Robert 303

Ellinger, Abbot 251, 255
Elton, Charles 74
Evans, Clifford 2, 50

Faull, M.L. and Stinson, M. *Domesday Book: Yorkshire* 236
Fell, John 245–6
Floyd, A.G. 315
Francis, Jane *172*
Francis, W.D. 315

Frater, Mark 322
Frodin, D.G. 310–11
Fukamachi, K. ix, 148, Chapter 22

Galloway, J.A. 262
Gammage, B. 314, 317; *The Biggest Estate on Earth* 317
Gardener, Lauren 370
Gardner, A.R. 263
Gelling, Margaret 37
Ghigliotti, Antonio Maria 226
Gilbert, Oliver 271
Girardclos, O. 265, 266–7
Girella, Luca 187
Godwin, Sir Harry 2–3
Gordon, A.M. 159–60
Govigli, Valentino Marini x, Chapter 13
Grant, M.J. 263–4
Green, Ted 10–11, 280, 320, 323, 324, 327
Grendi, Edoardo 219
Grove, Alfred Thomas (Dick) 7, 36, 139–40, 153–4, 155, 156–8, 159, 219
Grove, Jean 139
Grubb, Peter Foreword, x, 6, 382

Hallager, Erik and Birgitta 187
Hampton, Marc 59
Haneca, K. 266
Harrison, George 172
Harrison, John 240–1
Harvey-Samuel, Tim 368
Hassall, J.S. 308
Hoffmann, Richard x, Chapter 16
Hong, Sun-Kee 148
Hooke, Della x, Chapter 2, 37, 277; *The Anglo-Saxon Landscapes of the West Midlands* 37
Hopkins, Amos Lawrence 351, 352
Hoskins, W.G. 27, 31, 271, 283; *The Making of the English Landscape* 271, 283
Hughes, Donald x, Chapter 10
Hughes, Lucy 372

Johann, Elisabeth x, Chapter 8
Johnson, Richard 308
Jones, Huw 372
Jones, Melvyn xi, 4–5, 8, Chapter 15, 272, 273, 281

Kalantzi, F. 204
Kamitani, Tomohiko xi, Chapter 23
Karetsou, Alexandra 187
Keen, Paula 377, 378
Keene, D. 262
Keyser, Richard 262
Khatzegake, A.K. 184
Kilvert, Francis 43
Kirby, Keith xi, Chapter 4, 187, 368
Kizos, Thanasis xi, Chapter 11
Kral, F. 127

Lagomarsini, Sandro 218, 219, 222
Laing, Stuart and Sibella 368
Lamb, Henry 156
Lambrick, Camilla 368
Leatherdale, Simon 368
Lebret, T. *Wild Geese in the Netherlands* 103–5
Leeds, Duke of 242
Levine, Adrian 172
Linnaeus 307
Lloyd, Phillip 64
Lowdermilk, Walter C. 156
Lund, H.G. 125

Mabey, Richard 78, 374; and Evans, Tony *The Flowering of Britain* 78
Macnae, W. 306
Makhzoumi, Jala 143, *144*, 147, 148
Marguerie, D. 265
Markoulaki, Stavroula 187
Marsh, George Perkins 155–6
Martin, Michael 2
Mathieu, Jon 127
McKenzie, Euan 370
McKibben, Bill 350
Meiggs, Russell 156
Melville, Elinor 251
Merchant, Carolyn 350
Merton, Francis 18
Miguchi, Hideo xi, Chapter 23
Miller, Philip 307
Mlinsek, Dusan 125
Monbiot, George *Feral* 53
Moody, Jennifer (Jenny) xi, 6–8, 139, Chapter 12, *173*, *175*, *178*, 378; computer files 370; Crete 36; digital images 370; *The Friends of Oliver Rackham (FOR)* newsletter 374; *The Making of the Cretan Landscape* 153; memories of Oliver Rackham 12
Moravčik, M. 134
Moreno, Diego xii, 6, Chapter 14, 218, 219, 222, 228; *Dal documento al terreno* 228; *Quaderni Storici* (journal) 218
Morfitt, David xii, Chapter 5, 377; *Ecology of the Moor* 83; *The historical ecology of the woods of Binley, Warwickshire* 81n6, 85, 94
Morrish, Sarah (Royal Horticultural Society Botanical Art Show 2016) 375

Nakashizuka, Tohru xii, Chapter 23
Newman, S.M. 159–60
Newton, Adrian xii, Chapter 3, *178*
Ntinou, Maria 187

Oku, O. 148
Orchard, A.E. 302
Oswald, Philip 377, 382
Otto II, Emperor 251
Out, Welmoed 267

Papanastasis, V.P. 156
Parker, John 239
Paterson, G. 309–10
Perlman, Paula *172*
Pescott, E.E. 308
Peterken, George xii, Chapter 1, 17, 64, 65, 82, 276, 283, 367; *Wye Valley* 82
Peters, R.H. 50
Phillip, Arthur 116
Pigott, C. Donald 1–2, 10, 18, 271, 273, 275, 281
Pinto, Bruno 156
Plato *Critias* 154
Plieninger, T. 161
Pliny the Elder *Natural Histories* 184, 260
Poleggi, Ennio 219
Poorter, Ernst 105
Preston, Christopher 368, 382
Prosperetti, L. 377
Pungetti, Gloria xiii, 8, Chapter 9, 139–40, *144*, 147, 148; *Island Landscapes* 146

Rackham, Oliver (OR) *ii*, 1, *145*, *146*, *147*, *173*, *176*, *177*, *180*, *192*, *194*, *282*, *321*, *380*; accessible approaches 73; aphorisms 62; approaches to scholarship 377–9; archives 368; book dedications 377; categories and diversity 379; conferences in his memory 374; continuing influences 362; face-to-face communication 33; field trips 51; global influences 299; herbarium collection 370–2, *371*; Honorary Professor of Historical Ecology 36; humanity 377; Keeper of the College Plate 374; legacy 213–14, 374–7; linguistic gifts 6; love of natural world 379; maestro 140, 148–9; and 'meaning' 52, 379; OBE 1; obituaries 10; PhD 2; recording field data 359–60; research questions 214; stories 52; studies of ground flora 65–8; tree dedications *169*, 375–7, *378*; unfinished manuscripts 375, 377
Ranson, Susan 368, 382
Read, Helen 322
Ricciardi, Francisco, Count of Camalduli 300
Ričkienė, Aurika xiii, Chapter 7
Rockingham, Marquis of 236
Rose, Francis 11, 23
Rosenburg, Arthur E. 353
Rotherham, Ian D. xiii, 10, 24, Chapter 18, *272*, 273; *Arboricultural Journal* 10; *Shadow Woods* 272; *The Woodland Heritage Manual* 273; 'Trouble in the Woods' conference (2024) 374
Runnels, Curtis 156

Satterthwaite, Sheafe 356
Schimper, A.W.F. 311
Schweingruber, F.H. 265
Scott, Meryn *178*
Sears, Paul B. *Deserts on the March* 156
Shvidenko, Anatoly 125
Sinclair, A.R and Norton-Griffiths, M. *The Serengeti: Dynamics of an Ecosystem* 106–7
Smith, Humphrey 85
Smith, John 323
Stara, Kalliopi xiii, 7, Chapter 13, 191, *194*, 215; and Vokou, D. *The Ancient Trees of Zagori and Konitsa* 215
Stephanoudakis, Stelios 186–7, *187*
Stylianos, Father (Lákkoi, Crete) *169*
Swallow, Richard 245–6
Symonds, H.H. *Afforestation in the Lake District* 64
Szabó, Péter xiii, Chapter 17

Tanner, Edmund *375*
Tansley, Arthur 99–101, 103, 105, 106, 108
Tench, Watkin 307
Theophrastus *Enquiry into Plants* 172, 179, 182, 184, 258
Thorpe, Harry 37
Tsiakiris, Rigas xiii, 7, Chapter 13, 191
Tubbs, Colin 19
Tyrrell, Henry 240

Varro *De Rustica* 184
Vavilov, Nikolai Ivanovich 59–60
Veen, Harm van de 106–7
Vera, Frans xiii–xiv, Chapter 6, 271, 327; *De Oostvaardersplassen* 105; *Grazing Ecology and Forest History* 99, 109 *see also* Vera theory and debate
Vrška, T. 265, 281

Wakefield, E.M. and Dennis, R.W.G. *Common British Fungi* 51

Waller, M. 263–4
Walters, S. Max 2–3, 6
Warren, Peter M. 6, 165
Waterman, Thomas 288–9, *289*
Watkins, Charles xiv, 19, 64, 65, Chapter 14, 218
Ważny, Olga *173*, *177*
Ważny, Tomasz *173*, 187
Wentworth, Thomas, first Marquis of Rockingham 242, 243
White, John *Journal of a Voyage to New South Wales* 310
Williams, P.A. 159–60
Williamson, Tom xiv, Chapter 19, 374
Willis, Robert 377
Wilson, E.O. 50
Witney, K.P. 291
Wong, Jennifer L.G. xiv, Chapter 13
Wysmułek, Jacek 121

Zacharin, R.R.F. 300